Superplasticity—Current Status and Future Potential

MATERIALS RESEARCH SOCIETY
SYMPOSIUM PROCEEDINGS VOLUME 601

Superplasticity—Current Status and Future Potential

Symposium held November 29–December 1, 1999, Boston, Massachusetts, U.S.A.

EDITORS:

Patrick B. Berbon
Rockwell Science Center
Thousand Oaks, California, U.S.A.

Min Z. Berbon
Rockwell Science Center
Thousand Oaks, California, U.S.A.

Taketo Sakuma
University of Tokyo
Tokyo, Japan

Terence G. Langdon
University of Southern California
Los Angeles, California, U.S.A.

Materials Research Society
Warrendale, Pennsylvania

This work was supported in part by the Army Research Office under Grant Number DAA019-99-1-0359. The views, opinions, and/or findings contained in this report are those of the author(s) and should not be construed as an official Department of the Army position, policy, or decision, unless so designated by other documentation.

Single article reprints from this publication are available through University Microfilms Inc., 300 North Zeeb Road, Ann Arbor, Michigan 48106

CODEN: MRSPDH

Published by:

Materials Research Society
506 Keystone Drive
Warrendale, PA 15086
Telephone (724) 779-3003
Fax (724) 779-8313
Web site: http://www.mrs.org/

Library of Congress Cataloging-in-Publication Data

Superplasticity—current status and future potential : symposium held
    November 29–December 1, 1999, Boston, Massachusetts, U.S.A. / editors,
    Patrick B. Berbon, Min Z. Berbon, Taketo Sakuma, Terence G. Langdon
        p.cm.—(Materials Research Society symposium proceedings,
        ISSN 0272-9172 ; v. 601)
        Includes bibliographical references and indexes.
        ISBN 1-55899-509-9
        1. Superplasticity—Congresses. 2. Metals—Plastic properties—Congresses.
    3. Ceramics—Plastic properties—Congresses. 4. Alloys—Plastic properties—
    Congresses. I. Berbon, Patrick B. II. Berbon, Min Z. III. Sakuma, Taketo
    IV. Langdon, Terence G. V. Series: Materials Research Society symposium
    proceedings ; v. 601

TA418.14.S875 2000
620.1'1233—dc21   00-025161

Manufactured in the United States of America

CONTENTS

*Invited Paper

*Invited Paper

SUPERPLASTICITY IN INDUSTRY

HIGH STRAIN RATE SUPERPLASTICITY

*Invited Paper

DEVELOPMENTS USING SEVERE
PLASTIC DEFORMATION

*Invited Paper

PREFACE

This volume contains papers presented at Symposium HH, "Superplasticity—Current Status and Future Potential," at the 1999 MRS Fall Meeting in Boston, Massachusetts. Superplasticity refers to the ability of a crystalline material to exhibit large strains when pulled in tension. This phenomenon is of academic interest, but it also has a considerable industrial potential because it provides the capability for the forming of complex parts from sheet metals. The only other Materials Research Society symposium dealing exclusively with superplasticity was held at the 1990 MRS Spring Meeting in San Francisco. The proceedings of that symposium were published as Volume 196 in the Materials Research Society symposium proceedings series, and was edited by Merrilea J. Mayo, Masaru Kobayashi, and Jeffrey Wadsworth.

The incentive for organizing the present symposium was the recognition that there have been several significant developments in the field of superplasticity over the last decade. New techniques have recently become available for the production of materials with ultrafine grain sizes, typically in the submicrometer or nanometer range. These materials provide at least the potential for utilizing superplastic forming capabilities at much faster strain rates, and thereby expanding the technology from the current fabrication of low-volume high-value components into high-volume commercial applications. Furthermore, there have been several major advances in the techniques available for microstructural characterization including, for example, the use of high-resolution electron microscopy and computer-aided electron backscatter pattern analysis. Several papers in this symposium describe results obtained using these new procedures.

It is a pleasure to acknowledge the very considerable cooperation we have received from all of the authors participating in this symposium. We are especially grateful to the many authors who assisted in reviewing the papers during the three-day symposium. These reviews were undertaken in an efficient and professional manner, and they resulted in approximately one-half of the papers being returned to the primary authors for at least minor revision of their manuscripts. These revisions were undertaken very promptly with the result that, despite the Christmas holidays and the world-wide excitement of moving into a new millennium, all of the manuscripts were received in a form ready for publication by the first week of January 2000. This remarkable cooperation has resulted in a published proceedings containing all except three of the papers presented in the symposium.

This Meeting attracted papers on all aspects of superplasticity, with individual contributions from ten different countries. Several of the invited speakers came from overseas, and in this respect we are grateful for the financial support provided by the Army Research Office and by the Lawrence Livermore National Laboratory. It is a pleasure also to acknowledge the secretarial assistance provided by Ms. Carolyn Sax in preparing the manuscripts for publication. Finally, the success of this Meeting owes much to the dedicated staff of the Materials Research Society, who have provided thorough and speedy responses to our many inquiries during our organization of the symposium.

Patrick B. Berbon
Min Z. Berbon
Taketo Sakuma
Terence G. Langdon

January 2000

MATERIALS RESEARCH SOCIETY SYMPOSIUM PROCEEDINGS

MATERIALS RESEARCH SOCIETY SYMPOSIUM PROCEEDINGS

Prior Materials Research Society Symposium Proceedings available by contacting Materials Research Society

Superplasticity in Metals
and Intermetallics

COMPUTER SIMULATION OF GRAIN BOUNDARY CHARACTER IN A SUPERPLASTIC ALUMINUM ALLOY

T.R. MCNELLEY*, M.T. PERÉZ-PRADO**
*Department of Mechanical Engineering, Naval Postgraduate School, 700 Dyer Road, Monterey, CA 93943-5146; tmcnelley@nps.navy.mil
**Materials Science Group, Department of Mechanical and Aerospace Engineering, University of California at San Diego, 9500 Gilman Drive, La Jolla, CA 92093-0411; tpprado@ames.ucsd.edu

ABSTRACT

High-angle grain boundaries are generally deemed necessary for superplasticity in metals. In polycrystalline materials the grain boundary character must be described in terms of a probability distribution rather than by a single parameter, and little has been reported on the relationship between this distribution and fine-grain superplasticity. For aluminum alloys that exhibit continuous recrystallization the results of computer-aided electron backscatter diffraction analysis have shown that bimodal grain boundary disorientation distributions are present in as-processed material and persist during subsequent annealing. Such distributions may be simulated by computer methods based on a model of the microstructure which assumes that deformation banding occurs during deformation processing. High-angle boundaries ($\geq 30°$) develop in association with deformation banding while boundaries of lower disorientation ($<30°$) develop by dislocation reaction within the bands. Improved understanding of the grain boundary types associated with various microstructural transformation mechanisms will aid the design of processes to produce superplastic microstructures.

INTRODUCTION

The exceptional tensile ductility of superplastic materials allows near-net-shape forming of complex components. Benefits include reduced weight, cost and energy consumption coupled with improved component and system reliability [1]. The recognized requirements for superplasticity in aluminum alloys include a high value of m, the strain rate sensitivity coefficient of the flow stress, due to deformation at temperatures and strain rates where grain boundary sliding (GBS) is predominant during deformation. GBS must be accommodated by either slip or stress-directed diffusion and together these processes constitute the mechanism of superplasticity. For this, grains must be equiaxed, highly refined, reasonably stabile in size and with mobile boundaries that can slide while resisting tensile separation [2-5].

The mechanical behavior of fine-grained superplastic materials may be described by a constitutive law involving the independent, additive contributions of the dislocation creep mechanism and GBS [2,4,5]. Dislocation creep is usually envisioned to occur within the core regions of the grains and the associated deformation rate is generally thought to be independent of grain size. Then, GBS and its accommodation mechanisms are envisioned to occur within the grain boundaries and in adjacent, mantle-like regions of the grains; accordingly, the deformation rate during GBS increases strongly as grain size becomes finer [6]. This leads to a grain-size dependent transition from dislocation creep to GBS, and superplastic response at higher deformation rates (and lower deformation temperatures) as grain size is refined [6,7]. Indeed, this is the basis for the requirement that the grain size, d, must be finer than about 10 μm for useful superplastic response in engineering materials.

3

It is often asserted that high-angle grain boundaries are also necessary for superplasticity because the lattice registry across boundaries of low disorientation makes GBS difficult or impossible [e.g. 1-8]. However, few studies have been published on the relationship between grain boundary characteristics, such as the disorientation distribution, and GBS in polycrystalline materials. In Figure 1a, the size of equiaxed grains in a polycrystal may be represented by a single parameter, e.g. the mean linear intercept grain size, d_0. In contrast, the disorientation between adjacent grains must be described in terms of a probability distribution. Mackenzie [9] has determined this distribution when random processes determine the orientation of adjacent grains in a cubic material and shown that it is of the form depicted in Figure 1b. Here, the angle of disorientation, θ, is taken as the minimum among all crystallographically equivalent rotations that would bring two adjacent cubic lattices into coincidence. For cubic materials, $0° \leq \theta \leq 62.8°$ and the distribution exhibits a maximum at $45°$, as also shown in Figure 1b [9]. Lattice reorientation during thermomechanical processing (TMP) and post-TMP treatments involving annealing and deformation typically result in varying degrees of preferred orientation, which may, in turn, alter the distribution of disorientation angles.

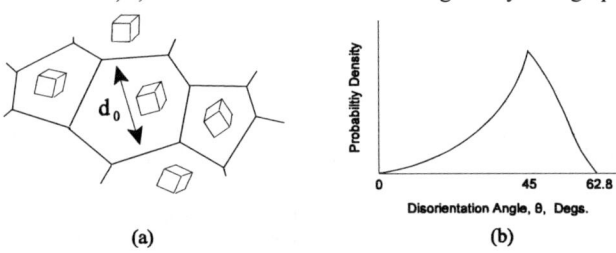

(a)   (b)

Figure 1. In (a), a schematic representation of an equiaxed grain structure, of size d_0 and in a material having randomly oriented grains, is shown. The corresponding distribution of disorientation angles is shown in (b) [9].

For example, the peak superplastic ductility was shown to improve dramatically when TMP parameters were adjusted to increase the population of boundaries of disoriented by $5°-15°$ in an Al-Mg-Zr alloy [10]. Such distributions can now be readily evaluated and so the interactions among TMP parameters, grain boundary characteristics and superplasticity may now be quantified.

Attaining the necessary microstructural prerequisites for superplasticity is greatly facilitated in alloys of iron or titanium by the phase transformations that occur on cooling of these metals. For most other alloys of engineering interest, including those based on copper and nickel as well as aluminum, the required grain refinement can only be attained by deformation and recrystallization treatments. The present understanding of deformation processing and recrystallization in engineering alloys is largely empirical in nature. Current developments have been recently reviewed [11]. The relatively low values of stored strain energy due to deformation preclude the formation of recrystallization nuclei in a severely deformed metal by thermal fluctuation. Instead, new grains form from cells or subgrains within the deformation microstructure and this accounts for the observation that the orientations of new grains generally are not random but, instead, evolve from orientations present within the deformation microstructure. For this reason, further progress in understanding of recrystallization and related phenomena requires improved understanding of the deformation-induced structure in severely deformed metals. Recent work has begun to examine such structures in pure metals and some solid solution alloys [e.g. 12]

Grains subdivision has been identified as one of the mechanisms of microstructural evolution during severe plastic deformation. During this process adjacent regions separated by dislocation boundaries of medium to high disorientation rotate by slip toward different end orientations of the texture. As strain is increased an elongated, banded or ribbon like structure develops and the prior boundaries may no longer be distinguishable. The deformed state consists of fine cells or subgrains containing few dislocations surrounded by dislocation boundaries of varying disorientation and typically exhibits a strong deformation texture [13]. The dislocation boundaries often exhibit disorientations $\geq 15°$. Upon annealing, such cellular microstructures will be unstable and exhibit discontinuous growth of isolated cells of high disorientation if the fraction of high-angle boundaries (disorientation $\geq 15°$) is less than about 0.7 [13,14]. This corresponds to normal, or primary, recrystallization. On the other hand, when the fraction of such high-angle boundaries is greater than 0.7 - 0.75, the structure becomes stable against discontinuous growth and exhibits instead stable, homogeneous coarsening. Then, recovery processes apparently predominate and this evidently corresponds to the continuous recrystallization process. A dispersion of fine particles may facilitate the retention of a fine grain structure.

The development of grain boundary character during annealing of Supral 2004, a highly superplastic aluminum alloy (nominally Al - 6.0 wt. pct. Cu - 0.5 wt. pct. Zr), has been studied within the foregoing framework. The presence of a fine dispersion of Al_3Zr particles is intended to retard the formation an migration of high-angle boundaries and this leads to a predominance of recovery processes, i.e. continuous recrystallization, in the development of a superplastically enabled microstructure. However, many important features of this process remain to be determined, including the nature of the grain boundaries that support superplasticity. Recently developed computer-aided electron backscatter diffraction (EBSD) methods have been used to evaluate grain boundary character during annealing of as-processed Supral 2004 [15,16]. The experimentally observed disorientation distributions will simulated by a computer-based model that incorporates deformation banding of the microstructure during severe plastic deformation, and also includes a fine, cellular deformation-induced structure within the bands.

EXPERIMENTAL DETERMINATION OF THE DISORIENTATION DISTRIBUTION

Computer-aided EBSD analysis was used for experimental evaluation of disorientation distributions [15,16]. A thorough background on EBSD has been provided by Randle [17]. Specific procedures employed here have been described previously [10,15,16,18]. Essentially, diffraction patterns are obtained along a linear traverse on a selected plane of examination as illustrated in the schematic of Figure 2; the electron beam of the SEM, operating in the spot mode, serves as a probe of the local orientation on a highly inclined sample of the material. Patterns can be obtained from grains as small as 0.5 μm in a conventional SEM. The beam is initially positioned and a diffraction pattern is captured, automatically indexed and stored by the system software (TexSEM Laboratories, Inc., Provo, UT). The data are in the form of the three Euler angles, φ_1^1, Φ^1, and φ_2^1, that specify the orientation of crystal 1 relative to axes associated with the final rolling stage of the prior TMP. These angles are defined according to Bunge's format [19]. The electron beam is then moved, pixel by pixel, until a new diffraction pattern is observed, collected, indexed and the corresponding data also stored; this process is repeated until K (typically 300 to 500) orientations have been acquired. The distance between successive patterns is also recorded The resulting microtexture data may be displayed in various forms including discrete pole figures or orientation distribution functions.

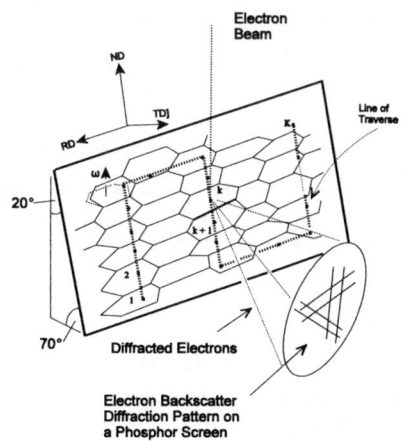

Figure 2. A schematic of EBSD for collection of grain-specific orientation data along a linear traverse through the microstructure.

The successive diffraction patterns correspond to the orientations of *adjacent* grains and so the disorientation of the intervening boundary (highlighted in Figure 2) may be calculated as the minimum rotation angle about a common axis to bring the neighboring crystal lattices into coincidence. Complete specification of the boundary would also require the determination of the orientation of the boundary plane (denoted by ω in Figure 2) but the additional difficulties in such measurements would severely restrict the quantity of related orientation data. For the kth grain, the matrix

$$A_k = \begin{pmatrix} a_{11}^k & a_{12}^k & a_{13}^k \\ a_{21}^k & a_{22}^k & a_{23}^k \\ a_{31}^k & a_{32}^k & a_{33}^k \end{pmatrix} \tag{1}$$

describes the orientation of the cubic lattice with respect to the reference axes of the processed material. The matrix elements are functions of the Euler angles φ_1^k, Φ^k, and φ_2^k for this grain [7].

A similar matrix is obtained for in terms of the Euler angles for an *adjacent* grain, i.e. A_{k+1}, in terms of φ_1^{k+1}, Φ^{k+1}, and φ_2^{k+1}, and the disorientation, or orientation difference, between the two grains, may be calculated in terms of the rotation matrix

$$M_{k,k+1} = A_k^{-1} A_{k+1}, \quad k = 1.....K - 1 \tag{2}$$

In general, there will be several $M_{k,k+1}$ rotation matrices, each corresponding to one of the symmetry elements of the crystal lattice. The convention for determination of the disorientation angle is to perform a minimization operation among the rotation angle values for each of the crystallographically equivalent rotation matrices, i.e

$$\theta_{k,k+1} \equiv \min abs \left(\cos^{-1} \left(\frac{\left(m_{11}^{k,k+1} + m_{22}^{k,k+1} + m_{33}^{k,k+1} - 1 \right)}{2} \right) \right) \tag{3}$$

Here, the $m_{ii}^{k,k+1}$ represent the elements of the trace of the corresponding $M_{k,k+1}$ matrix. The rotation axis, $R_{k,k+1} = (r_1^{k,k+1}, r_2^{k,k+1}, r_3^{k,k+1})$, that corresponds to $\theta_{k,k+1}$ may also be obtained, following Randle [17], from the elements of the $M_{k,k+1}$ matrix as

$$r_1^{k,k+1} = m_{23}^{k,k+1} - m_{32}^{k,k+1}$$
$$r_2^{k,k+1} = m_{31}^{k,k+1} - m_{13}^{k,k+1} \tag{4}$$
$$r_3^{k,k+1} = m_{12}^{k,k+1} - m_{21}^{k,k+1}$$

For K orientation measurements, there will be K-1 disorientation matrices and therefore a set of K-1 disorientation angle/axis pairs. When defined according to equation 2, these are referred to as *correlated* disorientations because the $M_{k,k+1}$ are always the orientation differences between *adjacent* grains.

Orientation differences between the kth grain and all other grains in the assemblage may be calculated as

$$M_{k,l} = A_k^{-1} A_l, \quad k,l = 1.....K \quad (k \neq l) \tag{5}$$

and the corresponding angles, $\theta_{k,l}$, and axes, $R_{k,l}$, calculated as in equations 8 and 9. For K individual orientation measurements, this procedure would result in $(K-1)^2$ disorientations although only $(K(K-1))/2$ are distinct. The disorientations calculated according to equation 5 are *uncorrelated* disorientations because the grains generally do not share a common boundary. Differences between the correlated and uncorrelated disorientation distributions may reflect grain-to-grain interactions during deformation and annealing. The distribution of the uncorrelated disorientations, $M_{k,l}$, for K

6

orientation measurements is the discrete analogue to the continuous disorientation distribution function (MDF), g, which may be calculated form the texture as described by Adams [20] and Pospiech, et al. [21]. The disorientation distributions of the grain boundaries (both correlated and uncorrelated) may then be represented by histograms of relative number versus disorientation angle.

ANNEALING OF SUPRAL 2004

Details of the composition, microstructural effects of annealing and data on the mechanical behavior of this material have been presented previously [15]. Discrete pole figures from as-received material are shown in Figure 3a. This is a deformation texture comprising mainly orientations that lay along the β-fiber; the brass component (B, {110}<112>) and nearby orientations rotated about the sheet normal are the most prominent, from analysis of the orientation distribution function. Corresponding discrete pole figures for material annealed 6 hr at 450°C are shown in Figure 3b, and for material annealed 12 hr at 450°C are shown in Figure 3c.. The only significant change from the

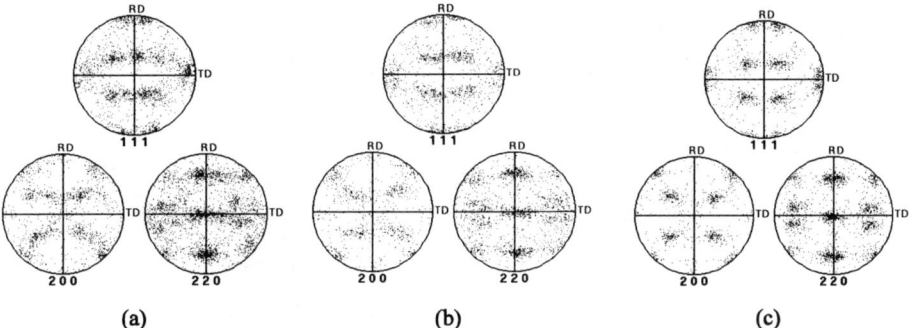

(a)   (b)   (c)

Figure 3. Discrete pole figures showing a deformation texture in as-processed Supral 2004 (a) and retention of this deformation texture following either 6 hr of annealing (b) or 12 hr of annealing (c) at 450°C.

as-received condition is the sharpening of the texture when compared to the as-received condition. The B orientation remains the most prominent component, and the sharpening suggests reduced a spread of orientations about it.

Disorientation distribution histograms for as-received material and for samples annealed 30 min and 6 hr as well as 12 hr are shown in Figure 4 a-d. During the traverse in the as-received material (Figure 4a) successive orientation measurements were obtained at 0.5 μm spacings, which is much less than the reduced thickness of the original grains. Thus, these measurements reflect the disorientation distribution in a fine, cellular deformation-induced structure. The correlated disorientation distribution for as-received material (shown by the bars) is bimodal in nature while the uncorrelated (texture) disorientation data (represented by the dotted curve) exhibit a single peak. Thus, the populations of boundaries disoriented by 0°-15° and 55°-62.8° exceed those predicted by texture considerations while the population of boundaries of disorientation angle 20°-55° is less than predicted by the texture. During annealing for times up to 12 hr the bimodal nature of the disorientation distribution is retained. The populations of 0°-15° and 55°-62.8° always exceed those predicted by texture considerations. For the longest annealing time, the uncorrelated distribution has become bimodal as the texture (Figure 3c) has become more sharply defined.

Typical times involved in heating and superplastic forming of this material range from 30 mins. to perhaps two hours. It can be surmised from these data that the boundary character at the initiation superplastic straining is essentially that developed by the prior deformation processing, i.e. that of the deformation-induced cellular microstructure of the as-processed condition. Recovery

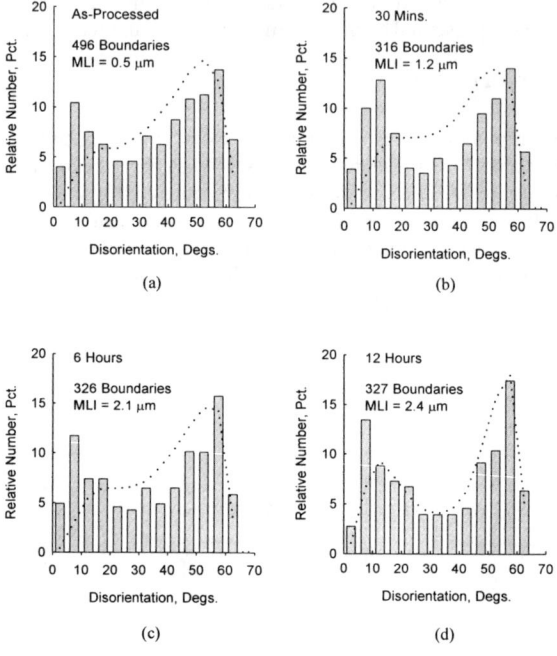

Figure 4. Histograms representing disorientation data for as-received material (a), and for material annealed 30 min (b), 6 hr (c) and 12 hr (d). Correlated data are shown by the bars while the uncorrelated distributions are shown by the dotted curves.

processes during continuous recrystallization in this alloy lead to the sharpening and retention of some cells and elimination of others from the deformation microstructure, but in a manner consistent with retention of the initial deformation texture of the material. In the following sections a computer-based simulation of these effects is described.

SIMULATION OF THE GRAIN BOUNDARY CHARACTER IN SUPRAL 2004

Model of the Microstructure

The following simulation of the grain boundary character of as-processed and processed and annealed Supral 2004 is based on the deformation banding model for microstructural evolution during severe straining of an FCC metal [22-24]. Deformation banding is a process of grain subdivision during deformation that results in lattice rotation in different senses in adjacent regions of the grain; its potential importance in severe deformation has long be recognized [25,26]. Its occurrence during rolling deformation may allow a grain as a whole to deform in plane strain even though plane-strain conditions may not be met in each individual band. Less work is done by slip within the bands than would be the case for homogeneous deformation but this requires a particular arrangement of the bands so that the net strain for a group of bands meets the overall strain.

Such a model for Supral 2004 is depicted in the schematic of Figure 5. The microstructure in Figure 5a comprises band-like features that are elongated in the rolling direction, and flattened in the thickness direction reflecting the plane strain deformation conditions of the rolling process. An idealization of this structure is shown in Figure 5b while details of the crystallography for such a structure are illustrated in Figure 5c. The elongated bands are here shown as block-like features, each of orientation near that of the B texture component. For the B component, there are two crystallographically distinct variants, hereafter denoted as B_1 and B_2. The lattice orientations for these variants are included in Figure 5c. According to the deformation banding model these variants would be arranged in an alternating pattern [22,23]. A shear, $\gamma_{RD,TD}$, is illustrated by the shape of each band (in both Figures 5 b and c); this shear would develop by slip on two systems for plane strain deformation in a crystal of the B orientation. The alternating arrangement of B_1 and B_2 shown allows plane strain conditions to be maintained for a group of bands despite the tendency to develop this shear. Of course, this added shear must be accommodated at the ends of the bands; Hirsch and Lücke [27] have pointed out that difficulties in accommodation of this shear often leads to instability the B orientation in many circumstances.

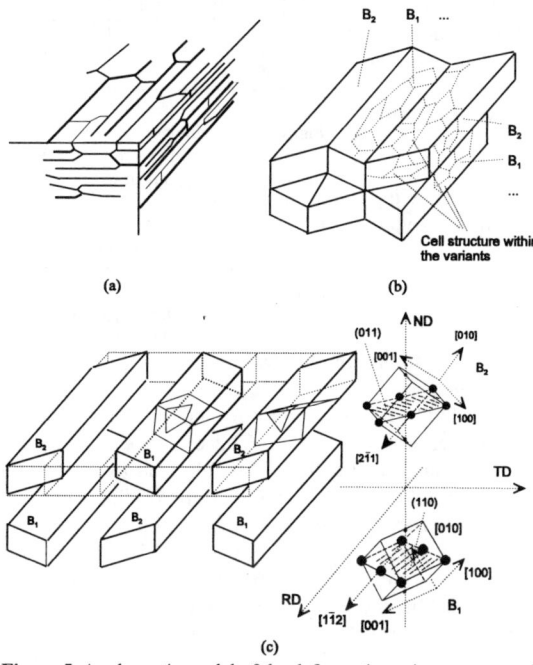

(a)   (b)

(c)

Figure 5. A schematic model of the deformation microstructure of Supral 2004 (a), and an idealization of the structure in terms of deformation banding as shown in (b). Crystallography for this model is included in (c).

Here, each band is assumed to contain a cellular dislocation structure such that the lattice orientation within a band is distributed about its respective B orientation. The fundamental zone of Euler space for a cubic material is shown in Figure 6. A linear traverse through the microstructure would encounter successive orientations associated with the cell structure that are near one or the other of the distinct B_1 or B_2 variants, as well as orientations alternating between these variants. The cell structure would give relatively small disorientation values; large disorientations would be associated with boundaries between the variants. Looking down the TD, the arrangement of bands might appear as shown in Figure 7a. The diagram at the upper left of this figure represents the alternating bands of B orientation and shows the trace of a <111> plane. A more detailed representation of the cellular structure within one of the variants is shown at the lower left in Figure 7a. The cell walls are envisioned to be regions of high dislocation density and thus pronounced lattice curvature due to accumulation of like-signed dislocations during deformation processing. This is indicated by the curvature of the solid diagonal line through the cell walls. The cell interiors are shown as relative free of dislocation structure. The lattice curvature is of opposite sense on each side of a cell so that the lattice orientation varies about the mean orientation of the variant, which is indicated by the dashed line. In this manner, the cell structure has an orientation, on the average, which remains that of the variant and slip planes within variants may be thought of as corrugated sheets.

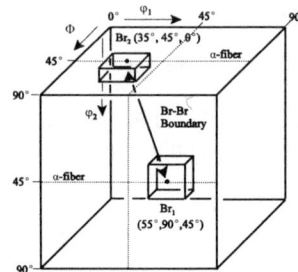

Simulation of the boundary disorientation distributions expected by traversing through such a microstructure may be accomplished by assuming that the Euler angles for the kth orientation (Figure 2) are each normally distributed about the corresponding exact angle values for a variant. This is shown on the right in Figure 7a. Thus, the distribution of each of the Euler angles, φ_1, Φ and φ_2, for the cellular structure within B_1 may be assumed to be normally distributed about their respective mean values; i.e. $\mu_0 = 55°$ for the distribution of φ_1, $\mu_0 = 90°$ for the distribution of Φ, and $\mu_0 = 45°$ for φ_2. Then, e.g., the probability density distribution for $x = \varphi_1$ would be given by

Figure 6. The fundamental zone of Euler space showing regions near the two distinct B variants in which most orientations are distributed in the texture of Supral 2004. A correlated disorientation distribution may be calculated by assuming that adjacent cells in a banded structure have orientations alternately laying in locations distributed throughout the B_1 or B_2 regions.

$$P(x = \varphi_1) = \frac{1}{\sqrt{2\pi}} \exp\left(\frac{(x - \mu_0)^2}{\sigma^2}\right) \qquad (6)$$

where the standard deviation σ is a measure of the spread in orientation about the mean for the angle in question. The probability distributions for Φ and φ_2 are assumed to be of the same form. The Euler angles within B_2 are likewise distributed about the exact angles for this variant. The effect of annealing is illustrated in Figure 7b. Migration of variant interfaces and cell walls would lead to coarsening; furthermore, recovery would result in sharpening of cell walls while annihilation of dislocations with the walls would result in reduced orientation spread about the mean orientation within the variant. In this model, this is reflected only in a reduced value of the standard deviation σ. The simulations were conducted by assuming that, during a traverse, n cell orientations will first be encountered within B_1, and then n cells orientations within B_2, as illustrated in Figure 8. The result is a sequence of Euler angles from which discrete pole figures and disorientation distributions may be determined.

Simulation Results

A series of simulations were conducted by computing a total of N orientations, each one consisting of a set of Euler angles. Typically, N = 500 for each simulation. At the beginning of a simulation the total number of orientations, N, and the number of cells within each variant, n, were specified (Figure 8). The values of the Euler angles for each variant were also entered, along with an assumed standard deviation, σ, for the distribution about the exact orientation of the variant. The individual Euler angles for an orientation were each computed by random selection from a normal distribution with a specified mean value, μ_0, and standard deviation, σ.

The results of simulations with N = 500 and n = 2 are shown in Figure 9. Each result represents a different value of the standard deviation, σ, so that the data of Figure 9 represent a simulation of the annealing of as-processed material. Assumed values were $\sigma = 10°$ in Figure 9a and b; 8° in Figure 9c and d; and 6° in Figure 9e and f. Comparison of these results with the experimental data of Figures 3 and 4 indicates that these simulations capture the essential features of the annealing response of this alloy. The simulation in Figure 9a and b would correspond to as-processed material in which a high dislocation density would lead to lattice curvature and large spread about the mean orientation of the main B texture component. The agreement between

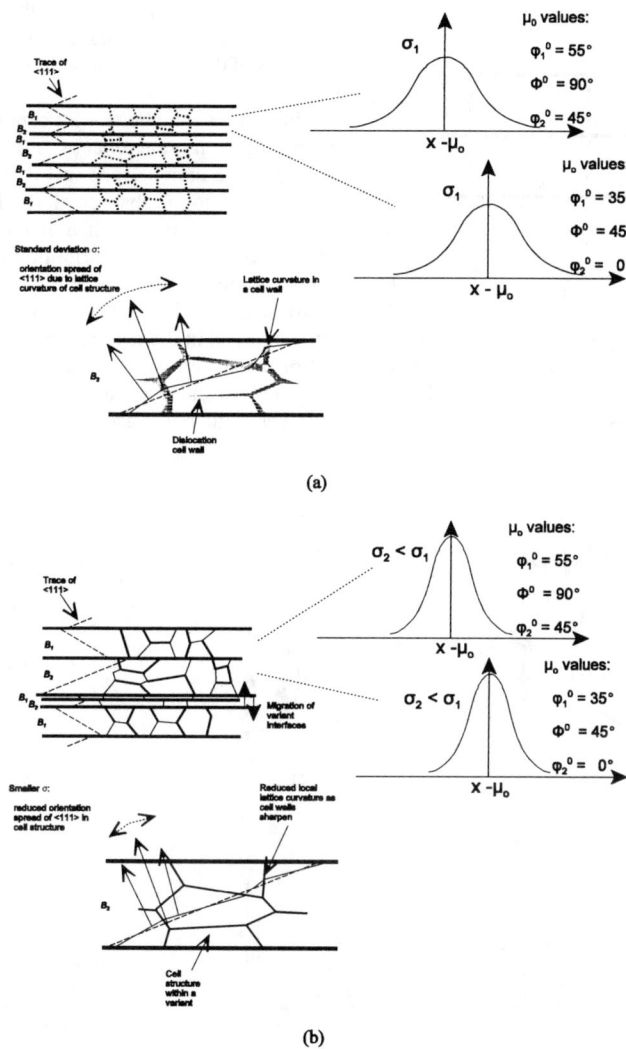

Figure 7. A model of the microstructure of Supral 2004 showing the alternating bands, of orientation near B, and a cellular structure within the bands for as-processed material, (a), and annealed material, (b). Orientations within each band are normally distributed and the distribution narrows upon annealing.

predicted (Figure 9a and b) and experimental (Figures 3a and 4a) textures and, most importantly, the disorientation distributions for $35° < \theta \leq 62.8°$ is excellent. The peak in the distribution of the disorientation angles for $0° \leq \theta \leq 35°$ lays at a larger angle in the predicted data when comparison is made to the experimental results. The simulations capture the sharpening of texture and associated changes in the disorientation distribution during annealing (Figures 9c and d; e and f). The population of boundaries of $55°$-$62.8°$ disorientation is predicted to increase during annealing,

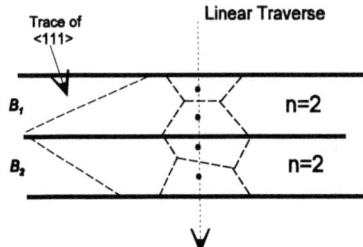

Figure 8. For n=2, two orientations are determined for each variant in sequence.

although the simulations overestimate this effect; similarly, the simulations indicate a gradual shift in the populations of the boundaries of $0° \le \theta \le 35°$ toward lower disorientation angles.

Additional simulations were conducted to examine the influence of the number of cells, n, within each variant and the results are summarized in Figure 10. For n = 1, which corresponds to the absence of a cell structure within the variants, a simulation with a standard deviation, σ, value of $8°$ shows only the distribution of the high-angle boundaries associated with the variant interfaces (Figure 10a and b). This emphasizes the extent of the spread in disorientation angle associated with such an angular spread in the texture. On the other hand, for n = 3 but also with $\sigma = 8°$, the relative heights of peaks at high and low disorientation angles are reversed as comparison of Figures 10c and d with and 9c and d reveals. Thus, as n increases, the structure comprises an increasing population of lower angle boundaries.

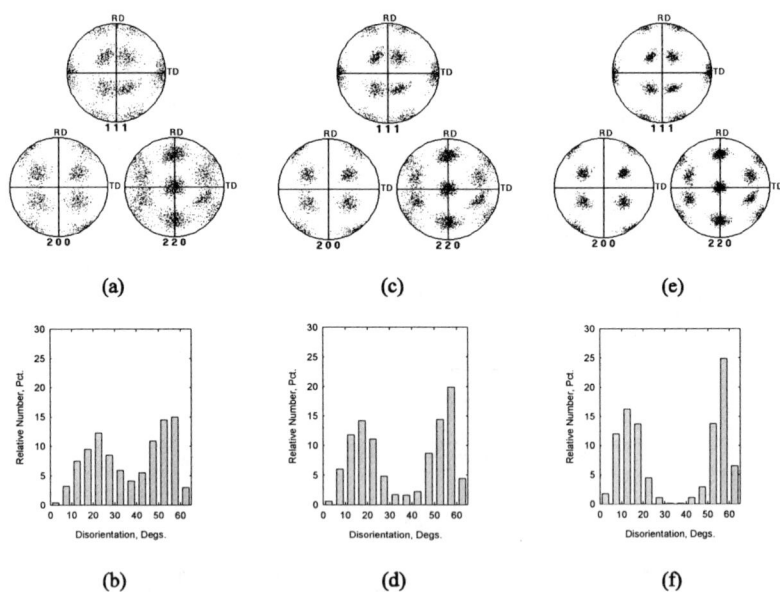

(a)   (c)   (e)

(b)   (d)   (f)

Figure 9. Results of simulations for n = 2 and N = 500. Data in (a) and (b) correspond to $\sigma = 10°$; the data of (c) and (d) are for $\sigma = 8°$; and the data of (e) and (f) are for $\sigma = 6°$. Altogether, these data correspond to annealing of as-processed material.

CONCLUSIONS

Computer simulation based on a model for the deformation-induced microstructure of Supral 2004 which includes a single, B-type texture and deformation banding during severe rolling deformation of the TMP reproduces the essential features of the grain boundary disorientation distribution in the as-processed material. This approach is also

capable of simulating the evolution of the grain boundary structure during elevated temperature annealing. These results suggest new insights into the development of microstructure during processing of aluminum alloys such as Supral 2004 to enable superplastic response. Factors which control of the process of deformation banding and the cellular dislocation structures within the bands may be amenable to control by adjusting the appropriate TMP parameters. This, in turn, may lead to improved processes to enable superplasticity in this alloy and insight into the design of processes for other alloys of engineering interest.

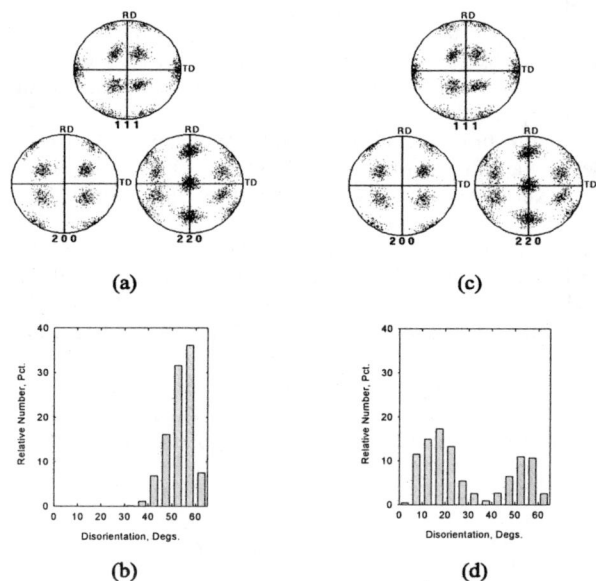

Figure 10. Results of simulations for n = 1 (a) and n = 3 (b). In both cases, N = 500 and σ = 8° (compare to Figure 10c and d.

REFERENCES

1.　　R. Grimes: Advances and Future Directions in Superplastic Materials, NATO-AGARD Lecture Series, 1988, No.168, pp. 8.1-8.16.
2.　　O.D. Sherby and O.A. Ruano: in Superplastic Forming of of Structural Alloys, N.E. Paton and C.H. Hamilton, eds., TMS-AIME, New York, 1982, pp. 241-254.
3.　　T. G. Langdon: Metall. Trans. A, 1982, vol. 13A, pp. 689-701.
4.　　O.D. Sherby and J. Wadsworth: in Deformation Processing and Microstructure, G. Krauss, ed., ASM, Materials Park, OH, 1984, p. 355
5.　　O.A. Ruano and O.D. Sherby: Revue Phys. Appl., 1988, vol. 23, pp. 625-637..
6.　　R.C. Gifkins: Metall. Trans. A, 1976, vol. 7A, p. 1225-32.

7.   J.P. Poirier: in Creep of Crystals, Cambridge University Press, 1985, pp. 79
8.   A. Ball and M.M. Hutchison: Met. Sci. J., 1969, vol. 3, p. 1-7.
9.   J.K. Mackenzie: Biometrica, 1958, vol. 45, pp. 229-240
10.   T.R. McNelley and M.E. McMahon: Metall. Mater. Trans. A, 1996, vol. 27A, pp. 2252-2262.
11.   R.D. Doherty, D.A. Hughes, F.J. Humphreys, J.J. Jonas, D. Juul-Jensen, M.E. Kassner, W.E. King, T.R. McNelley, H.J. McQueen, and A.D. Rollet: Mater. Sci. Eng. A, 1997, vol. A238, pp. 219-274.
12.   D.A. Hughes and N. Hansen: Acta Mater., 1997, vol. 45, pp.3871-3886.
13.   F.J. Humphreys, P.B. Prangell and R. Priestner: in Proceedings of the Fourth International Conference on Recrystallization and Related Phenomena, T. Sakai and H. Suzuki (eds.), Japan Inst. of Metals, Sendai, 1999, pp. 69-78
14.   F.J. Humphreys: Acta Mater., 1997, vol. 45, p. 5031
15.   T. R. McNelley and M. E. McMahon: Metall. Mater. Trans. A, 1997, vol. 28A, pp. 1879-87.
16.   T. R. McNelley, M. E. McMahon and M. T. Pérez-Prado: Phil. Trans. R. Soc. Lond. A, 1999, vol. 357, pp. 1683-1705.
17.   V. Randle: Microtexture Determination and Its Applications, The Institute of Metals, 1992.
18.   T.R. McNelley and M.E. McMahon: J. of Metals, 1996, vol. 48, no. 2, pp. 58-60.
19.   H.J. Bunge, Texture Analysis in Materials Science, Butterworths, London, 1982.
20.   B.L. Adams:Metall. Trans. A, 1986, vol. 17A, pp. 2199-2207.
21.   F. Haessner, J. Pospiech, and J. Sztwiertnia: Mater. Sci. Engng., 1983, vol. 57, pp.1-14.
22.   C.S. Lee, B.J. Duggan and R.E. Smallman: Acta Mater., 1993, vol. 41, pp. 2265-2270.
23.   C.S. Lee and B.J. Duggan: Acta Mater., 1993, vol. 41, pp. 2691-2699.
24.   S.S. Kulkarni, E.A. Starke and D. Kujlmann-Wilsdorf: Acta Mater., 1998, vol. 46, pp. 5283-5301.
25.   C.S. Barrett: Trans. Am. Inst. Min. Engrs., 1939, vol. 135, p. 296.
26.   C.S. Barrett and L.H. Levenson: Trans. Am. Inst. Min. Engrs., 1940, vol. 137, p. 112-127.
27.   J. Hirsch and K. Lücke: Acta Metall., 1988, vol. 36, pp. 2883-2904.

SUPERPLASTICITY IN Nb$_3$Al/Nb IN-SITU COMPOSITES

S. HANADA and W. FANG
Institute for Materials Research, Tohoku University, Sendai, 980-8577, Japan

ABSTRACT

Microstructures of a binary Nb-15.8at%Al alloy ingot were controlled by isothermal forging and heat treatment to produce equiaxed, fine grains of Nb$_3$Al and Nb solid solution (Nb$_{ss}$). Nb$_3$Al/Nb$_{ss}$ two phase alloy (in-situ composite) is found to exhibit superplasticity especially when one of the constituent phases, Nb$_{ss}$, is supersaturated. During superplastic deformation Nb$_{ss}$ transforms to Nb$_3$Al, and Al content in Nb$_{ss}$ decreases. After superplastic deformation the microstructure consisting of equiaxed grains is left unchanged, although a slight grain growth is observed. It is suggested that stress induced by grain boundary sliding is effectively accommodated through dislocation glide and climb in the soft Nb$_{ss}$.

INTRODUCTION

Current studies on the replacements of nickel-base superalloys at very high temperatures have focused on refractory intermetallics and their composites [1-5]. Among them, Nb$_3$Al/Nb$_{ss}$ in-situ composites are considered to be one of the most promising candidates because of their high melting point, low density and favorable mechanical properties. However, poor deformability of Nb$_3$Al/Nb$_{ss}$ in-situ composites has been pointed out to be one of major obstacles in practical application. As generally believed, superplastic forming will provide a technological breakthrough in solving these problems. Superplasticity has been found in many ordered intermetallics when initial microstructures are fine and grain growth is sluggish during deformation, except for some bcc-ordered intermetallics. However, no information is available concerning superplasticity in refractory intermetallic alloys. Recently, superplasticity in Nb$_3$Al/Nb$_{ss}$ in-situ composites, which were prepared by HIPing of mechanically alloyed Nb-Al powders, has been investigated. Unexpectedly, superplastic elongation was not obtained even though initial grain sizes of the constituent phases, Nb$_3$Al and Nb$_{ss}$, were less than a few micron. Scanning electron microscopy on fractured surfaces indicated that many cavities were produced around niobium carbide particles which were generated by a reaction of process control agent (methanol) with niobium. Therefore, small elongation was suggested to be due to the cavity formation and its coalescence. However, superplasticity did not appear even when the carbide formation was suppressed by mechanical alloying without using methanol and the grain sizes were still fine. At least two explanations are possible for these experimental results. One is that the cohesion of grain boundaries or interface boundaries is not strong in samples prepared by powder metallurgy. The other is that no suitable accommodation process, which is accompanied with grain boundary sliding, is activated in fine-grained Nb$_3$Al/Nb$_{ss}$ in-situ composites.

The objective of this work is to investigate the possibilities experimentally and to clarify whether refractory Nb$_3$Al alloys exhibit superplasticity or not. To check the former possibility, fine-grained samples are prepared by thermomechanical processing of cast ingots. The latter possibility is checked by measuring tensile elongation as a function of strain rate at a fixed temperature, since stress accommodation in Nb$_{ss}$ is considered to be a rate-controlling process.

EXPERIMENTAL PROCEDURE

Referring to the Nb-Al binary phase diagram in Fig. 1 [6], Nb-16at%Al alloy was arc melted to obtain two phase alloy consisting of Nb_3Al and Nb_{ss}. In addition, Nb-7at%Al, Nb-6at%Al and Nb-5at%Al alloys were arc melted to evaluated high temperature strength of monolithic Nb_{ss}. The arc-melting was carried out under an argon atmosphere using a non-consumable tungsten electrode in a water-cooled copper hearth. Arc-melted Nb-16at%Al alloy buttons of 40 mm in diameter by 8 mm in thickness were flipped and remelted five times to improve their homogeneity. They were homogenized at 2173 K for 3 h in the range of a single phase Nb_{ss} in a vacuum of better than 8×10^{-4} Pa, and isothermally forged to 70% reduction in thickness at 1573 K in the two phase region and at an initial strain rate of about 10^{-4} s^{-1}. This process has been recently found to yield equiaxed dual phase microstructure. The Nb_{ss} alloys were homogenized at 2173 K for 3 h and electro-discharge machined to produce compression samples with dimensions of 4 mm × 4 mm × 5 mm. The chemical composition of the as-isothermally forged button is Nb-15.8 at%Al (this alloy will be abbreviated as Nb-15.8Al hereafter) and compositions of the Nb_{ss} alloys are 6.60, 5.60 and 4.04 at%Al (abbreviated as Nb-6.60Al, Nb-5.60Al and Nb-4.04Al, respectively).

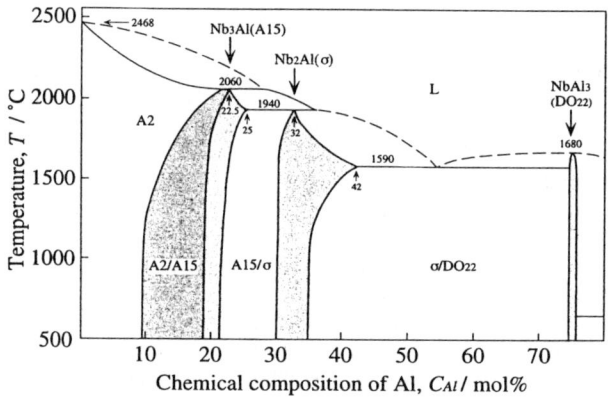

Figure 1   Binary Nb-Al phase diagram [6].

Tensile specimens of 1.5 mm × 2 mm × 10.5 mm in gage sections were electro-discharge machined from the forged blocks. Tensile tests were carried out using an Instron testing machine in a vacuum of better than 6×10^{-4} Pa and below 1623 K in a furnace equipped with a tungsten mesh heater. Tensile specimens were heated to deformation temperature at a heating rate of 15 K/min and maintained at the temperature with an accuracy of ± 1 K during deformation. They were pulled isothermally at a constant cross head speed from 0.02 to 0.5 mm/min, corresponding to initial strain rates ($\dot{\varepsilon}$) from 3.2×10^{-5} to 8×10^{-4} s^{-1}, until final fracture occurred. The obtained stress-strain curves were converted to true stress-strain curves by assuming homogeneous deformation according to the method employed by Oomori et al. [7]. In some cases, deformation was interrupted at a given elongation and specimens were cooled as fast as possible, in order to retain the microstructure evolved during high temperature deformation. Specimens were mechanically ground with 1500 grade SiC emery paper and lapped using lapping tape up to 4000 grade. The samples were electro-polished using a solution of 5% H_2SO_4 in methanol at 227 K and 18 mA. Electro-polished samples were chemically etched using a solution of $5HNO_3 + 10HF +$

$15H_2SO_4$ + $70H_2O$ (volume percent). Metallographic and fractographic observations were conducted by means of scanning electron microscopy (SEM) and transmission electron microscopy (TEM). Identification of existing phases was carried out by X-ray and electron diffraction, and electron probe microanalysis (EPMA) was performed to identify chemical compositions of the constituent phases. The volume fraction of phases was determined from SEM micrographs. Thin foils for TEM were taken from the center part of the gage section and observed using a HITACHI-8100 microscope operated at 200 kV.

RESULTS AND DISCUSSION

Tensile properties

It has been found that needle-like Nb_3Al precipitates in Nb_{ss} when a Nb-Al alloy with a composition of Nb-10~18%Al is annealed in the two phase region after homogenization in the Nb_{ss} single phase region at high temperature. Recently, Tabaru and Hanada [8] have observed that when precipitation of Nb_3Al occurs during isothermal forging of homogenized alloys, the two phase structure consisting of equiaxed Nb_3Al and Nb_{ss} grains is formed. Equiaxed microstructure similar to their result was obtained in this study after isothermal forging at 1573 K. Figure 2 shows SEM micrographs of Nb-15.8Al annealed at 1573 K for 60 h after isothermal forging. Quantitative analysis of the micrograph indicated that equiaxed grains have an average grain size of about 4 μm for both the phases, and volume fraction of Nb_{ss} is 0.35. These micrographic features seem to satisfy the conditions for fine-grained superplasticity. Actually a maximum elongation of about 400% was obtained in the annealed Nb-15.8Al samples with the equiaxed microstructure, which was greater than the samples prepared by powder metallurgy. Since the samples used in this experiment were prepared by thermomechanical processing of ingots, grain boundaries and phase boundaries would not be contaminated, as compared with the samples prepared by powder metallurgy processing. Thus, this result will not rule out the first possibility that the weak cohesion of grain boundaries or interface boundaries is responsible for the small elongation of powder metallurgy-processed samples.

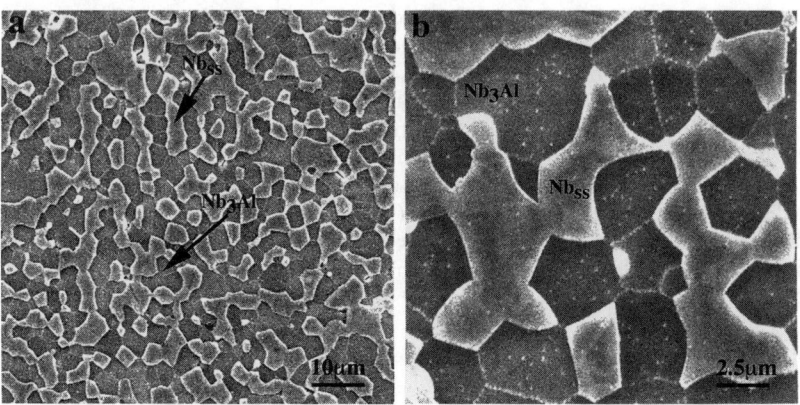

Figure 2   (a) SEM micrographs of Nb-15.8Al annealed at 1573 K for 60 h
after isothermal forging. (b) at a high magnification.

Next, the second possibility concerning a stress accommodation process is discussed. In superplastic α+β titanium alloys it has been revealed that β phase plays an important role in a stress accommodation process of grain boundary sliding, thereby leading to a large superplastic elongation at a suitable volume fraction of β phase. We have found in the annealing experiment described above that the volume fraction of Nb_{ss} changes from 0.48 to 0.35 on annealing at 1573 K after isothermal forging, although grain size and grain shape are left almost unchanged. To investigate the effect of volume fraction of Nb_{ss} on elongation, tensile properties were examined using as-isothermally forged samples with the Nb_{ss} volume fraction of 0.48. Microstructure of the as-isothermally forged sample is shown in Fig. 3. Although particles of Nb_3Al and Nb_{ss} appear to be slightly elongated in the SEM micrograph in Fig. 3(a), the TEM micrograph in Fig. 3(b) shows equiaxed microstructure consisting of Nb_3Al and Nb_{ss}. This implies that a few or several Nb_3Al or Nb_{ss} grains are clustered to form elongated particles in SEM.

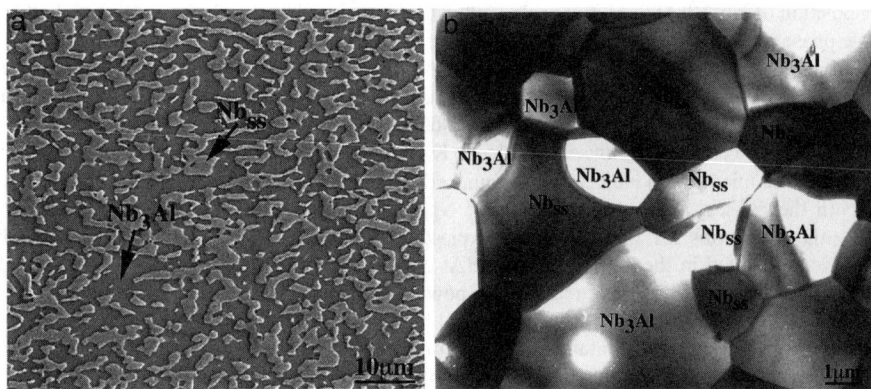

Figure 3   (a) SEM and (b) TEM micrographs of as-isothermally forged Nb-15.8Al.

True stress-strain curves at 1573 K are shown in Fig. 4 as a function of strain rate. Observed are weak work hardening at low strain rates and weak work softening at high strain rates. Flow stresses at ε = 0.1, 0.5, 1.0 and 1.5 are plotted against strain rate in Fig. 5. It is obvious that large m values from 0.5~0.4 are obtained depending on strain. Tensile specimens after deformation are shown in Fig. 6. Evidently, superplastic elongation without necking appears especially at lower strain rates. The three specimens at the lower strain rates did not fracture within the limitation of cross head movement of the tensile testing machine, which corresponded to 900% elongation. To know fracture strain under superplastic deformation conditions, the shape of tensile specimens was modified to have a shorter gage length of 5 mm (the original gage length is 10.5 mm) and tested at 1573 K and at various strain rates. The obtained results are summarized in Fig. 7. A maximum elongation of 950% is obtained for the specimen with the gage length of 5 mm at the initial strain rate of $1.6 \times 10^{-4} s^{-1}$.

Microstructures

Corresponding to tensile properties, deformation microstructure is sensitive to strain rate. Figure 8 shows SEM micrographs of specimens fractured at three strain rates, where the surfaces

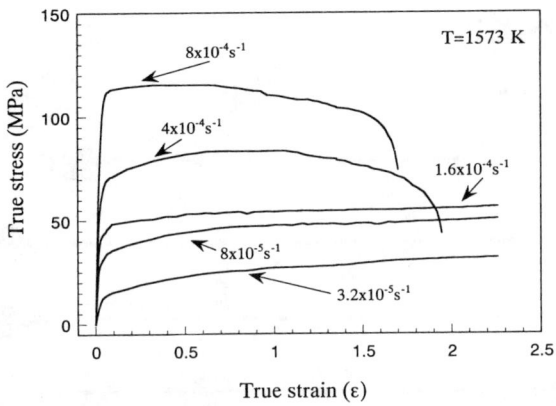

Figure 4   Stress-strain curves of as-isothermally forged Nb-15.8Al at 1573 K.

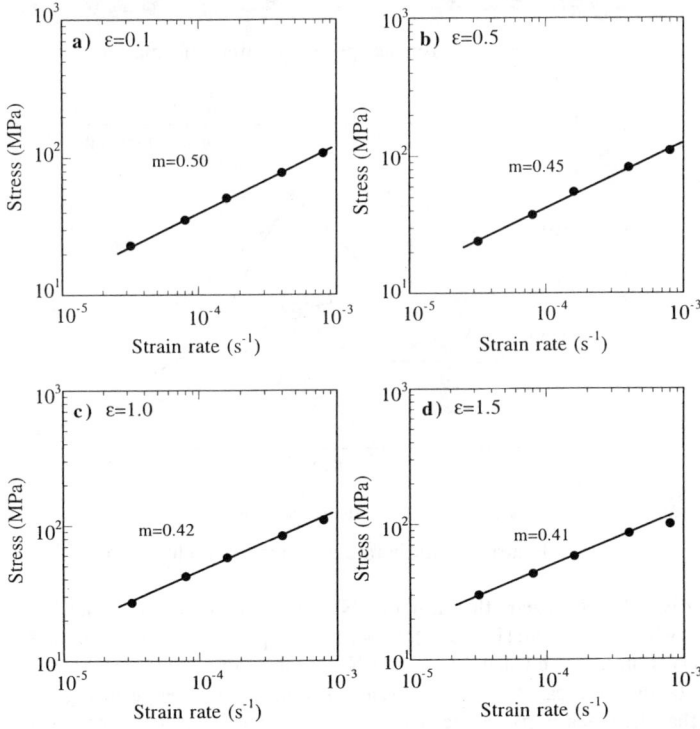

Figure 5   Strain rate dependence of flow stress for as-isothermally forged Nb-15.8Al at 1573 K.

of the fractured specimens were mechanically removed, polished with emery paper and finally electro-polished. Tensile direction is vertical and the side surface of the specimens is in the left edge

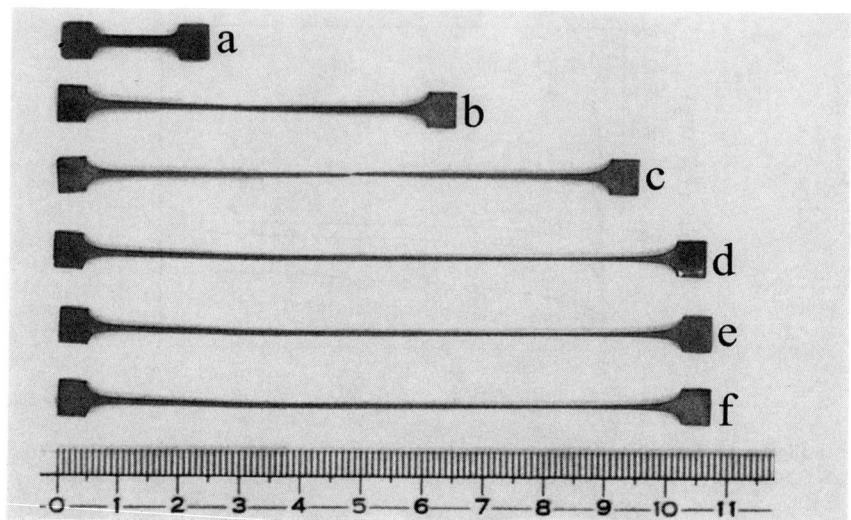

Figure 6　Tensile specimens after deformation.

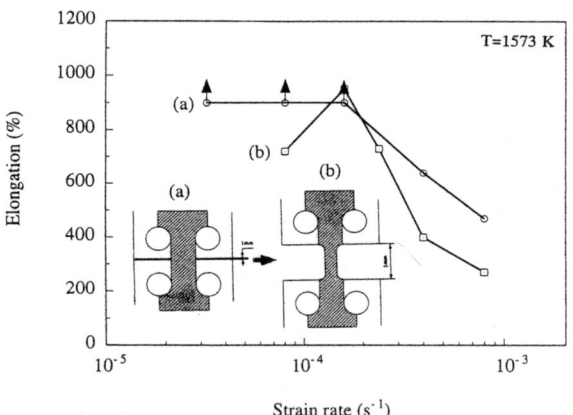

Figure 7　Strain rate dependence of elongation.

of the micrographs. It appears that although Nb_{ss} clusters are slightly elongated toward the tensile direction at a high strain rate (Fig. 8(a)), they remain equiaxed at low strain rates (Fig. 8(b) and (c)). TEM observation indicated that Nb_3Al and Nb_{ss} grains are still equiaxed and their grain sizes are less than 10 μm. One can see black particles around the side edge in Fig. 8(b) and (c). EPMA indicated that the black particles are Al_2O_3. Since the specimens deformed at a low strain rate are hold for a long time at high temperature, it is considered that their surfaces are oxidized and Al_2O_3 is formed in spite of testing in vacuum. Another characteristic feature in Fig. 8 is cavity formation. Coalescing cavities are frequently observed especially at the lowest strain rate (Fig. 8(c)), suggesting that fracture is controlled by cavity formation and coalescence.

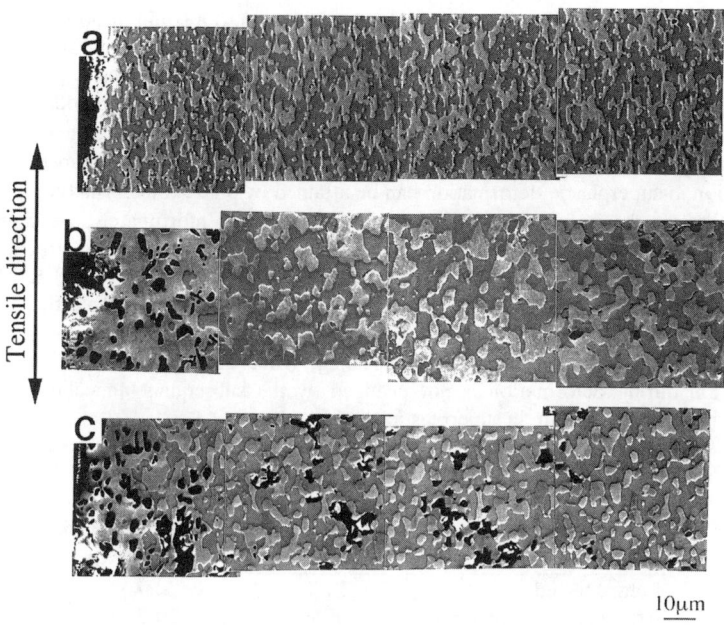

Figure 8   SEM micrographs of specimens fractured at (a) $8 \times 10^{-4}\text{s}^{-1}$,
(b) $1.6 \times 10^{-4}\text{s}^{-1}$ and (c) $8 \times 10^{-5}\text{s}^{-1}$.

Changes of volume fraction of Nb_{ss} during deformation at 1573 K are plotted in Fig. 9, where data for the static heat treatment (annealing without tensile stress) are included for comparison.

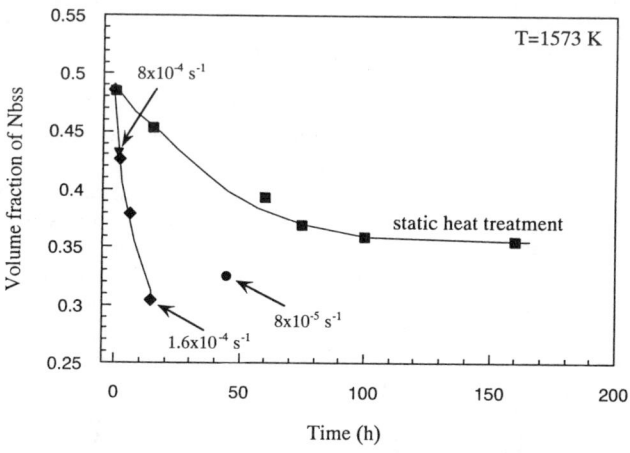

Figure 9   Changes of Nb_{ss} volume fraction during deformation.

The volume fraction of Nb_{ss} during superplastic deformation changes rapidly compared with that during static heat treatment, and after superplastic deformation to a strain of 900% at 1.6 x 10^{-4}s^{-1} it reaches about 0.3, which is smaller than that under the static heat treatment. Changes in volume fraction at other strain rates are also accelerated by deformation. These drastic decreases in Nb$_{ss}$ during superplastic deformation may be resistant to the operation of an accommodation process of grain boundary sliding. Recently, Koike et al. [9] have found that β volume fraction in α + β titanium alloy increases during superplastic deformation. They claimed that stress accommodation in superplastic deformation can be attained by phase transformation from the α to the β phase through the isotropic expansion in atomic volume. Unfortunately, we cannot discuss the atomic volume change accompanied with phase transformation from Nb$_{ss}$ to Nb$_3$Al, because at present there are no available data on lattice constants of Nb$_3$Al and Nb$_{ss}$ at 1573 K. However, it should be noted that although phase transformation is associated with superplasticity in the both alloys, its role will be different. In the α + β titanium alloy, the increased β volume fraction during deformation is reduced by the subsequent annealing, while in Nb-15.8Al, the decreased Nb$_{ss}$ volume fraction during deformation is not changed by the subsequent annealing. Therefore, a different mechanism will operate in superplastic Nb-15.8Al.

EPMA results show that Al concentration in Nb$_{ss}$ in the gage section decreases from 11.7 to 4.8 at% after deformation to 900%, as shown in Table 1. The analyzed Al content of 4.8 at% is much

Table 1   Chemical compositions of Nb$_{ss}$ and Nb$_3$Al before and after tensile tested.

	Nb$_{ss}$ (at%Al)	Nb$_3$Al (at%Al)
Before tested	11.4	20.9
After tested (gage area)	4.8	20.7
After tested (grip area)	6.6	20.9

lower than the solubility of Al in Nb$_{ss}$ at 1573 K, which was reported by Jorda et al. [6], but it is in good agreement with the result recently presented by Semboshi et al. [10]. This discrepancy may be due to different heat treatment conditions to attain equilibrium state; Jorda et al. extrapolated the solubility limit of Al to Nb$_{ss}$ around 1573 K using a sample annealed at 1273 K for 24 h, while Semboshi et al. did at 1473 K for 24 h. Therefore, the latter sample seems to approach the equilibrium state more closely. If this is the case, it is concluded that deformation enhances Al diffusion in supersaturated Nb$_{ss}$ and the equilibrium state is reached more quickly during tensile testing than during static annealing.

TEM observations of superplastically deformed Nb-15.8Al revealed that Nb$_3$Al grains remain essentially equiaxed and contain no dislocation, suggesting that grain boundary sliding is occurring at grain boundaries between Nb$_3$Al and adjacent Nb$_3$Al or Nb$_{ss}$ grains. By contrast, Nb$_{ss}$ grains contain a high density of dislocations, especially around grain boundaries, as shown in Fig.10, although the equiaxed structure is still maintained and no precipitation is observed inside Nb$_{ss}$ grains. The formation mechanism of the high dislocation density areas near grain boundaries is not understood at present, but one possible explanation is that they are produced by grain boundary sliding between Nb$_3$Al and Nb$_{ss}$ grains. That is, these dislocations will be introduced to accommodate stress generated by grain boundary sliding. This is simply because Nb$_{ss}$ grains are not elongated, but remain equiaxed. At the highly strained regions, grain boundary migration from Nb$_3$Al to Nb$_{ss}$ will be promoted through enhanced Al diffusion. That is, phase transformation always occurs at stress concentrated regions near grain boundaries. Such grain boundary migration is needed to decrease the volume fraction of Nb$_{ss}$ without changing the equiaxed grain microstructure. This explanation is consistent with the TEM observation that Nb$_3$Al does not precipitate within Nb$_{ss}$ grains.

The proposed idea to explain the present superplasticiy is concerned with the introduction of

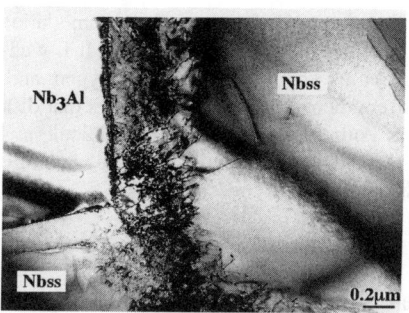

Figure 10   TEM micrograph indicating a grain boundary between Nb_3Al and Nb_{ss}.

dislocations in Nb_{ss}, which is accompanied with grain boundary sliding. The ease of dislocation glide and climb in Nb_{ss}, which is closely related to stress accommodation, can be evaluated by flow stress of Nb_{ss} at the superplastic deformation condition. The decrease in the volume fraction of Nb_{ss} during superplastic deformation necessarily decreases Al content of Nb_{ss}. Therefore, solid solution strengthening of Nb_{ss} was investigated as a function of Al content. As shown in Table 1, Al contents of Nb_{ss} are 10.7 at% before tested, 4.8 at% after tested (gage area) and 5.1 at% after tested (grip area).   If Nb_3Al precipitation from supersaturated Nb_{ss} occurs during deformation at 1573 K, the composition dependence of solid solution strengthening in Nb_{ss} cannot be evaluated precisely. Then, deformation was carried out at 1773 K and at $1.6 \times 10^{-4} s^{-1}$ using monolithic Nb-4.04 at%Al, Nb-5.60at%Al and Nb-6.60 at%Al. Fig.11 shows the stress-strain curves of these alloys. Solid solution strengthening is clearly seen; flow stress is remarkably decreased with decreasing Al content. The decreased flow stress is expected to facilitate dislocation glide and climb as an accommodation process. As a result, the accommodation process will work well in low Al containing alloy, even though the Nb_{ss} volume fraction is decreased during superplastic deformation.

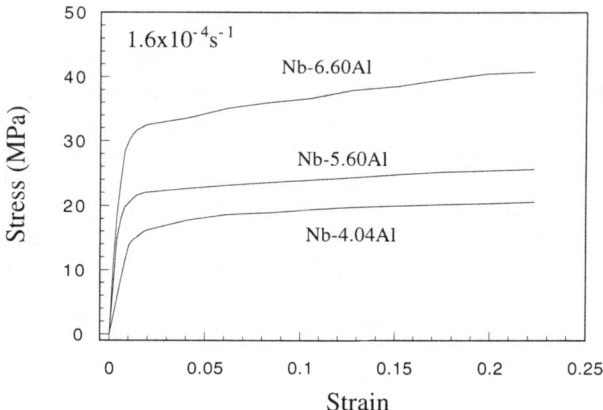

Figure 11   Stress-strain curves of monolithic Nb-Al alloys.

Thus, the second possibility concerning a stress accommodation process was related to impurity contamination during powder metallurgy processing. It is well known that niobium has a strong tendency to absorb gaseous impurities such as oxygen and nitrogen during powder processing and is remarkably strengthened by them. This means that dislocation glide and climb for stress accommodation of grain boundary sliding become difficult in Nb_{ss}, which results in the smaller elongation of powder metallurgy-processed Nb_3Al/Nb_{ss} in-situ composite with fine grains, as compared with ingot metallurgy-processed ones.

CONCLUSION

1. Superplastic elongation over 900% is obtained in Nb_3Al/Nb_{ss} in-situ composite (Nb-15.8Al alloy) on deformation at 1573K and at strain rates of 3.2×10^{-5} to 8×10^{-4} s^{-1} under the m values of 0.5~0.4.
2. Initial equiaxed grain structure remains unchanged after superplastic deformation, although grain growth occurs slightly.
3. Volume fractions of Nb_{ss} (Nb solid solution) and Nb_3Al change during superplastic deformation; the volume fraction of Nb_{ss} decreases from 0.48 to 0.3 during superplastic deformation to 950%.
4. Decrease in Nb_{ss} volume fraction during deformation is enhanced with increasing strain rate, i.e., with increasing flow stress.
5. Nb_3Al grains contain no dislocation after superplastic deformation, while a high density of dislocations is observed in Nb_{ss} especially near grain boundaries.
6. Grain boundary migration from Nb_3Al to Nb_{ss} occurs probably at highly strained Nb_{ss} regions near grain boundaries.
7. Strength at 1773 K of monolithic Nb-Al solid solution alloys decreases with decreasing Al content, which facilitates stress accommodation via dislocation glide and climb.
8. Based on the obtained results, it is suggested that superplasticity in Nb_3Al/Nb_{ss} in-situ composite is associated with grain boundary sliding and grain boundary migration accompanied with phase transformation.

REFERENCES

1. M.R.Jackson and K.D.Jones, Refractory Metals: Extraction, Processing and Applications, K.C.Liddell, D.R. Sadoway and R.G.Bautista, eds., TMS, Warrendale, PA, 1991, p. 311.
2. M.G.Mendiratta and D.M.Dimiduk, High Temperature Ordered Intermetallic Alloys III, C.T.Liu A.L.Taub, N.S.Stoloff and C.C.Koch, eds., MRS, Pittsburgh, PA, 133, 237 (1991).
3. M.G.Mendiratta, J.J.Lewandowski and D.M.Dimiduk, Met. Trans. A. 22A, 1573 (1990).
4. M.G.Mendiratta and D.M.Dimiduk, Met. Trans. A. 24A, 501 (1993).
5. D.M.Shah and D.L.Anton, Intermetallic Matrix Composites II, D.B.Miracle, D.L.Anton, J.A.Graves, eds., MRS, Pittsburgh, PA, 273, 385 (1993).
6. J.L.Jorda, R.Flukiger and J.Muller, J. Less-Common Metals, 75, 227 (1980).
7. T.Oomori, T. Yoneyama and H. Oikawa, Trans. JIM. 29, 399 (1988).
8. T. Tabaru and S. Hanada, Intermetallics, 6, 735 (1998).
9. J. Koike, Y.Shimoyama, T.Okamura and K.Maruyama, Mater. Sci. Forum, 304-306, 183 (1999).
10. S.Semboshi, T.Tabaru, H.Hosoda and S.Hanada, Intermetallics, 6, 61 (1998).

A TEM STUDY OF DYNAMIC CONTINUOUS RECRYSTALLIZATION IN A SUPERPLASTIC AL-4MG-0.3SC ALLOY

L.M. DOUGHERTY*, I.M. ROBERTSON*, J.S. VETRANO**
*Dept of Materials Science & Engineering, Univ of Illinois at Urbana-Champaign, Urbana, IL
**Pacific Northwest National Laboratory, Richland, WA

ABSTRACT

An Al-4Mg-0.3Sc alloy, aged at 280°C for 8 hours and cold rolled to a 70% reduction, exhibited dynamic recrystallization during superplastic forming at 460°C and at a strain rate of 10^{-3}sec^{-1}. To understand the progression of recrystallization during forming, specimens were deformed under these same conditions to 0.1, 0.2, 0.4 and 0.8 true strain and studied post-mortem using optical microscopy, transmission electron microscopy and orientation imaging microscopy. The microstructural evolution that occurred between each strain state was directly observed during deformation experiments at a nominal temperature of 460°C in the transmission electron microscope. These in-situ experiments showed the migration, coalescence, disintegration and annihilation of subboundaries. This combination of post-mortem analysis of specimens strained in bulk with real time observations made during these in-situ experiments allows the mechanisms operating during dynamic continuous recrystallization to be ascertained.

INTRODUCTION

The occurrence of dynamic continuous recrystallization during superplastic forming is well documented. Several mechanisms have been proposed to account for the microstructural changes that accompany this process. These proposed mechanisms include the migration of subboundaries and their coalescence to form higher-angle boundaries or their annihilation to reduce subboundary density [1]. This is basically the same mechanism believed to occur during static continuous recrystallization except the process is accelerated by the applied stress. Another proposed mechanism is the generation of dislocations at sources within the grains or grain boundaries. These dislocations migrate individually until they are incorporated into subboundaries, thereby increasing misorientation incrementally [2]. However, this mechanism alone cannot account for the decrease in subboundary density with dynamic recrystallization. A third proposed mechanism is the rotation of subgrains along their low-angle boundaries, in a process similar to grain boundary sliding [3]. This is based on research by Weinburg [4], which shows that shear can occur along boundaries with misorientations as low as 5 degrees. Hales et al. [5] later modified this theory by suggesting that sliding occurs on high-angle grain boundaries remaining from casting while subboundaries are eliminated through dislocation emission. Since grains from casting are large, typical accommodation mechanisms for grain boundary sliding break down and dislocations are generated at triple junctions and other points of stress concentration.

These mechanisms are based on post-mortem examination of the microstructure and suffer from the drawback that the mechanisms producing the structure are not observed. In this paper we report preliminary observations from elevated temperature straining experiments performed in-situ in the transmission electron microscope (TEM). This technique allows the path that terminates with the dynamically recrystallized microstructure of superplastically deformed specimens to be observed at high spatial resolution and in real time.

Mat. Res. Soc. Symp. Proc. Vol. 601 © 2000 Materials Research Society

EXPERIMENT

The alloy used in this study is Al-4Mg-0.3Sc (in weight percent) produced by Kaiser Aluminum. The chemical composition is 95.13Al-3.40Mg-0.239Sc-0.0597Si-0.262Fe with trace amounts of Cu, Zn and Mn. The alloy was cast in an 8x10x3-inch book mold with surfaces scalped after cooling. To assure consistency in initial structure and phase state, the alloy was then solution treated at 575°C for 4 hours.

A thermomechanical processing treatment was developed to produce a microstructure that would resist static recrystallization during the system stabilization hold prior to superplastic forming but permit dynamic recrystallization during forming. This treatment consists of aging for 8 hours at 280°C to attain peak hardness followed by cold rolling to a 70% reduction. This peak hardness condition was determined by performing a series of aging treatments at 250, 280, 300, 320 and 350°C and measuring the Vickers Hardness number for each condition. The processed alloy was found to statically recrystallize at temperatures above 510°C but not at 460°C; therefore, 460°C was selected as the superplastic testing temperature.

Bulk tensile specimens were strained at 460°C and at a strain rate of 10^{-3}sec^{-1} in a Sintech 1/G MTS tensile testing apparatus controlled by MTS Testworks software with crossheads enclosed by an Applied Test Systems, Series 300, 3-zone, cylindrical furnace. One specimen was strained to failure and 4 others were deformed to true strains of 0.1, 0.2, 0.4 and 0.8 and quenched with a liquid nitrogen spray to freeze in the microstructure.

The specimens were sectioned, mounted and then polished to study grain boundary distribution by using orientation imaging microscopy (OIM) and optical microscopy. The OIM samples were first mechanically polished down to a 0.25μm diamond paste, then polished with a dilute silica slurry in a vibratory polisher, and then chemically polished with 5% HF in water for 25 seconds. OIM was used to define texture, the size and shape of grains, the crystallographic orientation of grains and subgrains, and the misorientation of boundaries in the gage and grip of each specimen.

TEM samples were prepared by punching disks from material ground to a 150μm thickness. These disks were jet-polished with 5% perchloric acid in methanol. Post-mortem TEM study was used to examine subboundaries and dislocation structures. In-situ, heating and straining, TEM experiments were performed with a Gatan sample holder in a Hitachi 9000 at Argonne National Laboratory on material deformed to true strains of 0.2 and 0.4. Results were captured on Super VHS video tape for later analysis.

RESULTS

Material aged at 280°C for 8 hours and cold-rolled to a 70% reduction attained 375% engineering strain when deformed at 460°C and at a strain rate of 10^{-3}sec^{-1}, as shown in Fig. 1. Optical microscopy and OIM revealed partial recrystallization in the gages of the specimens strained to 0.4 and 0.8 true strain but not in the grips. This proves that dynamic recrystallization occurred during superplastic forming while static recrystallization was completely suppressed.

The TEM micrographs shown in Fig. 2 illustrate the progression of the microstructure during superplastic testing. After a system stabilization hold at 460°C for 17 minutes, the microstructure is partially recovered, as shown in Fig. 2(a). At $\varepsilon_T = 0.1$, Fig. 2(b), recovery appears to be complete as most lattice dislocations have disappeared, although some remain pinned at the Al_3Sc particles. The subboundaries are now ordered with uniform spacing between the dislocations. In addition, there is very little evidence of lattice dislocations interact-

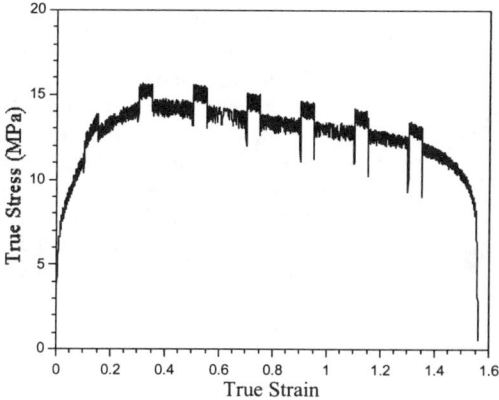

Fig. 1. True stress vs. true strain curve of a sample aged for 8 hours at 280°C, cold rolled to a 70% reduction and deformed at 460°C and at a strain rate of 10^{-3}sec^{-1}. The bumps in the curve are due to slight variations in the strain rate which are used to determine the strain rate sensitivity of the alloy.

ing with them. Between 0.1 and 0.8 true strain, the lattice dislocation density increases and the average subgrain size increases. In addition to this increase in dislocation density with increasing strain, the complexity of dislocation interactions with Al_3Sc particles and grain and subgrain boundaries increases. Dislocation lines become more convoluted as they now interact extensively with the fine Al_3Sc particles and with other dislocations. The dislocation arrays are not as well ordered, with irregularities in the dislocation spacing. At 0.8 true strain, Fig. 2(c), most boundaries are high-angle and the grain structure is very fine. No dislocation arrays are evident, however, entire grains full of dislocation tangles still exist. These qualitative observations, made using the TEM, of increasing high-angle boundary density and grain size with increasing strain were confirmed by optical microscopy and OIM.

To understand the path followed by the evolving microstructure between true strains of 0.2 and 0.4, samples taken from the specimen strained to 0.2 were deformed at a nominal temperature of 460°C in the TEM. These experiments were used to reveal the mechanisms operating during dynamic continuous recrystallization. In the following, we show an example of migration, coalescence, disintegration and annihilation of subboundaries. To aid in describing the events, a schematic of the area is presented in Fig. 3 in which the boundaries (numbers), particles (letters) and direction of motion of the subboundaries (hollow arrows) are indicated.

A series of images captured from videotape are presented in Fig. 4. Fig. 4(a) shows a low magnification image of the area illustrated schematically in Fig. 3. The first event is the migration and coalescence of the dislocation arrays 1 and 2. The creation of array 2 was not ob-

Fig. 2. TEM micrographs of the subgrain structure at (a) •T = 0, (b) •T = 0.1 and (c) •T = 0.8.

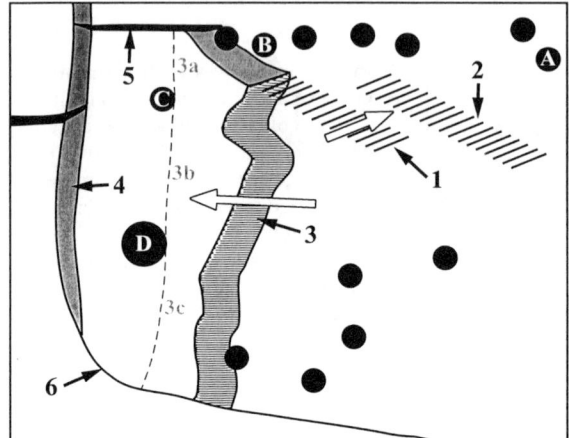

Fig. 3. A schematic of the initial state of the area observed during the in-situ heating and straining experiment in the TEM. Particles are indicated with letters, boundaries with numbers, and initial directions of boundary motion with hollow arrows. The thin line divided into segments 3a, 3b and 3c marks the position of boundary 3 once arrays 1 and 2 have coalesced.

served and array 1 appears to be associated with boundary 5 (it will become apparent later that array 1 was originally part of boundary 5 and is created as subboundary 3 migrates towards boundary 4). Array 1 migrates as a unit towards array 2, Fig. 4(b), and is eventually fully accommodated within array 2, Figs. 4(c) and 4(d). This unit migrates in the direction indicated and interacts with the large Al$_3$Sc particles, Fig. 4(e). Eventually the array disintegrates and the dislocations move independently. These dislocations are individually accommodated at a grain boundary (not shown).

The migration of subboundary 3 can be seen by comparing Figs. 4(f) and 4(g). In Fig. 4(f), the subboundary is between the Al$_3$Sc particles B and C. The position of subboundary 3 at this time is indicated by the dashed line in Fig. 3. In Fig. 4(g), it has moved past particle C. In addition, subboundary 3 also interacts with particle D (not shown). The subboundary does not break free of particle D as a unit but breaks in two instead. The upper segment is mobile while the lower segment (segment 3c in Fig. 3) remains pinned between particle D and boundary 6. Figs. 4(h)-(j) show the fate of the mobile segment of subboundary 3. Initially, only the lower portion is mobile as seen in Fig. 4(h). The upper portion is still pinned at particle C. The subboundary again divides at particle C, leaving an immobile segment (segment 3a in Fig. 3) between particle C and boundary 5. The mobile portion (segment 3b in Fig. 3) continues to move towards boundary 4 where it is eventually absorbed. During this stage, numerous dislocations emerge from the vicinity of particle D, Fig. 4(i), although the exact source was not observed. With further time (straining), segment 3a breaks free and migrates as a unit in the direction indicated in Fig. 4(j). Simultaneously, boundary 5 absorbs lattice dislocations and migrates in the direction indicated in Fig. 4(k). Finally, the last remnant of subboundary 3 (segment 3c) breaks free and moves as a unit in the direction indicated in Fig. 4(l). As with the other segments of the subboundary, the dislocations of this one are eventually individually accommodated at grain boundaries. This sequence provides direct evidence for how subboundaries are annihilated and a structure containing predominantly high-angle grain boundaries develops.

In the proceeding section we have focused on the migration and annihilation of subboundaries but there is also a significant amount of activity of lattice dislocations within the grains. Dislocations are observed being generated from and absorbed into grain boundaries, and interacting with the fine and with the large particles as they slip through the grains. Dislocation break-away from the particles is also observed.

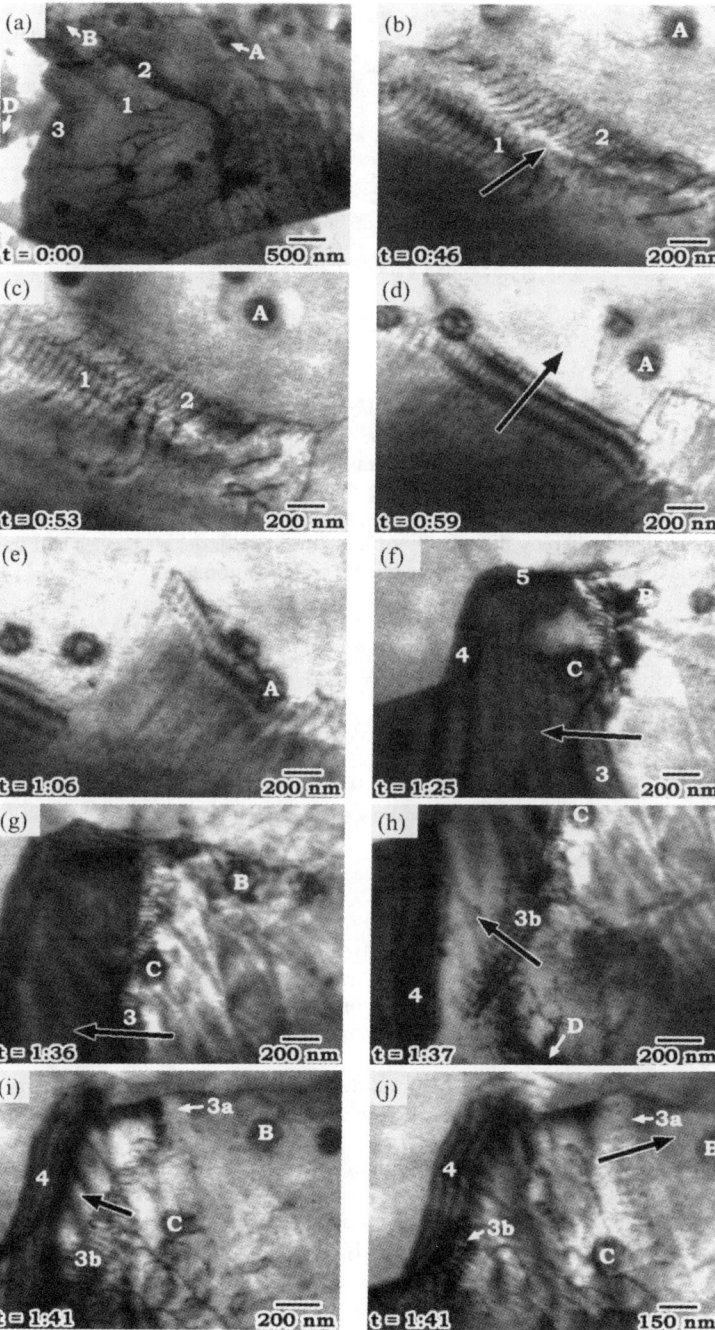

(a) t = 0:00 500 nm
(b) t = 0:46 200 nm
(c) t = 0:53 200 nm
(d) t = 0:59 200 nm
(e) t = 1:06 200 nm
(f) t = 1:25 200 nm
(g) t = 1:36 200 nm
(h) t = 1:37 200 nm
(i) t = 1:41 200 nm
(j) t = 1:41 150 nm

Fig. 4.
Images from video tape captured during an in-situ, heating and straining, TEM experiment. Letters and numbers correspond to the labels used in Figure 3. Arrows indicate the direction of subboundary or dislocation array migration. (a) Low magnification view showing the initial configuration of the region of interest. (b)-(d) Migration of subboundary 1 and coalescence with subboundary 2.
(e) Migration of new subboundary and interaction with large Al₃Sc particles.
(f)-(g) Migration of subboundary 3 as a whole.
(h) Breaking of subboundary 3 into 2 parts at particle D. (i) Division of upper part into segments 3a and 3b. Segment 3a remains fixed at particle C as segment 3b continues to migrate. (j) Accommodation of segment 3b at boundary 4. Unpinning and migration of segment 3a.

29

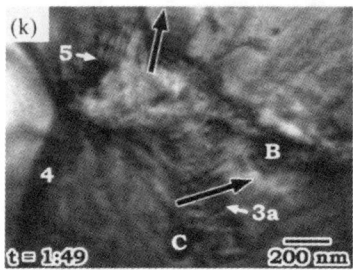

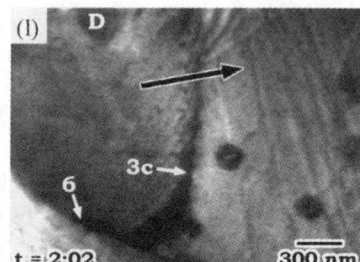

Fig. 4. (continued) (k) Initiation of boundary 5 migration. (l) Unpinning of segment 3c from particle D and migration in the direction indicated.

CONCLUSIONS

Dynamic continuous recrystallization has been shown to occur in Al-4Mg-0.3Sc when aged for 8 hours at 280°C, cold-rolled to a 70% reduction and superplastically formed at 460°C and at a strain rate of 10^{-3}sec^{-1}. In-situ heating and straining experiments in the TEM provide the first direct evidence for mechanisms operating during dynamic continuous recrystallization. Direct observations were made of subboundary migration, coalescence, disintegration and annihilation. These are mechanisms whereby the density of subgrain boundaries decreases and the grains become more equiaxed during dynamic continuous recrystallization

ACKNOWLEDGMENTS

The authors gratefully acknowledge sponsorship by the United States Department of Energy under contract 1-5-49183 and the Office of Basic Energy Sciences through the Pacific Northwest National Laboratory. Additional Support is provided by the Center for Excellence in the Synthesis and Processing of Advanced Materials. The use of the facilities in the Center of Microanalysis of Materials in the Frederick Seitz Materials Research Laboratory at the University of Illinois, at PNNL and Argonne National Laboratory are appreciated.

REFERENCES

[1]   A. M. Diskin and A. A. Alalykin, "Superplasticity of duralumin and magnalium type alloys with initial non-recrystallized structure," *Tsvetnye Metally (English translation)*, vol. 28, pp. 86-88, 1987.

[2]   X. Zhang and M. J. Tan, "Dislocation model for continuous recrystallisation during initial stage of superplastic deformation," *Scripta Metallurgica et Materialia*, vol. 38, pp. 827-831, 1998.

[3]   S. J. Hales and T. R. McNelley, "Microstructural evolution by continuous recrystallization in a superplastic Al-Mg alloy," *Acta Metallurgica et Materialia*, vol. 36, pp. 1229-1239, 1988.

[4]   F. Weinburg, "Grain Boundary Shear in Aluminum," *Transactions of the Metallurgical Society of AIME*, vol. 212, pp. 808-817, 1958.

[5]   S. J. Hales, T. R. McNelley, and H. J. McQueen, "Recrystallization and superplasticity at 300°C in an Al-Mg alloy," *Metallurgical Transactions A*, vol. 22, pp. 1037-1047, 1991.

EFFECT OF SN ADDITIONS ON SUPERPLASTICITY IN AL-MG-MN-SC ALLOYS

C. H. Henager, Jr., J. S. Vetrano, V. Y. Gertsman, and S. M. Bruemmer
Pacific Northwest National Laboratory, Richland, WA 99352, chuck.henager@pnl.gov

ABSTRACT

Identical Al-Mg-Mn-Sc alloys without and with 0.034-wt% Sn additions were fabricated, heat-treated, and tensile tested in a fine-grain (d < 6 μm) condition at four strain rates from 10^{-2} to 10^{-4} s^{-1} and at temperatures from 723K to 823K. Alloys with Sn additions exhibited reduced failure strains at 723K but higher failure strains at 823K for the slowest strain rates. The effect of Sn on flow stress, activation energy for flow stress, and strain rate exponent was explored and was found to be small. The main effect of Sn was suggested to be in reducing cavitation by allowing a redistribution of stress at critical hetero-junctions in the alloys.

INTRODUCTION

Grain boundary sliding (GBS) and concurrent accommodation control superplastic deformation in metals by diffusion and dislocation motion[1,2]. The process of GBS is considered to occur by diffusion coupled with movement of extrinsic grain boundary dislocations (EGBDs) along the interface[3]. The ease of motion and absorption of these grain boundary dislocations depends on the boundary characteristics, boundary chemistry, and boundary diffusivity. The process of lattice dislocation incorporation into the grain boundary can be influenced significantly by alloying elements[4,5]. Such solutes can also change local diffusion coefficients and, thus, rates of boundary sliding through grain boundary segregation.

Vetrano et al.[6] measured segregation of Sn (present as an impurity) to grain boundaries in a superplastic Al-Mg-Mn-Zr alloy. Though the role of Sn could not be positively ascertained, it was noted that the alloy with Sn cavitated less than one without under deformation conditions where GBS dominated. Research at our laboratory has shown that Sn can influence GB properties in Al-Mg-Sn alloys[5] by demonstrating that 80 appm Sn suppressed the spreading or dissociation of dislocations that had impinged on a grain boundary to a higher temperature than without Sn. This is consistent with a decrease in grain boundary diffusivity due to Sn[4,5], but is also consistent with solute pinning of grain boundary dislocations at Sn-containing boundaries[5]. Trace amounts of Sn have also been shown to increase sintering rates of Al-Cu-Mg powder processed alloys by binding to Al-vacancies[7] with a Sn-vacancy binding energy of 0.45 eV. Sn is expected to collect at vacancy sinks and to decrease the grain boundary diffusion rates relative to pure Al for boundaries at which Sn has segregated.

The role of Sn on deformation was investigated using a series of Al alloys (modified 5083 type) containing Mg, Mn, and Sc with and without Sn in solid solution. These were subjected to deformation under conditions that lead to extensive GBS to explore effects of Sn on activation energies, flow stresses, strain rate exponents, and elongations. Transmission electron microscopy (TEM) was performed on specimens quenched under load at specific strain levels to examine the deformed microstructure and microchemistry.

EXPERIMENTAL

The Al-Mg-Mn-Sc(-Sn) alloys were produced by the Kaiser Center for Technology and supplied as 25 cm x 20 cm x 5-cm book molds (Table 1). Following warm rolling at 573K and a recrystallization anneal for 2 h. at 773K, the material was cold-rolled 80% and given a recrystallization treatment of 10 m. at 673K to create a grain size of approximately 6 μm.

31

Standard dog-bone tensile samples were machined with a gage length of 25 mm and a width of 6.3 mm. Mechanical tensile testing using a 444.8 N (100 lb.) load cell was performed at temperatures from 723K to 823K and strain rates from 1×10^{-4} s^{-1} to 1×10^{-2} s^{-1} in a servo-motor electromechanical tensile tester[†] equipped with feedback software to maintain a constant true-strain-rate and to provide constant loads for the quench studies. A vertical split-chamber furnace with a 30-cm constant-temperature-heated-zone length was attached to the frame such that a total crosshead travel of 25.3 cm was available.

Table 1. Alloy compositions in wt. %, balance is Al.

Alloy	Mg	Mn	Sc	Sn	Si	Fe
A4M5S	4.20	1.00	0.52	-	0.029	0.039
A4M5S-3	4.00	1.00	0.52	0.034	0.027	0.027

Computer control of the crosshead allowed the implementation small strain-rate increases (bumps) to study the strain-rate sensitivity. The strain-rate increases were programmed to occur at true-strains of 0.1, 0.3, 0.5, 0.7, 0.9, 1.1, and 1.3. The magnitude and duration of the strain rate increase is shown in Table 2. A quench rate of approximately 75K s^{-1} was achieved by blowing liquid/gaseous N_2 over the sample while the load was held constant.

Table 2. Strain Rate and Strain Rate Increase Information.

Strain Rate (s^{-1})	Base Rate (s^{-1})	Increased "Bump" Rate (s^{-1})	Duration (Strain)
1×10^{-4}	1×10^{-4}	1.5×10^{-4}	0.05
5×10^{-4}	5×10^{-4}	7.5×10^{-4}	0.05
1×10^{-3}	1×10^{-3}	1.2×10^{-3}	0.05
1×10^{-2}	1×10^{-2}	1.2×10^{-2}	0.05

RESULTS

The alloys exhibited increasing failure strains with increasing temperature and decreasing strain rate, with the largest failure strains at 823K and 1×10^{-4} s^{-1} (Figure 1). The Sn alloy exhibited increased failure strains compared to the non-Sn alloy at 823K and 5×10^{-4} and 1×10^{-4} s^{-1}. The Sn alloy exhibited similar or reduced failure strains at 773K and at 723K compared to the non-Sn alloys (Figure 2). The effects of Sn on failure strains can be summarized as mildly detrimental for 723K and 773K and beneficial at 823K and 5×10^{-4} and 1×10^{-4} s^{-1}.

The strain rate exponent, or m-value, was evaluated both from the overall response of the series of alloys tests (Figure 3) using the base strain rate data at strains of 0.1 and 0.3 and from the instantaneous strain rate change (bump) data obtained over the strain range 0.1 to 1.3 (Figure 4). These data indicate that the m-value increases from 0.3 at 723K to 0.46 at 823K using the flow stress at strains of 0.1 and 0.3 and the base strain rate data for the Sn alloys whereas the m-value at 823K for the non-Sn alloy is slightly lower at 0.44. The strain rate exponent, when computed from the strain rate bumps (Table 2), reveals that the m-value varies as a function of strain (Figure 4). The m-values decrease with increasing strain for 1×10^{-2} and 1×10^{-3} strain rates, remain essentially constant for 5×10^{-4}, and increase with increasing strain for 1×10^{-4}, the strain rate for which the maximum elongation occurs. The only systematic difference between the Sn and non-Sn alloys is that the m-value for 1×10^{-2} is smaller for the Sn alloy compared to the non-Sn alloy and slightly higher at the slowest strain rates.

[†] Sintech 1G Test Frame running TestWorks 3.05, MTS Corporation, Eden Prairie, MN, USA.

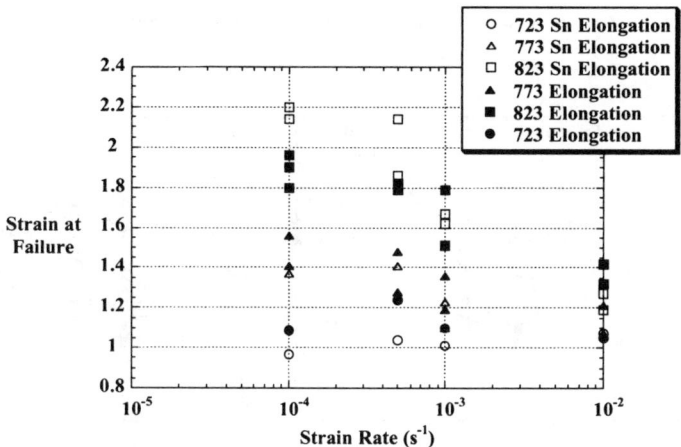

Figure 1 Strain at failure for all alloys as function of strain rate and test temperature (denoted by different symbols) and as a function of Sn (open symbols) or non-Sn (closed symbols).

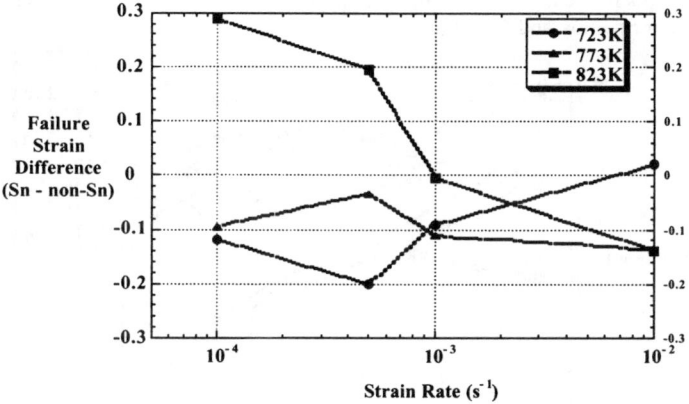

Figure 2. Difference in failure strains for Sn alloys compared to non-Sn alloys as a function of strain rate. Sn alloys exhibit improved properties at 823K and slowest strain rates.

Activation energies were computed from plots of the log strain rate as a function of reciprocal temperature at constant flow stress (Table 3). There is a transition from the low activation energies for the high strain rates and lower temperatures to the increased activation energies for the slower strain rates and higher temperatures. An m-value of 0.3 (or stress exponent, n, > 3) suggests a solute drag regime[8] while an m-value of 0.45 (n ~ 2) suggests a grain boundary sliding regime[2]. There is little difference between the Sn and non-Sn alloys at the low strain rates (GBS regime) but an appreciable difference at the high strain rates (solute drag regime).

TEM examination of material prepared from quenched specimens has revealed that small pockets or regions of impurities form at aluminum grain boundaries where they meet the eutectic constituent (EC) particles, which are micron-sized intermetallic particles that form during casting.

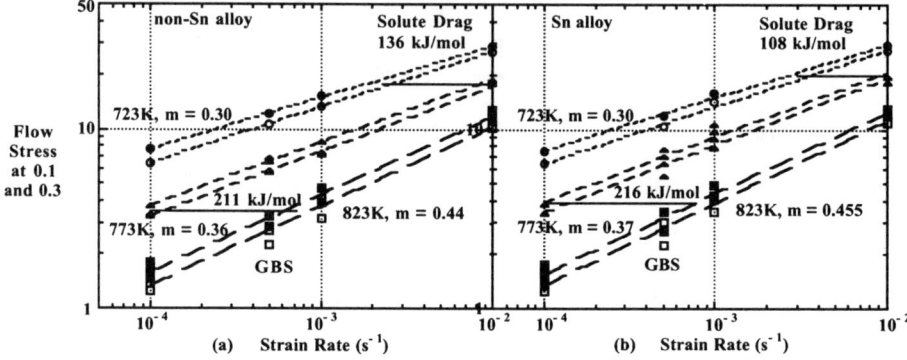

Figure 3. Log flow stress as a function of log strain rate showing variation of strain rate exponent with temperature and Sn-content at strains of 0.1 (open symbols) and 0.3 (closed symbols). (a) non-Sn alloy and (b) Sn alloy.

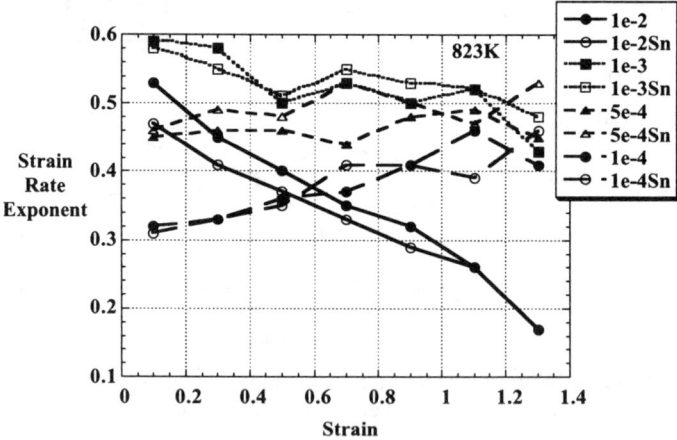

Figure 4. Strain rate exponent as a function of strain determined by periodic strain rate increases as described in Table 2. This data is for 823K for the four strain rates.

Table 3. Activation energies and m-values for each alloy in solute drag and GBS regimes.

Strain Rate Regime	Sn alloy E_{act} (kJ/mol) and m-value	Non-Sn alloy E_{act} (kJ/mol) and m-value
Low strain rates High T (GBS)	216 m = 0.46	211 m = 0.44
High strain rates Low T(Solute Drag)	108 m = 0.3	137 m = 0.3

These hetero-junctions, or hetero-triple-junctions, are often the sites at which cavities nucleate and are likely places of intense sliding stresses during deformation. These pockets were observed to contain Sn, Si, and O using EDS and also appear to be less dense than the surrounding Al (Figure 5a) as evidenced by a lighter contrast but higher-Z components. The pockets observed in the non-Sn alloys were not as extensive and did not contain any Sn. Optical microscopy (OM) was performed on polished sections of two specimens, one of each alloy, tested under identical conditions at 823K and 1×10^{-4} s^{-1} and stopped prior to failure at a strain of 1.86, to study cavitation. While more detailed tests remain to be completed, it appears that the cavity shapes were different between the Sn and non-Sn alloys (Figures 5b and 5c). The non-Sn alloy has cavities that are smoothly shaped while the Sn alloy cavities are irregular.

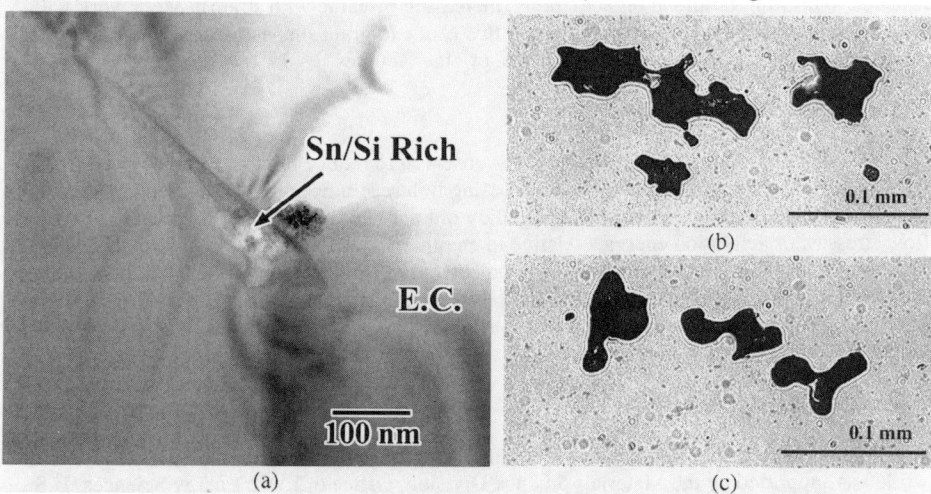

Figure 5. (a) Sn/Si/O-rich pocket imaged at an EC/Al-grain boundary interface in Sn alloy following quenching at 823K after a strain of 0.7 at 1×10^{-4} s^{-1}. (b) Irregular cavities in Sn alloy and (c) smooth-shaped cavities in non-Sn alloy illustrating shape differences.

DISCUSSION

The use of selected impurities to probe mechanisms of superplasticity in metals shows promise by allowing companion studies on related alloys to reveal differences that can be attributed to the action of the impurity species. In this case, Sn was selected based on its ability to segregate to grain boundaries and because of its previously observed effects (serendipitous) on cavitation. Its effects on grain boundary dislocation spreading kinetics may also play a subtler role. However, in this study examining the more macroscopic properties it is seen that Sn has some observable effects on elongation and properties but that these effects are small except for the conditions at 823K and slowest strain rates (GBS regime). Here the evidence seems to point to reduced cavitation since neither m-value nor activation energy differences, which are noticeable but small, can readily explain the improved properties. Rather, the TEM evidence suggests that Sn-containing pockets form and may allow a redistribution of stresses at the critical EC/GB interfaces where cavitation occurs.

Sn binds to Al-vacancies (~0.45 eV) and would thus tend to collect at vacancy sinks preferentially. Although it probably does this under all the test conditions it is likely that until a critical temperature is reached at which the Sn-containing regions soften relative to the surrounding Al that little stress redistribution occurs. At lower temperatures and higher strain rates the alloys are in the solute-drag deformation regime and Sn acts to retard deformation by dislocation pinning. The activation energy for the non-Sn alloy of 137 kJ/mol matches quite well with that reported for creep of Al-Mg alloys[8]. That, coupled with the m-value of 0.3, places the high strain rate deformation in the solute drag regime. The reason for the decreased activation energy of the Sn alloy in the solute drag regime is unknown.

The results reported in Fig. 4 showing a relationship between m-value and strain seem to be an important observation made during this study (independent of the Sn-effects) and suggest that more should be done to investigate the microstructural changes that occur during deformation. The implications are that under certain strain rates the material becomes more resistant to necking failure at higher strains (increasing m-value with strain). More work will have to be performed to understand whether this is due to grain shape changes, cooperative grain motion resulting in optimal grain alignments, or other sources.

CONCLUSIONS

A small amount of Sn (0.034 wt%) was observed to reduce failure strains at low temperatures and fast strain rates while increasing failure strains at higher temperatures (823K) and slower strain rates. The Sn-containing alloy did not have significant differences in m-values, flow stresses, or activation energies relative to an otherwise identical non-Sn alloy. Instead, evidence seems to point towards Sn effects on cavitation and final fracture via the formation of Sn-containing pockets at hetero-triple-junctions (particle/grain boundary) in the alloys. These pockets are suggested to allow for a redistribution of stresses at these hetero-junctions, which would slow the growth of the cavities.

ACKNOWLEDGEMENTS

The authors would like to thank Konrad Lasota for assistance with the microscopy. This work was supported by the Materials Science Division, Office of Basic Energy Sciences, U.S. Department of Energy (DOE) under Contract DE-AC06-76RLO 1830.

REFERENCES

1. M. F. Ashby and R. A. Verrall, *Acta Metall.*, 1973, **21** (2), pp. 149-163.

2. O. D. Sherby and J. Wadsworth, *Prog. Mater. Sci.*, 1989, **33** (3), pp. 169-221.

3. A. P. Sutton and R. W. Balluffi, Interfaces in Crystalline Materials, 1995, Oxford University Press, New York, NY.

4. W. Lojkowski, J. W. Wyrzykowski, J. Kwiecinski, D. L. Beke, and I. Godeny, *Diffus. Defect Data, Pt. A*, 1990, **66-69** (Diffus. Met. Alloys-DIMETA 88, Pt. 2), pp. 701-712.

5. S. G. Song, J. S. Vetrano, and S. M. Bruemmer, *Mater. Sci. & Engr. A: Structural Materials: Properties, Microstructure and Processing*, 1997, **A232** (1-2), pp. 23-30.

6. J. S. Vetrano, J. C. H. Henager, E. P. Simonen, S. G. Song, and S. M. Bruemmer, in *Boundaries and Interfaces in Materials: The David A. Smith Memorial Symposium*, 1998, Indianapolis, IN, TMS, Warrendale, OH, pp. 205-212.

7. T. B. Sercombe and G. B. Schaffer, *Acta Mater.*, 1999, **47** (2), pp. 689-697.

8. T. R. McNelley, D. J. Michel, and A. Salama, *Scripta Metall.*, 1989, **23**, pp. 1657-1662.

SUPERPLASTIC DEFORMATION BEHAVIOR OF Zn-Al ALLOYS

Tae Kwon Ha, Hyun Woo Koo, Tae Shin Eom and Young Won Chang
Center for Advanced Aerospace Materials (CAAM)
Pohang University of Science and Technology, Pohang, Kyungbuk 790-784, South Korea

ABSTRACT

The superplastic deformation behavior of a quasi-single phase Zn-0.3 wt.% Al and a microduplex Zn-22 wt.% Al eutectoid alloy has been investigated in this study within the framework of the internal variable theory of structural superplasticity (SSP). Load relaxation and tensile tests were conducted at room temperature for quasi-single phase alloy and at 200 °C for eutectoid alloy. The flow curves obtained from load relaxation tests on superplastic Zn-Al alloys were shown to consist of the contributions from interface sliding (IS) and the accommodating plastic deformation due to dislocation activities. The IS behavior could be described as a viscous flow characterized by the power index value of $M_g = 0.5$. In the case of quasi-single phase Zn-0.3 wt.% Al alloy with the average grain size of 1 μm, a large elongation of about 1400 % was obtained at room temperature. In a relatively large-grained (10 μm) single-phase alloy, however, grain boundary sliding (GBS) was not expected from the analysis based on the internal variable theory of SSP.

INTRODUCTION

Structural superplasticity (SSP) is the ability of a polycrystalline material to produce a large elongation of more than several hundred percents when deformed under a tensile loading. Interface sliding (IS) is widely accepted as the major deformation mechanism of SSP [1-4]. The prerequisites for SSP are well known as a stable, equiaxed and extremely fine microstructure, an optimum strain rate and test temperature normally above the half of the absolute melting temperature T_M [1-4]. There have been an extensive number of reports that very large elongation could be obtained in the various classes of materials including ceramics [5-7] as well as amorphous alloy [8].

Both the eutectoid and dilute Zn-Al alloys have been reported to exhibit excellent superplasticity [9-14]. As mentioned above, the stable and fine-grained microstructure is very important for superplastic deformation. In the present work, it has, therefore, been attempted to produce a stable and fine-grained microstructure in an eutectoid and a dilute Zn-Al alloy through proper thermomechanical treatment process (TMTP). The grain sizes thus obtained were about 2 μm in the eutectoid alloy and 1 μm in the Zn-0.3 wt.% Al alloy. A relatively coarse-grained material with the grain size of 10 μm was also prepared through the subsequent aging treatment of the fine-grained material to investigate the grain size effect. A series of load relaxation and tensile tests have been conducted at room temperature for dilute Zn-Al alloy and at 200 °C for eutectoid alloy. The internal variable theory of SSP [15] has been employed to characterize the superplasticity of the Zn-Al alloys.

INTERNAL VARIABLE THEORY OF STRUCTURAL SUPERPLASTICITY

A brief summary is described here and details of the development of this theory are given elsewhere [15]. Figure 1 shows the rheological and physical models for SSP adopted in this study.

Mat. Res. Soc. Symp. Proc. Vol. 601 © 2000 Materials Research Society

The IS is mainly accommodated by a dislocation process, giving rise to an internal strain ($\underset{\sim}{a}$) and plastic strain ($\underset{\sim}{g}$). The stress variables $\underset{\sim}{\sigma}^{I}$ and $\underset{\sim}{\sigma}^{F}$ represent the internal and the frictional resistance to the dislocation motion [15]. In the temperature range where SSP can be expected, $\underset{\sim}{\sigma}^{F}$ is, in general, very small compared to $\underset{\sim}{\sigma}^{I}$ and the time rate of internal strain $\underset{\sim}{\dot{a}}$ can be neglected if the relaxation test is performed uni-axially at a steady state. The total inelastic strain rate ($\dot{\varepsilon}_{T}$) at a given stress level can, therefore, be expressed as the sum of the unrecoverable plastic strain rate ($\underset{\sim}{\dot{g}}$) and the strain rate due to IS ($\underset{\sim}{\dot{g}}$);

$$\dot{\varepsilon}_{T} = \underset{\sim}{\dot{g}} + \underset{\sim}{\dot{g}} \tag{1}$$

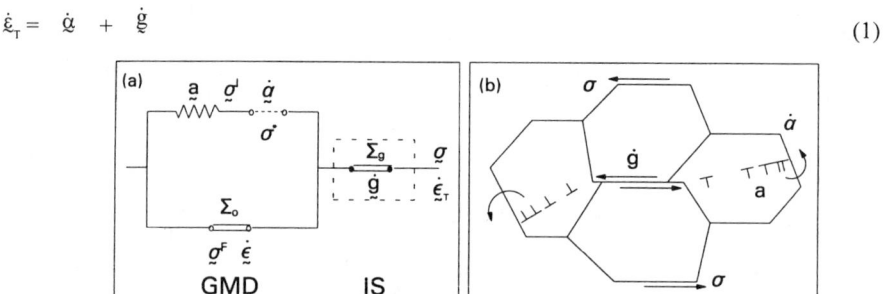

Figure 1. (a) Rheological and (b) physical models representing the IS accommodated by grain matrix deformation (GMD) [15].

In this circumstance, we can describe the superplastic deformation fully with the constitutive relations for $\underset{\sim}{\dot{g}}$ and $\underset{\sim}{\dot{g}}$ elements, which are prescribed as follows;

$$(\sigma^{*}/\sigma^{I}) = \exp(\dot{\alpha}^{*}/\dot{\alpha})^{p} \tag{2}$$

$$(\dot{g}/\dot{g}_{o}) = (\sigma/\Sigma_{g} - 1)^{1/M_{g}} \tag{3}$$

where p and M_{g} are material constants, σ^{*} and $\dot{\alpha}^{*}$ are the internal strength variable and its conjugate reference strain rate, while Σ_{g} and $\dot{g}_{o}$ are the static friction stress for IS and its reference rate, respectively. Interface sliding is considered here as a viscous drag process similar to the frictional glide process of dislocation [15].

EXPERIMENTAL PROCEDURE

A Zn-22 wt.% Al alloy and a Zn-0.3 wt.% Al alloy used in this study were prepared by melting the commercial purity elements in an induction furnace. The melts were then cast into a steel-mold. The chemical compositions of the materials are given in Table 1. After the homogenization at 300 °C for 10 hrs, extrusion with the area reduction ratio of 25:1 on the eutectoid Zn-Al alloy and warm rolling for the thickness reduction from 50 to 10 mm on the dilute Zn-Al alloy were performed at 220 °C. The grain size of eutectoid alloy was about 2 μm after annealing at 220 °C for 30 min. The plates of the dilute Zn-Al alloy were finally cold-rolled into a sheet of 4 mm in thickness followed by aging at room temperature for 168 hrs and the resulting grain size was 1 μm. To obtain a coarse-grained specimen in the Zn-0.3 wt.% Al alloy, an additional annealing treatment was carried out at 220 °C for 3 hrs to produce the grain size of 10 μm. The tensile specimens for mechanical testing were then machined from these sheet materials.

Table 1. Chemical compositions of Zn-Al alloys used in this alloy. (wt.%)

	Al	Cu	Fe	Mg	Pb	Ti	Zn
Zn-22 wt.% Al	21.94	0.02	0.11	0.07	0.003	0.004	Bal.
Zn-0.3 wt.% Al	0.28	0.002	0.01	-	0.001	-	Bal.

Load relaxation tests have been conducted at room temperature for the dilute Zn-Al alloy and at 200 °C for the eutectoid alloy. Load relaxation test provides a broader range of strain rates with a small plastic straining [16]. The specimens for load relaxation tests had a square cross-section (4 mm × 4 mm) with the gauge length of 27 mm. In the load relaxation, the load was recorded as a function of time at the fixed extension and the flow stress σ and the inelastic strain rate $\dot{\varepsilon}$ were then calculated from the load-time data following the usual procedure described in the literature [17]. The strain level undergone by the specimen before the onset of load relaxation was less than 5 % in all cases. Tensile tests were also performed on the Zn-Al alloys at the various initial strain rates ranging from 2×10^{-4}/s to 5×10^{-2}/s under a constant ram speed condition.

RESULTS AND DISCUSSION

The typical microstructures of materials used in this study are shown in Fig. 2(a) and (b) for Zn-0.3 wt.% Al and (c) for Zn-22 wt.% Al alloys. After the cold rolling of 60 % reduction at room temperature, Zn-0.3 wt.% Al alloy is observed to have a very fine and uniform grain size of about 1 μm from Fig. 2(a). This fine-grained microstructure was found to be very stable at room temperature and further aging did not cause any grain growth. Annealing treatment of the cold-rolled sheets at 220 °C for 3 hrs has been found to provide a grain size of about 10 μm as can be seen from Fig. 2(b). The grain size of Zn-22 wt.% Al alloy was evaluated as about 2 μm.

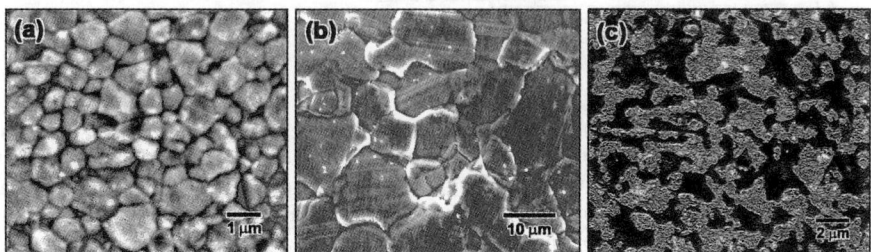

Figure 2. Typical microstructures of materials used in this study. (a) Fine-grained Zn-0.3 wt.% Al, (b) coarse-grained Zn-0.3 wt.% Al, and (c) fine-grained Zn-22 wt.% Al alloys.

The flow curves obtained from load relaxation tests are given in Fig. 3. It is easily noted from this figure that the flow curves of fine-grained (d = 1 μm) dilute Zn-Al and eutectoid alloy show a typical sigmoidal shape, whereas that of coarse-grained dilute Zn-Al alloy exhibits a monotonically convex shape in the entire strain rate range. The sigmoidal shape of flow curve is typically exhibited in the various superplastic materials [1-4, 18].

The flow curves in Fig. 3 have been analyzed based on the internal variable theory of SSP described in the earlier section and the results are given in Fig. 4. The thin solid lines marked as GMD in these figures represent the contribution from grain matrix plasticity, $\overset{\cdot}{\alpha}$ element of Eq. (2) and the dotted lines contribution from IS, $\overset{\cdot}{g}$ element of Eq. (3). The flow curve of coarse-grained Zn-0.3 wt.% Al alloy has no IS contributions at all. The bold lines, composite curves constructed by Eqs. (1), (2), and (3), show good agreements with the experimental data. The constitutive parameters used in the construction of the predicted curves are listed in Table 2. It is very interesting to note that the value of the parameter M_g is 0.5 for both quasi-single phase and microduplex superplastic Zn-Al alloys. The parameter M_g has been reported to represent the characteristics of interface or grain boundary [15]. In the case of microduplex superplastic materials including the Zn-Al eutectoid alloy used in the present study, M_g is well known as 0.5 [19, 20]. The value of M_g in quasi-single phase superplastic Al alloy, on the other hand, has been obtained as unity suggesting that grain boundary sliding (GBS) be the Newtonian viscous flow [21, 22]. Recently, grain boundary character of some superplastic Al alloys has been reported to change with the accumulation of plastic deformation such that the value of M_g was also changed from unity to near 0.5 [23]. The value of M_g obtained as 0.5 in the fine-grained Zn-0.3 wt.% Al alloy indicates that the grain boundary character of this quasi-single phase superplastic alloy is different from other superplastic Al alloys and more systematic investigation is necessitated.

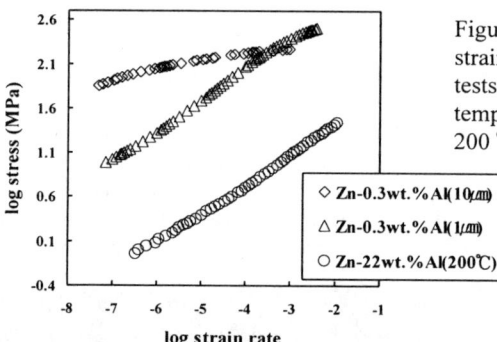

Figure 3. Flow curves of log stress vs. log strain rate constructed from load relaxation tests for Zn-0.3 wt.% Al alloy at room temperature and for Zn-22 wt.% Al alloy at 200 °C.

The tensile elongations at room temperature for Zn-0.3 wt.% Al alloy and at 200 °C for Zn-22 wt.% Al alloy are plotted as a function of the initial strain rates ranging from 2×10^{-4}/s to 5×10^{-2}/s in Fig. 5. It can be easily noted that the elongation to failure in fine-grained dilute Zn-Al alloy and eutectoid Zn-Al alloy appear to be very sensitive to the initial strain rate. The elongation of the fine-grained dilute Zn-Al alloy increases with decreasing strain rate and maximum elongation of about 1400 % was obtained at the strain rate of 2×10^{-4}/s, which is clearly in the strain rate range where IS is predominant deformation mechanism as illustrated in Fig. 4. The room temperature elongation of 1400 % in dilute Zn-Al alloy is the largest one ever reported in the open literature. The specimens with a coarser grain size (d = 10 μm), on the other hand, showed the elongation less than 100 %, seemingly not superplastic as expected in Fig. 4. The tensile elongation of the eutectoid Zn-Al alloy appeared to reach a maximum value of about 1400 % at the initial strain rate of 5×10^{-3}/s and then to decrease with decreasing initial strain rate at 200 °C. The tendency is consistent with the result of Mohamed et al [10].

Table 2. The constitutive parameters determined from the analyses of the flow curves of Zn-Al alloys based on the internal variable theory of SSP.

| Specimens | $\log \sigma^*$ | GMD | | | IS | |
		$\log \dot{\alpha}^*$	P	$\log \Sigma_g$	$\log \dot{g}_o$	M_g
Zn-22 wt.% Al	1.932	-1.989	0.15	0.086	-5.013	0.5
Fine-grained Zn-0.3 wt.% Al	2.883	-2.782	0.15	0.968	-6.263	0.5
Coarse-grained Zn-0.3 wt.% Al	2.402	-6.627	0.15	-	-	-

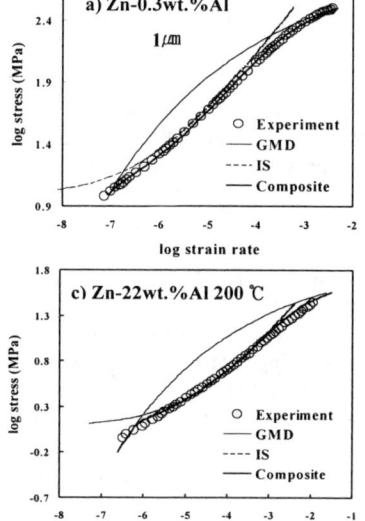

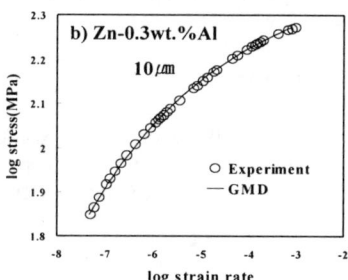

Figure 4. Results of application of the internal variable theory of SSP to the flow curves obtained from load relaxation test on Zn-Al alloys.

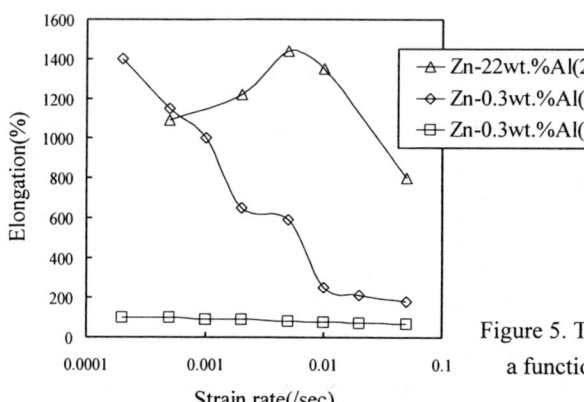

Figure 5. Tensile elongation to failure as a function of the initial strain rates.

CONCLUSIONS

A remarkably large elongation of 1400 % was obtained at room temperature in a Zn-0.3 wt.% Al alloy with a very fine and stable grain size of 1 μm, which is the largest room temperature elongation ever reported in the open literature for dilute Zn-Al alloys. By applying the internal variable theory of structural superplasticity, the flow curves of superplastic Zn-Al alloys could effectively interpreted as consisting of contributions from IS ($\dot{g}$) and the accommodating grain matrix plastic deformation ($\dot{\alpha}$). The IS behavior appeared to be a viscous flow process characterized by the power index value of 0.5 in both the microduplex and quasi-single phase superplastic Zn-Al alloys.

ACKNOWLEDGMENTS

This work was financially supported by Non-directed Research Fund from Korea Research Foundation, 1997 – 99. The authors wish to thank the Advanced Materials Division of Research Institute of Industrial Science and Technology (RIST) for the provision of experimental facilities.

REFERENCES

1. K. A. Padmanabhan and G. J. Davis, *Superplasticity*, Springer-Verlag, New York, 1980, p. 11.
2. O. D. Sherby and J. Wadsworth, Prog. Mater. Sci., **33**, p. 169 (1989).
3. J. W. Edington, K. N. Melton, and C. P. Cutler, Prog. Mater. Sci., **21**, p. 61 (1976).
4. A. H. Chokshi, A. K. Mukherjee, T. G. Langdon, Mater. Sci. Eng., **R10**, p. 237 (1993).
5. Y. Yoshizawa and T. Sakamura, J. Am. Soc. Cer., **73**, p. 3,069 (1990).
6. D. J. Schissler, A. H. Chokshi, T. G. Nieh and J. Wadsworth, Acta Matell., **40**, p. 581 (1992).
7. A. H. Chokshi, Mater. Sci. Eng., **A166**, p. 119 (1993).
8. Y. Kawamura, T. Shibata, A. Inoue and T. Masumoto, Scripta Mater., **37**, p. 431 (1997).
9. H. Naziri and R. Pearce, Acta Metall., **22**, p. 1321 (1974).
10. F. A. Mohamed, M. M. I. Amed and T. G. Langdon, Metall. Trans., **8A**, p. 933 (1977).
11. O. A. Kaibyshev, B. V. Rodionov and R. Z. Valiev, Acta Metall., **26**, p. 1,877 (1978).
12. P. Malek, Scripta Metall., **19**, p. 405 (1985).
13. M. Furukawa, Y. Ma, Z. Horita, M. Nemoto, R. Z. Valiev and T. G. Langdon, Mater. Sci. Eng., **A241**, p. 122 (1998).
14. A. Arieli, A. K. S. Yu and A. K. Mukherjee, Metall. Trans., **11A**, p. 181 (1980).
15. T. K. Ha and Y. W. Chang, Acta Mater., **46**, p. 2,741 (1998).
16. E. W. Hart, in *Stress Relaxation Testing*, edited by A. Fox (ASTM Special Technical Pub. No. **676**, Baltimore, Md, 1979) pp. 5-20.
17. D. Lee and E. W. Hart, Metall. Trans., **2A**, p. 1,245 (1971).
18. D. A. Woodford, Metall. Trans., **7A**, p. 1,244 (1976).
19. T. K. Ha and Y. W. Chang, Scripta Metall., **35**, p. 1,317 (1996).
20. J. S. Kim, Y. W. Chang and C. S. Lee, Metall. Mater. Trans., **29A**, p. 217 (1998).
21. T. K. Ha and Y. W. Chang, Scripta Metall., **32**, p. 809 (1995).
22. T. K. Ha, H. J. Sung, K. S. Kim and Y. W. Chang, Mater. Sci. Eng., **A271**, p. 160 (1999).
23. Y. N. Kwon and Y. W. Chang, in *Light Weight Alloys for Aerospace Applications V*, edited by E. W. Lee, N. J. Kim, K. V. Jata and W. E. Frazier (TMS-AIME Annual Meeting, San Diego, CA 1999), In-print.

SUPERPLASTICITY OF NICKEL-BASED ALLOYS WITH MICRO- AND SUB-MICROCRYSTALLINE STRUCTURES

V.A. VALITOV[1], B.P. BEWLAY[2], Sh. Kh. MUKHTAROV[1], O.A. KAIBYSHEV[1] and M.F.X. GIGLIOTTI[2]
[1]Institute for Metals Superplasticity Problems, Ufa 450001, Russia
[2]General Electric Corporate Research and Development, Schenectady, NY 12301, USA

ABSTRACT

This paper will describe the generation of micro- and sub-microcrystalline structures in two Ni-based alloys that are typically strengthened by phases, such as γ' and $\gamma''+\delta$. The relationship between the superplastic behavior and microstructure will be discussed. High strain deformation processing in the temperature range of $0.9T_m$ to $0.6T_m$ results in reduction of the initial coarse-grained structure (>100 μm) to a range of structures including micro-crystalline (MC) (grain size <10 μm) and sub-microcrystalline (SMC) (grain size <1 μm) with increasing deformation. The influence of alloy chemistry and constituent phases on dynamic and static recrystallization is considered, and their effect on grain refinement is described. Low-temperature and high strain rate superplasticity can be observed in dispersion-strengthened alloys with SMC structures. It was established that in dispersion-hardened Ni alloys with SMC structures, superplasticity can be observed at temperatures 200-250°C lower than in alloys with MC structure.

INTRODUCTION

There is presently substantial research in materials with sub-microcrystalline structures because they display unique physical and mechanical properties [1-3]. In particular, they offer expansion of the temperature-strain rate range of superplasticity (SP) into regimes of both low temperature and high strain rate SP. In the case of high-temperature Ni-based alloys, there is interest in decreasing the temperature of SP deformation and increasing the strain rates for metal forming. In Ni-based alloys that possess a micro-duplex microstructure (4-10 μm) SP is usually observed at high homologous temperatures (0.8-0.85T_m) and low strain rates (10^{-3}-$10^{-4}s^{-1}$), depending on the alloy chemistry and constituent phases [4]. Several papers describing the mechanical behavior of Ni-based alloys with SMC structure have already been published [5,6]. However, these previous data do not permit sufficiently reliable predictions about specific features of SP behavior of dispersion-hardened 718 and EP962 Ni-based alloys with SMC structure. The goal of this paper is to identify processing regimes for the generation of MC and SMC structures in Ni-based alloys and to examine their influence on SP properties.

EXPERIMENTAL

The alloys used in this investigation (Table I) are typical of commercial Ni based alloys: dispersion-hardening alloys – alloy 718 ($\gamma''+ \delta$) and EP962 (γ'). The samples for investigations of alloy 718 and EP962 were fabricated from hot worked rods, 200 mm in diameter. The mechanical behaviors of these alloys were investigated in both compression and tension over the temperature range of 600-1270°C at strain rates of 6.67×10^{-5}-$1.7 \times 10^{-1} s^{-1}$.

Table I. Chemical compositions of Ni based alloys used in the present study (wt. %).

Alloy	C	Cr	Co	W	Mo	Fe	Al	Ti	Nb	Ni	
Alloy 718	0.05	19	-	-	3.1	18	0.5	1.0	5.1	base	
EP962	0.1	13.1	10.7	2.8	4.6	0.6	3.2	2.6	3.4	base	0.01 B

43

Initially, cylindrical samples 10 mm in diameter and 15 mm in height were compressed. On the basis of MC and SMC processing regimes that were developed using these samples together with high-strain deformation techniques [4], larger SMC samples were produced. Tensile tests of flat samples, 10x5x2 mm in the gauge section, and cylindrical samples, with gauge sections 5 mm in diameter and 25 mm length, were performed. Transmission electron microscope investigations were conducted using a JEM-2000EX.

RESULTS

In the initial condition, alloy 718 has a completely recrystallized structure with a grain size of ~40 μm. Within the γ- phase grains there are disk-like γ''-phase precipitates which are uniformly distributed. The diameter of γ''-phase disks is about 60 nm, and their thickness is 20 nm. The γ''-phase precipitates are also observed on grain boundaries.

In the initial condition, the microstructure of the EP962 is also coarse-grained with a grain size of ~80 μm. Within the grains there are coherent dispersed particles (0.1 μm) of γ'-phase which are uniformly distributed. The grain boundary precipitates of γ'-phase are 2-3 times coarser than intragranular precipitates, and they have partially coherent interphase boundaries.

Formation of submicrocrystalline structure

The alloys 718 and EP962 were subjected to high-strain thermo-mechanical processing (TMP) including isothermal deformation with intermediate annealing and gradual, step-by-step reduction of the TMP temperature. Differences in the formation of the alloy 718 and EP962 SMC structures are due to differences in the alloy chemistries and constituent phases.

Alloy 718 : A specific characteristic of alloy 718 is precipitation of lamellar δ-phase particles during heating for deformation. The morphologies of the δ and γ''-phase precipitates after annealing can be distinguished from γ'-precipitates [4]. The large-scale transgranular δ-phase plate-like precipitates and the disk-shaped γ'' precipitates are not favorable for the formation of the MC structure during hot deformation and its subsequent transformation to a SMC structure.

In the initial stages of deformation (5-10%) over the temperature interval 975-850°C, elongated sub-grains are formed in the vicinity of initial grain boundaries, and there are separate dislocations decorated by γ''-phase particles precipitated inside of grains. Further increase in strain results in dynamic recrystallization in regions near grain boundaries, and this causes softening.

The near-boundary sub-grains transform to grains with high angle grain boundaries, and δ-phase particles form during deformation at grain boundaries. Deformation to a strain of 75% did not generate complete recrystallization. The post-deformation annealing conducted at a deformation temperature 900°C for 2 hours contributes to formation of a completely recrystallized structure with a matrix grain size of 1-2 μm. The mean δ-phase particle size was 0.6 μm in length and 0.15 μm in width. Additional deformation-thermal processing of the alloy at temperature below 850°C leads to further structure refinement. The matrix grain size decreases to 0.3 μm (Fig. 1a). Elongated δ-phase particles (0.2-0.6 μm long) were observed distributed primarily at grain boundaries. The microstructure of the SMC alloy produced by TMP exhibits high dislocation density both within the grains and at the grain boundaries. Grain boundary contrast typical of high angle grain boundaries was not observed.

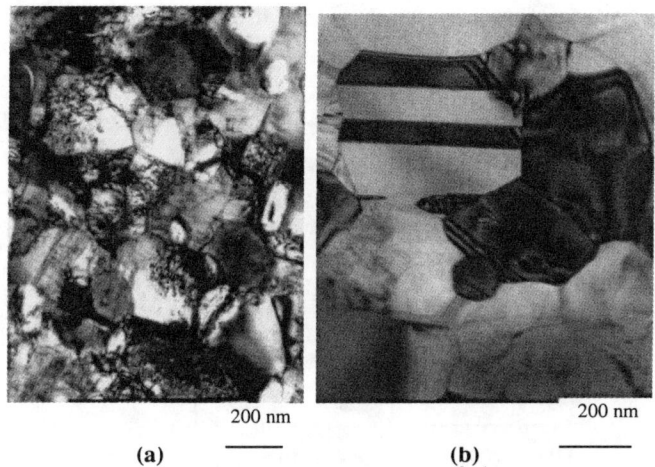

(a) ——— (b) ———

Figure 1. Microstructures after forging at strain rate $5.5 \times 10^{-3} s^{-1}$ of (a) alloy 718 at 700°C, (b) EP962 deformed and annealed at 850°C.

EP962: In order to improve the ductility and create the most favorable conditions for recrystallization, the EP962 was annealed by heating to a temperature above that of the gamma prime solvus temperature (~1150°C) and then cooled slowly in a furnace. This treatment did not change the matrix grain size (γ-phase) (100 μm), but the γ'-phase particle size was increased to 0.5 μm, and interphase boundaries became coherent. Deformation of the EP962 at 1075°C to a total strain of 80% led to intense dynamic recrystallization. TEM analysis of microstructural changes established the mechanism of transformation of the coarse-grained matrix structure to a microduplex structure. Formation of a microduplex structure during hot deformation occurred via transformation of partially coherent γ/γ' interphase boundaries to non-coherent type boundaries, and transformation of γ-phase subgrains to grains with high angle grain boundaries. The grain size of γ and γ'-phases is 5.5 and 2.5 μm, respectively. Further reduction in the deformation temperature from 975 to 850°C led to precipitation of γ'-phase particles, 0.1-0.2 μm in size, within the γ grains (2.5 μm). The total volume of the γ'-phase was 45%.

After a deformation strain of 65% at 850°C and strain rate $10^{-3} s^{-1}$, the SMC structure with γ and γ'-phase grain sizes of 0.1-0.4 μm and a recrystallized volume fraction of 20% was generated from samples with MC structures (d=2.5 μm). A high density of dislocations was observed in non-recrystallized areas both in the matrix and the γ'. This observation, as well as the refinement of the initial γ and γ'-phase grains, suggests that recrystallization occurs both in the γ and the γ'. Post deformation annealing at 850°C for 8 hours increased the volume fraction of recrystallized grains to 70%; the mean γ and γ'-phase grain size was ~0.25 μm (Fig. 1b).

Microstructure and properties of alloys during SP deformation

In this section the SP behaviors of alloy 718 and EP962 with SMC and MC structures are compared. Although these alloys have similar grain sizes in the SMC condition, the distinctions in chemistry and constituent phases lead to different superplastic behaviors.

45

Alloy 718. Analysis of the mechanical property data from samples with MC and SMC structures shows that the mechanical behavior of alloy 718 (Fig. 2) correlates with the microstructural changes.

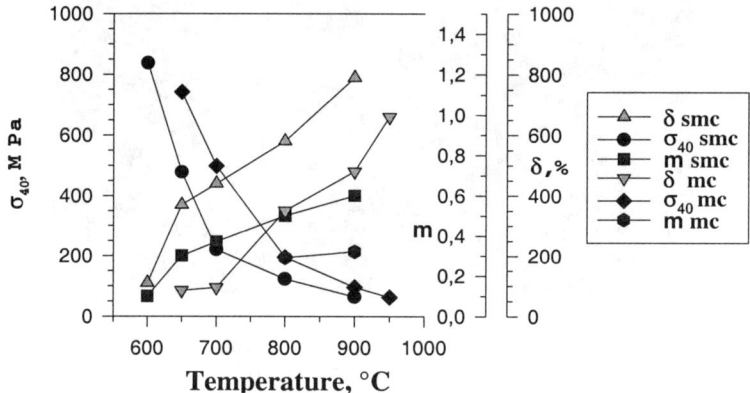

Figure 2. SP properties (stress, strain and strain rate sensitivity of the flow stress (m value)) of SMC and MC alloy 718 for a strain rate of $5.5 \times 10^{-4} s^{-1}$. The stress, σ_{40}, is that at a strain of 40%.

Fig. 2 indicates that the decrease in the grain size in the MC state from 6 to 1-2 μm results in a decrease in the lower temperature limit of SP behavior. Additional refinement of the micro-structure to sub-micron sizes contributes to low-temperature superplasticity in the alloy 718. Alloy 718 with SMC structure displays SP properties even at 650°C. The lowest level of flow stress is observed in the SMC alloy 718.

The analysis of microstructural changes after deformation of the MC structure has shown that after deformation at a strain rate of $7 \times 10^{-3} s^{-1}$ in the temperature range 800-1050°C the microstructure remains fine-grain. The mean grain size increases from 3.8 μm at 900°C up to 8.3 μm at 1050°C. On increasing the strain rate from 7×10^{-4} up to $2 \times 10^{-2} s^{-1}$ at 925°C the grains remain equiaxed; their mean grain size (~5 μm) does not change significantly.

The microstructure of the SMC alloy 718 also remains equiaxed after deformation. However, the increase in the temperature of deformation leads to grain growth, such that above 900°C the grain size exceeds 1 μm. In addition, after superplastic deformation the grain boundaries possess a more stable configuration.

EP962 alloy: The mechanical properties with both microduplex (5.5 μm) and SMC (0.25 μm) structures are given in Table II for a range of temperatures and strain rates. For the SMC structure the low temperature limit for SP behavior is reduced. In addition, the strain rate range for SP becomes wider for the SMC alloy, as shown at 950°C. At 800°C the samples with SMC structure did not fail even after elongations of ~550% at the strain rate $3.3 \times 10^{-4} s^{-1}$.

The changes in SMC structure after SP deformation were investigated. TEM studies of samples deformed at 800°C and strain rate $3.3 \times 10^{-4} s^{-1}$ to 550% strain show that during deformation the grains remain equiaxed.

Table II. Characteristics of SP properties of the EP962

Grain/	Particle Size (µm)	Tempera-ture	Strain Rate	Stress, σ_{40}	Strain, δ,	m
γ-phase	γ'-phase	°C	s⁻¹	MPa	%	
5.5	2.5	1075	$1.33\cdot10^{-3}$	50	>550	0.6
		950	$1.33\cdot10^{-3}$	120	>500	0.3
		950	$1.33\cdot10^{-2}$	551	76	0.16
		875	$1.33\cdot10^{-3}$	378	60	0.13
0.25		950	$1.33\cdot10^{-2}$	144	252	0.43
		875	$1.33\cdot10^{-3}$	187	337	0.39
		800	$1.33\cdot10^{-3}$	435	98	0.21
		800	$3.3\cdot10^{-4}$	350	>550	0.3

The character of dislocation structures in γ and γ' grains is different. In γ-phase grains, only random dislocations are observed, while in γ'-phase grains the dislocation density is significantly higher. During SP deformation pore formation is observed and the density decreases on approaching a failure zone. Moreover, only random pores, (<1 µm) can be revealed in sections far from the fracture zone, while near the failure zone, pores are aligned along the deformation axis, and coalescence has formed coarser pores. Pore formation occurs mainly near carbide particles both in the SMC sample and in the sample with microduplex structure (5.5 µm) after SP deformation under conventional conditions (1075°C, strain rate 1.33×10^{-3} s⁻¹, δ>550%). The size and volume fraction of carbide particles are similar in both cases. However, in the sample with microduplex structure the mean size of pores is coarser and is equal to 3.6 µm, but the volume fraction of pores is similar to that in SMC structure. Density measurements are given in Table III. The density was determined by the Archimedes method.

Table III. Density of MC and SMC EP962 after deformation at 800°C and a range of strain rates

	Sample Microstructure			
	d=5 µm	d=0.25 µm		
Strain rate, s⁻¹	$1.33\cdot10^{-3}$	$1.33\cdot10^{-3}$	$3.3\cdot10^{-4}$	$6.67\cdot10^{-5}$
Elongation, %	72	100	550	450
Density of the tensile sample head, g/cm³	8.225	8.178	8.225	8.224
Density of the gauge section, g/cm³	8.223	8.146	8.094	7.846
Density changes Δρ/ρ, %	0.03	0.39	1.59	4.60

The bulk density of EP962 with SMC structures is less than the density in samples with microduplex structure, and no cracks and pores are present in SMC samples. Annealing of shoulders of specimens during deformation restores the density in these sections to the density of the alloy with microduplex structure. At the same time, during SP deformation a further decrease of density in the gauge section of the specimen occurs. The lower the strain rate, the larger the reduction in the sample density. This is probably caused by pore formation.

For equivalent microstructural states (MC or SMC) alloy 718 has lower flow stresses at a given temperature than EP962, and superplasticity can be achieved at lower temperatures in alloy 718 than EP962. Microduplex EP962 exhibits a heterogenous distribution of dislocations; there is a high dislocation density in the γ, and a low dislocation density in the γ'.

DISCUSSION

The results of this investigation indicate that in Ni-based alloys with different chemical and phase composition, high-strain TMP generates dynamic recrystallization; this is accompanied by precipitation and coagulation of second phases, which results in formation of microduplex structures. The SMC microstructure is characterized by features that distinguish it from the usual state of a material. Significant elastic stresses are observed, and the density of the material in the SMC state is less than in the coarse-grained state. This result is explained by a significant growth of intercrystalline grain boundary areas and a change in the structure of regions near boundaries. The reduction in the low temperature limit for SP in alloys with SMC structure is generated by both the reduction in grain size and the formation of non-equilibrium grain boundaries.

The investigation of SP of dispersion-hardened Ni-based alloys EP962 and alloy 718 shows that they can display high ductility at relatively low deformation temperatures. The same ductility is observed in Ni alloys with a microduplex structure at high temperatures but at lower strain rates and at higher flow stresses. As temperature increases, the SP strain rate range of the SMC structure shifts to higher strain rates. Structural changes occurring in these alloys during low-temperature SP are similar to those observed at usual temperatures for SP. At the same time the nature of pore formation is different in the MC and SMC structures. The size of pores is significantly smaller in the SMD than in the microduplex alloy deformed under conventional conditions. In the case of the SMC structure, the level of peak stresses on grain boundaries is significantly less, because there is more effective relaxation of stresses caused by grain boundary sliding [7], dislocation pile-ups near grain boundaries, and triple junctions. However, because of the lower temperatures of SP deformation, the stress relaxation due to diffusion is lower, and as a result the volume fraction of pores remains almost the same.

CONCLUSIONS

1. The generation of SMC structures in Ni-based alloys with different chemical and phase characteristics can provide both low-temperature and high strain rate superplasticity.
2. In comparison with the MC structure, the SMC structure reduces the lower temperature limit of SP by 200-250°C. The SMC structure also decreases the flow stress by a factor of 1.5-2, and allows the maximum strain rate for SP to be increased by 1-1.5 orders of magnitude.

REFERENCES

1. G.A. Salishchev, O.R. Valiakhmetov, V.A. Valitov, Sh. Kh. Mukhtarov, Mater. Sci. Forum. ICSAM 94 **170-172**, 121 (1994).
2. G. Salishchev, R. Zaripova, R. Galeev, O. Valiakhmetov, Nanostructured Materials **6**, pp.913-916 (1994).
3. H. Gleiter, Prog. Mat. Sci. **33**, 223 (1990).
4. O.A. Kaibyshev, *Superplasticity of Alloys, Intermetallides and Ceramics* (Springer Verlag, Berlin, 1992).
5. *Superplastic Forming of Structural Alloys*, edited by N.T. Paton and C.H. Hamilton (The Metallurgical Society of AIME, 1982).
6. J.K. Gregory, J.C. Gibeling, and W.D. Nix, Metall. Trans. **16a**, 777 (1985).
7. O.A. Kaibyshev, A.I. Pshenichniuk, V.V. Astanin, Acta Mater. **46** (14), 4911-4916 (1998).

TEXTURE EVOLUTION DURING LOW TEMPERATURE SUPERPLASTICITY IN 5083 AND 5052 Al-Mg ALLOYS

S.W. SU, I.C. HSIAO, AND J.C. HUANG
Institute of Materials Science and Engineering, National Sun Yat-Sen University, Kaohsiung, Taiwan 804, R.O.C.

ABSYRACT

Low temperature superplasticity (LTSP) at 250 °C and 1×10^{-3} s^{-1} was observed in the 5083 Al-Mg base alloy after thermomechanical treatments (TMT). With increasing TMT rolling strain, the high angle grain boundary fraction increased, more favorable for the further operation of grain boundary sliding and LTSP. The strong texture components and bimodal misorientation distributions present after TMT were not affected by static annealing at 250 °C, but evolved gradually into a random orientation distribution during LTSP straining from 30% to 100%. When the LTSP elongation was greater than 150%, the macro-deformation anisotropy R ratio would finally reach a stable level. It seems that the LTSP performance was controlled by a large fraction of high angle boundaries, but not by the special coincidence site lattice boundaries.

INTRODUCTION

In our pervious studies [1,2] a simple rolling-type TMT was applied to process an inexpensive commercial Al-Mg alloy, resulting in low temperature superplasticity at around 250 °C and 1×10^{-3} s^{-1}, with an optimum tensile elongation to 400%. The TMT processed thin sheet contained grains and subgrains measuring around 0.3-0.5 μm. At temperatures lower than 300 °C, the grains grew slightly and maintained LTSP, with failure by cavity coalescence and slight necking. The flow stress of the LTSP specimens dropped to nearly one half as compared with the as-received (AR) non-LTSP samples, and the strain rate sensitivity increased from 0.15-0.2 of the AR specimens to above 0.4 of the LTSP ones. In this report, the evolution of texture and grain misorientation during TMT processing and LTSP straining are characterized and discussed. It is of most interest in this study to explore three aspects. (1) Whether grain boundary sliding (GBS) could occur at a low temperature of 250 °C for Al base alloys? (2) Whether there is a critical transition strain level that the microstructure would transform from ill-defined subgrains characteristic of anisotropic slip deformation into well-defined equiaxed grains favorable for GBS? (3) What could be the most crucial microstructure parameter that governs the LTSP behavior?

EXPERIMENT

The 5083 (Al-4.7wt%Mg-0.7%Mn) and 5052 (Al-2.6%Mg) alloys were obtained from China-Steel Aluminum Corp., Taiwan, in the form of hot-rolled thick plates of 30 mm thickness. The AR thick plates possessed elongated grains measuring 500x80x8 μm^3 and did not exhibit any LTSP. Simple TMT processes were applied, including annealing at 500 °C for 1 h, followed by air cooling and a series of low temperature rolling (25-250 °C) to a final thickness of 0.5-3 mm. The rolling reduction during TMT varied from 90-98.3% (true strain of 2.3-4.1). Some AR or TMT processed specimens were prepared for exploring the thermal stability of grain structures. The TMT sheet specimens were heated under controlled heating rate, i.e., 50 min from room temperature to the set temperatures such as 250, 280, and 300 °C. Once the desired temperature was reached, the specimens were water quenched and ready for TEM foil specimen preparation. Another group of specimens were all heated to 250 °C during 50 min, followed by further 10-60 min holding time, in order to study the effect of static annealing on the grain-structure thermal stability at this very temperature.

Constant crosshead speed tensile tests were conducted using an Instron 1125 universal testing machine, with the loading direction parallel to the rolling direction. It usually took 50 min for the tested specimens to reach in the desired loading temperature without thermal fluctuations. The specimen gauge length was 8 mm. The grain structure and grain orientation distribution were examined using a Jeol 6400 SEM, equipped with an Oxford Link Opal electron backscattered diffraction system. The surface of the examined specimens were first mechanically polished, followed by electrochemical polishing in 6% perchloric acid-80% ethanol-14% distilled water at room temperature and a DC voltage of 30 V. The spatial resolution limit of EBSD is ~0.3 μm.

RESULTS

Macro-properties

Figure 1 shows the typical true stress-strain curves for the AR and TMT processed specimens loaded at 250 °C. The TMT processed one revealed LTSP to 400% in the optimum loading condition of 250 °C and 1×10^{-3} s^{-1}. The ultimate tensile stress (UTS) typically occurred at a true tensile strain of ε=0.4. After transformed into the true stress and true strain curve, the highest true stress was usually located at ε=0.5. It seemed that once the subgrains formed during TMT transformed successfully into high angle grain boundaries during the critical initial straining over 0<ε<0.5, the material could further deform smoothly to larger strains.

The plastic anisotropy behavior in SP samples can be evaluated in terms of the R ratio of sample width reduction to thickness reduction [3,4]. If the grain structures become equiaxed and randomly oriented, this ratio will be stabilized to a fixed value close to one. Figure 2 shows the increasing trend of R with increasing strain level. The low but increasing R ratios at low superplastic strains were due to the structure evolution which gradually changed the as-TMT anisotropic structures during the initial stage into equiaxed and randomly oriented structures. It seems that the specimen needs at least a local plastic true strain (the local true strain at the necked position under examination) of ε'=1.3, or a tensile elongation of ~150%, for the grain structure evolution to be completed, so as to exhibit a stabilization of R ratios near 0.8.

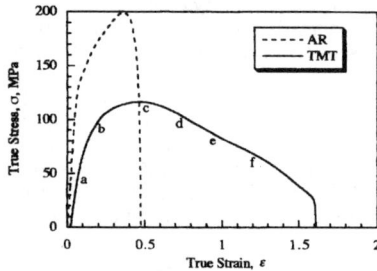

Fig. 1 The true stress and strain curves for the AR and TMT 5083 alloys. The strain levels for EBSD examinations are marked as a to f.

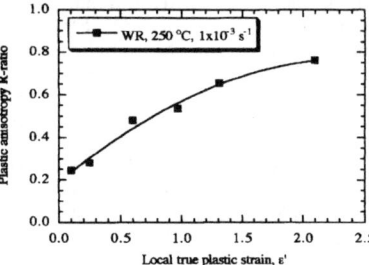

Fig. 2 The variation of the plastic deformation anisotropy R-ratio as a function of true local strain ε' .

Evolution of texture during TMT processing

Based on the EBSD results, the texture under the AR+annealing at 500 °C condition was relatively weak. The highest intensity contour with respect to background was around 4-8x. When subjected to TMT processes, preferred orientation texture components were developed. As the TMT was performed using cold rolling (CR) at 25 °C, a strong texture with a high portion of near brass {110}<112> type was observed. With increasing TMT temperature from 25 °C to 100-200 °C, the texture contained a mixture of Cu {112}<111>, S {123}<634> (also its association R {123}<412>), and brass components. The highest texture intensity contour in the as-TMT specimens scattered within 10-25x. The LTSP tensile elongation became higher for a warm rolling (WR) TMT. For example, the highest LTSP elongations for the 5083 alloy at 250 °C for the cases of CR, or WR at 150 °C and 200 °C were ~150%, 250% and 400%, respectively. At the same rolling strain, CR seemed to result in a higher fraction of the Brass texture component than WR. This can be understood since cold rolling tends to involve less cross slip and produces a higher portion of brass texture. Meanwhile, CR also induced a much higher portion of LAB and a much lower fraction of HAB than WR, resulting in poor performance of LTSP in the CR specimens.

The texture evolution during different TMT stages has been examined for the WR route. The 5083 thick plates were rolled to four different rolling strains of 2.4, 2.7, 3.4, and 4.1. Figures 3a and 3b shows the representative inverse pole figures for the sheet normal and longitudinal directions obtained from specimens with TMT $\varepsilon=3.4$. The sheet normal tended to align along a band from <112> through <123> to <110>; and the sheet longitudinal direction pointed from <112> through <634> to <111>. The misorientation angle between two neighboring grains has an important impact on the activation of cooperative or individual grain boundary sliding (CGBS or GBS). We have defined the boundaries with a misorientation angle less than 10° as low angle boundaries, LAB, and these are commonly treated as the ones unlikely to motivate GBS. Those angles greater than 30° are termed as high angle boundaries, HAB, most favorable for GBS. The rest within $10-30^o$ are classified as medium angle boundaries, MAB. After CR, the fraction of LAB increased up to 60%, compared with 20-35% for the WR counterparts. An apparent bimodal distribution was consistently observed for CR and WR as-TMT sheets; and the former was particularly noticeable.

The bimodal grain boundary misorientation distribution was persistently present after the WR process. Figures 3c and 3d show the misorientation distributions for $\varepsilon=2.4$ and 4.1. With increasing rolling strain, the LAB seemed to decrease and the HAB to increase. The WR specimen with $\varepsilon=3.4$, which was the one exhibiting the highest LTSP elongation, possessed a high fraction of HAB (62%) under the as-TMT condition. The thinnest WR specimens with $\varepsilon=4.1$ (rolled down to 0.5 mm in thickness), though with an even higher HAB fraction of 69%, resulted in a lower LTSP elongation due to the detrimental surface effects from surface oxidation and surface defects which would affect most severely the LTSP properties for the thinnest sheet. The variation of the relative percentages of LAB, MAB and HAB versus the true rolling strain during TMT is shown in Fig. 4a. The obvious trends are the gradual decrease in LAB and gradual increase in HAB with increasing TMT strain, with a similar portion of MAB. It appeared that the grain structures evolved into a more and more mature state during TMT. With a higher TMT rolling strain, the grain boundaries with $\theta>30^o$ constituted a greater fraction, favorable for facilitating GBS and LTSP. Finally, around $11\pm2\%$ of the boundaries were classified as the coincidence site lattice (CLS) Σ boundaries. Usually, 40-60% of the total CLS boundaries belonged to the $\Sigma=3^n$ twin-related categories, which accounted for about 5% of the total grain boundary population.

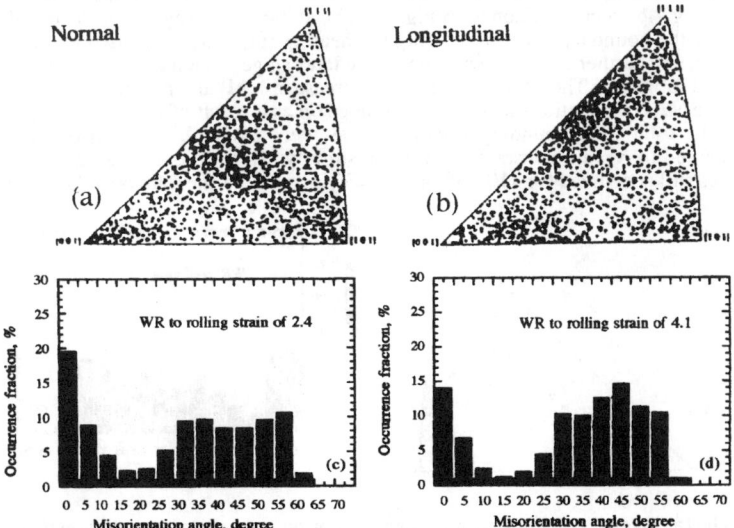

Fig. 3 The inverse pole figures for the sample (a) normal and (b) longitudinal directions of the WR 5083 sheet with $\varepsilon=3.4$, and the misorientation distributions for TMT $\varepsilon=$ (c) 2.4 and (d) 4.1.

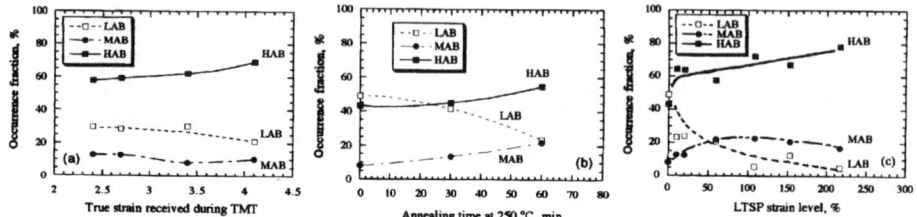

Fig. 4 The variation of LAB, MAB and HAB occurrence fraction as a function of (a) true strain received during TMT, (b) static annealing time at 250 °C, and (c) LTSP strain level.

Evolution of texture during static annealing

Upon heating to 250 °C, the inverse pole figures remained similar to that of the as-TMT specimens. With increasing holding time at 250 °C, the texture contour intensities became slightly lower, but the major texture components remained unchanged. Figure 5 presents the inverse pole figure and misorientation distribution for the 5083 WR specimens heated to 250 °C and held for 60 min. On heating to 250 °C, the boundaries first suddenly became sharpened. Numerous fine grains or subgrains with a low angle boundary appeared, resulting in an increase of LAB fraction from 30-40% in the as-TMT condition to ~50% after heated to 250 °C. With further holding time at 250 °C, the bimodal tendency became weaker, so that the LAB fraction decreased to 24% and the MAB fraction increased to 22% in the specimen held for 60 min. The concurrent grain growth effect from 0.5 to 1.4 μm would also lower the fraction of LAB, eliminating most cells or subgrains with boundary misorientation angles less than 5°.

Distinctly different behaviors were observed from the CR and WR specimens upon static annealing. Due to the suppression of recovery during CR at room temperature, the abundant subgrains and cells (with LAB constituting 60%) proceeded rapid recovery and recrystallization during heating and static annealing, resulting in a much lower LAB fraction of 33%. In contrast, there were continuous recovery and partial recrystallization during WR. The variation of the relative fractions of LAB, MAB and HAB in the 5083 WR sample versus annealing time is presented in Fig. 4b, in comparison with Fig. 4a. The static annealing at 250 °C mainly induced sharpening of the boundary structures during the heating stage, which caused an increase in the LAB fraction, and further grain growth during the later stage, which occurred at the expense of small retained subgrains. The latter resulted in a decrease of LAB and an increase of HAB. In the meantime, the MAB population continuously climbed up as a result of the weakening of bimodal behavior. The total CLS boundary population, as well as the $\Sigma = 3^n$ twin-related boundary fraction, seemed to remain constant or increase slightly after static annealing. Basically, total Σ boundaries still accounted for ~10-13% and $\Sigma = 3^n$ for 5-7% of the total boundary population.

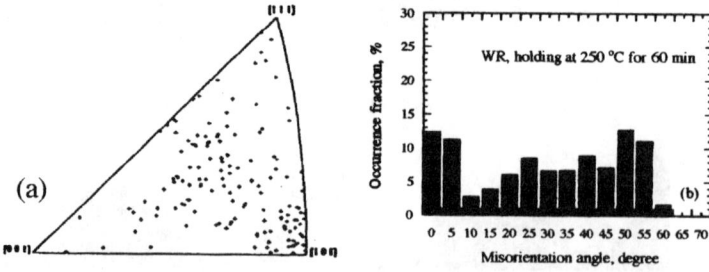

Fig. 5 The (a) inverse pole figure for the sample normal, and (b) misorientation distribution for the 5083 alloy after static annealing at 250 °C for 60 min.

Evolution of texture during LTSP straining

Systematic characterizations on the LTSP strained specimens have been performed for six cases indicated in Fig. 1 with ε'=0.10-2.21 or tensile elongations of 10-216%. At ε'=0.97 or above, a much more uniform orientation distribution was observed, as shown in Fig. 6 for ε'=2.21. There was no sharp transition for the texture transformation; the evolution proceeded in a continuous and gradual way. The highest texture contour level decreased from $17\pm7x$ under the as-TMT condition to ~3x after ε'=2.21. The preferred orientation and the small misorientation between neighboring subgrains or grains were gradually evolved during the initial superplastic straining over $0<\varepsilon'<1.0$. At $\varepsilon'\sim1.0$ (or $\varepsilon\sim0.7$, or a tensile elongation of 100%) the grain structure and grain orientation have both transformed into a mature state favorable for overall GBS. At $\varepsilon'>1.3$ (or $\varepsilon>0.9$, or a tensile elongation >150%), the effective GBS operating at most grain boundaries would result in a nearly isotropic deformation characteristic of an R ratio near 0.8.

The bimodal misorientation distribution sustained over the initial straining from ε'=0-0.25. Significant changes occurred within $0.3<\varepsilon'<1.0$ (i.e. tensile elongation 30-100%). Above $\varepsilon'>1.0$, the distribution was transformed mostly into semi-normal behavior, with a LAB portion only around $8\pm4\%$ and a high HAB population of 70-80%. Such a distribution of grain boundary mutual misorientation can be considered to be a result of extensive CGBS, GBS and grain rotation. This result confirms that, even at a low temperature of 250 °C, the fine-grained Al materials can still deform with GBS the same as those at high temperatures. The variation of LAB, MAB and HAB as a function of LTSP strain level has also been presented in Fig. 4c. It is evident that HAB increased at the expense of LAB. The MAB increased initially in reflection of the disappearance of bimodal distribution, and then remained a stable level of ~20%. After a LTSP elongation of 100%, these three categories of grain boundaries would reach their steady relative fractions. Figure 7 compares the inverse pole figures and misorientation distributions for the 5052 and 5083 alloys LTSP strained to an equivalent strain level of ~100%. With the help from the Mn-containing particles which suppressed grain growth and facilitated CGBS and GBS, the evolution in the 5083 alloy proceeded more rapidly and successfully than the 5052 alloy. The total Σ and $\Sigma=3^n$ boundary fractions in the WR specimens during LTSP continuously remained stable within 9-14% and 3-8%, respectively, similar to the one seen in the as-TMT specimens.

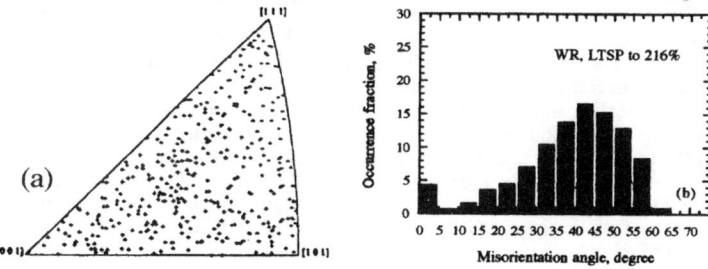

Fig. 6 (a) The inverse pole figure for sample normal, and (b) misorientation distribution of the 5083 sheet after 216% LTSP elongation.

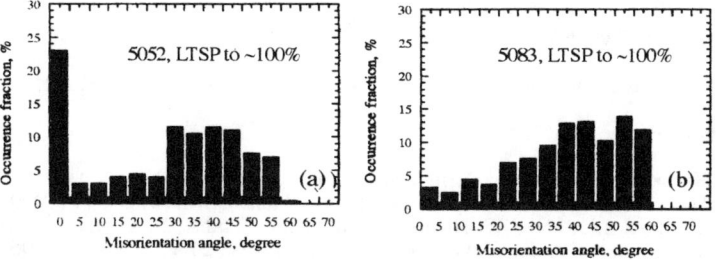

Fig. 7 The misorientation distributions of the (a) 5052 and (b) 5083 alloys to ~100% elongation.

53

Relationship between EBSD and LTSP results

It has been postulated that the special twin related boundaries would affect the GBS operation and superplastic performance [5]. In this study, we did not notice conclusive trends for the total CLS or $\Sigma=3^n$ occurrence fraction during TMT, static annealing and LTSP straining (mostly scattering at $11\pm3\%$ and $5\pm2\%$, respectively). On the other hand, the fractions of LAB and HAB followed an apparent opposite trend as LTSP straining proceeded. Based on the limited data on the current Al-Mg alloys in Fig. 8a, the variation of CLS and $\Sigma=3^n$ boundaries would not correspond to any direct indication of the LTSP elongation performance. Whether the tensile specimens that would exhibit a LTSP elongation of 50% or 400%, the CLS and $\Sigma=3^n$ boundary fractions were all scattered within a narrow band. It seems unlikely that a ±3 or $\pm2\%$ difference in the population of total CLS or $\Sigma=3^n$ boundaries would account for the primary responsibility of the drastically different behavior in LTSP elongation. In contract, a higher LTSP elongation would result if the as-TMT specimens possessed a higher fraction of HAB, as shown in Fig. 8b. It seems that the prevalence degree of HAB could affect the LTSP performance.

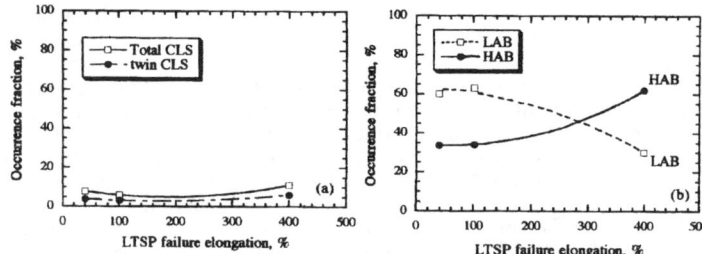

Fig. 8 The dependence of occurrence fractions of (a) total Σ and $\Sigma=3^n$ boundaries, and (b) LAB and HAB versus the LTSP performance in terms of the tensile failure elongation.

CONCLUSIONS

(1) Multiple texture components of the type of brass, S and Cu texture components were observed in the as-TMT specimens. With a higher TMT rolling strain, the high grain boundaries partitioned a greater fraction, more favorable for facilitating GBS and LTSP.

(2) Static annealing at 250 °C itself could not alter the existing texture much, but it reduced the fraction of low angle boundaries through recovery, recrystallization and limited grain growth.

(3) At the peak stress (at $\varepsilon\sim0.5$ or 60% elongation), the structure and texture had not yet completely evolved and GBS had not overwhelmed the deformation. After superplastic straining of 100% or higher, a much more random orientation and misorientation distributions were formed, resulting from extensive GBS and grain rotation occurred at a low temperature at 250 °C. The evolution of microstructure and microtexture proceeded in a gradual and continuous manner. After 150%, the evolution had nearly completed, and exhibited an R-ratio near 0.8.

(4) The amount of HAB (not the special coincidence site lattice boundaries) could affect the LTSP performance. The EBSD layout might be used as one of the indexes in predicting the characteristics of LTSP.

ACKNOWLEDGMENTS

The work was sponsored by National Science Council of ROC (NSC 87-2216-E-110-018).

REFERENCES

1. I. C. Hsiao and J. C. Huang: Scripta Mater., **40**, p. 697 (1999).
2. I. C. Hsiao, J. C. Huang, and S. W. Su, Mater. Trans. JIM, **40**, p. 744 (1999).
3. H. P. Pu, F. C. Liu, and J. C. Huang: Metall. Mater. Trans., **26A**, p. 1,153 (1995).
4. P. A. Friedman and A. K. Ghosh: Metall. Mater. Trans., **27A**, p. 3,827 (1996).
5. T. R. McNelley and M. E. McMahon, Metall. Mater. Trans., **27A**, p. 2,252 (1996).

ASSESSING FAILURE MECHANISMS DURING TRANSFORMATION SUPERPLASTICITY OF Ti-6Al-4V

C. SCHUH, D. C. DUNAND
Department of Materials Science and Engineering, Northwestern University, Evanston, IL, USA

ABSTRACT

During thermal cycling through the α/β phase transformation under the action of a small external biasing stress, Ti alloys exhibit an average deformation stress exponent of unity and achieve superplastic strains. We report tensile experiments on Ti-6Al-4V with an applied stress of 4.5 MPa, aimed at understanding the failure processes during transformation superplasticity. The development of cavities is assessed as a function of superplastic elongation, and macroscopic neck formation is quantified at several levels of elongation by digital imaging techniques. The effects of thermal inhomogeneity on neck initiation and propagation are also elucidated experimentally.

INTRODUCTION

Transformation superplasticity (TSP) is a deformation mechanism observed in polymorphic materials upon cyclic polymorphic transformation [1]. Internal transformation mismatch strains are biased in the direction of an externally-applied stress, resulting in deformation with a low stress sensitivity. In this manner, repeated transformations (by, e.g., thermal cycling) can produce superplastic strains. With the demonstration of TSP in advanced metallic materials (i.e., intermetallic alloys [2] and metal-matrix composites [3]), as well as the extension of this mechanism to multiaxial modes of deformation [4], TSP is of current interest as a prospective shape-forming technology.

Studies of TSP to date have focussed primarily on the mechanics of steady-state deformation, with some limited studies on post-deformation mechanical properties [5, 6]. However, to our knowledge there have been no systematic studies describing the mechanisms of failure during TSP, in uniaxial tension or otherwise. For implementation of TSP in shape-forming processes, knowledge of damage evolution, including cavitation and macroscopic plastic instability development, is essential for process design.

The purpose of the present work is to describe preliminary experiments ultimately aimed at understanding failure processes during TSP of titanium alloys and composites. The procedures and results of experiments on a single tensile specimen of Ti-6Al-4V are described in detail, experimental and mechanistic issues are discussed, and some preliminary conclusions are drawn.

EXPERIMENT

Powder metallurgical Ti-6Al-4V was supplied by Dynamet Technology (Burlington, MA), and machined into a cylindrical tensile specimen with a 20 mm gauge length. The specimen was deformed in tension by TSP in a custom creep frame described in detail elsewhere [7], under an atmosphere of purified argon. All deformation of the specimen took place under a nominal uniaxial true tensile stress of $\sigma_n = 4.5 \pm 0.2$ MPa, which was computed assuming uniform gauge deformation and volume constancy. The α/β transformation was produced by performing triangular thermal cycles between 840 and 1030° C with an 8 min. period. These thermal cycles were chosen to maximize the volume of transformation product (~90% [8]). Temperature was controlled by a BN-coated type-K thermocouple (1.6 mm diameter) positioned

Mat. Res. Soc. Symp. Proc. Vol. 601 © 2000 Materials Research Society

at the gauge surface, approximately 7 mm from the upper specimen head. A second thermocouple was positioned at the specimen head. The fillet radius between the specimen gauge and head was sharp (about 0.5 mm).

The experiments described in the following were designed to elucidate issues central to damage evolution during TSP. The development of internal cavities and external necks was assessed by performing interrupted deformation experiments, in which the specimen was deformed to a selected strain by TSP (with thermal cycles and stress conditions as described above) and the experiment interrupted by an excursion to room temperature. Following density determination and dimensional measurements at room temperature, the specimen was again loaded into the creep frame for an additional increment of deformation. Prior to the first stage of deformation, the specimen and load train were allowed to come to equilibrium at the upper cycling temperature (1030° C) for 75 min., after which the load was applied and thermal cycling initiated. This procedure is used to ensure a coarse β-grain size ($\sim$ 0.5-1 mm), thereby minimizing subsequent grain growth during thermal cycling. A similar equilibration procedure was performed prior to each subsequent stage of deformation, but at the lower cycling temperature (840° C) to minimize thermal sintering of any internal cavities.

During excursions to room temperature, the following two measurements were made:
- Specimen density was determined using the Archimedes method in distilled water. The water temperature was measured with an accuracy of 0.1 K, and the water density corrected for temperature according to Ref. [9]. The density of the gauge section was calculated assuming that the specimen heads remained at full density for all stages of deformation. After the specimen had failed, the heads were removed from the gauge with a diamond saw, and their density measured. This procedure indeed verified that the specimen heads remained at full theoretical density. This technique was found to give reliable gauge densities to $\pm 3 \cdot 10^{-3}$ g/cm^3 ($\pm 0.07\%$ for Ti-6Al-4V).
- The macroscopic profile of the gauge section was assessed by the following methods. A digital image of the specimen was first captured, and an edge-finding algorithm employed to

Table I: Density and porosity of the gauge section of Ti-6Al-4V as a function of superplastic elongation

elongation [%]	gauge density [g/cm^3] $\pm 3 \cdot 10^{-3}$ g/cm^3	porosity [%] $\pm 0.07\%$
0	4.422	0
32	4.409	0.31
66	4.409	0.31
101	4.414	0.18
136	4.415	0.17
172	4.406	0.35
203	4.405	0.38
260	4.405	0.38

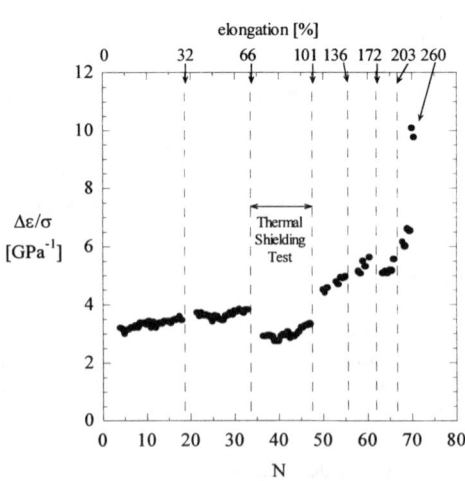

Figure 1: Deformation history of Ti-6Al-4V; stress-normalized strain increment $\Delta\varepsilon$ as a function of thermal cycle number N.

obtain the profile of the gauge section. The projected diameter was then measured as a function of axial position using image processing software. Two images were analyzed after each stage of straining, representing two perpendicular projections of the specimen. This method was found to give accurate measurements, with precision of about 60 μm on the diameter measurements. In addition, calipers (with accuracy of 10 μm) were used to measure the specimen diameter at six to twelve axial positions, for comparison with the image analysis results. The deformation was interrupted six times; together with data collected before deformation and after fracture, this allowed for density and profile measurements at eight elongations between 0 and 260% (Table I).

In addition to the above experiments, heat transfer effects were considered in a single experiment. During the third stage of deformation (from 32 to 66% elongation), the second sheathed type-K thermocouple (1.6 mm diameter) was extended from the top of the creep frame to the lower head of the specimen. This arrangement served two purposes. First, by comparing the temperature measured at the gauge surface to that measured at the lower head, thermal uniformity along the specimen length was verified. Second, since heating is performed by four symmetrically-positioned radiation heat lamps, the thermocouple acted as a radiation barrier, blocking an estimated 10-20% of the incident radiation from the heaters.

RESULTS

The measured displacements during thermal cycling are composed of thermal expansion and contraction of the specimen and load train, as well as deformation of the specimen by TSP and creep. Thermal expansion displacements are fully recovered after a complete thermal cycle, allowing for the measurement of a true strain increment, $\Delta\varepsilon$, after each thermal cycle, which represents deformation of the specimen, averaged over the gauge length. In Fig. 1, the full deformation history of the interrupted experiment is shown, in which each strain increment $\Delta\varepsilon$ is normalized by the nominal applied stress, σ_n, and plotted against the cycle number, N. Vertical dotted lines are shown where the test was interrupted. At the beginning of each stage of deformation, the measured displacement is complicated by thermal transients, which disappear after a dynamic steady-state is established (typically requiring about 2-3 thermal cycles); data points associated with transient cycles have been removed from Fig. 1.

As seen in Fig. 1, the normalized strain increment, $\Delta\varepsilon/\sigma_n$, was found to be ~3.1 GPa^{-1} at the beginning of the test, rising to larger values over the course of the experiment. For the stage of deformation in which thermal shielding was investigated, $\Delta\varepsilon/\sigma_n$ was reduced by about 25% relative to the data points of the immediately preceding and following stages.

After all stages of straining, the specimen was found to exhibit a circular cross section, with only slight deviations from circular symmetry (~5%). Thus, the two independent digital images yielded specimen profiles which overlapped to within 20 μm in most cases, 70 μm in the worst case. Additionally, the manual diameter measurements made with a caliper were found to agree well with the data collected from the digital image processing. Therefore, the diametral measurements from the two images were averaged, to yield the specimen profiles shown in Fig. 2. This figure also shows the distribution of cross-sectional area (calculated from the average diameter) after each stage of straining. Beginning at the early stages of the test, the specimen exhibited a tendency to deform homogeneously, except near the upper head (left side of Fig. 2), where deformation proceeded more slowly. This inhomogeneity resulted in a neck (arrows in Fig. 2) which ultimately thinned until the specimen failed.

Finally, the volume fraction of cavities measured after each stage of deformation is shown in Table I, and was found to be less than 0.4% for all stages of deformation.

DISCUSSION

Transformation Superplasticity

Greenwood and Johnson [10] developed a continuum-mechanics theory of TSP, in which an external uniaxial stress σ biases internal strains produced by the transformation volume mismatch. For a material in which the weaker phase creeps with a power law, they derived a linear relationship between the strain increment produced on each thermal cycle, $\Delta\varepsilon$, and σ:

$$\Delta\varepsilon = C\cdot\sigma \qquad (1)$$

where C incorporates characteristics of the internal stress and strains produced by the transformation, as well as the creep behavior of the weaker phase.

The linear relationship of Eq. (1) has been observed by several authors [6, 11, 12] for thermal cycling of Ti-6Al-4V. Although the value of the constant C is characteristic of the Ti-6Al-4V α/β transformation, it depends on the thermal cycle amplitude, since different amplitudes produce transformations of differing completeness. In a previous study [11], we examined the relationship of Fig. 1 for Ti-6Al-4V with thermal cycles identical to those used in the present work, and reported $C \approx 3.1$ GPa^{-1}. The value observed at the beginning of the present experiment, $C \approx 3.1$ GPa^{-1} (Fig. 1) is thus in agreement with our previous work, as well as that of Kot *et al.* [6], who found $C \approx 3.2$ GPa^{-1} for thermal cycles in the range 760-980° C. The observed increase of C over the course of deformation is attributed to the development of plastic instabilities, as observed in Fig. 2, and will be the subject of future publications.

Thermally-Initiated Flow Instability

The importance of uniform thermal conditions for TSP flow stability is emphasized by the data in Fig. 1. When the specimen was slightly shielded from the radiant heaters by a thermocouple probe, the measured strain increments were about 25% lower than expected. Given a lower thermal flux, the specimen may not have completely transformed on heating, which would reduce the deformation expected from TSP. Furthermore, for the period of time during each cycle when the specimen is fully in the β-field (> 1000° C), power-law creep occurs

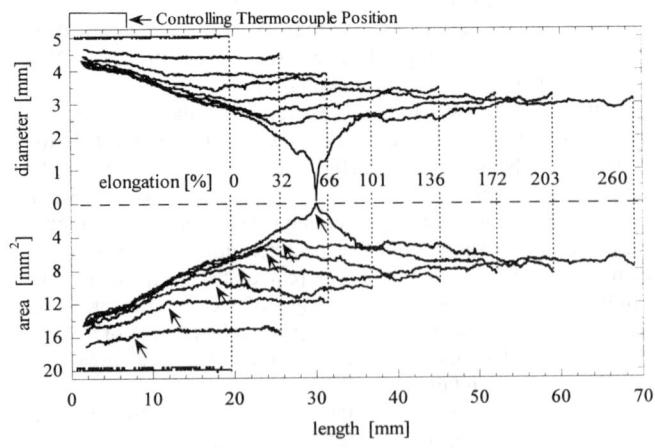

Figure 2: Axial profiles of specimen diameter and cross-sectional area at several superplastic elongations. Controlling thermocouple position is indicated by the shaded region, and the arrows follow the progression of a plastic instability.

58

instead of TSP. If the thermal shielding effect reduces the length of time spent in the β-field or the maximum temperature achieved during the cycling, the contribution of power-law creep in the β phase is reduced.

The thermal effect described above was, in the thermal shielding experiment, uniform over the gauge length of the specimen, and thus the deformation rate of the entire gauge section was affected (Fig. 1). The implications of this experiment for non-uniform thermal conditions are clear; local hot or cold spots can experience substantially different deformation behavior, which may lead to the nucleation and/or growth of plastic instabilities. This is illustrated in the following paragraph.

For each stage of deformation, the controlling thermocouple was positioned on the specimen gauge section, about 7 mm from the upper specimen head. This thermocouple thus provided partial thermal shielding for the upper 7 mm of the specimen, while the remainder of the specimen gauge was unshielded. As shown in Fig. 2, the position of the controlling thermocouple can be directly correlated with the onset of inhomogeneous deformation. The upper 7 mm of the specimen experienced only a small radial contraction, even after the specimen was deformed to 260% elongation. In contrast, sections of the gauge which were far from the controlling thermocouple deformed in a generally homogeneous manner. Furthermore, a diameter inhomogeneity initiated near the thermocouple position in the early stages of deformation spread and developed into a significant profile gradient over the course of superplastic elongation. The site of fracture is associated with this inhomogeneity, as shown by the arrows in Fig. 2.

Damage Accumulation

Barring environmental degradation or microstructural coarsening, the two most common modes of tensile creep fracture are by the accumulation of internal damage (cavity growth and coalescence), or macroscopic damage (necking or loss of section) [13]. In the present case of TSP in coarse-grained Ti-6Al-4V, almost no cavitation damage was detected (Table I), and the specimen was pulled to failure at a fine point. This result is similar to that for fine-grained Ti-6Al-4V deformed isothermally in the grain- or phase-boundary sliding regime, in which superplastic flow without significant cavitation has been observed up to elongations over 1000% [14]. In contrast, β-transformed Ti-6Al-4V hot-worked in the α+β field exhibits extensive cavitation, leading to premature fracture at elongations of less than 300% [15]. This result suggests that TSP could be useful in shape-forming applications where suppression of internal damage is of more interest than high deformation rates, particularly when a fine-grained structure cannot be achieved (e.g., cast alloys).

There is one experimental factor which may affect the cavitation measurements in the present experiments. After any given stage of TSP straining, the excursion to room temperature allows measurement of cavities developed during that particular stage. However, before the following deformation increment the specimen is allowed to equilibrate at 840° C for 75 min. (without applied stress) prior to re-initiating the TSP experiment. This time may allow for sintering of the cavities, and thus may somewhat heal, or rejuvenate, the material. Since cavity growth rates are cavity size-dependent, repetition of this annealing procedure multiple times (as in the present work) would tend to amplify the error due to this phenomenon. This problem will be addressed in future experiments.

Neglecting potential errors from the mechanism described above, it is apparent from Table I and Fig. 2 that macroscopic plastic instabilities govern the tensile ductility during TSP of Ti-6Al-4V, and that cavitation plays a negligibly small role. Furthermore, as described in the previous section, thermal variations appear responsible for initiation of the critical plastic instability. For industrial implementation of TSP as a forming method, these issues may govern the selection of part size or thermal cycling rate (through the requirement of uniform dynamic thermal conditions), or may alternatively restrict the achievable forming ductility. However, without the development of internal cavitation damage, the formed parts should retain the mechanical properties of the bulk material, as also found in post-deformation mechanical testing [5, 6].

CONCLUSIONS

Failure of Ti-6Al-4V during uniaxial tensile deformation by transformation superplasticity has been investigated during thermal cycling through the α/β transformation range under a uniaxial stress of 4.5 MPa. By periodically interrupting the superplastic straining with excursions to room temperature, internal cavity development and external plastic instabilities could be assessed. Despite plastic elongation to ~260% engineering strain, less than 0.4% cavities were detected in the Ti-6Al-4V specimen at any stage of deformation. Furthermore, successive profiles of the specimen gauge section illustrated the progressive growth of a neck. Thermal variations are found to have a significant impact on the deformation rate, and are associated with the initiation of the inhomogeneity which ultimately lead to fracture.

ACKNOWLEDGEMENTS

This study was primarily funded by NSF SBIRs #9760593 and #9901850, monitored by Dr. R. Coryell, through subcontracts from Dynamet Technology. Helpful discussions with W. Zimmer of Dynamet Technology are gratefully recognized, and C.S. acknowledges support of the U.S. Department of Defense through a National Defense Science and Engineering Graduate Fellowship.

REFERENCES

1. T.G. Nieh, J. Wadsworth, and O.D. Sherby, *Superplasticity in Metals and Ceramics*. (Cambridge University Press, Cambridge, 1997).
2. C. Schuh and D.C. Dunand, Acta Mater. **46**, 5663 (1998).
3. D.C. Dunand and C.M. Bedell, Acta Mater. **44**, 1063 (1996).
4. D.C. Dunand and S. Myojin, Mater. Sci. Eng. **230A**, 25 (1997).
5. C. Schuh, W. Zimmer, and D.C. Dunand. in *Creep Behavior of Advanced Materials for the 21st Century* edited by R.S. Mishra, A.K. Mukherjee, and K.L. Murty (Warrendale PA, TMS, 1999) p. 61.
6. R. Kot, G. Krause, and V. Weiss. in *The Science, Technology and Applications of Titanium* edited by R.I. Jaffe and N.E. Promisel (Oxford, Pergamon, 1970) p. 597
7. P. Zwigl and D.C. Dunand, Metall. Mater. Trans. **29A**, 2571 (1998).
8. W. Szkliniarz and G. Smolka, J. Mater. Proc. Tech. **53**, 413 (1995).
9. D.W. Green and J.O. Maloney, eds. *Perry's Chemical Engineer's Handbook*, (New York, McGraw-Hill, 1984).
10. G.W. Greenwood and R.H. Johnson, Proc. Roy. Soc. Lond. **283A**, 403 (1965).
11. C. Schuh and D.C. Dunand, Int. J. Plastic. (in print).
12. K. Sato, T. Nishimura, and Y. Kimura, Mater. Sci. Forum. **170-172**, 207 (1994).
13. M.F. Ashby and B.F. Dyson. in *Advances in Fracture Research, Fracture 84,* edited by S.R. Valluri, *et al.* (Oxford, Pergamon, 1984) p. 3.
14. M.T. Cope and N. Ridley, Mat. Sci. Technol. **2**, 140 (1986).
15. S.L. Semiatin, V. Seetharaman, A.K. Ghosh, E.B. Shell, M.P. Simon, and P.N. Fagin, Mater. Sci. Eng. **A256**, 92 (1998).

CAVITATION IN A UNIAXIALLY DEFORMED SUPERPLASTIC Al - Mg ALLOY

D. H. BAE and A. K. GHOSH
Department of Materials Science and Engineering, The University of Michigan, Ann Arbor, MI 48109

ABSTRACT

Cavitation caused by superplastic straining of a fine-grained Al-Mg-Mn-Cu alloy under uniaxial tension has been systematically evaluated. Tensile tests were conducted in the strain-rate range of $10^{-4}s^{-1}$ to $10^{-2}s^{-1}$ and in the temperature range of $450^{\circ}C$ to $550^{\circ}C$. Measurements of the number and size of cavities were made by image analysis through optical microscopy on tested specimens. With increasing imposed strain, the cavity population density increases. Cavity growth has been found to be primarily due to the plastic deformation of the matrix. These results are characterized by the total volume fraction of cavities which is found to increase exponentially with strain. However, the dependencies of cavity volume fraction on strain-rate and temperature are not straightforward and the notion of just a few large cavities controlling the total cavity volume is not always true. Attempts to explain these complex dependencies have been carried out based on the concepts of debonding between the matrix and non-deformable particles, the continuous nucleation of new cavities, and plasticity-based cavity growth for large cavities.

INTRODUCTION

Superplastic cavitation is generally quantified by the total volume fraction of voids. Volume fraction, V, can be expressed as $V = \sum N_i \times v_i$ where N_i is the number of cavities per unit volume having a cavity volume v_i. It has been found that the level of cavitation increases with strain not only due to the increasing cavity size but also due to the increasing cavity population [1]. During complex part fabrication using superplastic forming processes, the strain, strain-rate, and stress-state vary widely in different regions of the part, causing varying levels of cavitation. In addition, cavitation behavior of different superplastic alloys also varies significantly with test temperature and strain-rate [2, 3]. An understanding of the principles behind the different cavitation tendencies for different materials, microstructures, and test conditions is difficult without knowledge of the population of cavities as well as their growth kinetics.

However, existing cavitation models (particularly the diffusional growth model [4]) grossly underestimate the level of internal damage developed during superplastic deformation for fine-grained metals [3, 5]. Many of the cavitation models [4, 6, 7] were originally developed for the prediction of creep life in coarse microstructures. Later these models were applied to fine-grained superplastic alloys exhibiting several hundred to several thousand percent elongations, where the test temperature and strain-rate are much higher [8-10]. The predictions of void size and volume based on existing creep cavity growth models are less than adequate for superplastic deformation conditions, where diffusional accommodation is very high.

For a particle-containing alloy, submicron scale interface defects can cause matrix decohesion from particle. This early cavity growth is rapid due to constraint from the non-deformable particle interface. The growth process, discussed in Ref. 11 shows how submicron size defects grow large enough to be visible under the optical microscope. After the entire particle interface is debonded, the size of the cavity suddenly rises to the size of particles (1 ~ 4μm). Beyond this, cavities may grow with continued plastic deformation of the matrix, in which the particle's interface constraint effect is essentially removed. Detailed and quantitative information on the development of cavities is required to understand the progress of cavitation damage beyond the initial development. In this study, cavitation under uniaxial tension has been systematically investigated in a fine grain Al-Mg-Mn-Cu alloy. Based on a recent study of the plasticity based cavity growth model [5], the growth of single cavities after debonding from particles has been simulated to explain the dependencies of cavitation on superplastic forming parameters. New insight into the cavitation problem has been gained by combining the growth of individual cavities with the evolution of cavity population density with strain.

MATERIAL AND EXPERIMENTAL

An aluminum alloy containing 4.7%Mg, 0.8%Mn, and 0.4%Cu by weight was used in this study. The alloy was provided by Kaiser Aluminum Company in the DC cast condition. It was homogenized at 500°C for 12 hours and subsequently hot forged. To obtain a fine-grained microstructure, it was cold rolled to 1.7mm final thickness ($\sim$ 90% reduction), and then annealed at 500°C for 0.5 hour in a salt bath. The resulting grain structure showed the average grain size of 7.9μm. Two types of particles have been observed in the SEM sample [12]: (i) aging precipitates containing Mg-Al-Cu (< 0.1μm), which are in solid solution at the test temperature (> 450°C) and (ii) dispersed intermetallic particles, such as Al_6Mn containing Cu ($0.5 \sim 5$μm), which are located at the grain boundaries.

Uniaxial tensile tests were carried out on this material (specimen gage length = 12.7mm) under constant strain-rate. The strain-rate range for the tests was within $10^{-4}s^{-1}$ - $10^{-2}s^{-1}$ and the temperature range was within 450°C - 550°C. Maintaining a constant $\dot{\varepsilon}$ in superplastic tensile tests is not straightforward due to flow of soft material from the grip. To correct for this flow problem, an improved cross-head speed schedule for a constant strain-rate [13] was used. Cross-head speed was controlled by computer through a digital interface board on an Instron machine. Test interruptions were made to produce samples with a variety of strain levels in the range of 0.15 to 1.2 and the samples quenched for metallographic observation.

To measure the number and size of cavities under the microscope, the tested samples were sectioned along three orthogonal planes: L-S, L-T, and T-S planes, taken from the uniform part of the gage region, and then mechanically polished. Local plastic strain for each section was determined from its cross-sectional areas before and after the test. The measurements of cavity size and distribution were made by using digitized images (1400x) taken with a CCD camera (640 x 480 pixels) through an optical microscope. The dimension of each image was 273.5 x 205.13μm with a pixel size of 0.427μm. Two-dimensional measurements of cavities on the L-S, L-T, and T-S planes were conducted using NIH-Image software. The cavity images were computer-fitted to ellipses (best fit) of equivalent area, which provided the dimensions of the major and minor axes of the ellipse, and the orientation of the major axis, and an equivalent diameter for each cavity.

RESULTS

Optical Micrographs of Cavities

Fig. 1 shows the micrographs of specimens deformed at T = 500°C and $\dot{\varepsilon}$ = $10^{-3}s^{-1}$ for strain levels of 0.46 and 0.84. Cavities are found to be randomly distributed and cavity size varies over a broad range at a fixed strain. As the strain increases, the population and average size of cavities are found to increase but new small size cavities also emerge. Some extension in the loading direction (L) is also observed. They also grow in the transverse directions, and their shape is anisotropic in the T-S plane (not shown); the length in the T direction is larger than that in the S direction, which may be related to the anisotropic grain morphology of this alloy.

Number Density of Cavities

Fig. 2 shows the total number of cavities (expressed as density = number per volume; the conversion of 2-D data to 3-D values required a special program, described in Ref. 5) for three different strain-rates as a function of strain at the two different temperatures: (a) 500°C and (b) 550°C. The cavity density at zero strain was obtained from the specimen just before the start of deformation at the corresponding temperature. At 550°C, this level is low possibly due to damage recovery during annealing (~10min) to stabilize the isothermal test condition. With increasing strain, the number density of cavities increases for all test conditions, i.e. new cavities are continuously formed and grow during deformation [1, 5, 11].

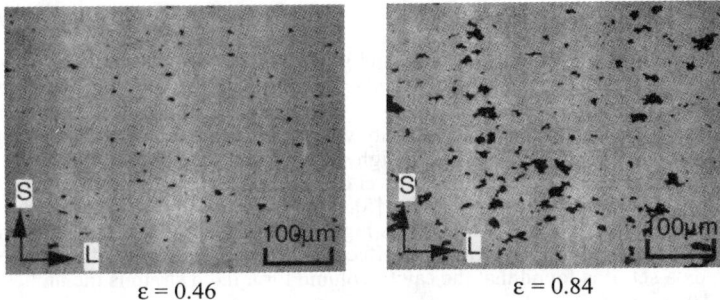

ε = 0.46   ε = 0.84

Fig. 1. Micrographs of cavities in an Al-Mg-Mn-Cu alloy developed during superplastic deformation at T = 550°C and $\dot{\varepsilon}$ = 10⁻³s⁻¹ in the L-S plane. Tensile tests were interrupted at the strain of 0.46 and 0.84. Tensile axis is horizontal.

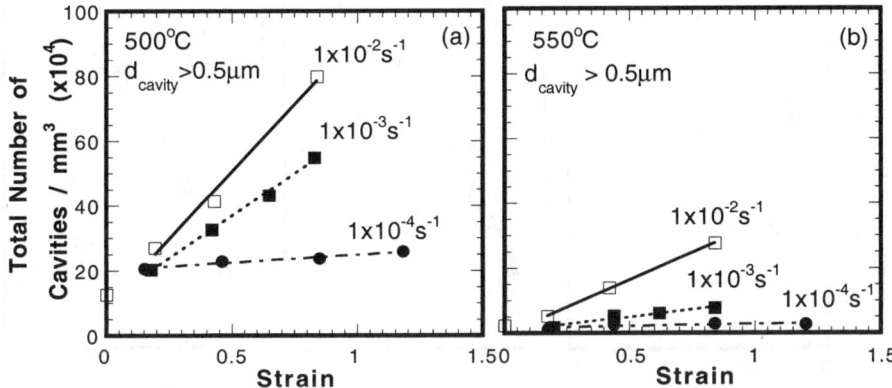

Fig. 2. Cavity population (number/mm³) for three different strain-rates as a function of strain at the test temperature of (a) 500°C and (b) 550°C in the superplastic Al-Mg-Mn-Cu alloy deformed under uniaxial tension.

Recent work [11] has shown that the nucleation of cavities in these alloys involves a process of debonding of the matrix from the non-deformable second phase particles followed by a complete separation via interface-constrained plasticity growth. This process is strongly influenced by strain-rate. Typically, large particles and large initial defects induce more rapid decohesion with increasing strain, while smaller particles have a smaller constrained zone and require more strain to produce complete debonding. We believe that this initial growth process causes more small (nanoscale) voids to become visible continuously with increasing strain (continuous "nucleation"). The frequency of cavity formation events increases with increasing strain-rate. This strain-rate effect on continuous nucleation is believed to be related to a higher stress concentration around the particles and lower diffusional accommodation at the interface.

Volume Fraction of Cavities

The volume fraction of cavities, which is equivalent to the area fraction of cavities, is related to the population and size of the various cavities. The measured results of cavity volume fraction are plotted in a semi-logarithmic scale as a function of strain in Fig. 3 for three different temperatures: (a) 450°C, (b) 500°C, and (c) 550°C, in which arithmetic average value of volume fraction obtained from the L-S, L-T and T-S planes is plotted. The volume fraction of cavities increases nearly exponentially with strain, which is in agreement with the data of other superplastic alloys [2, 9], indicating that void growth is generally controlled by plasticity [6]. In Fig. 3, the slope of these plots $\eta = d\ln V/d\varepsilon$ termed the cavity growth rate factor was determined in the strain range showing a linear region of Fig. 3. At the near optimum superplastic temperature of 550°C, the cavity volume is found to be dependent on the strain-rate. A high strain-rate ($10^{-2}s^{-1}$) produces a high level of cavitation, and the level of cavitation decreases at lower strain-rates. The low level of cavitation at a low strain-rate ($10^{-4}s^{-1}$) is due to the slow initial growth of the new cavities [5]. At 500°C, the strain-rate effect on cavity volume is more complicated. The cavity volume is higher at an intermediate strain-rate ($10^{-3}s^{-1}$). At the test temperature of 450°C, the strain-rate effect on the cavity volume is found to be weak. From the entire data set, it is found that the cavity volume for a fixed strain is the highest at T = 500°C and $\dot{\varepsilon} = 10^{-3}s^{-1}$.

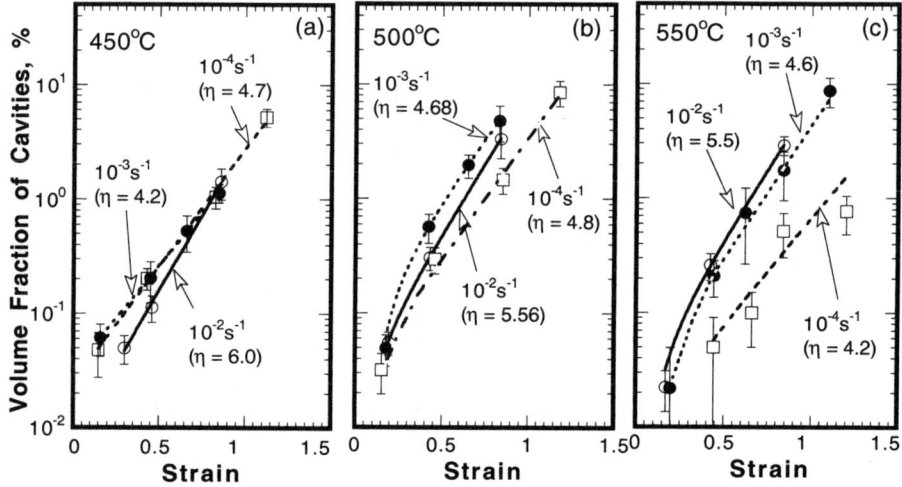

Fig. 3. The variation of volume fraction of cavities (plotted logarithmically) with strain at the test temperature of (a) 450°C, (b) 500°C, and (c) 550°C.

DISCUSSION

While it is a common belief that the total cavity volume is determined by a few large cavities only, the real situation is more complex as seen above. The level of cavitation increases with strain not only due to the increasing cavity size but also due to the increasing cavity population (continuous nucleation) [1, 5]. The knowledge of the population of cavities as well as their growth kinetics is required to understand the different cavitiation tendencies. The rate of continuous nucleation was empirically found to increase with increasing strain-rate and decreasing temperature for such an alloy, both of which can be attributed to the reduced diffusional accommodation and higher flow stress of the matrix [5].

After initial debonding from particles [11], the rate of cavity growth becomes much slower. A more detailed model of this process has now been developed [5]. In this model, a perturbed flow field near the cavity is considered because the cavity carries no stress (Fig. 4). The higher rate of deformation in the perturbed region leads to faster growth of the cavity. This was estimated as functions of several parameters such as the strain-rate sensitivity (m), the zone size of strain concentration (k_1), and the number density of cavities (ρ) (i.e. average cavity spacing (L_o)), etc.

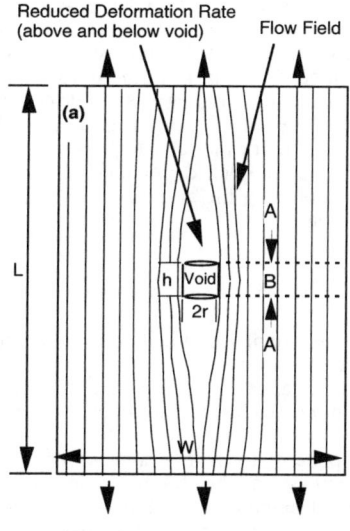

Reduced Deformation Rate
(above and below void)   Flow Field

(a)

L

h | Void
2r

A
B
A

W

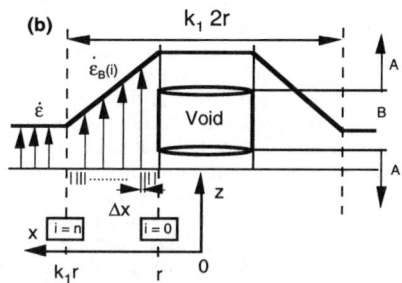

(b) $k_1 2r$

$\dot{\varepsilon}_{B(i)}$

$\dot{\varepsilon}$ Void   A

B

A

Δx z

x $\boxed{i=n}$ $\boxed{i=0}$

$k_1 r$ r   0

1) Force balance between region A and region B

$$\bar{\sigma}_B = \frac{\bar{\sigma}_A}{1-f} \qquad \text{where } f = 1/k_l^2$$

2) Applied extension rate along the axial direction

$(L-h)\dot{\bar{\varepsilon}}_A + h\dot{\bar{\varepsilon}}_B = L\dot{\varepsilon}$

3) Transverse contraction rates in regions A and B are equal to $-W\dot{\varepsilon}/2$

4) Strain-rate continuity at the periphery of the zone

$$r_{new} = [(k_1 r)^2 \exp(-\varepsilon) - \sum_{i=0}^{n-1}[(k_1 r - i\Delta x)^2$$
$$- (k_1 r - (i+1)\Delta x)^2]\exp(-\varepsilon_B(i))]^{(1/2)}$$

Fig. 4. Schematic representation of (a) unit cell with a cavity showing assumed flow field around a cylindrical void of radius = r and height = h within a unit cell of the dimensions of height = L and width = W and (b) the assumed strain-rate distribution near the void.

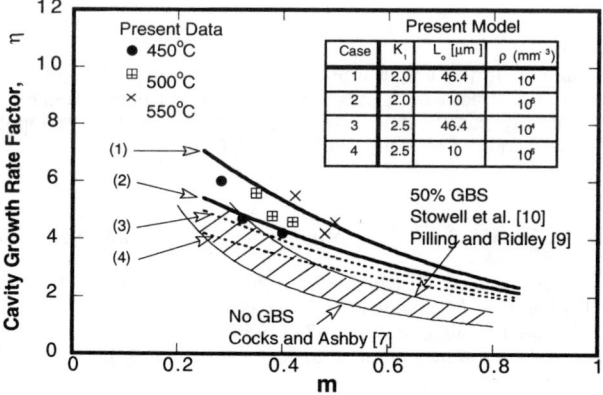

Fig. 5. The variation of cavity growth rate factor, η, with the strain-rate sensitivity (m).

The η-values calculated from this model versus m are plotted in Fig. 5 ((1) - (4)) by varying zone sizes, k_l, and/or the initial number density of cavities. The experimentally obtained η-values (Fig. 3) are also plotted along with theoretical estimates from existing models [7, 9, 10]. Using the present plasticity based model, the zone size k_l = 2 reasonably fit the experiment data for the actual cavity number density in the range of 10^4 - $10^6/mm^3$ (Fig. 2), indicating that strain field is localized near the void in this region. However, existing models again show an underestimate of cavity growth in this material, even after inclusion of the 50% GBS effect. The effect of strain-rate sensitivity on the cavity growth is significant and reasonably predictable.

Another important effect is the continuous decrease in the size of the unit cell as more voids emerge. When the cavity population is small, the strain contribution from the uncavitated region to total strain is large. This causes faster enlargement of the few cavities as seen at the low $\dot{\varepsilon}$. At higher $\dot{\varepsilon}$, many cavities nucleate but they grow slightly because the unit cell is much smaller in size. Based on this plasticity based single cavity growth model and the study of continuous nucleation of new cavities during deformation, the complex dependencies of the various parameters that are involved can be understood by combining the growth of individual cavities with the evolution of cavity population density with strain.

CONCLUSIONS

1. In order to characterize cavitation in formed parts, the size of individual voids and their distribution must be known in addition to the total void volume.
2. The notion that only a few large cavities control the total cavity volume is not always true. Observed cavity density increases with increasing strain, i.e., cavities continuously emerge and become visible (continuous "nucleation").
3. The total volume fraction of cavities increases exponentially with strain. Around the optimum superplastic temperature, the level of cavitation decreases as the strain-rate decreases. At other temperatures and strain-rates, its dependencies are not monotonic. This may be related to the combined roles of grain boundary sliding causing strain concentration and diffusional processes providing accommodation.
4. Based on an improved plasticity based cavity growth model, it is found that the growth of cavities is more for low strain-rate sensitivity and lower number density of cavities. The complex dependencies of strain-rate and temperature that are involved can be understood by combining the growth of individual cavities with the evolution of cavity population density with strain.

ACKNOWLEDGMENTS
This work was performed under support from US Dept of Energy under grant FG02-96ER45608-A000, and a contract from General Motors R & D Center. Acknowledgement is also due to the US Air Force Contract F33615-94-C-5804 for sabbatical leave appointment of A. K. Ghosh at the Air Force Research Laboratory at WPAFB, OH.

REFERENCES
1. A. K. Ghosh and D. H. Bae, Mat. Sci. Forum, Trans. Tech. Publications, Switzerland, 243-245, 89 (1997).
2. N. Ridley, D. W. Livesey and A. K. Mukherjee, J. Mat. Sci., 19, 1321 (1984).
3. J. Pilling and N. Ridley, Res. Mechanica, 23, 31 (1988).
4. D. Hull and D. E. Rimmer, Philos. Mag., 4, 673 (1959).
5. D. H. Bae and A. K. Ghosh, to be submitted to Acta Mater., (1999).
6. J. W. Hancock, Metal Sci., 10, 319 (1976).
7. A. C. F. Cocks and M. F. Ashby, Metal Sci., 14, 395 (1980).
8. M. J. Stowell, Metal Sci., 17, 1 (1983).
9. J. Pilling and N. Ridley, Acta Metall., 34, 669 (1986).
10. M. J. Stowell, D. W. Livesey and N. Ridley, Acta Metall., 32, 35 (1984).
11. A. K. Ghosh, D. H. Bae and S. L. Semiatin, Mat. Sci. Forum, Trans. Tech. Publications, Switzerland, 304-306, 609 (1999).
12. H. Watanabe, K. Ohori and Y. Takeuchi, Trans., ISIJ, 27, 730 (1987).
13. P. A. Friedman and A. K. Ghosh, Metall. Trans., 27A, 3030 (1996).

CAVITATION BEHAVIOR OF COARSE-GRAINED AL-4.5MG ALLOY EXHIBITING SUPERPLASTIC-LIKE ELONGATION

H. Iwasaki[a], T. Mori[a], H. Hosokawa[b], T. Tagata[c], M. Mabuchi[d],   and K. Higashi[e]

a) College of Engineering, Department of Materials Science and Engineering,
Himeji Institute of Technology, Shosha, Himeji, Hyogo 671-2201, Japan
b) Graduate Student of Department of Metallurgy and Materials Science,
  Osaka Prefecture University, Gakuen-cho, Sakai, Osaka 599-8531, Japan
c) SKY Aluminum Co., Ltd, 1351 Uwanodai, Fukaya, Saitama 366-0801, Japan
d) National Industrial Research Institute of Nagoya,
Hirate-cho, Kita-ku, Nagoya 462-8510, Japan
e) College of Engineering, Department of Metallurgy and Materials Science,
Osaka Prefecture University, Gakuen-cho, Sakai, Osaka 599-8531, Japan

ABSTRACT

Cavitation behaviors related to ferrous primary crystals have been investigated at a temperature of 653 K and a strain rate of 10^{-3} /s for Al-4.5%Mg-0.05%Fe and Al-4.5%Mg-0.2%Fe alloys which have a grain size of 50 μm. The alloys constantly exhibited a large elongation-to-failure above 300% at the temperature of 653 K and strain rate of 10^{-3}/s. Cavitation was increased by increasing the iron content. Most cavities were nucleated at the interface between the ferrous primary crystal and matrix and elongated parallel to the tensile direction. The experimental critical diameter of the primary crystal, above which cavity is nucleated, was 1.5 μm at the grain boundary and 0.5 μm at grain interior, which were very close to double the critical diffusion length.

INTRODUCTION

Al - high Mg alloys exhibit a low stress exponent value of 3, or a high strain rate sensitivity value of 0.33 for dislocation creep due to viscous dislocation glide [1- 11].   Recently, large elongations above 300 % have been attained in coarse-grained Al - high Mg alloys [9 - 11].   It should be noted that large elongations are attained without significant grain boundary sliding. The plastic stability increases with increasing strain rate sensitivity. Therefore, the large elongation for the Al-Mg alloys is likely attributed to the high plastic stability due to the high stain rate sensitivity value of about 0.3.   However, elongation at elevated temperatures depends not only on the plastic stability, but also on cavitation and/or crack formation.   Taleff et al. [8] investigated the ductility in dislocation creep for coarse-grained Al-Mg alloys with a variety of Mg contents and they showed that the maximum elongation is about 120 % for the Al-Mg alloys in spite of the high stain rate sensitivity of about 0.3.   This is probably due to cavitation development.   Therefore, it is important to investigate not only the deformation behavior, but also cavitation in order to understand the origin of the large elongation in the Al - high Mg alloys.
In general, commercial Al-Mg alloys contain impurities of Si, Fe and so on. The effect of Si on cavitation has been investigated[12] and it has been clarified that Mg_2Si particles promoted extensive cavity nucleation.   The aim of this paper is to investigate the effects of the ferrous primary crystal on the cavitation behavior of Al - 4.5 Mg alloys.

EXPERIMENT

The Al-4.5Mg-0.05Fe alloy   and Al-4.5Mg-0.2Fe alloy were used.   The chemical compositions of the alloys are shown in Table 1. These alloys were fabricated under the same processing conditions.   The alloy ingots were homogenized at 800 K for 10 h. The alloys were reheated to 750 K and hot-rolled to a thickness of 6 mm.   The alloys were then cold-rolled from 6 mm to 1 mm.   These sheets were annealed at 723 K for 25 s in a salt-bath.   Tensile specimens with a gage length of 10 mm and a gage width of 5 mm were prepared with the tensile axes parallel to the rolling direction.   Two types of tensile tests were conducted.   The first type of test was the

Table 1.The chemical compositions of the alloys (mass%)

Alloy	Mg	Si	Fe	Ti	Al
0.05Fe	4.49	Tr	0.05	0.01	bal.
0.2Fe	4.53	Tr	0.20	0.01	bal.

strain-rate-change test. The strain-rate-change tests were performed at temperatures of 613, 653 and 693 K. The specimens were pulled at a series of discrete constant true strain rates ranging from 3×10^{-5} to 3×10^{-2} s^{-1}. The second type of test was a constant-strain-rate tensile test. The tests were performed at 653 K and at 10^{-3} s^{-1} to strains of 0.2, 0.7 and 1.0. For both tests, the specimens were held in the furnace for 1.8 ks before pulling. The temperature variation during the tensile tests was not more than 1 K.

A metallographic investigation was carried out using optical and scanning electron microscopies. The grain size of the alloys was measured by optical micrography. The second-phase particles were observed using the scanning electron microscope (SEM). Furthermore, cavity formation at the particle/matrix interfaces was investigated by SEM. Quantitative metallographic measurements were conducted on mechanically polished specimens. The size distributions of the second-phase particles and the cavities were analyzed by assuming that all particles and cavities were spherical. The volume of the cavities in the gage length of the deformed specimens was investigated by hydrostatic weighing in water with the corresponding gauge head being used as the density standard.

RESULTS

Microstructure

The microstructures of the two Al-Mg alloys are shown in Fig. 1 after the specimens were annealed at 653 K for 1.8 ks. The grains were almost equiaxial and the grain size was about 50 μm in both alloys. Second-phase particles were observed at the grain boundaries and in the interior of the grains. The chemical composition analysis revealed that the second-phase particles are Al$_m$Fe. The particle sizes were quantitatively measured for both alloys. The results are shown in Fig. 2. The particle sizes were less than 7 μm and the peak particle size was in the range of 1 ~ 2 μm for both alloys.

Fig. 1 Optical micrographs of the Al-4.5Mg alloys annealed at 653 K for 1.8 ks: (a) 0.05 Fe alloy; (b) 0.2 Fe alloy.

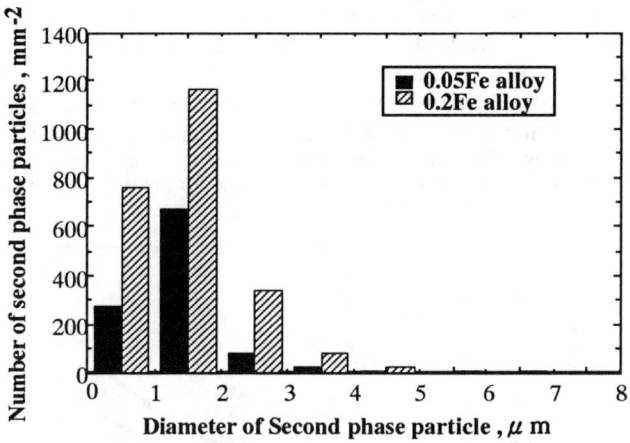

Fig. 2 Particle size distributions for the 0.05 Fe alloy and the 0.2 Fe alloy.

Deformation Behavior at 613 ~ 693 K

The variation in flow stress as a function of strain rate at 613, 653 and 693 K for the 0.05Fe alloy is shown in Fig. 3. These data were obtained from the strain-rate-change tests. The strain rate sensitivity, m, is approximately 0.3 at all strain rates and temperatures investigated for both alloys. Therefore, the deformation mechanism for both Al-Mg alloys is likely viscous dislocation glide creep similar to many other solid-solution alloys[8-11].

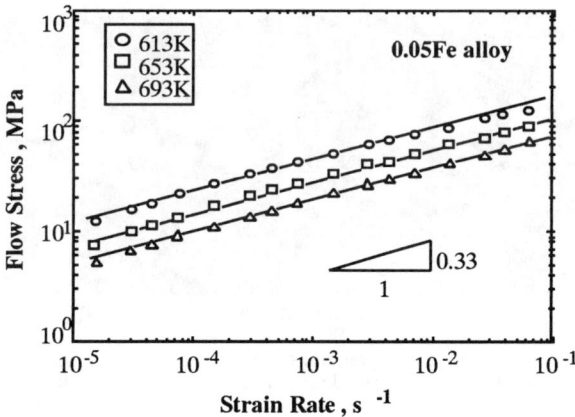

Fig. 3 The variation in flow stress as a funtion of strain rate for 0.05 Fe alloy.

Cavitation

The variations in the volume fractions of cavitation as a function of strain are shown in Fig. 4, where the specimens are pulled at a strain rate of 10^{-3} s^{-1} at 653 K. Figure 4 shows that the cavity

69

volume fraction for the 0.2 Fe alloy is larger than that for the 0.05 Fe alloy, and that the Fe addition significantly affects cavitation. Figure 5 shows the microcavities nucleated at the interface of the particle/matrix, which developed parallel to the tensile direction in the specimen, deformed to $\varepsilon =1.1$ at a strain rate of 10^{-3} s^{-1} at 653 K.

The cavity diameter distribution is shown in Fig. 6, where the specimens are deformed to ε =0.2 and ε =1.1 at a strain rate of 10^{-3} s^{-1} at 653 K.   For both alloys, the total number of cavities increases with strain, suggesting that continuous cavity nucleation occurs during deformation. The number of cavities for the 0.2 Fe alloy was greater than that for the 0.05 Fe alloy.   Clearly, the large addition of Fe promotes cavity nucleation.

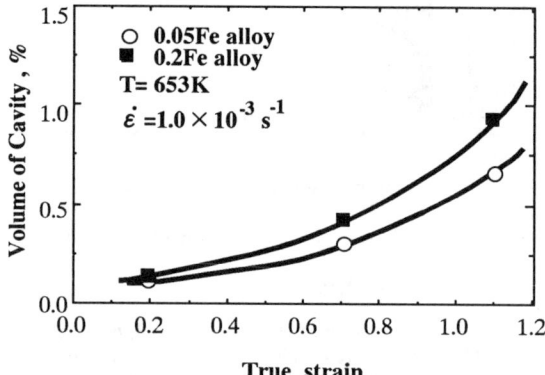

Fig. 4 The variation in cavity volume fraction at a function of true strain for the 0.05 Fe alloy and the 0.2 Fe alloy.

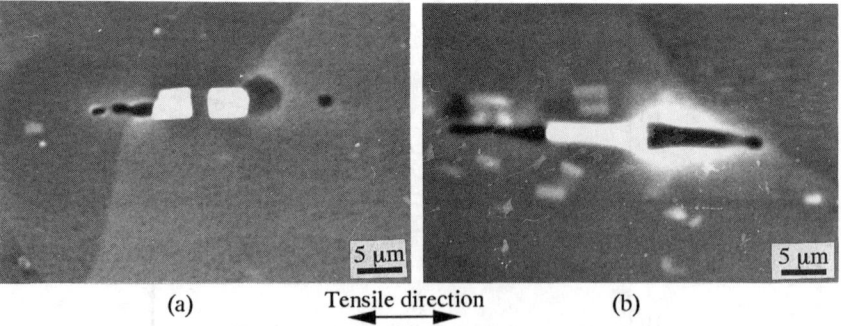

(a)   Tensile direction   (b)

Fig. 5 SEM micrographs showing intergranular cavities (a) and intragranular cavities (b) nucleated at the interface for the 0.2 Fe alloy deformed up to a strain of 1.1 at 653 K at 1.0×10^{-3}s^{-1}.

The critical particle size for cavity nucleation

The sites for cavity nucleation were experimentally investigated for two cases of the intergranular particles and intragranular particles in the specimen deformed to ε =0.2 at 653K and at 10^{-3} s^{-1} for the 0.2 Fe alloy.   These results are shown in Fig. 7.   It can be seen that no cavities are nucleated when the particle size is less than 1.5 μm for the intergranular particle, and no cavities are nucleated when the particle size is less than 0.5 μm for the intragranular particle. Therefore, it is likely that the critical particle size for cavity nucleation is 1.5 μm for the

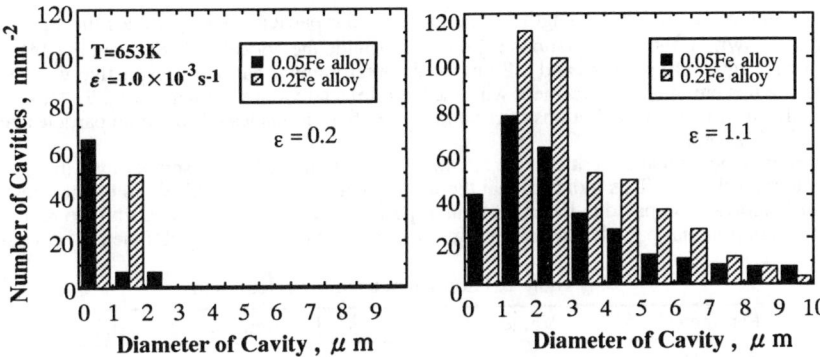

Fig. 6 Cavity size distributions for the 0.05 Fe alloy and the 0.2 Fe alloy at ε =0.2 (left) and ε = 1.1 (right).

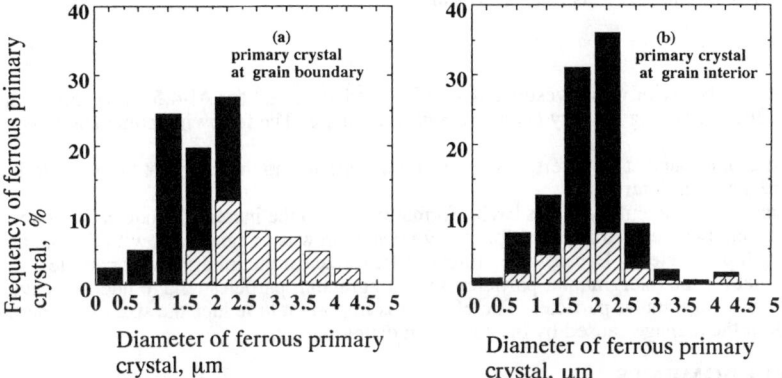

Fig. 7 Size distribution of ferrous primary crystals with cavity (☐) and with no cavity (■) for the 0.05 Fe alloy deformed up to a strain of 1.1 at 653 K at $1.0 \times 10^{-3} s^{-1}$. (a) grain boundary and (b) grain interior.

intergranular particle and 0.5 μm for the intragranular particle.

In general, stress concentrations may be relaxed by diffusion at elevated temperatures. Needleman and Rice [13] developed the diffusion length parameter. When the particle size is less than the diffusion length parameter, the stress concentrations due to the particles can be relaxed by diffusion and no cavitation is formed at the particles. In the case of the intergranular particles, the critical diffusion length is given by

$$\Delta_{gb} = \left(\frac{\Omega \delta D_{gb} \sigma}{kT\dot{\varepsilon}}\right)^{1/3} \qquad (1)$$

where Δ_{gb} is the critical diffusion length for the intergranular particles, Ω is the atomic volume, δ is the grain boundary width, D_{GB} is the coefficient for grain boundary diffusion and σ is the applied stress. In the case of intragranular particles, the critical diffusion length is given by

$$\Delta_l = \left(\frac{\Omega D_l \sigma}{\pi kT\dot{\varepsilon}}\right)^{1/2} \qquad (2)$$

71

where Δ_i is the critical diffusion length for the intragranular particles. Using values for the parameters shown in Table 2, equations (1) and (2) provide the critical diffusion length (2Δ) of 1.6 μm for the intragranular particles and 0.2 μm for the intergranular particles, respectively. These calculated values are in good agreement with the experimental values shown in Fig. 7. This suggests that no stress concentrations occur and no cavities are nucleated for small particles below the critical diffusion length.

It should be noted that the critical size for the intragranular particles is smaller than that for the intergranular particles. This indicates that the intragranular particles are likely to be the sites for cavity nucleation as compared to the intergranular particles. Therefore, it is important to reduce the size of the intragranular particles in order to limit the damage due to the development of cavitation.

Table 2. Parameters used for calculations

Parameter	Value	Parameter	Value
Ω (m³)	1.7×10^{-29}	T (K)	653
D_l (m²s⁻¹)	$1.7 \times 10^{-4}\exp(-142000/RT)$	σ (MPa)	29
δD_{gb}(m³s⁻¹)	$1.7 \times 10^{-14}\exp(-84000/RT)$	$\dot{\varepsilon}$ (s⁻¹)	10^{-3}

* R(8.3J mol⁻¹K⁻¹) is the gas constant.

CONCLUSIONS

The cavitation behavior was investigated at 653 K and at 10^{-3} s⁻¹ for Al-4.5Mg-0.05Fe and Al-4.5Mg-0.2Fe containing primary crystal particles of Al$_m$Fe. The following conclusions were obtained:
1. Most cavities nucleated at the interfaces between the particle and the matrix at both the grain boundary and the grain interior.
2. The increase in Fe content promotes cavity formation due to the increase in nucleation rate.
3. The experimental critical particle size for cavity nucleation was in agreement with double the diffusion length parameter. Because the critical particle size for the intragranular particles is smaller than that for the intergranular particles, the intragranular particles tend to nucleate cavities compared to the intergranular particles. Therefore, it is important to reduce the size of the particles in order to limit the damage caused by the cavitation development.

ACKNOWLEDGMENTS
H.I gratefully acknowledges the financial support of a Grant-in-Aid from the Ministry of Education Science and Culture of Japan.

References
1. O.D. Sherby and P.M. Burke, Progr. Mater. Sci. **13**, p.325 (1969).
2. F.A. Mohamed, *Scripta Metall.* **12**, 99 (1978).
3. T.R. McNelly, D.J. Michel and A. Salama, Scripta Metall. **23**, p.1657 (1978).
4. W.C. Oliver and W.D. Nix, *Acta Metall.* **30**, 1335 (1982).
5. P. Yavari, F.A. Mohamed and T.G. Langdon, Acta Metall. **29**, p.1495 (1981).
6. P. Yavari, and T.G. Langdon, Acta Metall. **30**, p.2181 (1982).
7. M.S. Mostafa and F.A. Mohamed, Metall. Trans. **17A**, p.365 (1986).
8. E.M. Taleff, D.R. Lesuer, and J. Wadsworth, Metall. Mater. Trans. **27A**, p.343 (1996).
9. S.S. Woo, Y.R. Kim, D.H. Shin and W.J. Kim, Scripta Mater. **37**, p.1351 (1997).
10. M. Otsuka, S. Shibasaki and M. Kikuchi, Mater.Sci. Forum **233-234**, p.193 (1997).
11. J.K. Kim, D.H. Shin and W.J. Kim, Scripta Mater. **38**, p.991 (1998).
12. H. Hosokawa, H. Iwasaki, T. Mori, M. Mabuchi, T. Tagata and K. Higashi, Acta Mater. **47**, p.1859 (1999).
13. A. Needleman and J.R. Rice, Acta Metall. **28**, p.1315 (1980).

ANALYSIS OF CAVITATION IN A NEAR-γ TITANIUM ALUMINIDE DURING HIGH TEMPERATURE/SUPERPLASTIC DEFORMATION

Carl M. Lombard*, Amit K. Ghosh**, and S. Lee Semiatin*
*Air Force Research Laboratory, Materials and Manufacturing Directorate, AFRL/MLLM, Wright-Patterson Air Force Base, OH, 45433;
**Univ. of Michigan, Dept. of Materials Science & Engineering, Ann Arbor, MI, 48109

ABSTRACT
The superplastic flow behavior of a near-γ titanium aluminide (Ti-45.5Al-2Cr-2Nb) is determined under uniaxial tension in as-rolled or rolled-and-heat treated conditions (1177°C/4 hr or 1238°C/2 hr). Cavitation characteristics, including cavity growth rates, are established via isothermal, constant strain rate tests conducted at 10^{-4} to 10^{-2} s^{-1} and temperatures between 900°C and 1200°C. Differences in cavitation as a function of initial structure, strain, strain rate and temperature are noted. Cavity growth is found to be largely plasticity controlled. Experimental growth rates are compared with equations that predict rates as a function of strain rate sensitivity. Although the equations assume no coalescence and no nucleation of new cavities, which are experimentally observed, they are useful in predicting actual growth rates.

INTRODUCTION
The potential for producing superplastic near-γ titanium aluminides exists because relatively stable microstructures containing fine two phase mixtures of α_2-Ti$_3$Al and γ-TiAl can be developed via thermomechanical processing [1]. Results show these materials exhibit good superplastic characteristics, i.e. they can achieve tensile elongations of several hundred percent or more and have strain rate sensitivity values (m) greater than 0.3 [2-3]. m is defined as

$$m = \left(\frac{d(\log \sigma)}{d(\log \dot{\varepsilon})}\right)_{T,\varepsilon} \tag{1}$$

where σ is stress, $\dot{\varepsilon}$ is strain rate, T is temperature, and ε is strain. A number of near-γ compositions (ranging from 43 to 49.5 at.% Al) have been tested; the superplastic behavior is dependent on composition and grain size in addition to temperature and strain rate. While parallels can be drawn to other superplastic two-phase titanium alloys, such as Ti-6Al-4V, one notable exception is that near-γ cavitates under appropriate superplastic testing conditions, which leads to premature failure [3].

The current research centers on understanding cavitation behavior in near-γ under tensile superplastic forming conditions. Previous research shows that cavity growth is predominantly plasticity controlled in these alloys. For a cavity of radius r, the relationship

$$\frac{dr}{d\varepsilon} = \frac{\eta}{3}\left(r - \frac{3\Gamma}{2\sigma_E}\right) \tag{2}$$

is proposed for growth by deformation of the surrounding matrix [4]. Γ is the surface energy and σ_E is the Von Mises equivalent stress. Equations for η, the cavity growth rate of a single cavity, are dependent on the applied stress state, strain, strain rate, temperature, geometry of deformation, and in two phase alloys, the phase volume fractions. They are of the form

$$\eta = \frac{3}{2}\left(\frac{m+1}{m}\right)\sinh\left(2\left(\frac{2-m}{2+m}\right)\left(\frac{k_s}{3} - \frac{P}{\sigma_E}\right)\right) \tag{3}$$

as developed in [5] by following an analysis of Cocks and Ashby [6], or

$$\eta = \frac{3}{2}\left((1 + 0.932m - 0.432m^2)^{1/m}\left(\frac{k_s}{3} - \frac{P}{\sigma_E}\right)\right) \tag{4}$$

as developed by Suery [7] and Budiansky, et al. [8]. k_s is a constant dependent on geometry of deformation and extent of grain boundary sliding (gbs) while P is the superimposed hydrostatic stress. The range of k_s is $1 \leq k_s \leq 2$ where $k_s = 1$ represents no gbs while $k_s = 2$ indicates free sliding. That portion of the equation that contains k_s and P defines the triaxiality of stress local to the grain boundary. Davenport proposes a simpler expression to predict η [9]

$$\eta = 1/m \qquad (5)$$

Relationships describing the bulk behavior of cavity growth have been developed. Simplifying assumptions are usually made: cavities are the same size and regularly distributed along the specimen axis, and η is a constant for a given material and test condition. It is also assumed that η_v, the volumetric cavity growth rate is identical to η. The equation developed is

$$C_v = C_{vo} \exp(\eta_v \varepsilon) \qquad (6)$$

where C_{vo} represents the initial cavity volume in the specimen and C_v is the volume at a strain ε. Eq. (6) is valid only for non-interacting cavities. Lian and Suery [10] and Nicolaou, et al. [11], show that the value of m is a major influence on the strain to failure when η_v is small ($\eta_v \leq 1$). In these cases, flow localization, not cavitation/fracture events, controls failure. By contrast, when η_v is greater than ≈ 3, strain to failure is almost independent of m for $m \geq 0.25$. Thus, a high m value is a necessary but not a sufficient condition for large elongations to failure.

MATERIALS AND PROCEDURES

Initial Microstructures The input material for this study is hot rolled sheet with composition Ti-45.5Al-2Cr-2Nb. Three microstructures are selected for an assessment of the influence of grain size and phase volume ratios on superplastic forming characteristics. Conditions tested are the as-rolled (AR) or rolled-and-heat treated (1177°C/4 hr or 1238°C/2 hr). In the AR condition, the microstructure is two-thirds recrystallized equiaxed γ and α_2 grains of 3-5 μm diameter. Regions of untransformed ($\gamma + \alpha_2$) lamellae constitute the remainder of the structure. Volume ratios of α_2 and γ are 25/75, respectively. A small amount of β phase (< 5%) is also present in the alloy. The 1177°C/4 hr condition has an equiaxed microstructure with a 10-12 μm γ grain size. α_2 is found in a plate morphology at γ grain boundaries. Blocky ($\gamma + \alpha_2$) lamellar colonies of 5 μm, formed during heat treatment, are also present. Volume ratios of α_2 and γ are 25/75, respectively. The 1238°C/2 hr condition has a fine equiaxed microstructure with a 10-12 μm γ and α grain sizes and approximately 50/50 phase ratio.

Tensile Testing/Metallography Sheet specimens are tested with the bulk being 12.70 mm gage length specimens with 3.18 mm width. For 1200°C tests, specimens with 6.35 mm gage length and 1.59 mm width are employed. Both types have 1.52 mm thickness. Tension testing is done in high purity (gettered) argon with a tungsten wire mesh furnace. Prior to pulling, each specimen is allowed to equilibrate at the test temperature for 30 min. Tests are conducted at constant true axial strain rates. Select interrupted tests are also conducted. Metallographic sections containing the axial through-thickness are prepared using standard techniques. They are examined using optical and backscattered electron microscopy.

Cavity Measurements Cavity populations, equivalent diameters, and aspect ratios are determined using a montage of photos taken at 500x magnification at a minimum of four sections along the length. The montage is computer scanned and analyzed using the NIH *Image* program [12]. Typical individual sections analyzed are of 500 μm height and 200 μm width. Cavities less than five pixels in size (equivalent diameter of 1.6 μm) are not considered because of potential errors due to film defects, spotting of sample, etc. Volume fractions are associated with local axial (longitudinal) strain, ε_l, determined from measurements of local thickness strain, ε_t, and local width strain, ε_w, ($\varepsilon_l \approx -\varepsilon_w - \varepsilon_t$, neglecting the effects of cavitation on the change in volume). Equivalent diameters are grouped in 2 μm increments.

Comparison of Data with Plasticity Based Models The formulation of Eq. (6) assumes pre-existing cavities, no cavity nucleation once deformation commences, and no cavity coalescence. In the present work, because continuous nucleation and coalescence occurs [13], only an overall cavity volume growth rate (η_{app}), which represents a composite value of growth, coalescence, and nucleation, can be fitted to the measurements. Thus, these values cannot be compared directly with those generated in [14]; however, qualitative statements about the trends found can be made. Using the assumptions that η approximates η_v and void area is proportional to void volume, η_{app} values are determined by plotting C_v versus ε and fitting an exponential curve through the data. Measured cavity volume fractions are fitted to Eq. (6) using a C_{vo} value of 10^{-4}. There are no physical measurements to back this selection. However, past research on determining values of C_{vo} show extensive evidence that justifies the use of this number [14].

RESULTS

Cavity Growth – Microstructural For the AR specimen tested at 1000°C 10^{-3} s^{-1}, growth and coalescence of cavities is a function of strain. Numerous small cavities, on a scale and spacing commensurate with the grain size, are observed in the material after straining. Cavities elongate in the direction of the tensile axis, indicating their growth is plasticity controlled, and align in stringers. In this alloy, the cavities are generally found at α_2/γ grain boundaries [13]. Cavities at these boundaries are expected because of the greater mismatch in flow stress between two different phases as compared to two like phases. With increasing strain, significant void opening is observed. Finally, near the fracture area, linkage of adjacent void stringers occurs and a significant fraction of the cross-sectional area contains voids. At 1200°C, cavity stringers are encountered less often. Cavities still generally occur at α_2/γ grain boundaries.

Examination of cavitation found in 1238°C/2 hr specimen tested at 1000°C 10^{-3} s^{-1} shows that, at low strains, cavity size is also on the order of grain size and that cavities are generally located at α_2/γ interfaces. Note that this means that cavities are approximately 3x the size of that observed in AR specimens. Comparison with AR specimens of similar strains show that cavities in 1238°C/2 hr material are less numerous but they grow and/or coalesce to a greater extent. The decrease in cavity density is related to fewer boundaries present and better chemical homogeneity in the heat treated conditions. Faster cavity growth rates associated with heat treated specimens are related to lower m values (Table I). Results like those of the 1238°C/2 hr tests are found for 1177°C/4 hr condition.

Cavity Growth – Quantitative Figs. a-c present cavity distribution results for AR specimens deformed at 1000°C 10^{-3} s^{-1} to tensile elongations of 60%, 120% and failure. These correspond to true strains of 0.47, 0.79, 1.03. In the specimen deformed to 60%, the distribution of equivalent 2 µm size cavities is nearly equal for the four strain levels (Fig. a). Fig. b shows that, after 120% elongation, the number of 2 µm cavities is still increasing, thus providing evidence for continuous nucleation of cavities. The largest size cavities increase from 12 µm to 18 µm in going from 60-120% elongation. At failure, the largest cavity is 88 µm (Fig. c). More 2 µm cavities nucleate, as the frequency is approximately 2000/mm^2 near the fracture tip.

It should be noted that overlapping of the Figs. a-c produces some discrepancies, i.e. a slightly lower strain region from the specimen strained 120% has a higher density of cavities than a similarly strained region from the specimen strained 60%. Such individual inconsistencies are expected because of the statistical nature of cavitation. However, the overall trends are consistent with the intuition that cavities should continuously nucleate and grow with strain.

Fig. d shows the effect of increasing temperature from 1000 to 1200°C at 10^{-3} s^{-1} in AR specimens. Cavity distributions for approximately equivalent strains indicate that number and size decreases with increasing temperature. This is to be expected because of decreasing flow stress, increasing diffusion accommodation, and higher m values at higher temperatures.

Effect of strain rate on cavity distribution at 900°C is shown in Fig. e for 1238°C/2hr specimens. The order of magnitude decrease in strain rate results in one-half as many 2 µm cavities for approximately the same strain and a decrease in the largest cavity observed from 30 µm to 14 µm. Note that a strain of 0.69 at 10^{-3} s^{-1} produces an almost equivalent cavity distribution as a strain of 1.42 at 10^{-4} s^{-1}. Lower strain rates result in lower flow stresses, more time for accommodation and diffusion processes to occur, and higher m values, thereby reducing cavity nucleation and growth.

Influence of Strain on Cavity Aspect Ratio It is noted that with increasing strain the largest aspect ratio found increases, which supports the conclusion of plasticity controlled cavity growth. The probability of finding a low aspect ratio cavity (< 3) is approximately equal at all strain levels. It is also found that the largest cavities do not always have the highest aspect ratios. Similar observations of cavity aspect ratio being nearly independent of size in near-γ titanium aluminides are made by Sneary, et al. [15].

Cavity Growth Rates - Experimental Results Values of η_{app} for the AR and heat treated conditions are summarized in Table I. For 900-1000°C at 10^{-3} s^{-1}, the value of η_{app} for the AR specimen is less than that of the heat treated specimens. As seen in the table, heat treated specimens tested at 900-1000°C have low m and η_{app} greater than 3. Thus, cavitation and fracture control failure and cause it prematurely than what would happen if failure is solely due to localized thinning, e.g. that predicted for non-cavitating titanium and zirconium alloys [14]. Cavity growth is expected to be slower for the fine-grained, AR microstructure because of the greater ability to relieve stress concentrations at grain boundaries by diffusional processes and higher m values as compared to the coarser grained heat treated materials. Additionally, although Eq. (5) is not dependent on the applied stress, the role of hydrostatic stresses in causing cavity opening cannot be neglected. The generally higher flow stresses of the heat treated specimens will lead to higher local hydrostatic tensile stresses and hence higher cavity growth rates.

Table 1. Values of η_{app}, m and peak flow stress.

Temp. (°C)	η_{app} 10^{-2} s^{-1}	η_{app} 10^{-3} s^{-1}	η_{app} 10^{-4} s^{-1}	m 10^{-2} s^{-1}	m 10^{-3} s^{-1}	Peak 10^{-2} s^{-1}	Stress 10^{-3} s^{-1}	(MPa) 10^{-4} s^{-1}
AR 900	-	3.0	2.6	-	0.20	-	412	261
1000	-	2.4	--	-	0.43	-	157	59
1100	-	2.5	2.4	-	0.56	-	54	18
1200	2.1	1.8	--	0.68	0.80	90	22	3.5
HA 900	-	7.7	2.6	-	0.14	-	348	255
1000	-	5.8	2.6	-	0.25	-	201	114
1100	-	2.9	3.0	-	0.44	-	90	33
1200	2.7	2.0	3.0	0.29	0.72	98	50	9.5
HB 900	-	7.6	2.7	-	0.19	-	428	278
1000	-	3.8	2.0	-	0.30	-	204	102
1100	-	2.8	2.5	-	0.49	-	95	31
1200	3.2	--	-	0.43	-	95	35	-

AR = as-rolled. HA = 1177°C/4hr. HB = 1238°C/2hr. - = not tested. -- = not measured.
Note that m values are slightly different than in [13] because of the use of improved curve fitting techniques in the current work.

As the test temperature is increased, higher m values and increased diffusion kinetics lead to lower values of η_{app} for all the three initial microstructures. In fact, for the three conditions the measured value of η_{app} drops to approximately 2.0 at 1200°C 10^{-3} s^{-1}. Thus, it can be concluded that as the temperature increases, failure by cavitation becomes less critical while flow localization dominates as the failure mode. At 10^{-4} s^{-1}, the η_{app} values are approximately 2-3 for all temperatures and material conditions, indicating that the growth rate is independent of temperature and initial microstructure. The cause of this approximately constant growth rate can be best explained by the expected higher m values at 10^{-4} s^{-1}. Since η_{app} values fall in the regime of being greater than 1 and less than 3, both m and η influence strain to failure.

Fitting of η_{app} Data to Phenomenological Relations η_{app} data are plotted along with calculated values of η per Eqs. (3-5) in order to determine the best equation to predict η for this system (Fig. f). Eqs. (3) and (4) are solved using upper and lower bound values of k_s for a range

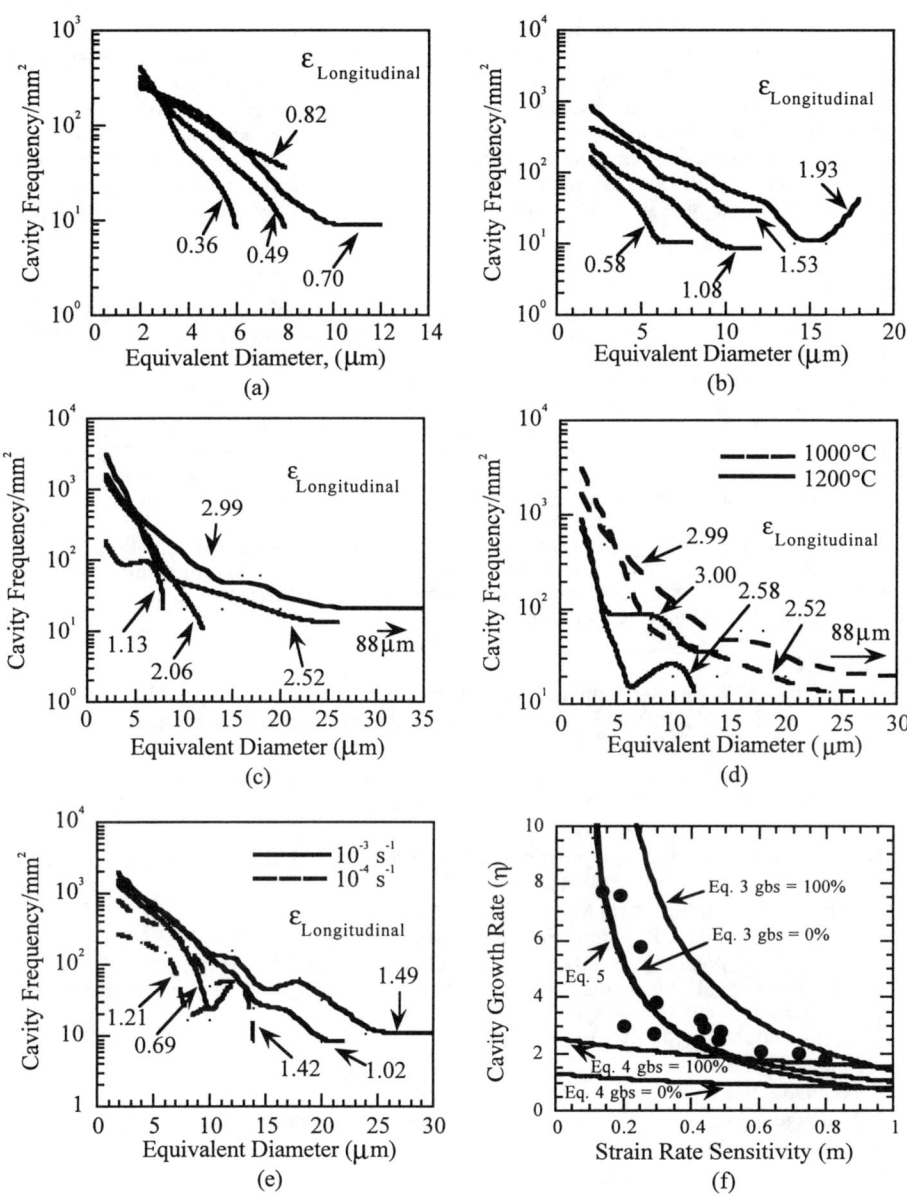

Figs. a-e. Cavity frequency/mm^2 as a function of size for various test conditions.
Fig. f. η as a function of m. Experimentally determined η_{app} (blackened circles) plotted for comparison.

of m values. Values of η derived from the three equations approach 0.75-1.50 for $m = 1.0$. From Eq. (4), η is weakly dependent on m while Eqs. (3) and (5) show considerable increases in η for low m values. The η_{app} data closely follow Eq. (3) with $k_s = 1$ (0% gbs) and Eq. (5). (Note that m values are calculated using the peak flow stress values.) This is not to say that the gbs mechanism is not operative in the test regime used in this study, for the amount of gbs is expected to increase with increasing temperature in this material. The high values of m are indicative of gbs and their increasing value with temperature provides evidence of the increasing role that gbs plays. For the range $m = 0.35$-0.65, the experimental data closely follows Eq. (4) with $k_s = 2$ (100% gbs).

CONCLUSIONS

1. For all three microstructural conditions, cavity growth is plasticity controlled. Also, the cavity growth rate decreases with an increase in test temperature or a decrease in the strain rate indicating that cavitation plays a less important role while localized thinning becomes critical in causing final failure at higher temperatures/lower strain rates. The growth rate decreases because of increasing accommodation by diffusion and increasing m values.

2. At lower test temperatures and strain rate equal to 10^{-3} s^{-1}, η_{app} values for the heat treated conditions are consistently higher than those for the as-rolled material. Fine grained as-rolled material has 1) lower flow stress, thus reducing cavity growth rate due to hydrostatic tensile stresses, and 2) higher m values and greater ease of accommodating stress concentration/cavity growth due to gbs.

3. Apparent cavity growth rates closely follow a hyperbolic sine equation ($k_s = 1$, 0% gbs) as well as the simpler relation $1/m$. For $m = 0.35$-0.65, the experimental data follows a power law relation ($k_s = 2$, 100% gbs). The amount of gbs occurring in this system, however, should be further investigated experimentally.

REFERENCES

1. S. L. Semiatin, V. Seetharaman, and V. K. Jain, Met. Trans. A, **25A**, pp. 2753-2768 (1994).
2. W. B. Lee, H. S. Yang, Y.-W. Kim and A. K. Mukherjee, Scripta Met. Mater., **29**, pp. 1403-1408 (1993).
3. C. M. Lombard, A. K. Ghosh and S. L. Semiatin, *Proceedings of Gamma Titanium Aluminides Symposium*, eds. Y.-W. Kim, R. Wagner and M. Yamaguchi (TMS, Warrendale, PA 1995) pp. 579-586.
4. N. Ridley and J. Pilling, *Superplasticity*, eds. B. Baudelet and M. Suery; (CNRS, Paris, 1985), Ch. 8.
5. J. Pilling and N. Ridley, Acta Metall., **34,** pp. 669-679 (1986).
6. A. C. F. Cocks and M. F. Ashby, Progress in Matl. Sci., **27**, pp. 189-244 (1982).
7. M. Suery, *Superplasticity*, eds. B. Baudelet and M. Suery; (CNRS, Paris, 1985), Ch. 9.
8. B. Budiansky, et al., *Mechanics of Solids*, eds. H. G. Hopkins and M. J. Sewell; (Pergamon Press, Oxford, 1982).
9. M. J. Stowell, Metal Sci., **14**, pp. 267-271 (1980).
10. J. Lian and M. Suery, Mat. Sci. Tech., **2**, pp. 1093-8 (1986).
11. P. Nicolaou, S. L. Semiatin, and C. M. Lombard, Met. Trans. A, **27A**, pp. 3112-9 (1996).
12. NIH Image public domain, image analysis computer program (developed at the U.S. National Institutes of Health and available from the Internet at zippy.nimh.nih.gov).
13. C. M. Lombard, A. K. Ghosh and S. L. Semiatin, *Superplasticity and Superplastic Forming 1998*, eds. A. K. Ghosh and T. R. Bieler, (TMS, Warrendale, PA, 1998), pp. 267-275.
14. J Lian and M. Suery, Mat. Sci. Tech., **2**, pp. 1093-1098 (1986).
15. P. R. Sneary, R. S Beals and T. R. Bieler, Scripta Mat., **34**, pp. 1647-1654 (1996).
16. D. Lee and W. Backofen, Trans. Met. Soc. AIME, **239**, pp. 1034-1040 (1967).

Superplasticity in Ceramics

GRAIN BOUNDARY ENGINEERING OF HIGHLY DEFORMABLE CERAMICS

MARTHA L. MECARTNEY
University of California, Department of Chemical and Biochemical Engineering and Materials Science, Irvine, CA 92697-2575 martham@uci.edu

ABSTRACT

Highly deformable ceramics can be created with the addition of intergranular silicate phases. These amorphous intergranular phases can assist in superplastic deformation by relieving stress concentrations and minimizing grain growth if the appropriate intergranular compositions are selected. Examples from 3Y-TZP and 8Y-CSZ ceramics are discussed. The grain boundary chemistry is analyzed by high resolution analytical TEM is found to have a strong influence on the cohesion of the grains both at high temperature and at room temperature. Intergranular phases with a high ionic character and containing large ions with a relatively weak bond strength appear to cause premature failure. In contrast, intergranular phases with a high degree of covalent character and similar or smaller ions than the ceramic and a high ionic bond strength are the best for grain boundary adhesion and prevention of both cavitation at high temperatures and intergranular fracture at room temperature.

INTRODUCTION

The challenges for superplasticity in oxide ceramics are significant, especially when compared to most metallic systems. Higher temperatures are required due to the slower diffusion of oxide ions than metallic species. Often only relatively low strain rates are possible to achieve in ceramic systems even at high temperatures. A fine grain size is needed for superplastic deformation. This value is generally accepted to be approximately 1 μm or less for ceramics compared to 10 μm for most metal alloy systems. The long time at high temperature make it difficult to prevent grain growth and retain the fine grain size required for superplastic forming of ceramics. The ceramic 3 mol% yttria tetragonal zirconia polycrystals (3Y-TZP) is one of a few ceramics that has an intrinsically slow grain growth rate and was the first fine grained ceramic to be used to demonstrate extensive superplasticity [1]. Even with a fine grain size, in some ceramic systems there appears to be an intrinsic lack of tensile ductility. Defect formation (cavitation and crack formation) can occur during extensive deformation, degrading the mechanical properties of the material and significantly lower the strength of brittle ceramics. The mold materials for shape forming operations must be able to withstand high temperatures and thermal cycling while not reacting with the ceramic, and these requirements leave only a few covalent materials (SiC, for example) that can be used. This type of material is extremely hard and difficult to machine, adding additional costs to the forming.

In order to make superplastic forming of ceramics a useful process, these challenges must be addressed. The approach described in this paper is the addition of small amounts of intergranular amorphous silicate phases. In many oxide ceramic systems, silicate impurities are common result from processing, and small amounts do not significantly degraded the required mechanical behavior. These silicate phases soften and in some cases become liquid at the deformation temperature and can assist in stress accommodation and minimize cavitation at multiple grain junctions. This approach has been used by the group of Sakuma at the University of Tokyo and in 3Y-TZP extensive deformations over 1000% have been achieved with the addition of 5 wt% silica [2]. It is also well established in metals that small amounts of liquid phase that that form at the grain boundaries with low melting point eutectic composition can assist in superplastic deformation and generate high strain rate superplasticity [3]. Another example is low melting point copper oxide that has been added to some ceramic systems such as 3Y-TZP and 8 mol% yttria stabilized zirconia to assist in high temperature deformation [4,5].

The question posed is whether the composition of the silicate grain boundary phase can be manipulated in order to not only assist in achieving easy deformation at high temperatures but also used to prevent grain growth. The materials studied and presented in this paper include 3Y-TZP, with a very slow grain growth rate, and 8Y-CSZ, a chemically similar ceramic but with a much more rapid grain growth rate. Cubic 8Y-CSZ has a resultant order of magnitude larger grain size that makes superplastic deformation difficult. If silicate phases can be used to limit grain boundary mobility with respect to grain growth, while relieving stress concentrations, then it may be possible to achieve superplasticity in other oxides by this route.

The addition of second phase silicates to the material can assist in limiting grain growth by several mechanisms. If the second phase is discretely distributed and has a low solubility for the oxide component, Zener pinning of the grain boundaries can occur. If the second phase wets the grain boundaries at high temperatures, grain growth can be limited by reducing the driving force for grain growth in the presence of the lower energy liquid/ceramic interfaces. In addition, the mobility of the grain boundaries with respect to grain growth can be decreased by increasing the effective grain boundary width, which will occur if liquid phases extend intergranularly along the grain boundaries. [6,7].

EXPERIMENTAL PROCEDURE

3Y-TZP and 8Y-CSZ powders from Tosoh were used to prepare ceramic samples. 3Y-TZP samples were mixed with glass powders (see Table I for composition) in a slurry of isopropanol, dried, sieved, and cold isostatic pressed at 60 MPa prior to sintering in air at 1400°C for two hours. For grain growth studies, these samples were hot-isostatically pressed at 1650°C. 8Y-CSZ samples were attritor milled with the glass powders or colloidal silica in isopropanol, dried, sieved, and cold isostatically pressed at 60 MPa. For some grain growth studies, these samples were sintered two hours at 1400C and then hot isostatically pressed at 1650°C. For other grain growth studies and for superplastic deformation, 8Y-CSZ samples were sintered at 1350°C for 6 minutes then hot isostatically pressed at 1400°C for 15 minutes in order to densify the material while retaining a fairly fine grain size. Grain growth experiments on these samples were conducted by annealing at 1400°C in air. The glass composition were chosen to include a range of solubilities and viscosities. The solubility of Zr and Y in the glasses is highest for the bariumsilicate (BaS), less for the borosilicate (BS), lesser for the aluminosilicate (AS), and the lowest for pure silica. The viscosity of the intergranular phases scales inversely with the solubility of the ceramic in the intergranular phase.

Additive	SiO_2	Al_2O_3	B_2O_3	Na_2O_3	K_2O	BaO	SrO	As_2O_3
BaS	45.8	2.1	21.6	----	----	29.2	0.7	0.5
BS	83.3	1.5	11.2	3.6	0.4	----	----	----
AS	79.8	20.2	----	----	----	----	----	----
Silica	100	----	----	----	----	----	----	----

Table I. Chemical Composition of Intergranular Silicate Additives. .

High temperature mechanical deformation was conducted both in tension and compression. Samples were heated to the testing temperature within 3 hours. Testing temperatures ranged from 1200°C - 1500°C. In compression appropriate loads were calculated assuming constant volume and uniform diameter and loads were varied to keep the stress constant with increasing sample diameter during testing. The tension experiments were conducted at the University of Tokyo using a constant strain rate. Room temperature mechanical properties of hardness (H) , elastic modulus (E) and fracture toughness (K1C) were measured using high load indentation techniques. Samples were scored and fractured with a high impact load in order to characterize the fracture path.

Samples were examined by field emission gun scanning electron microscopy using a Philips XL30. The analytical transmission electron microscopy was conducted on a Philips CM200FEG TEM – STEM at Oak Ridge National Laboratory with a 200 KeV field emission gun, Oxford super-ATW energy dispersive X-ray spectrometer (EDS) and XP3 pulse processor, and a EMiSPEC Vision acquisition system. The electron probe size was selected to maximize the EDS signal while considering beam damage effects and best spatial resolution. Typically a probe size of 1.2 nm FWHM at 0.5 nA was used. EDS spectrum lines were acquired in the STEM mode across edge-on grain boundaries, usually for 20 points with a 0.5 - 2.0 nm spacing and a 10 second dwell/point.

RESULTS

Grain Growth

The grain growth of 3Y-TZP is much more sluggish than 8Y-CSZ. It was found, however, that certain additives, such as a borosilicate glass (BS) would reduce grain growth in both 3Y-TZP and 8Y-CSZ (Figure 1). The borosilicate phase was tested to determine if it could limit grain growth in 8Y-CSZ during deformation by conducting annealing tests at 1400°C. While the borosilicate phase successfully limited grain growth, it was found that a fine dispersion of colloidal silica was even more effective and reduced the starting grain size to well below 1 µm for samples with 5 wt% silica (Figures 2 and 3). The improved reduction of grain growth with pure silica is not surprising since silica has a lower solubility for yttrium and zirconium and has a higher viscosity and thus a lower diffusion coefficient for transport of yttrium and zirconium.

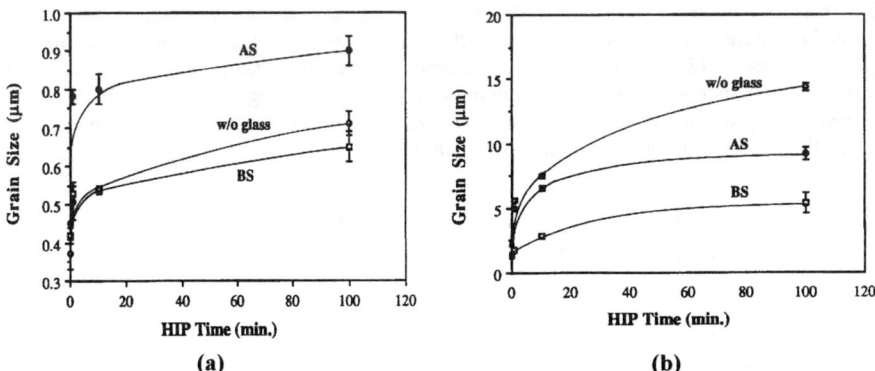

(a)   **(b)**

Figure 1. Grain size versus HIP time at 1650°C in (a) 3Y-TZP and (b) 8Y-CSZ without glass additions, with 5 wt% aluminosilicate glass (AS), and with 5 wt% borosilicate glass (BS).

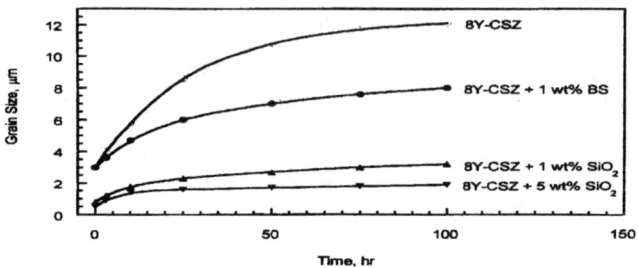

Figure 2. Grain growth at 1400C for 8Y-CSZ with different intergranular additives.

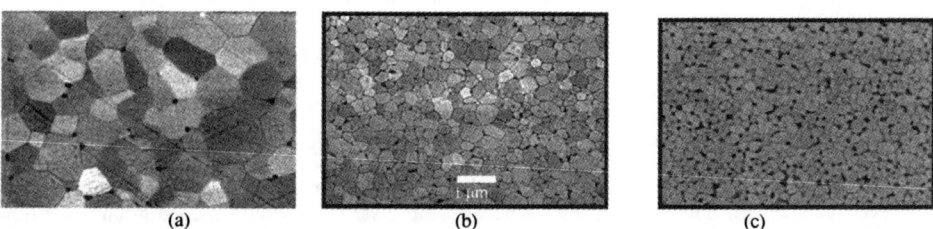

(a)   (b)   (c)

Figure 3. SEM images of initial microstructures of (a) pure 8Y-CSZ, (b) 8Y-CSZ + 1 wt% SiO, and (c) 8Y-CSZ + 5 wt% SiO₂.

Deformation

Additives of the high viscosity and low solubility borosilicate phase and a low viscosity, high solubility bariumsilicate phase were added to 3Y-TZP in order to enhance superplastic deformation. In compression, samples with 5 wt% additives behaved similarly, with a slightly higher strain rate for the lower viscosity phase (Figure 4). When tested in tension, samples had a similar stress exponent close to 2 at 1300°C and the flow stress was similar for the two intergranular phases, with a slightly lower flow stress for the low viscosity bariumsilicate intergranular phase (Figure 5). However, differences between the two intergranular phases became very apparent when the tensile ductility was measured (Figure 6). The bariumsilicate phase had less tensile ductility than the pure 3Y-TZP, but the higher viscosity borosilicate phase enhanced the ductility for temperatures above 1300°C.

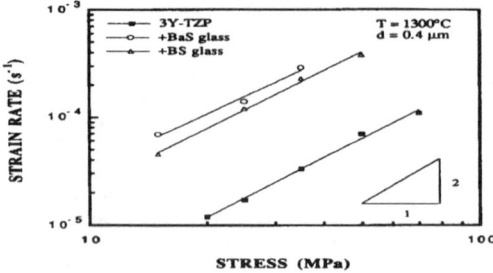

Figure 4. Strain rate versus true compressive stress for 3Y-TZP with various amounts of glass.

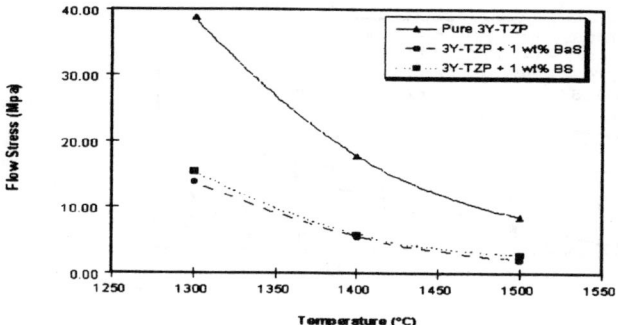

Figure 5. Graph of flow stress versus temperature for pure 3Y-TZP, 3Y-TZP + 1 wt% BaS, and 3Y-TZP+ 1 wt% BS at a strain rate of 1.3 X 10^{-4} /s in tension.

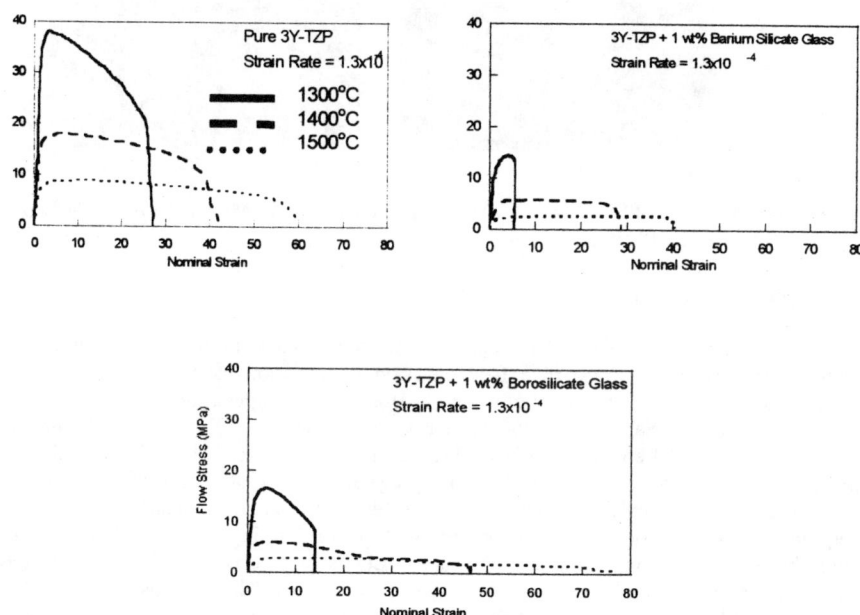

Figure 6. Stress - strain curves obtained at 1300°C, 1400°C and 1500°C with a strain rate of 1.3x10^{-4} /s for pure, BaS added, and BS added 3Y-TZP tested in tension.

Samples of 8Y-CSZ with different intergranular glass phases were tested in compression. (Figure 7). The highest strain rates, approaching 0.001, were obtained for the fine grain material with silica additions. Some Ostwald ripening was observed during deformation but no internal cavitation was observed. Figure 8 illustrates the grain growth found during deformation at the various temperatures. The dark regions are silica pockets.

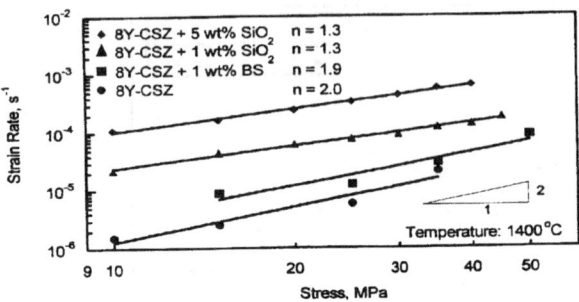

Figure 7. Comparison of the strain rates of various 8Y-CSZ samples at 1400°C.

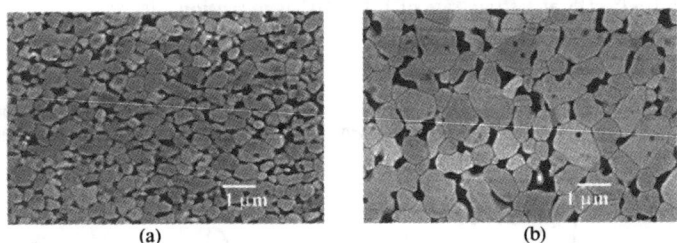

Figure 8. SEM images of 8Y-CSZ + 5 wt% SiO_2 deformed at 45 MPa (a) 1300°-- 50% true strain and (b) 1500° -- 180% true strain.

Grain Boundary Analysis

The grain boundaries of the materials were examined using high resolution energy dispersive spectroscopy (EDS) in a field emission transmission electron microscope (TEM). This approach allowed a very fine spot size to be used so that segregation at the grain boundaries could be determined. Samples of 3Y-TZP that were pure (Figure 9), or had added glass phases (borosilicate Figure 10 and bariumsilicate Figure 11) were anlyzed. It was found that that all samples had a high concentration of yttrium segregated to the grain boundaries and a depletion of zirconium. Surprisingly, no silicon was detected at the grain boundaries for either of the intergranular compositions. The glass appeared to remain in triple junctions and not be uniformly distributed along the grain boundary. (Limits of detectability are approximately one atomic percent and possible for elements heavier than boron.) The bariumsilicate sample had a high degree of barium segregation at the grain boundary (Figure 11). These samples also had a lower tensile ductility during high temperature deformation (Figure 6).

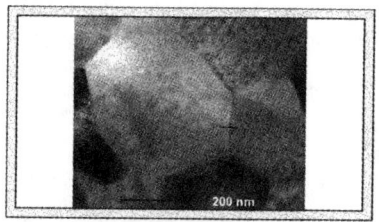 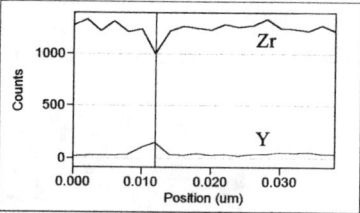

Figure 9. STEM image and corresponding line scan for pure 3Y-TZP.

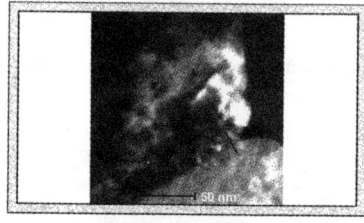 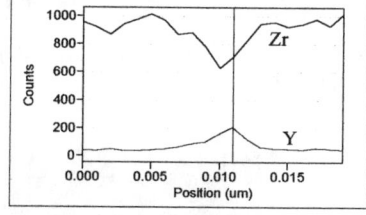

Figure 10. STEM image and line scan for 3Y-TZP + 1 wt% borosilicate glass.

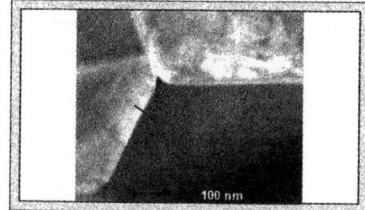 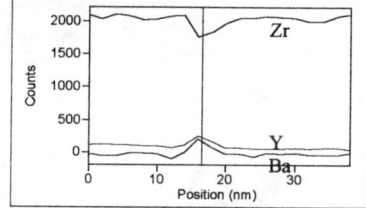

Figure 11. STEM image and line scan for 3Y-TZP + 1 wt% barium silicate glass.

Grain boundary analysis of 8Y-CSZ materials showed different trends. The pure samples still had a depletion of zirconium at the grain boundaries, but no strong increase in yttrium concentration compared to the grain interior was noted (Figure 12). This still means, however, that the ratio of the yttrium to zirconium is higher at the grain boundary than in the grain interior. The borosilicate samples still had no detectable silicon at the grain boundaries (Figure 13). The samples containing pure silica, however, had significant segregation of silicon along the grain boundaries (Figure 14). A similar segregation has been seen in 3Y-TZP doped with silica [8]. This indicates a very different wetting behavior and distribution for the various intergranular phases. It should be noted, however, that these measurements were taken at room temperature and not at the temperatures of deformation, where the wetting behavior could be quite different.

The different segregants could have a dramatic effect on the grain boundary cohesion. Ikuhara and Sakuma [9] have proposed that grain boundaries with a more covalent character undergo more extensive high temperature deformation. For the case of barium silicate additions, the barium segregated to the grain boundary is highly ionic and so would be expected to reduce the ductility. In addition, the barium ion is very large and the strain associated with accommodation of this defect can weaken the grain boundaries. Lastly, the ionic bond strength, calculated from Pauling's rules as the formal charge on the cation divided by its coordination number, is much lower for barium-oxygen than for silicon-oxygen ion pairs [10]. A summary of these calculation is presented in Table II.

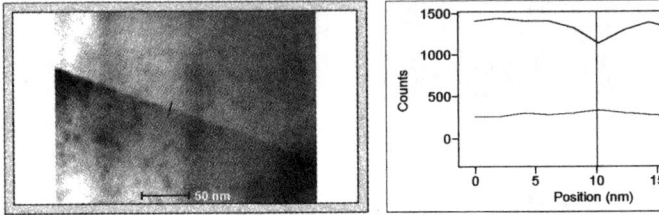

Figure 12. STEM image and corresponding line scan for pure 8 mol % yttria cubic stabilized zirconia (8Y-CSZ).

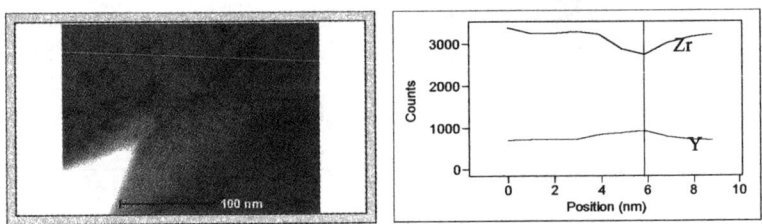

Figure 13. STEM image and corresponding line scan for 8Y-CSZ + wt% borosilicate glass.

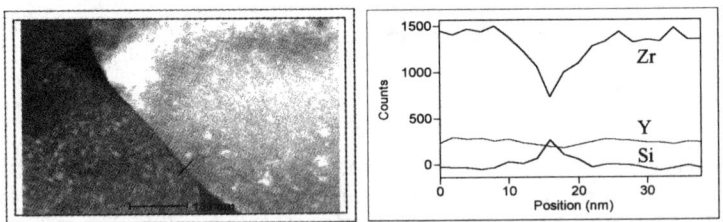

Figure 14. STEM image and corresponding line scan for 8Y-CSZ + 1 wt% SiO_2.

Ion Pair	Ionic character	Cation Size(Å)	Ionic Bond Strength
Ba-O	82% Ionic	1.36	1/3
Y-O	70% Ionic	0.89	1/2
Zr-O	67% Ionic	0.72	1/2
Al-O	55% Ionic	0.53	1/2
Si-O	43% Ionic	0.40	1
B-O	40% Ionic	0.20	1

Table II. Ionic character, cation size, and ionic bond strength for various ion pairs.

Room Temperature Fracture and Mechanical Properties

One concern about using silicate additives is how the presence of these additives will affect the mechanical behavior of the material. Earlier tests have found that while 1 wt% of glassy phases does not degrade the mechanical properties of 3Y-TZP, higher amounts of 5 wt.% are deleterious [11]. Samples of 3Y-TZP with one wt% glass additions were fractured and the fracture path characterized by SEM in order to determine if the presence of these phases promoted intergranular fracture (Figure 15). It can be seen that, consistent with the lack of segregation at the grain boundaries for the borosilicate added samples, the primary fracture mode is transgranular for both the pure and the borosilicate samples. The barium silicate samples, however, had primarily intergranular fracture, confirming that the presence of barium weakened the grain boundaries.

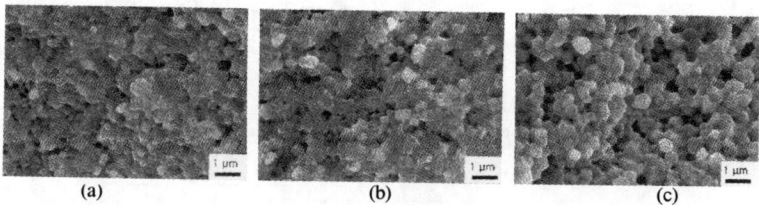

(a)   (b)   (c)

Figure 15. SEM images of room temperature fracture surfaces of (a) pure 3Y-TZP, (b) 3Y-TZP + 1 wt% borosilicate glass and (c) 3Y-TZP + 1 wt% barium silicate glass.

8Y-CSZ samples with silica additions fractured transgranularly at room temperature, despite the confirmed segregation of silicon (presumed to be silica) at the grain boundary (Figure 16a). The ability to maintain transgranular fracture even with 5 wt% (approximately 10 vol%) of intergranular silicate phase is quite remarkable. Fracture during high temperature deformation, however, indicated that intergranular fracture is dominant at high temperatures (Figure 16b). This may also indicate that the width of the segregation is different at high temperatures and a more uniform coverage of the grain boundaries may result.

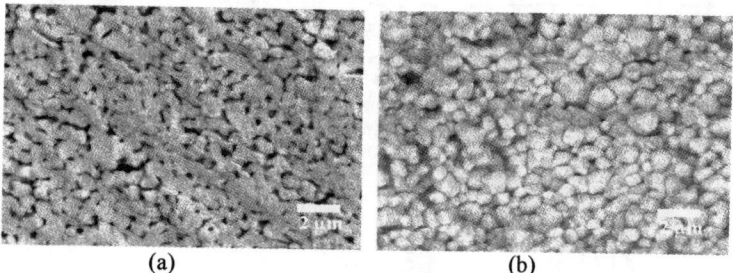

(a)   (b)

Figure 16. 8Y- CSZ + 5 wt% SiO_2 SEM images of (a) transgranular fracture at room temperature and (b) intergranular fracture at high temperature (1300°C).

Samples of 8Y-CSZ were tested for hardness, elastic modulus and fracture toughness as a function of the amount of silica added to the samples in order to determine if the addition of silica was detrimental and if so, the maximum amount that could be added without degrading the properties (Figure 17). From the data presented, 3wt.% of silica was an optimal amount that did not degrade the properties. Further studies were undertaken with 3 wt.% silica added to 8Y-CSZ to determine if deformation affected the mechanical behavior. Samples were deformed up to 100% and then tested for hardness, elastic modulus, and fracture toughness (Figure 18). The results indicate that deformations of 50% appeared to increase the fracture toughness in the material. Further testing to determine the bend strengths of the materials and to better detect defect formation are needed to ensure that highly deformable ceramics can still maintain their mechanical integrity after deformation.

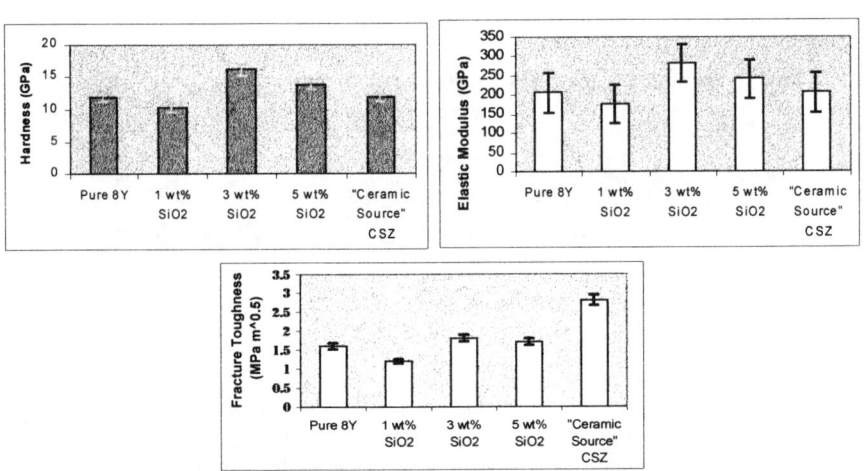

Figure 17. Hardness, elastic modulus and fracture toughness of pure 8Y-CSZ, 1 wt%, 3 wt%, and 5 wt% silica and compared values from the Ceramic Source '89.

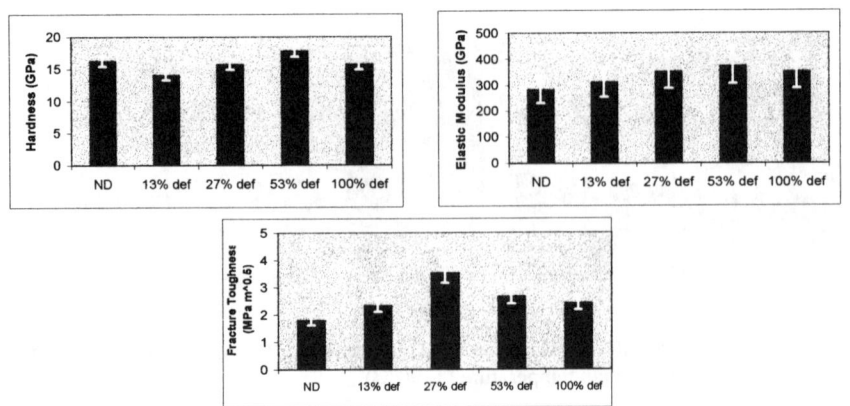

Figure 18. Hardness, elastic modulus and fracture toughness after compression at 1500°C.

CONCLUSIONS

Small amounts of amorphous silicate phase additions in ceramics can successfully be used to enhance superplasticity by (1) increasing the strain rate, (2) lowering the flow stress, (3) minimizing grain growth, and (4) increasing the total ductility. The most promising candidates to date for these intergranular phases are silicates with a low solubility for the ceramic component and a high viscosity. Characterization of the composition of the grain boundary phase and segregation along the grain boundaries is key to understanding how these intergranular phases can resist cavitation and fracture.

ACKNOWLEDGEMENTS

The author would like to acknowledge the work of the graduate students who have participated in this project and provided most of the research results discussed here: Phil Imamura, Adel Sharif, Maria Gust, and Yung-Jen Lin. In addition, the support and input from external collaborators including Dr. Neal Evans and Dr. Jim Bentley at Oak Ridge National Laboratory (SHaRE program), Dr. Terry Mitchell at Los Alamos National Laboratory (support of this work under a CULAR grant), and Professor T. Sakuma at the University of Tokyo (sponsor of Phil Imamura for a research summer on an NSF graduate student program) are also gratefully acknowledged.

REFERENCES

1. F. Wakai, S.Sakaguchi, and Y. Matsun, Adv. Ceram. Mater, 1 (1986) 259.
2. P. Thavarniti, Y. Ikuhara, and T. Sakuma, J. Am. Ceram. Soc., 81 (1998) 2927.
3. J. Koike, K, Miki, K, Maruyama, and H. Oikawa, Phil. Mag. A, 78 (1998) 599.
4. C.M.J. Hwang and I.W. Chen, J. Amer. Ceram. Soc., 73 (1990) 1626.
5. S. Tekeli and T.I. Davies, J. Mater. Sci., 33 (1998) 3267.
6. Turnbull, D., Trans. AIME 191 (1951) 227.
7. Kingery, W.D.; Bowen, H.K.; Uhlmann, D.R., Introduction to Ceramics, 2nd ed. John Wiley and Sons (1976).
8. Y.Ikuhara, P.Thavarniti, and T. Sakuma, Acta Mater. 45 (1997) 5275.
9. Sakuma, T; Ikuhara, Y; Takigawa, Y; Thavorniti, P, Mat. Sci. and Eng. A - Structural Materials Prop., Microstructure and Processing 234 (1997) 226.
10. L. Pauling, Nature of the Chemical Bond, 3rd. Ed. Cornell University Press (1996).
11. M. Gust, G. Goo, J. Wolfenstine, and M. L. Mecartney, J Am. Ceram. 76 (1993) 1681.

Creep deformation in a 3mol%Y_2O_3-Stabilized Tetragonal Zirconia

K. Morita, K. Hiraga and Y. Sakka

National Research Institute for Metals, 1-2-1 Sengen, Tsukuba-shi, Ibaraki 305-0047, Japan.

ABSTRACT

The high temperature deformation behavior of a 3mol% yttria-stabilized tetragonal zirconia (3Y-TZP) is examined. The analysis of creep strain rate, which was monitored directly with an optical extensometer and corrected for the current grain size, reveals a sigmoidal feature in the true stress-creep rate relationship. At lower stresses, a stress exponent n of ~1.6 incorporated with a grain size exponent of 2.0 suggests the intervention of a lattice diffusion creep mechanism. For higher stresses where n~2.7, evidence of intragranular dislocation activities suggests that dislocations may play an important role in the accommodation process for grain boundary sliding in 3Y-TZP.

1. INTRODUCTION

For the characterization of high temperature deformation behavior in yttria-stabilized tetragonal zirconia (Y-TZP), the following creep equation has often been applied

$$\dot{\varepsilon} = A\sigma^n / d^p , \qquad (1)$$

where $\dot{\varepsilon}$ is the steady state creep rate, σ is the stress, d is the grain size, n is the stress exponent, p is the grain size exponent and A is a material constant. According to Eq. (1), the deformation behavior of high purity Y-TZP has been characterized by n~2.0 at higher stresses and n~3.0 at lower stresses in earlier studies [1-8]. To characterize the creep behavior strictly, it is necessary to determine the creep parameters from the instantaneous relationship among the steady state creep rate, the stress and the grain size. In the earlier studies [1-8], however, the creep parameters have been evaluated without considering the influence of grain growth.

The present study was performed to (i) determine the creep parameters from the instantaneous relationship and (ii) discuss possible deformation mechanisms for fine-grained, high purity 3Y-TZP based on the evaluated creep parameters and observed deformed microstructures.

2. EXPERIMENTS
2-1. Material and Method

A high purity Y-TZP powder containing 3 mol% yttria (Tosoh Co., Japan : <50 Al_2O_3, 50 SiO_2, <50 Fe_2O_3, 220 Na_2O in wt ppm) was cold-isostatically pressed at 100 MPa and sintered at 1673 K for 3 h in air. The sintered material had an initial grain size of 0.35 μm, where the grain size was determined as 1.56L; L is the average intercept length of the grains in SEM micrographs. The relative density of the material was higher than 98%. From the sintered material, dog-bone-shaped flat specimens with a gauge portion of $^t3 \times ^w3 \times ^l20$ mm were machined.

Constant load creep tests were performed at 1673 K in air and in a stress range of 3-80 MPa. The tensile displacement was measured with an electro-optical extensometer which can directly monitor the length between targets made at both ends of the gauge portion [9].

For TEM observation, the conventional creep tests at constant stress were carried out by an Instron-type machine at 1673 K in a vacuum of 2×10^{-3} Pa. The tensile specimens were deformed up to prescribed strains and cooled at an initial rate of 5 Ks^{-1}. During the cooling, the creep load was not removed to freeze-in the deformed structures. TEM specimens were cut from the deformed gauge portion, mechanically polished to a thickness of 100 μm and further thinned by Ar ion-milling.

Mat. Res. Soc. Symp. Proc. Vol. 601 © 2000 Materials Research Society

2-2. Creep Data Analysis

The monitored creep rate, $\dot{\varepsilon}_m$, was corrected for the instantaneous stress and grain size as follows. The details is described elsewhere [9, 10]. In brief, for uniform deformation and constant volume in the gauge portion, the creep rate can be corrected for the instantaneous stress as

$$\dot{\varepsilon}_\sigma = \dot{\varepsilon}_m \exp(-n\varepsilon_a) , \qquad (2)$$

where the ε_a is the arbitrary true strain. When grain growth is not negligible, the influence of both stress and grain size can be corrected as

Table 1 Material parameters used for the correction by Eq. (7).

n	2.6 for	$\sigma_0 \geq 15$ MPa
	2.2 for	$\sigma_0 < 15$ MPa
p	3.0 for	$\sigma_0 \geq 50$ MPa
	2.0 for	$\sigma_0 < 50$ MPa
r	3	
k (m³/s)	5.86×10^{-25}	
α	0.45	

$$\dot{\varepsilon}_{\sigma d} = \dot{\varepsilon}_m \exp(-n\varepsilon_a)(d_0 / d_a)^{-p} , \qquad (3)$$

where d_0 is the initial grain size and d_a is the instantaneous grain size at ε_a.

Next, the instantaneous grain size d_a is evaluated as a function of time and creep strain as follows. The static grain growth in 3Y-TZP [11] has been shown to obey

$$d_s^{\ r} - d_0^{\ r} = kt , \qquad (4)$$

where d_s is the grain size at time t, r is the grain growth exponent and k is the rate constant for static grain growth. Grain growth during deformation is reported to be accelerated owing to an additional strain-dependent component which can be represent by the following form [12, 13]

$$\ln(d_a / d_s) = \alpha\varepsilon_a , \qquad (5)$$

where d_a is the grain size at ε_a, d_s is the grain size attained only by annealing for the duration of $[\exp(\varepsilon_a)-1]/\dot{\varepsilon}_0$ and α is the rate constant for the strain-induced grain growth. Replacing d_s in Eq. (5) with that of Eq. (4) gives the instantaneous grain size d_a at ε_a as

$$d_a = (d_0^{\ r} + kt)^{1/r} \exp(\alpha\varepsilon_a) . \qquad (6)$$

Substitution of Eq. (6) in Eq. (3) yields the creep rate corrected for both instantaneous stress and grain size as

$$\dot{\varepsilon}_{\sigma d} = \dot{\varepsilon}_m \exp(-n\varepsilon_a)\{[(d_0^{\ r} + kt)^{1/r} \exp(\alpha\varepsilon_a)]/d_0\}^p . \qquad (7)$$

The parameters for the correction were measured experimentally and listed in Table 1. Here, the apparent n-values were determined from $\dot{\varepsilon}_m$-curve at $\varepsilon =0.03$, where a quasi-steady state was attained at $\sigma =3-80$ MPa and grain growth is assumed to be negligible. The p-value was determined from the slopes of $\ln(\varepsilon_a)-\ln(d_a)$ plots [10].

3. EXPERIMENTAL RESULTS

3-1. Creep Behavior

Figure 1 shows the creep rate curves at 1673 K under 30 MPa. The thin solid, broken and thick solid lines represent the monitored creep rate $\dot{\varepsilon}_m$, the creep rate corrected for the stress $\dot{\varepsilon}_\sigma$ and the corrected for both stress and grain size $\dot{\varepsilon}_{\sigma d}$, respectively. For the $\dot{\varepsilon}_m$- and $\dot{\varepsilon}_\sigma$-curves, a primary creep region appears up to $\varepsilon \sim 0.02$ and it is followed by a quasi-steady state region. For the $\dot{\varepsilon}_{\sigma d}$-curve, the correction shows a well defined steady state region above $\varepsilon >0.05$. We should note that the creep rate in the $\dot{\varepsilon}_\sigma$-curve decreases gradually with strain and the discrepancy between the $\dot{\varepsilon}_\sigma$- and $\dot{\varepsilon}_{\sigma d}$-curves tends to increase with an increase in strain. Since the grain growth was accelerated owing to the strain-dependent component (Eq. (5)), the decrease in the creep rate for the $\dot{\varepsilon}_\sigma$-curve can be associated with grain growth.

The open circles in Fig. 1 represent the creep rates, $\dot{\varepsilon}_{\sigma c}$, obtained by the conventional constant stress test, where the strain rate was monitored between the specimen shoulders. In contrast to the above results, the primary creep region extended to $\varepsilon \sim 0.3$ and the transition from

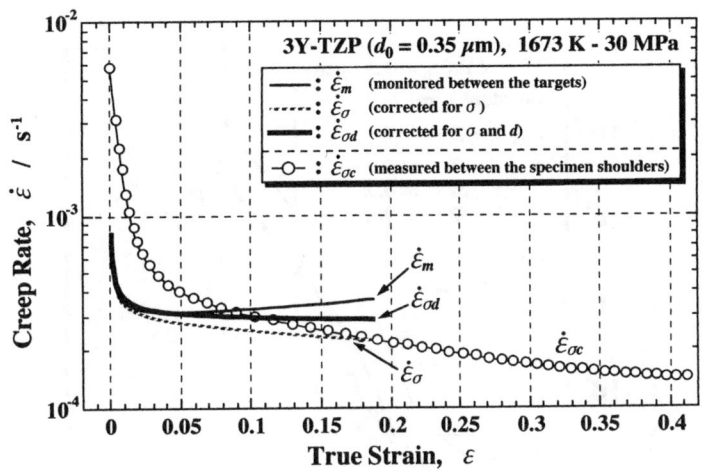

Fig. 1 Creep rate curves of 3Y-TZP at 1673 K, at 30 MPa.

the primary to the quasi-steady state creep regions becomes ambiguous in the $\dot{\varepsilon}_{\sigma c}$-curve. The difference from the $\dot{\varepsilon}_{\sigma d}$-curves can be attributed to some preferential deformation at around the specimen shoulders [9].

3-2. Evaluation of Stress Exponent, n

Figure 2 shows the corrected creep rate as a function of the true stress. The open circles represent the steady state creep rates, $\dot{\varepsilon}_{\sigma d}$, evaluated at ε =0.05~0.1. The open, filled and inverted triangles represent the corrected creep rate, $\dot{\varepsilon}_{\sigma}$, evaluated at ε =0.03, 0.05 and 0.1, respectively. For comparison, the typical data reported for a high-purity 3Y-TZP at 1673 K [3] are also shown by the open squares.

A sigmoidal feature similar to that of superplastic metals appears in the σ-$\dot{\varepsilon}_{\sigma d}$ relationship. The stress exponent evaluated from the $\dot{\varepsilon}_{\sigma d}$-curve takes a value of 2.7 at 15-80 MPa and 1.6 at stresses lower than 10 MPa. These regions are connected with a transition region with n>5. However, the transition of the stress exponent from 2.0 to 3.0 reported in the earlier studies [1-8] does not appear in the stress range of the present examination.

For the σ-$\dot{\varepsilon}_{\sigma}$ relationship, the n-value is almost the same at the higher stress region. Below 10 MPa, however, the stress exponent tends to increase with an increase in strain and takes a value of ~3.0 at ε =0.1 as shown by the broken line in Fig. 2. It is noted that the data for ε =0.1 parallel the earlier ones obtained without correcting the effect of grain growth [1-8].

3-3. Microstructural Evaluation

Figure 3(a) shows a bright field TEM image of the as-sintered material. The material had equiaxed grains with sharply faceted grain boundaries. No glass phases were found at both multiple and two grain junctions. Figure 3 (b) is a TEM micrograph after deformation at 30 MPa up to ε ~0.18. Although the grains were still equiaxed and surrounded by faceted grain boundaries after deformation, the strain contrast was more pronounced than in the as-sintered material.

Figures 3(c) and (d) show dark field TEM micrographs of the materials deformed at 30 and 50 MPa, respectively. The micrographs show that both dense alignment and tangle of

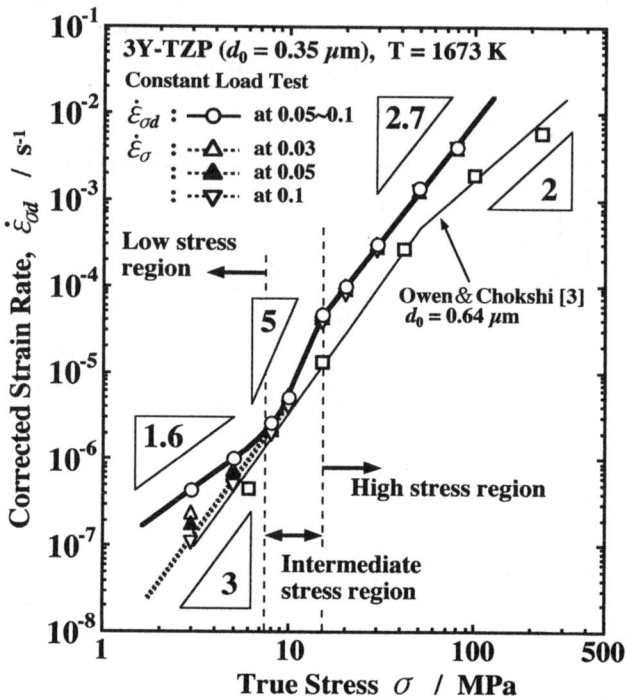

Fig. 2 Creep rates corrected for stress ($\dot{\varepsilon}_\sigma$) and for both stress and grain size ($\dot{\varepsilon}_{\sigma d}$) as a function of true stress. The open square shows the typical data reported by Owen and Chokshi [3] for a 3Y-TZP with an initial grain size of 0.64 μm.

dislocations exist in the grain interior. In the as-sintered material, intragranular dislocations were rarely observed, and hence such aligned nor tangled dislocations as seen in Figs. 3(c) and (d) did not preexist. Quantitative examination on dark field micrographs also confirmed that the numerical densities of intragranular dislocations were much higher in the deformed specimens than in the as-sintered material. It can thus be concluded that the pronounced strain contrast (Fig. 3(b)) and the dislocation structures (Figs. 3(b) and (d)) were not due to any artifacts during specimen preparation, but developed during deformation.

4. DISCUSSION
4-1. The *n*-value transition from ~2.0 to ~3.0

The transition of n~2.0 at higher stresses to n~3.0 at lower stresses has often been reported in earlier studies [1-8]. In the present study, however, n~2 did not appear at stresses up to 80 MPa. Such a difference cannot be explained by grain growth, because the effect of grain growth tends to decrease with increasing stress [9]. Interference from preferential deformation around the grip section may be concerned with the difference. In the earlier studies [1-5], the transition stress is reported to increase with decreasing the grain size. If this is true in Y-TZP, the n-value of ~2.0 may appear at stresses higher than 80 MPa. Additional experiments are necessary to confirm the existence of n~2.0.

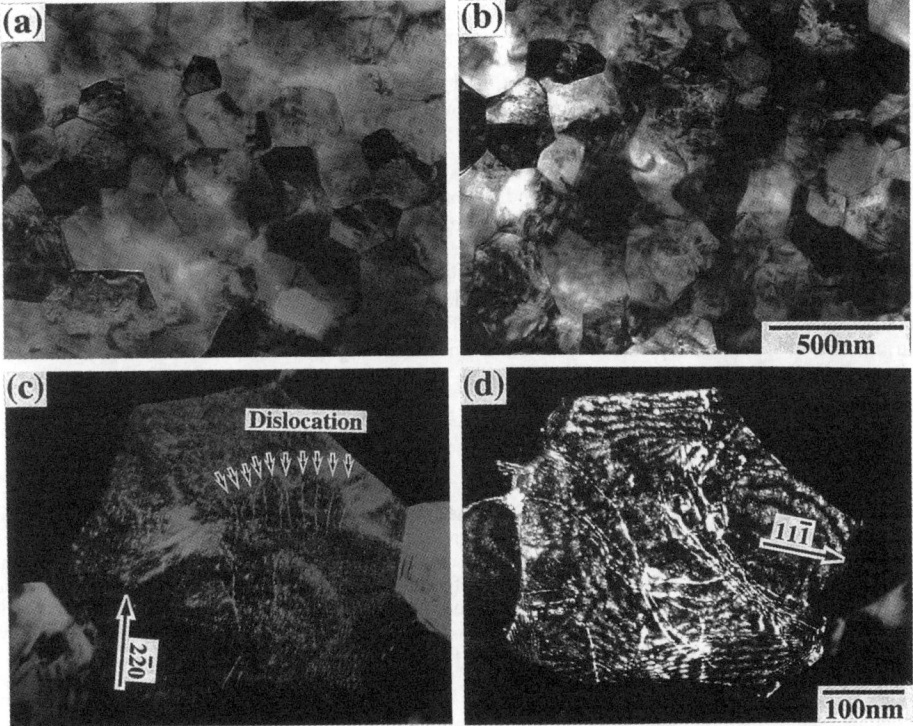

Fig. 3 TEM micrographs for 3Y-TZP of (a) before and (b)-(d) after deformation. The specimens were deformed up to $\varepsilon \sim 0.18$ at 1673 K, at stresses of 30 MPa ((b)-(c)) and 50 MPa ((d)).

4-2. High stress region of σ = 15-80 MPa

For the region with $n \sim 3.0$, an interface-reaction-controlled process [1-4] and a diffusional process incorporated with a threshold stress [5-8] have been proposed as the rate-controlling mechanisms for high purity Y-TZP. According to the combinations of creep parameters in existent creep models [14, 15], however, the observed creep parameters and dislocation structures (Figs. 3(c) and (d)) do not seem to be fully compatible with these deformation mechanisms.

Owing to a mismatch between the flow stress of fine-grained Y-TZP and the yield stress of yttria-stabilized zirconia single crystals, earlier studies have ruled out the possibilities of dislocation-related deformation mechanisms for Y-TZP. However, the present study revealed that dislocation motion is indubitably activated in this deformation region (Figs. 2 and 3). This result indicates that stresses much higher than the creep flow stresses should be concentrated at multiple grain junctions during grain boundary sliding (GBS) and be relaxed by the dislocation motion. The result also suggests that the activated dislocation motion may play an important role in the deformation process of fine-grained 3Y-TZP.

To confirm this suggestion, more detailed examinations on the deformed microstructures and stress concentrations are necessary at the multiple grain junctions. Within the scope of the present results, however, the following implications are possible. Irrespective of the observed dislocation activities incorporated with $n \sim 3.0$, conventional dislocation creep can be excluded as the deformation mechanism. This is because the creep rates of the present material depend upon grain size, *i.e.*, the p-values in Eq.(1) takes a value of 2.0 or 3.0. As mentioned before, the present observation typically shown in Fig. 3 suggests that intragranular dislocation motion is likely to contribute to the relaxation of the stress concentrations at multiple grain junctions. If this is the case, continuous removal of the dislocations is necessary for continuous relaxation of the stress concentration during deformation, and this process would control the creep rate as in the models proposed by Ball and Hutchison [16] and Mukherjee [17] for superplastic metals.

4-3. Low stress region of $\sigma < 8$ MPa

The comparison among $\dot{\varepsilon}_{od}$, $\dot{\varepsilon}_{\sigma}$ and the earlier data indicates that the region with $n \sim 3$ is spurious at stresses lower than 8 MPa. The comparison also indicates that such a spurious region arises from ignoring the effect of grain growth [9, 10]. The present results show that the smaller stress exponent of 1.6 is essential in this deformation region.

There has been little information available for discussing rate-controlling mechanisms in such a low stress region for fine-grained Y-TZP. It should be noted, however, that the creep parameters of $n=1.6$ and $p=2.0$ evaluated from the $\sigma - \dot{\varepsilon}_{od}$ plot is quite similar to those of coarse-grained Y-TZP, $i.e.$, $n \sim 1.2$ and $p \sim 2.2$ for $d_0 = 2.6$-9.5 μm [18]. The combination of the creep parameters suggests that lattice diffusional creep may intervene in this deformation region.

5. SUMMARY

In a fine grained, high purity 3Y-TZP, the present analysis of creep strain rate at 1673 K revealed a sigmoidal feature in the true stress-creep rate relationship. At stresses lower than 8 MPa, a stress exponent n of ~ 1.6 and a grain size exponent of 2.0 appears and this combination of creep parameters suggests the intervention of a lattice diffusion creep. At stresses between 15-80 MPa, $n \sim 2.7$ appears and microstructural observation revealed that intragranular dislocation motion were activated during deformation. This result suggest that dislocation motion may play an important role in the accommodation process for grain boundary sliding in 3Y-TZP.

REFERENCES

[1] A. H. Chokshi, Mat. Sci. Eng. **A166**, 119(1993).
[2] A. H. Chokshi, Mat. Sci. Eng. **A234-236**, 986(1997).
[3] D. M. Owen and A. H. Chokshi, Acta Mater. Mater. **46**, 667(1998).
[4] J. A. Hines, Y. Ikuhara, A. H. Chokshi and T. Sakuma, Acta Mater. Mater. **46**, 5557(1998).
[5] M. Jimenez-Melendo, A. Dominguez-Rodoriguez and A. Bravo-Leon, J. Am. Ceram. Soc. **81**, 2761(1998).
[6] A. Bravo-Leon, M. Jimenez-Melendo and A. Dominguez-Rodoriguez, Scripta Metall. **24**, 1459(1990).
[7] M. Jimenez-Melendo, A. Bravo-Leon and A. Dominguez-Rodoriguez, Mat. Sci. Forum **243-245**, 363(1997).
[8] A. Dominguez-Rodoriguez, A. Bravo-Leon, J. D. Ye and M. Jimenez-Melendo, Mat. Sci. Eng. **A247**, 97(1998).
[9] K. Morita and K. Hiraga, submitted to Scripta Metall.
[10] K. Morita, K. Hiraga and Y. Sakka, Proceedings of *"The 8th International Conference on Creep and Fracture of Eng. Materials and Stractures"*, 847(2000).
[11] J. Zhao, Y. Ikuhara and T. Sakuma, J. Am. Ceram. Soc. **81**, 2087(1998).
[12] E. Sato and K. Kuribayashi, ISIJ Int. **33**, 825(1993).
[13] J. R. Seidensticker and M. J. Mayo, Scripta Metall. **38**, 1091(1998).
[14] E. Arzt, M. F. Ashby and R. A. Verrall, Acta Mater. Mater. **31**, 1977(1983).
[15] M. F. Ashby and R. A. Verrall, Acta Mater. Mater. **21**, 149(1973).
[16] A. Ball and M. H. Hutchison, Met. Sci. J., **3**, 1(1969).
[17] A. K. Mukherjee, Mat. Sci. Eng. **8**, 83(1971).
[18] D. Dimos and D. L. Kohlstedt, J. Am. Ceram. Soc. **70**, 531(1987).

SUPERPLASTICITY AND JOINING OF ZIRCONIA-BASED CERAMICS

F. Gutierrez-Mora , A. Dominguez-Rodriguez and M. Jimenez-Melendo.
Dpto. de Física de la Materia Condensada, Universidad de Sevilla, 41080 Sevilla, Spain.

R. Chaim. and G. B. Ravi.
Dept. of Materials Engineering, Technion, Israel Institute of Technology, Haifa 32000, Israel.

J. L. Routbort
Energy Technology Division, Argonne National Laboratory.

Abstract

Steady-state creep and joining of alumina/zirconia composites containing alumina volume fractions of 20, 60, and 85% have been investigated between 1250 and 1350°C. Superplasticity of these compounds is controlled by grain-boundary sliding and the creep rate is a function of alumina volume fraction, not grain size. Using the principles of superplasticity, pieces of the composite have been joined by applying the stress required to achieve 5 to 10% strain to form a strong interface at temperatures as low as 1200°C

Introduction

The high-temperature plastic deformation of alumina ceramics has been extensively reported in the literature [1-3]. Achieving high ductility in pure alumina is difficult because of rapid static and dynamic grain growth that is accompanied by large strain hardening and severe cavitation. The relatively high grain-boundary energy (GBE) in pure alumina [1] provides a high driving force for grain growth. The high GBE is also responsible for the relatively low cohesive strength that results in severe cavitation [1].
Suppression of grain growth can be achieved through low temperature sintering or/and the use of additives. Although the solubility of a second phase in alumina is very limited, the use of additives such as MgO, ZrO_2 or Y_2O_3 inhibit the grain growth by segregation of these cations to the alumina grain boundaries [1,4].
In the case of alumina/zirconia, microstructure and mechanical properties has been widely studied [1, 4-7] and it has been shown that the equiaxed fine microstructure is preserved during deformation improving the ductility of this compounds [1,6,7].
In this paper we will characterize the mechanism controlling the plasticity as a function of the alumina volume fraction and of the phase distribution.
It is well known that grain boundary sliding (GBS) controls the macroscopic deformation of these compounds [6]. When two pieces of ceramics are compressed in the superplastic regime, the grains in the interface rotated with respect to each other producing interpenetration and a strong junction of both pieces [8-10]. Therefore, superplastic flow has been used to join alumina/zirconia ceramics and the conditions and strength of the junctions will be also discussed in this paper.

Experimental conditions

Alumina/zirconia composites with alumina volume fractions of 20, 60 and 85 %, and with different phase distribution, were investigated. The compositions are designated as ZT20A, ZT60A and ZT85A. Fabrication processes of the ZT60A and ZT85A were described in detail in references 11 and 12, respectively.
The ZT20A was fabricated as following: nanocrystalline ZrO_2-4 wt% Y_2O_3 (Y-TZP) powder with mean particle size of 10 nm was prepared by chemical precipitation from the

99

nitrate salt solutions, as previously described [13]. The commercial nanocrystalline alumina powder (Neomat) was composed of transition alumina (θ and δ) with a mean particle size of 10 nm. The Y-TZP and alumina powders were separately dispersed ultrasonically in ethanol; the pH of the dispersions were adjusted to 1.5 and 7 respectively. The two slurries were ball milled for 24 h before mixing; the resultant pH was between 1.5 and 2. At this stage two different mixing and sedimentation routes were adopted. In the case of the sedimentation (S designated samples), ammonia solution was added to the final mixed slurry to change its pH to 10.5. Then the slurry was left to sediment overnight; the latter was collected and washed with ethanol and dried at 200°C for 1 day. In the case of the milled (M designated samples), the mixed slurry was dried using a heating plate with continuous mixing to avoid sedimentation. Once the slurry was dried, the wet powder was dried at 200°C for 1 day. Cylindrical pellets of 15 mm in diameter and length were uniaxially cold pressed followed by cold isostatic pressing at 200 MPa. Finally, the pellets were sintered at 1600°C for 2 min using heating and cooling rates of 5 °C/min and 20 °C/min, respectively.

The average grain size of zirconia and alumina were determined by electron microscopy and the final density was determined by the Archimedes method. The phase content was determined by X-ray diffraction.

The samples were cut into parallelepiped shape having dimensions of 6x3x3 mm and compressed in air at temperatures between 1200 and 1350°C at constant load (creep experiments) or at constant cross-head speed in a Instron machine. In these latter experiments a steady-state stress, as defined by a zero work-hardening rate, was obtained after 1-2 % of deformation. The stress of the plateau is the flow stress for Instron experiments. The stress ranged between 5 and 500 MPa and the initial strain rate varied between 10^{-7} and 10^{-4} s^{-1}.

For the junction, pieces of ZT20AM and ZT60A were compressed in an Instron machine to a strain of 5 and 10 % at crosshead speed of 5 μm/min that corresponded to an initial strain rate, $\dot{\varepsilon} \sim 10^{-5}$ s^{-1}, for the dimensions of the pieces joined. The temperatures employed were between 1200 and 1300°C.

The junction strength was measured from cracks produced by a 100 N Vickers indentor whose diagonal was along the interface.

Experimental results and discussion

The ZT20AS specimens were characterized by a more homogeneous distribution of

Fig 1. Microstructure of the as-received ZT20AS

both alumina and zirconia grains compared with the ZT20AM specimens. Fig. 1 is a typical micrograph of the microstructure of a ZT20AS ceramics. Scanning electron microscopy (SEM) [11,12] revealed that the ZT60A and ZT85A consisted of a very homogeneous distribution of both phases. The characteristics of these ceramics are listed in Table I. No differences in the grain size of both phases were found in the two types of ZT20A.

The superplastic behavior of alumina/zirconia composites has been analyzed using the equation:

$$\dot{\varepsilon} = A\sigma^{n}d^{-p}\exp\left(\frac{-Q}{RT}\right) \tag{1}$$

where $\dot{\varepsilon}$ is the strain rate, σ is the applied stress, d the grain size, R and T have their usual meaning, and n, p and Q are the creep parameters characterizing the mechanism controlling the creep behavior.

TABLE I. Grain size of both phases of the materials used in this work.

Material	Al$_2$O$_3$ grains (μm)	3Y grains (μm)
ZT20A	0.8	0.6
ZT60A	0.7	0.4
ZT85A	1.3	0.5

Figs. 2a and 2b are creep curves typical of all materials tested. From stress and temperature changes it is possible to determine n and Q as shown. There are no

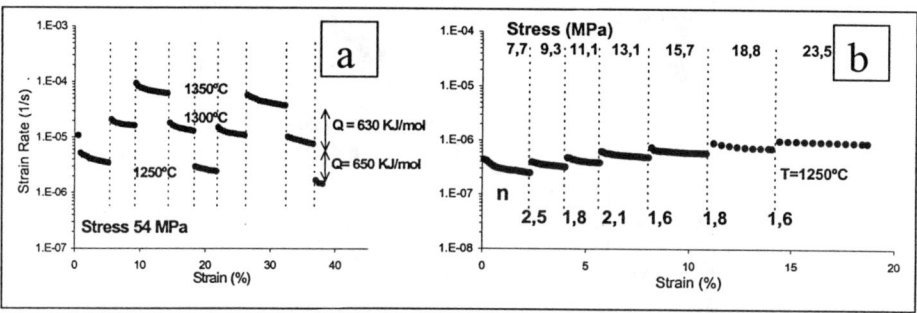

Fig 2. Typical creep curves to determine the activation energy Q (a) and the parameter n (b).

discontinuities in the extrapolation of the creep data for a given temperature, despite the fact that T was changed several times. Therefore, the creep data indicate that the microstructure remained constant during the test without grain growth. No grain-size exponent was determined because the grain size was not varied.

The effect of T and σ is shown in Fig. 3, which is a plot of log $\dot{\varepsilon}$ vs log σ for three different temperatures. The excellent agreement between the creep and Instron data is further evidence that microstructure was unchanged and that the tests were performed in steady state. That is, the final state is path independent. The slopes of these curves are close to 2 decreasing at the higher stresses. Q and n were determined from curves as in fig. 2. Values of n at low and intermediate stresses of 1.9 ± 0.1 and of 1.4 ± 0.1 at the higher stresses were measured. The values of n and σ are independent of the composition and phase distribution. Fig. 4 shows that there is no influence of the phase distribution. As shown in fig. 4, the value of 2 at stresses as low as 5 MPa indicates that the threshold stress is zero in these compounds. The same behavior has been reported in Y-TZP with low purity [14,15].

A n value of 2 has been also measured by Wakai et al. [6] for stresses below 80 MPa and by Clarisse et al [7]. However, these last authors found than n tends to 1 when the stress increases.

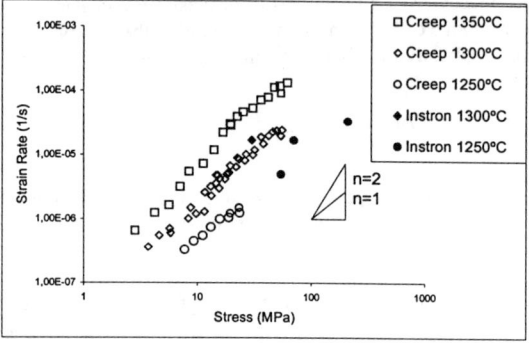

Fig. 3: Plot of log ε̇ versus log σ for different temperatures.

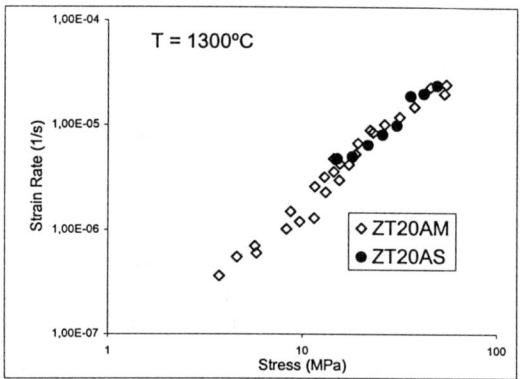

Fig.4: Influence of phase distribution in the mechanical behavior.

The Q value was weakly dependent on stress. A Q-value of 680 ± 20 kJ/mol at the lower stresses and of 640 ± 10 kJ/mol for stresses higher than 50 MPa were measured. Although the activation energy for non-doped alumina [7] and pure Y-TZP [14] ranged between 450-500 kJ/mol, a higher value of Q has been systematically found for the different alumina/zirconia composites [6,7,16]. The higher Q has been attributed to the interdiffusion of Zr to the Al_2O_3 phase resulting in the modification of the defect formation enthalpy [7].

As already mentioned, the suppression of grain growth in alumina can be achieved through the use of additives as zirconia. The zirconium, in alumina, precipitates to the interface [1,6,7] lowering the GBE and improving the stability of the microstructure and the superplastic formability of these compounds. At the same time, the segregation of Zr^{4+} at the grain boundaries reduces the aluminum diffusivity. For instance, the Al^{3+} in these compounds changes its diffusivity with respect to pure alumina, increasing the activation energy from 500 kJ/mol in pure alumina to 700 kJ/mol in alumina/zirconia composites [5,6]. No variation in Q with the zirconia/alumina composition has been detected. This result is compatible with stability of the microstructure of these compounds when the second phase is present at relatively low volume fractions [1,7], as a consequence of the precipitation of zirconium at the grain boundaries and the change of the diffusivity.

In table II the creep resistant for the three compositions used in this work are presented together with the results in the literature, for a stress of 50 MPa and two different temperatures of 1300 and 1350°C. It is obvious that the mechanical behavior of these compounds depend on the alumina volume fraction and not of the grain size. For instance, in our work, ZT20A and ZT60A have almost the same grain size, however the strain rate is one order of magnitude lower for ZT60A.

TABLE II. Strain Rates at a stress of 50 MPa for different compositions and two temperatures.

Material	Authors	Al_2O_3 grains (μm)	3Y grains (μm)	Strain Rate (1/s) 1300°C	Strain Rate (1/s) 1350°C
ZT20A	Present work	0.8	0.6	3×10^{-5}	1×10^{-4}
ZT20A	[7]	0.6	0.6	4×10^{-5}	2×10^{-4}
ZT50A	[7]	1.1	0.8	6×10^{-6}	3×10^{-5}
ZT50A	[6]	0.6	0.5	6×10^{-6}	2.5×10^{-5}
ZT60A	Present work	0.7	0.4	3×10^{-6}	-
ZT69A	[6]	1.0	0.6	1×10^{-6}	7×10^{-6}
ZT80A	[7]	1.4	0.7	-	7×10^{-6}
ZT85A	Present work	1.3	0.5	Broken	1×10^{-6}
ZT86A	[6]	1.0	0.5	1×10^{-6}	7×10^{-6}

After deformation, the microstructure remained unchanged (not shown here) with the same grain size and shape as measured by the form factor. These characteristics are typical when GBS is the main mechanism controlling plasticity.

Based on the GBS mechanism, two types of junction have been produced. In one case, two pieces of TZ20A were joined and in the other case one piece of TZ20A was joined with another of TZ60A. Fig. 5 shows the interface of the junction. The interface has almost

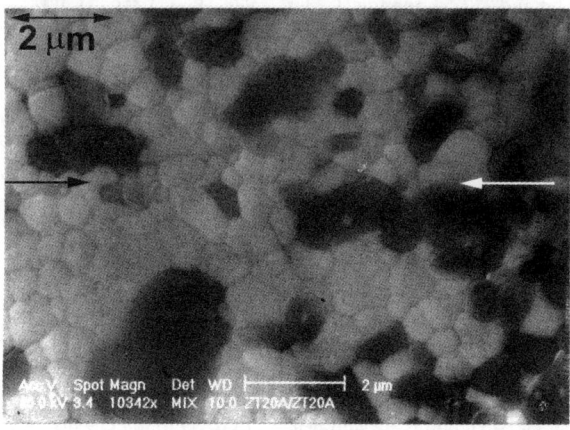

Fig. 5: Junction of two pieces of ZT20A. The arrows indicate the location of the interface.

disappeared, showing that the grains of one part interpenetrate the other. The junction is formed by rotation of the grains of both pieces. The crack lengths of the Vickers indentations developed along and perpendicular to the interface in the case of ZT20A/ZT20A are

280 ± 30 μm and 270 ± 20 μm. In the case of ZT20A/ZT60A, the cracks did not propagate along the interface but deviate into the brittle material (ZT60A). These results indicate that the pieces were perfectly joined.

Conclusion

It is shown that the ZTA composites due to the refinement of the microstructure of the alumina and to the presence of YTZP became more damage tolerant that pure alumina although the strength is controlled by this phase.

These compounds can be deformed superplastically at temperatures as low as 1250°C. The main mechanism of the plastic deformation at high temperature of these compounds is GBS accommodated by diffusion process, which are controlled at their turn by the precipitation of Zr at the grain boundaries.

Based on the GBS mechanism, different pieces of ZTA has been joined at temperatures as low as 1200°C for period of time as long as 30 min. The cracks develop along the interface after Vickers indentation show that the pieces were perfectly joined.

Acknowledgement:

This research was supported by CICYT under grant no MAT97-1007-C02 (Ministerio de Educación y Ciencia-Spain), by the Fund for the promotion of Research at the Technion (1999) and by the Israel Ministry of Science. J. Routbort is grateful to IBERDROLA for providing funding for his visit to the Universidad de Sevilla.

References

1. I.W. Chen and L. A. Xue J. Am. Ceram. Soc., **73**, p. 2585 (1990)
2. W. R. Cannon in Advances in Ceramics, Vol 10 *Structure and Properties of MgO and Al₂O₃ Ceramics*. Edited by W.D. Kingery. Am. Ceram. Soc. Columbus OH. 1984. p. 741-49.
3. W. R. Cannon and T. G. Langdon, J. Mater. Sci, **18**, p. 1 (1983). Ibidem, **23**, p. 1 (1998).
4. T. Sakuma, Y. Ikuhara, Y. Takigawa and P. Thavorniti, Mat. Sci. Eng. **A234-236**, p. 226 (1997).
5. H. Yoshida, K. Okada, Y. Ikuhara and T. Sakuma. Phil. Mag. Let. **76**, p. 9 (1997).
6. F. Wakai, T. Nagano and T. Iga. J. Am. Ceram. Soc. **80**, p. 2361 (1997)
7. L. Clarisse, R. Baddi, A. Bataille, J. Crampon, R. Duclos and J. Vincens. Acta Mater. **45**, p. 3843 (1997).
8. J. Ye and A. Dominguez-Rodriguez, Scripta Met. and Mater. **33**, p. 441 (1995).
9. A. Dominguez-Rodriguez, F. Guiberteau and M. Jimenez-Melendo. J. Mat. Res. **13**, p. 1631 (1998).
10. A. Dominguez-Rodriguez, E. Jimenez-Pique and M. Jimenez-Melendo. Scripta Mater. **39**, p. 21, (1998).
11. M. Jimenez-Melendo, C. Clauss, A. Dominguez-Rodriguez, G. De Portu, E. Roncari and P. Pinasco, Acta Mater. **46**, p. 3995 (1998).
12. M. Jimenez-Melendo, C. Clauss, A. Dominguez-Rodriguez, A. J. Sanchez-Herencia, and J. S. Moya, J. Am. Ceram. Soc. **80**, p. 2126 (1997).
13. G. Hare'l, B. G. Ravi, and R. Chaim, Mater. Lett. **39**, p. 63 (1999).
14. M. Jimenez-Melendo, A. Dominguez-Rodriguez and A. Bravo-Leon. J. Am. Ceram. Soc. **81**, p. 2761 (1998)
15. A. H. Chokshi, Mat. Sci. Eng. **A166**, p.119 (1993).
16. M. Jimenez-Melendo and A. Dominguez-Rodriguez, Phil. Mag. A, **79**, p.1591 (1999).

A DISCUSSION OF FLOW MECHANISMS IN SUPERPLASTIC YTTRIA-STABILIZED TETRAGONAL ZIRCONIA

MIN Z. BERBON[*], TERENCE G. LANGDON[**]
[*]Materials and Computational Science Division, Rockwell Science Center,
1049 Camino Dos Rios, Thousand Oaks, CA 91358, mzberbon@rsc.rockwell.com
[**]Departments of Materials Science and Mechanical Engineering,
University of Southern California, Los Angeles, CA 90089-1453

ABSTRACT

Flow data are reported for a superplastic yttria-stabilized tetragonal zirconia containing 3 mol % of Y_2O_3 (termed 3Y-TZP). The data are analyzed using a model for interface-controlled diffusion creep and the results are compared with published data for other samples of 3Y-TZP.

INTRODUCTION

The potential for achieving superplastic elongations in ceramic systems was first demonstrated in early reports of tensile elongations in excess of 100% in samples of yttria-stabilized tetragonal zirconia with 3 mol % yttria (termed 3Y-TZP) [1] and a composite containing 80 wt % of 3Y-TZP and 20 wt % Al_2O_3 [2]. Subsequently, there were several reports of superplasticity in 3Y-TZP and a detailed summary was presented by Chokshi [3].

The results from superplastic experiments are often plotted logarithmically in the form of the steady-state creep rate, $\dot{\varepsilon}$, against the applied stress, σ, where the slope of this plot is termed the stress exponent, n. Several of the more recent sets of data on 3Y-TZP reveal a possible transition from a value of n close to ~3 at the lower stresses to a value of n $\simeq$ 2 at higher stresses [4-8]. Two different approaches have been adopted in interpreting these results. First, it has been suggested that flow occurs through grain boundary sliding in the region with n $\simeq$ 2 but the value of n increases at the lower stress levels because of the increased segregation of impurities to the grain boundaries [4]. Second, the higher value of n at the lower stresses has been interpreted using the concept of a threshold stress [5-7]. Very recently, an alternative proposal was presented in which flow occurs through diffusion creep but with the advent of an interface-controlled process at the lower stresses [9]. In the present paper, new experimental results are reported for 3Y-TZP and these results, and other published data, are compared with this alternative approach.

EXPERIMENTAL MATERIAL AND PROCEDURE

The experiments were conducted using high purity samples of 3Y-TZP containing <0.005 wt % Al_2O_3 and <0.002 wt % SiO_2. The samples were prepared in an essentially fully-dense condition using isostatic pressing and a two-step sintering procedure [10]. The sintering times and temperatures were adjusted to give spatial grain sizes, d, of 0.67 and 1.5 μm, respectively, where d is defined as $1.74 \times L$ where L is the mean linear intercept grain size. Samples were pulled to failure in tension using an Instron testing machine operating at a constant rate of cross-head displacement. Tests were conducted at initial strain rates from 7.1×10^{-6} to 2.8×10^{-4} s^{-1} and at temperatures from 1623 to 1813 K.

EXPERIMENTAL RESULTS AND DISCUSSION

Figure 1 shows experimental data for samples tested with two different grain sizes: the stresses were taken from the stress-strain plots at a strain of 10% and the corresponding strain rates were estimated for this strain, where a strain of 10% was chosen to avoid any significant grain growth. These results suggest the possibility of a change in the value of n from ~3 with a grain size of 0.67 μm to ~2 with a grain size of 1.5 μm. It is interesting to note that a similar decrease in n with increasing grain size was reported in earlier experiments on samples of 3Y-TZP where the stress was also recorded at a strain of 10% [11].

As noted earlier, the advent of an increase in n at the lower stress levels in 3Y-TZP has been interpreted by invoking the occurrence of a threshold stress [5-7,12,13]. However, detailed analyses indicate that the concept of a threshold stress is not viable because, at least for some of the data, the predicted threshold stresses are negative [4,14] and, unlike metal matrix composites where threshold stresses are well-established [15], there is no evidence in 3Y-TZP for a stress exponent of n → ∞ at the lowest levels of the experimental stresses. It is necessary therefore to critically examine the alternative proposal that the region of n ≃ 2 is associated with grain boundary sliding as in the superplastic flow of metals [16]. Grain boundary sliding cannot occur in a polycrystalline matrix without a concomitant accommodation process in the form of some limited dislocation slip within the grains. The occurrence of limited intragranular slip has been demonstrated in superplastic metal alloys [17,18] but it appears unlikely in 3Y-TZP because of the very high Peierls stress which is estimated to lie in the range of ~260 - 910 MPa at a temperature of 1673 K [9].

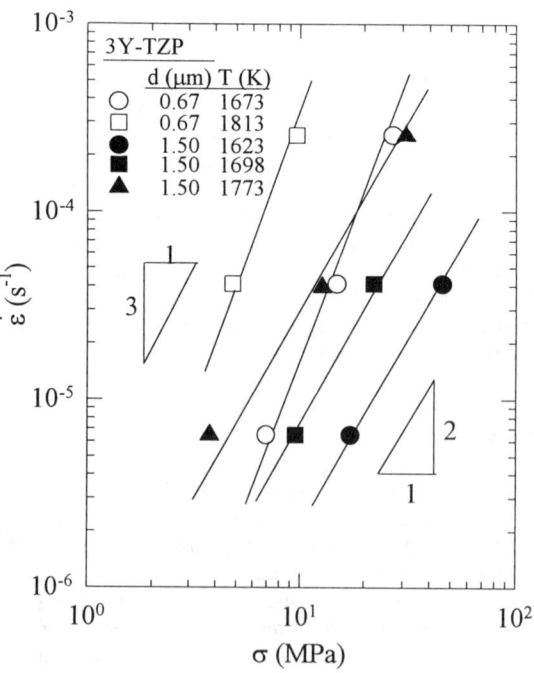

Fig. 1
Strain rate versus stress for samples
of 3Y-TZP tested with two
different grain sizes.

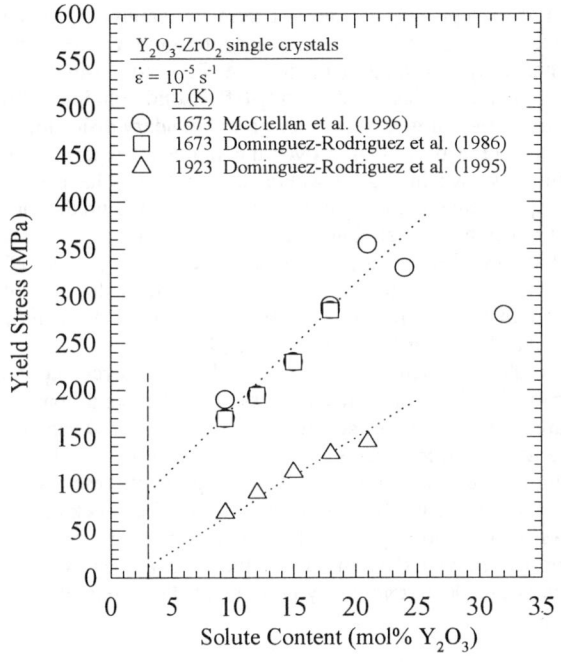

Fig. 2
Yield stress versus yttria content
for yttria-fully-stabilized cubic
zirconia single crystals [19-21]:
dotted lines show extrapolation to
3 mol % Y_2O_3.

The difficulty of initiating intragranular slip during superplasticity in 3Y-TZP is also evident from inspection of Fig. 2 where the measured yield stresses for single crystals of yttria-fully-stabilized cubic zirconia are plotted against the mol % of Y_2O_3 for tests conducted at a strain rate of 10^{-5} s^{-1} [19-21] and the dotted lines show a linear extrapolation to 3 mol % of yttria. Noting that the crystal structure of tetragonal ZrO_2 is very similar to that of cubic ZrO_2 except only for a minor distortion in the c-axis, it appears from Fig. 2 that the yield stress is ~90 MPa at 1673 K. The results in Fig. 2 are also consistent with the report of a yield stress of ~ 150 MPa in a pseudo-cubic ZrO_2 single crystal with 5.7 mol % Y_2O_3 tested at 1673 K with a similar strain rate of 6×10^{-5} s^{-1} [22]. Reference to Fig. 1 shows, however, that a stress of ~90 MPa at 1673 K is substantially higher than the anticipated stress for a strain rate of ~10^{-5} s^{-1}.

As a result of these difficulties, it is appropriate to examine the role of interface-controlled diffusion creep [23]. When Coble diffusion creep is dominant, the strain rate is given by a relationship of the form

$$\dot{\varepsilon} = \frac{AD_{gb}Gb}{kT} \left(\frac{\delta}{b}\right)\left(\frac{b}{d}\right)^3 \left(\frac{\sigma}{G}\right) \left[\frac{N^2}{N^2 + \frac{1}{2}}\right] \qquad (1)$$

where D_{gb} is the appropriate grain boundary diffusion coefficient, G is the shear modulus, b is the Burgers vector, k is Boltzmann's constant, T is the absolute temperature, δ is the effective grain boundary width for diffusion, N is the number of dislocations in a single grain boundary wall and A is a constant having a theoretical value of ~33.4. Equation (1) is the standard relationship for Coble diffusion creep [24] except only for the last term in parentheses which introduces interface-controlled behavior by assuming that vacancies are only emitted and absorbed at grain boundary dislocations [23]. Taking δD_{gb} for the Zr^{4+} ion using data determined experimentally for polycrystalline tetragonal CeO_2-ZrO_2-HfO_2 solid solutions [25] and expressing N as ($\sigma d/2Gb_{gb}$) where b_{gb} is the Burgers vector for grain boundary dislocations ($\sim b/2$), it is possible to estimate the variation of $\dot{\varepsilon}$ with σ over the range of experimental conditions given in Fig. 1. The result is shown in Fig. 3 where the solid lines have been adjusted to give a reasonable fit to the experimental datum points by multiplying the predictions of equation (1) by a factor of 40. It is apparent that, although there is an inconsistency, by a factor of 40, between the datum points and the theoretical predictions, nevertheless the overall trends are consistent with the experimental data. In the earlier comparison with the data of Owen and Chokshi [4], the predicted strain rates were also consistent with the data but they exceeded the measured rates by approximately one order of magnitude [9].

Figure 4 shows experimental results reported by Sato et al. [8] for polycrystalline samples of 3Y-TZP with a grain size of 0.4 μm tested in compression under conditions of constant displacement rate. These data also reveal, as in the earlier experiments of Owen and Chokshi [4], an increase in n from ~ 3 at the lower stresses to ~ 2 at higher stress levels. It is appropriate to consider whether equation (1) provides a reasonable description of these results in addition to the present data and the earlier results of Owen and Chokshi [4]. To check on the validity of this approach, Fig. 5 shows a logarithmic plot of the temperature and grain size compensated strain rate versus the normalized stress incorporating the term associated with interface-control. This plot shows there is excellent consistency between all three sets of data.

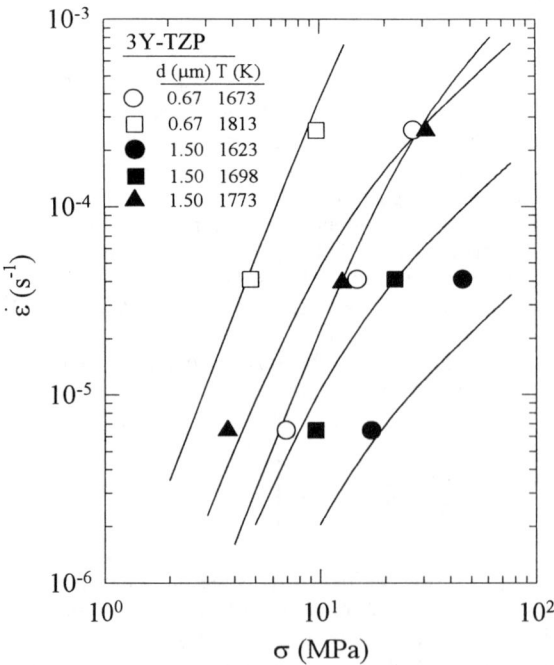

Fig. 3
Strain rate versus stress for
3Y-TZP showing the trends
predicted by equation (1) when A is
adjusted to give a good fit.

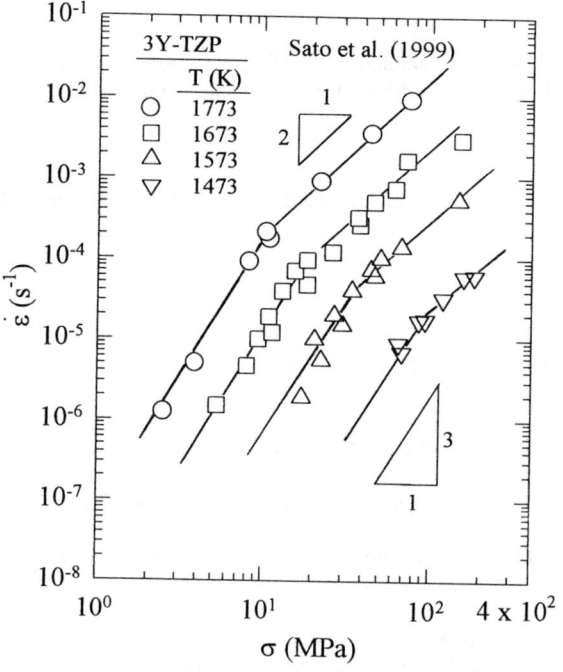

Fig. 4
Strain rate versus stress for
3Y-TZP tested in compression with
a grain size of 0.4 μm [8].

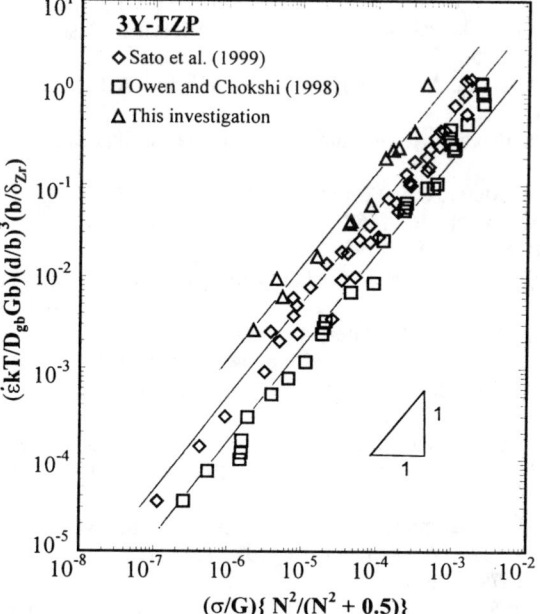

Fig. 5
Normalized plot of strain rate
against stress based on the
predictions of equation (1) for
interface-controlled Coble diffusion
creep.

CONCLUSIONS

The flow of superplastic 3Y-TZP can be explained by interface-controlled Coble diffusion creep. Good agreement is demonstrated for three different sets of experiments.

ACKNOWLEDGMENT

This work was supported by the United States Department of Energy under Grant No. DE-FG03-92ER45472.

REFERENCES

1. F. Wakai, S. Sakaguchi and Y. Matsuno, Adv. Ceram. Mater. **1**, 259 (1986).
2. F. Wakai and H. Kato, Adv. Ceram. Mater. **3**, 71 (1988).
3. A.H. Chokshi, Mater. Sci. Eng. **A166**, 119 (1993).
4. D.M. Owen and A.H. Chokshi, Acta Mater. **46**, 667 (1998).
5. M. Jiménez-Melendo, A. Domínguez-Rodríguez and A. Bravo-León, J. Amer. Ceram. Soc. **81**, 2761 (1998).
6. A. Domínguez-Rodríguez, A. Bravo-León, J.D. Ye and M. Jiménez-Melendo, Mater. Sci. Eng. **A247**, 97 (1998).
7. M. Jiménez-Melendo and A. Domínguez-Rodríguez, Phil. Mag. A **79**, 1591 (1999).
8. E. Sato, H. Morioka, K. Kuribayashi and D. Sundararaman, J. Mater. Sci. **34**, 4511 (1999).
9. M.Z. Berbon and T.G. Langdon, Acta Mater. **47**, 2485 (1999).
10. M. Zhou-Berbon, O.T. Sørensen and T.G. Langdon, Mater. Lett. **27**, 211 (1996).
11. J. Ye, A. Domínguez-Rodríguez, R.E. Medrano and O.A. Ruano in *Third Euro-Ceramics*, edited by P. Durán and J. Fernández (Faenza Editrice Ibérica, Castellón de la Plana, Spain, 1993), vol. 3, p. 525.
12. A. Bravo-León, M. Jiménez-Melendo, A. Domínguez-Rodríguez and A.H. Chokshi, Scripta Mater. **34**, 1155 (1996).
13. A. Bravo-León, M. Jiménez-Melendo and A. Domínguez-Rodríguez, Scripta Mater. **35**, 551 (1996).
14. M.Z. Berbon and T.G. Langdon, Mater. Sci. Forum **243-245**, 357 (1997).
15. Y. Li and T.G. Langdon, Scripta Mater. **36**, 1457 (1997).
16. T.G. Langdon, Acta Metall. Mater. **42**, 2437 (1994).
17. L.K.L. Falk, P.R. Howell, G.L. Dunlop and T.G. Langdon, Acta Metall. **34**, 1203 (1986).
18. R.Z. Valiev and T.G. Langdon, Acta Metall. Mater. **44**, 949 (1993).
19. K.J. McClellan, A.H. Heuer and L.P. Kubin, Acta Mater. **44**, 2651 (1996).
20. A. Domínguez-Rodríguez, K.P.D. Lagerlöf and A.H. Heuer, J. Amer. Ceram. Soc. **69**, 281 (1986).
21. A. Domínguez-Rodríguez, M. Jiménez-Melendo and J. Castaing in *Plastic Deformation of Ceramics*, edited by R.C. Bradt, C.A. Brookes and J.L. Routbort (Plenum, New York, 1995), p. 31.
22. D. Baither, B. Baufeld and U. Messerschmidt, J. Amer. Ceram. Soc. **78**, 1375 (1995).
23. E. Arzt, M.F. Ashby and R.A. Verrall, Acta Metall. **31**, 1977 (1983).
24. R.L. Coble, J. Appl. Phys. **34**, 1679 (1963).
25. Y. Sakka, Y. Oishi, K. Ando and S. Morita, J. Amer. Ceram. Soc. **74**, 2610 (1991).

CREEP BEHAVIOR OF A SUPERPLASTIC Y-TZP/AL₂O₃ COMPOSITE: AN EXAMINATION OF THE POSSIBILITY FOR DIFFUSION CREEP

SIARI S. SOSA, TERENCE G. LANGDON
*Departments of Materials Science and Mechanical Engineering
University of Southern California, Los Angeles, CA 90089-1453 ssosa@usc.edu

ABSTRACT

The creep characteristics of a high purity superplastic Y-TZP/Al₂O₃ composite were studied. The test samples had a uniform initial grain size of essentially equiaxed grains and the phase distribution was also uniform. Creep tests were conducted at elevated temperatures in tension and measurements were taken after testing to determine the average size and shape of the grains. The creep curves showed the existence of a primary stage of creep and then a transition into a steady-state region where the strain rate remained reasonably constant. Observations by scanning electron microscopy revealed an elongation of the alumina grains with the grain strain essentially matching the overall strain within the sample. There was some grain growth after testing at the lower stresses. Preliminary observations using an atomic force microscope revealed direct evidence for the occurrence of some grain boundary sliding within the samples. The results are compared with earlier data reported for Y-TZP/Al₂O₃ composites and they are analyzed using the interface reaction-controlled diffusion creep model.

INTRODUCTION

It is convenient to express the steady-state creep rate of a polycrystalline 3Y-TZP/20% Al₂O₃ composite using the standard relationship for high temperature deformation:

$$\dot{\varepsilon}_s = \frac{A\,D\,G\,b}{k\,T}\left(\frac{b}{d}\right)^p \left(\frac{\sigma}{G}\right)^n \qquad (1)$$

where A is a constant, D is the appropriate diffusion coefficient, G is the shear modulus, **b** is the magnitude of the Burgers vector, k is Boltzmann's constant, T is the absolute temperature, d is the grain size, σ is the flow stress and p and n are the exponents of the inverse grain size and the stress, respectively.

If a material deforms by diffusion creep [1-3], the value of n will be equal to 1 and it can be shown theoretically that a spherical grain will become an ellipsoid of revolution [4]. Therefore, any attempt to determine the significance of diffusion creep requires a careful evaluation of the grain shape, and especially the grain aspect ratio, after deformation. In the material used in this investigation, the possibility of diffusion creep was evaluated by taking separate measurements to determine the shapes and sizes of the alumina and zirconia grains.

EXPERIMENTAL MATERIAL AND PROCEDURES

The samples used in this investigation were prepared from high purity commercial powders of yttria-stabilized zirconia with 20% alumina obtained from the Tosoh Corporation (Nanyo Plant, Shin-nanyo, Yamaguchi-ken, Japan): this material will be designated 3Y20A. The composition of the samples was reported by the manufacturer as 3.8%Y₂O₃, 19.79% Al₂O₃, <0.002% SiO₂, 0.002% Fe₂O₃ and 0.020% Na₂O with the balance as ZrO₂.

The grain size of the samples was taken as the value of the mean linear intercept (L_o) measured in two orthogonal directions (L_1 and L_2) and calculated as

$$\overline{L}_o = (L_1 L_2^2)^{1/3} \qquad (2)$$

Separate values were calculated for the alumina and the 3Y-TZP grains. The resulting values for L_o were 0.41 and 0.45 μm for the alumina and zirconia grains, respectively.

Tensile creep tests were performed using a specially designed cam which maintained essentially a constant stress throughout the test [5]. In practice, there was a slight decrease in the stress with increasing elongation because of the counter-weight used to obtain the initial equilibrium condition. Calculations and experiments showed this decrease in stress was given by the following relationship:

$$\sigma = \sigma_o - \frac{W_o \, r_o \, (\exp \varepsilon - 1)}{A_o \, R_o} \qquad (3)$$

where σ_o is the initial stress, W_o is the counterweight added for initial equilibrium, r_o is the initial length of the loading arm, ε is the strain, A_o is the initial cross-sectional area of the sample and R_o is the length of the arm supporting the sample. A correction was made for this decrease in stress by periodically adding small weights to the initial load. A linear displacement transducer was used to continuously record the deformation of the sample. After heating the sample to the testing temperature of 1723 K, the selected load was applied smoothly using a hydraulic jack.

The surfaces of samples were examined after testing using a scanning electron microscope (SEM). A software package, Image Pro Plus™, was used to record the sizes and shapes of the individual grains in every image visible in the SEM. The calibration was carefully set for every area measured and grains in the border regions were excluded from the measurements. The grain aspect ratio (GAR) was determined for each grain, defined as the ratio of the projections of the grain in the horizontal (or tensile) axis and the vertical axis.

Some samples from the gauge sections were examined after testing using a Nanoscope-III atomic force microscope (AFM). The AFM was connected through a monitor to a computer equipped with software designed to manipulate the AFM data and images. This instrument was calibrated before each set of observations and surfaces were examined in the contact mode. The AFM images were modified for clarity using the "Plane Fit Auto" feature of the software.

EXPERIMENTAL RESULTS

Figure 1 shows plots of the instantaneous strain rate against the measured creep strain for several samples tested at a temperature of 1723 K using a range of constant stresses from 12 to 53 MPa.

All of these plots show a reasonable steady-state region although it is observed that the sample tested at the highest stress (53 MPa) shows no primary region and an immediate steady-state flow and some of the other samples exhibit creep rates which tend to decrease with increasing strain. This decrease may be due to an elongation of the grains through diffusion creep since it is known that a grain deforming by diffusion creep will elongate with increasing strain and there will be a continuous decrease in the creep rate after large strains [4]. The sample tested at the lowest stress shows a primary creep which may be associated with grain growth. This sample was tested at 12 MPa and the final grain sizes, determined after termination of the creep test, were measured as 0.75 and 1.12 μm for the alumina and the zirconia grains, respectively.

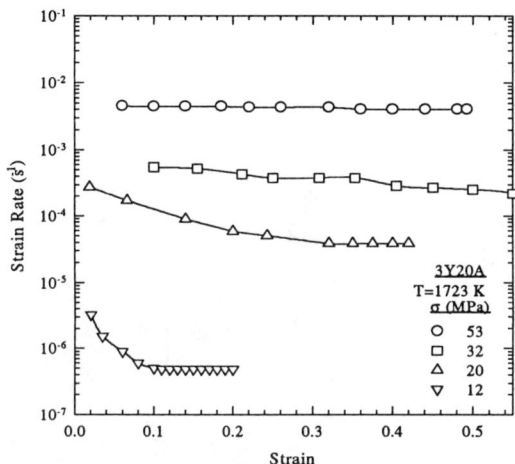

Fig. 1 Strain rate versus strain for 3Y20A samples tested at 1723 K.

An examination of the samples by SEM after creep testing showed that there was a tendency for the grains to become elongated along the tensile axis and this elongation was evident even at high stresses where there was little or no grain growth. Detailed measurements were taken to determine the grain aspect ratio for representative samples and typical results are shown in Fig. 2 for the alumina grains after testing at three different stress levels. For the sample untested and shown at zero on the stress axis, it is evident that the GAR for each individual grain clusters around an average value which is of the order of ~1.12. After tensile testing at 32 and 53 MPa, however, the distributions extend to significantly higher values for the GAR and the measured average values were ~1.60 and ~1.51, respectively, where the higher GAR for the sample tested at 32 MPa probably reflects the higher strain in this sample. Essentially similar distributions were also obtained for the zirconia grains, with an identical value of ~1.12 in the untested sample and values of ~1.45 and ~1.28 after testing at 32 and 53 MPa, respectively. The measurements of the GAR show that the alumina grains become more elongated than the zirconia grains during creep of the 3Y20A samples. Careful inspection confirmed these trends and showed also that some of the elongated alumina grains contained slip traces which suggested there was at least some deformation through the intragranular movement of dislocations. Figure 3 shows an example of an exceptionally elongated alumina grain in the sample tested at 53 MPa: the tensile axis in the photomicrograph is horizontal. Very careful inspection of the elongated grain in Fig. 3 reveals some slip traces which lie at approximately 45° to the stress axis.

The advent of superplasticity strictly requires that the grains move over each other in the process of grain boundary sliding [6]. However, recent analyses have suggested that the rate-controlling mechanism in the creep of Y-TZP ceramics may be associated with diffusion creep [7]. In practice, diffusion creep also leads to a relative displacement of adjacent grains where the displacement occurs as an accommodation process. These two types of grain displacement are generally termed Rachinger sliding [8] and Lifshitz sliding [9], respectively [10]. Grain boundary displacements may be revealed using an AFM [11] and an example is shown in Fig. 4 where two adjacent grains are displaced vertically by ~42 nm in the sample tested at 53 MPa.

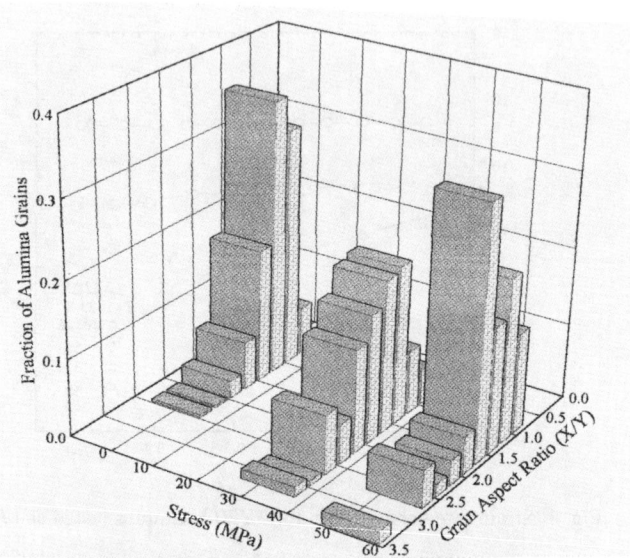

Fig 2. Variation of grain aspect ratio with stress for 3Y20A samples tested at 1723 K.

Fig. 3. SEM micrograph of a 3Y20A sample tested at 53 MPa and 1723 K.
The tensile axis is horizontal.

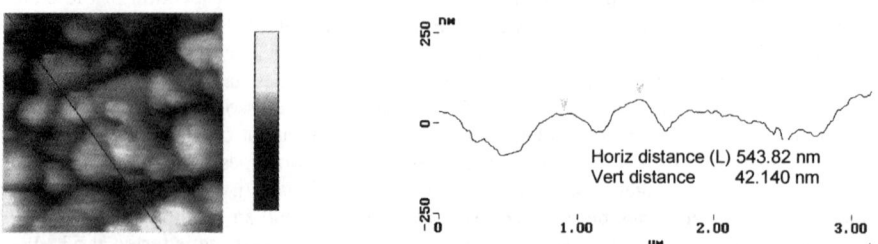

Fig. 4 . AFM 2-D Image of a 3Y20A sample tested at 1723 K and 53 MPa.

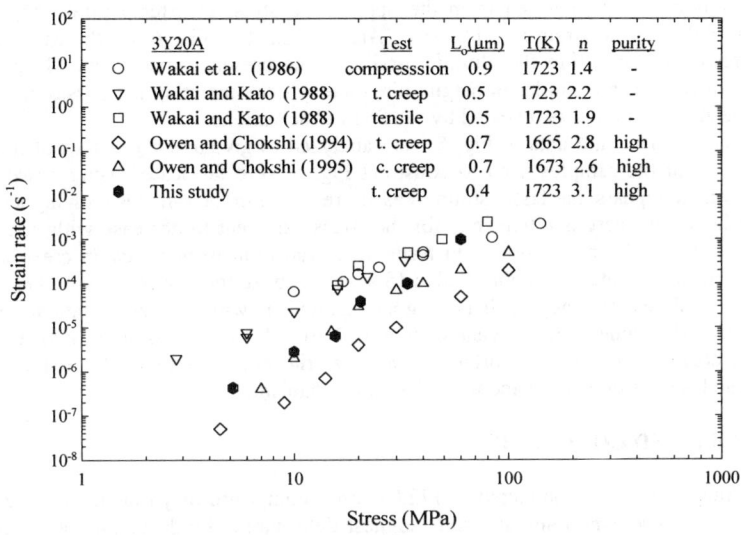

Fig. 5 Strain rate versus stress for samples of 3Y20A tested under different conditions [12-15].

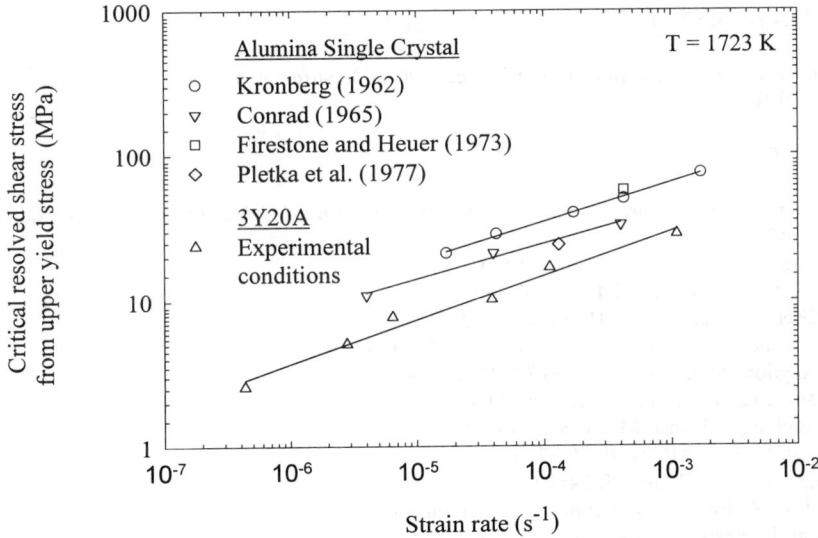

Fig. 6 Critical resolved shear stress versus strain rate for alumina [16-19].

DISCUSSION

The experimental creep date are plotted in Fig. 5 together with results reported by other investigators testing under different conditions [12-15]. In order to check on the possibility of intragranular slip within the alumina grains, Fig. 6 shows the critical resolved shear stress for alumina single crystals, derived from the upper yield stress at a temperature of 1723 K, as a function of the testing strain rate [16-19]. Also included in Fig. 6 are the estimated resolved shear stresses acting within the alumina grains in the present experiments. Inspection shows both sets of stresses are similar in magnitude, thereby demonstrating the possibility of some slip within the alumina grains as revealed by the SEM observations.

The experimental data in Fig. 5 were analyzed to give average values of n as indicated. Although all of the samples used to construct Fig. 5 were fabricated using powders from the same source, it appears that faster strain rates are recorded in the samples with higher purities. In practice, however, there is a tendency for the stress exponent to decrease with increasing stress and this trend has been interpreted in terms of a transition from diffusion creep at the higher stress levels to an interface-controlled diffusion creep at the lower stresses [7]. A detailed analysis has shown this approach is in good agreement with published data for the 3Y-TZP ceramic and the same trend is visible in Fig. 5 for 3Y20A. It is suggested, therefore, that diffusion creep is probably important in this material but with some slip within the alumina grains which serves only as an accommodation mechanism.

SUMMARY AND CONCLUSIONS

Creep tests were conducted at 1723 K on a composite of yttria-stabilized zirconia with 20% alumina. The experimental results suggest deformation by diffusion creep but with some intragranular slip within the alumina grains.

ACKNOWLEDGMENTS

This work was supported by the United States Department of Energy under Grant No. DE-FG03-92ER45472.

REFERENCES

1. F.R.N Nabarro in *Report of a Conference on Strength of Solids* (The Physical Society, London, 1948), p. 75.
2. C. Herring, J. Appl. Phys. **21**, 437 (1950).
3. R.L. Coble, J. Appl. Phys. **34**, 1679 (1963).
4. H.W. Green, J. Appl. Phys. **41**, 3899 (1970).
5. P. Yavari and T.G. Langdon, J. Testing Eval. **10**, 174 (1982).
6. T.G. Langdon, Mater. Sci. Eng. **A174**, 225 (1994).
7. M.Z. Berbon and T.G. Langdon, Acta Mater. **47**, 2485 (1999).
8. W.A. Rachinger, J. Inst. Metals **81**, 33 (1952-53).
9. I.M. Lifshitz, Soviet Phys. JETP **17**, 909 (1963).
10. W.R. Cannon, Phil. Mag. **25**, 1489 (1972).
11. L. Clarisse, A. Bataille, Y. Pennec, J. Crampon and R. Duclos, Ceram. Intl. **25**, 389 (1999).
12. F. Wakai, H. Kato, S. Sakaguchi and N. Murayama, J. Ceram. Soc. Japan **94**, 1017 (1986).
13. F. Wakai and H. Kato, Adv. Ceram. Mater. **3**, 71 (1988).
14. D.M. Owen and A.H. Chokshi, J. Mater. Sci. **29**, 5497 (1994).
15. D.M. Owen and A.H. Chokshi, Acta Mater. **46**, 667 (1998).
16. M.L. Kronberg, J. Amer. Ceram. Soc. **45**, 274 (1962).
17. H. Conrad, G. Stone and J. Janowski, Trans. AIME **233**, 889 (1965).
18. R.F Firestone and A.H. Heuer, J. Amer. Ceram. Soc. **56**, 136 (1973).
19. B.J. Pletka, A.H. Heuer and T.E. Mitchell, Acta Mater. **25**, 25 (1977).

SOME ASPECTS OF PLASTIC FLOW IN SILICON NITRIDE

T. ROUXEL , J. RABIER*, S. TESTU** and X. MILHET*
Laboratoire "Verres et Céramiques", UMR-CNRS 6512, Université de Rennes 1, Campus Beaulieu, 35042 Rennes cedex, France.
*Laboratoire de métallurgie physique, URA-CNRS 131, Université de Poitiers, Bd. 3 Téléport 2 BP 179, 86960 Futuroscope cedex, France.
** ENSCI, SPCTS, UMR-CNRS 6638, 47 Av. A. Thomas, 87065 Limoges cedex, France.

ABSTRACT

The different scales of plastic flow in silicon nitride were investigated either by indentation experiments and compression under hydrostatic pressure in the 20-850°C temperature range, and by stress relaxation and creep above 1350°C. [0001], 1/3<11-20> and 1/3<11-23> dislocations were evidenced by Transmission Electron Microscopy (TEM) in the low temperature range. Cross-slip events in {10-10} prismatic planes were observed at temperature as low as 20°C by Atomic Force Microscopy (AFM) on micro-hardness indents. By increasing the temperature, the deviation plane becomes {11-20} prismatic planes. The easiest slip system is by far the [0001]{10-10} system. Above 1350°C, the creep strain could be fitted by the sum of a transient component, $\varepsilon_t = \varepsilon^{\infty}[1-\exp-(t/\tau_c)^{b_c}]$, where τ_c reflects the duration of the transient creep stage, and b_c is between 0 and 1, and a stationary component, $\varepsilon_s = \dot{\varepsilon}_s t = A\sigma^n t$, where σ is the stress and n is the stress exponent. The increase of ε^{∞} with temperature is interpreted on the basis of the formation of liquid intergranulary phases above 1400°C by progressive melting of some of the grains. A creep exponent of 1.8 was determined. A single value could hardly be given to the activation energy since an S-shape curve was observed in the $\ln \dot{\varepsilon}_s$ versus 1/T plot, as for most glasses over large temperature ranges. The stress relaxation kinetics was found to follow the Kohlrausch-Williams-Watt expression: $\sigma/\sigma_0 = \exp[-(t/\tau_r)^{b_r}]$, where b_r ranges between 0 (solid state) and 1 (liquid state) and τ_r is a characteristic relaxation time constant. As in the case of glasses, τ_r decreases rapidly whereas b_r increases from about 0.2 to 0.7 as the temperature increases from 1400 to 1650°C. But again, it is very difficult to get a single value for the activation energy from the $\ln \tau_r$ versus 1/T plot.

INTRODUCTION

Silicon nitride-based polycrystalline ceramics have much evolved during the past thirty years. Whereas materials in the 70' were usually sintered with more than 10 vol.% additives and were developed for application at temperatures below 1250°C, the most advanced materials are now sintered with less than 3 vol.% - or no - sintering additives and exhibit impressive creep resistance up to 1350°C. This last generation of silicon nitride ceramics results from the continuous and intensive research efforts that were focussed on processing methods and on microstructure control. Nowadays, reliable creep resistant, tough, and resistant ceramics are available. However, the high temperature deformation mechanism is still not clearly understood. But it can be anticipated that besides mechanisms proposed in the 70' based on grain boundary sliding through viscous flow and solution-migration-precipitation processes, with the development of materials containing low amounts of

Mat. Res. Soc. Symp. Proc. Vol. 601 © 2000 Materials Research Society

intergranulary phases, the possibility for dislocation motion should also be considered. In other words, ceramic scientists are facing a situation that places them closer to metallurgists, and the exceptional intrinsic properties of covalent crystals such as Si_3N_4 should be of greater importance on the microscopic properties of polycrystalline ceramics.

DISLOCATION ACTIVITY

Quite a few study of dislocation activity in silicon nitride were reported so far. What seems clear is that even in highly deformed specimens, a relatively low dislocation density is observed. In β-Si_3N_4 crystals, most dislocations move in prismatic planes, and the [0001]{10-10} slip system is by far the most common. Dislocations with 1/3<11-20> and 1/3<10-10> Burgers vectors were also observed, but again, according to the values for the Peierls potential, slip occurs in 1st and in 2nd kinds prismatic planes [1,2].

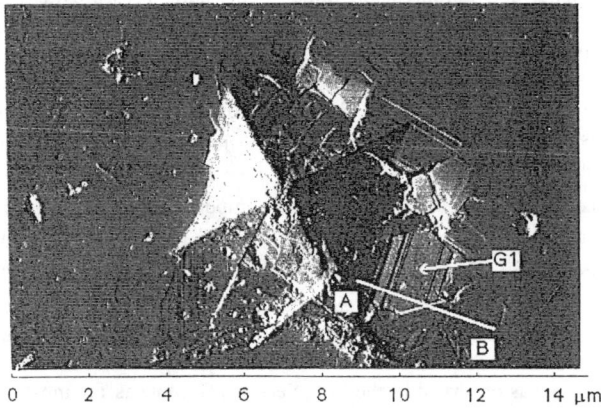

Figure 1 - Scanning electron micrograph of a Vickers indent performed under 50 g at room temperature on a mirror-polished surface of a polycrystalline silicon nitride with a coarse microstructure. Prismatic glide is observed at the surface of grain G1. An atomic force microscopy analysis along line AB reveals 30 nm high steps, along the c-axis, on the surface of the grain (from [3]).

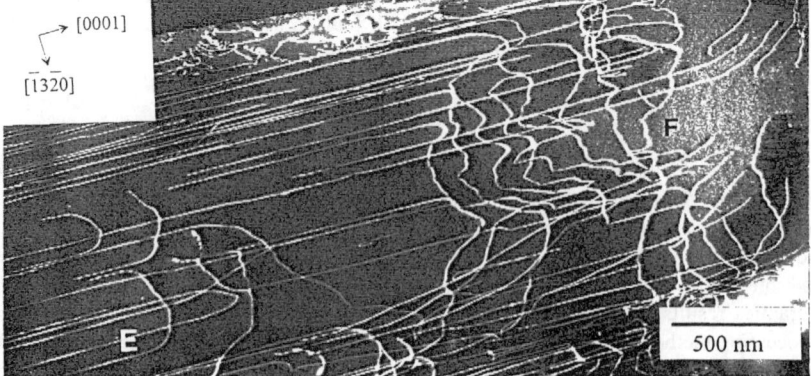

Figure 2 - Transmission electron microscopy image of an array of dislocations (weak beam, g=0002, B=[5-1-40]) observed in a specimen indented at room temperature. Note that almost all dislocations have a [0001] Burgers vector (from [3]).

118

Deformation experiments by high temperature micro-hardness or compression under hydrostatic pressure, show that, between 20 and 850°C, all the dislocation sub-structures resulting from plastic deformation are mainly built with screw dislocations with [0001] Burgers vector (Figs.1,2).

This type of dislocation activity should prevail up to 1550°C. It was recently proposed that the occurrence of ductility in compression for temperature higher than 1550°C, for imposed strain rates lower than $8.10^{-6}s^{-1}$ and $6.10^{-5}s^{-1}$ at 1550 and 1750°C respectively, correlates with the appearance of <-2113>{11-21} pyramidal slip [4]. Yield stresses between 50 and 250 MPa were measured at 1700°C by changing the strain-rate between $9.10^{-6}s^{-1}$ and $6.10^{-5}s^{-1}$. These stresses compare well with those applied in compression in the same temperature range to study superplasticity. Furthermore, a rough estimation from the general expression for the Peierls-Nabarro stress, $\tau_{P.N}$,

$$\tau_{P.N.} = \frac{2\mu}{1-\nu} \exp\left(-\frac{2\pi h}{(1-\nu)b}\right) \qquad (1)$$

where μ, ν, h and b are the shear modulus, Poisson's ratio, the spacing between the slip planes and the length of the Burgers vector respectively, also supports the possibility for dislocation motion during high temperature deformation in silicon nitride-based ceramics.

Considering the [0001]{10-10} slip system, assuming for μ a value of 105 GPa, as determined from the Young's modulus value at 1600°C (extrapolation from ref. [5] with $\nu=0.3$) and using h=2.91 Å and b=3.29 Å gives $\tau_{P.N.}=107$ MPa.

STRESS RELAXATION

Typical stress relaxation curves obtained in a material sintered without sintering additives, but using pre-alloyed powders [6], are plotted in Fig.3. With regards to the experimental duration, temperatures above 1350°C are required for relaxation to be observed.

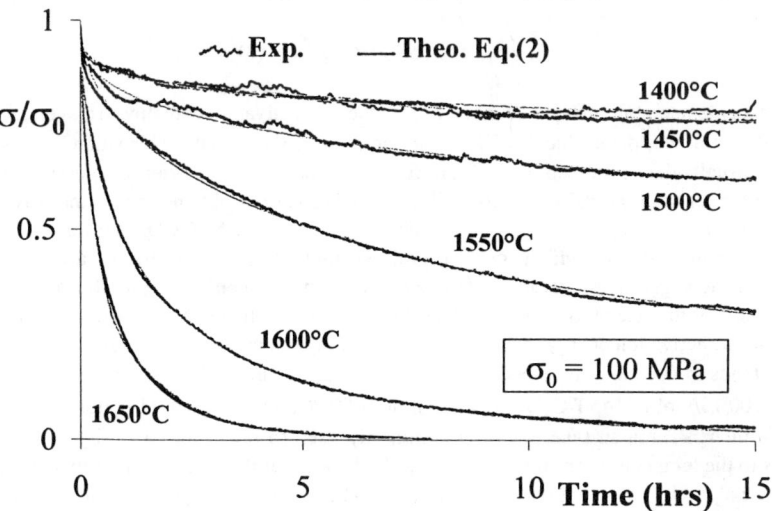

Figure 3- Stress relaxation curves obtained in compression, under nitrogen atmosphere.

The relaxation kinetics could be remarkably well modelled by the so-called Kohlrausch-Williams-Watt (KWW) equation [6] often quoted in glass science:

$$\sigma(t)/\sigma_o = \exp\left[-(t/\tau_r)^{b_r}\right] \qquad (2)$$

where τ_r is the characteristic relaxation time and, following previous works [7,8], parameter b_r is named correlation factor and can be tentatively related to the mobility of a structural unit at the atomic or molecular scale. b_r ranges between 0 (strong intermolecular coupling allowing for pure elastic deformation only) and 1 (weak correlation corresponding to the simple Debye - or Maxwell - relaxation of liquids). Accordingly, the increase of b_r with temperature, from 0.23 at 1400°C to 0.67 at 1650°C, would reflect a smooth transition between the elastic behavior in the solid state and the viscous behavior in the liquid state.

In silicon nitride-based polycrystalline ceramics, intergranular glassy films and pockets are supposed to play a major role in this transition. However, the presence of such films and pockets were not evidenced at room temperature in this material. It is therefore anticipated that the liquid phase that is necessary to promote sintering, crystallizes upon cooling, but partially reforms upon heating above 1150°C, as is evidenced during creep experiments (see next section), and that a significant amount of eutectic liquid may form during testing above 1400°C. Consistently, τ_r decreases rapidly with rising temperature, from thousands of hours at 1400°C to few minutes at 1650°C and a rapid softening is observed above 1150°C (Fig.4). Significant ductility does not show up until the relaxation time (τ_r) becomes smaller than the experimental duration at the usual experimental time scale for plastic shaping techniques, from few seconds to few hours. For the presently studied material with improved creep resistance, temperatures above 1750°C would be required (1550°C for fine-grained superplastic SiYAlON) [9,10].

Following the theory of thermally activated deformation, τ_r can be regarded as the waiting time of the activation process and depends on T according to:

$$\tau_r = \tau_{ro} \exp\left(\Delta G_a/(RT)\right) \qquad (3)$$

where τ_{ro} is temperature independent and ΔG_a is the free activation enthalpy. An average value of 1070 kJ/mol was determined by linear interpolation from the data. This value is consistent with the usual values for creep of silicon nitride or viscous flow of oxynitride glasses [11], but is much higher than the values of 406 kJ/mol and 470 kJ/mol obtained the same way in the same temperature range in superplastic ceramics, namely a SiYAlON and a Si_3N_4/SiC nanocomposite [10], for which the fine-grained microstructures allow for a much faster relaxation at a given temperature. However, as can be seen in Fig.4, the temperature dependence of the relaxation time constant does not follow the Gibb's equation with a single value for the activation energy over the entire range. In fact, as for most glasses, the activation energy tends to decrease on both sides of the temperature range. For instance, ΔG_a decreases from 1100 kJ/mol at 1450°C to 465 kJ/mol at 1650 °C. Thus, τ_r would better agree with a Vogel-Fulcher-Tamman type equation commonly used to relate the viscosity of inorganic glasses to the temperature from and above Tg. In the light of these results, it seems difficult to anticipate a physical meaning for the apparent activation energy averaged over the 1400-1650°C range. The high temperature value is close to 579 kJ/mol, corresponding to Si diffusion in SiO_2, and close to the value of 500 kJ/mol associated with creep anelasticity in silicon nitride and ascribed to the relaxation of local elastic stress fields at grain boundary asperities, through the redistribution of the viscous phase (SiO$_2$ films?) [6].

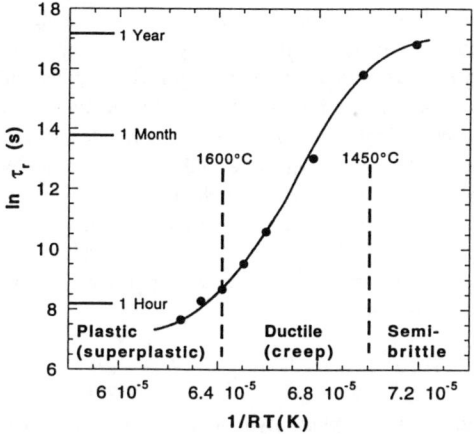

Figure 4- Temperature dependence of the relaxation time constant, τ_r.

CREEP

The creep behavior of the previously described glassy-phase free material, was characterized in compression under a stress of 100 MPa, in nitrogen atmosphere, for temperatures ranging between 1400 and 1650°C and stresses between 50 and 200 MPa. This unusual temperature range - creep experiments are usually conducted below 1400°C - was selected in order to reach the stationary creep regime more rapidly and with the aim to magnify the phenomena occurring during flow. A typical creep curve obtained at 1500°C under 100 MPa is shown in Fig.5 (from [5]).

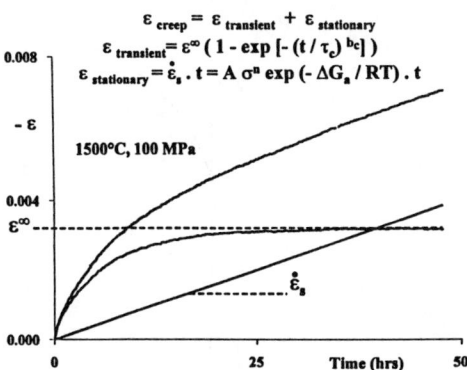

Figure 5- Typical creep curve obtained in compression under 100 MPa at 1500°C, in nitrogen atmosphere.

The creep strain can be visualized as the sum of a transient creep strain, which takes a form similar to that of the KWW relation for relaxation, and a stationary creep strain where flow appears to be non-Newtonian with n=1.8. As during relaxation, τ_c decreases and b_c increases with rising temperature. However, it is noteworthy that τ_c decreases with rising stress and that the asymptotical strain, ε^∞, increases with temperature (from 0.0009 at 1400°C to 0.045 at 1600°C). In silicon nitride, the primary creep could be explained by a redistribution of the glassy phase consequently to the bringing together of neighboring grains. Suppose now that some glassy phase forms at elevated temperature, the former observation, which were quite unexpected, strongly suggest that i) τ_c correlates with the time necessary for grains to get closer each to the other, which depends both on the temperature, through the grain boundary phase viscosity, and on the stress, through the intergranulary phase thickness, and that ii) the amount of glassy phase increases with increasing temperature above 1400°C. Therefore, it turns out that starting from a material with relatively clean grain boundaries to get insight into the deformation mechanisms, ends with a material containing some liquid phase during the deformation process!

SUPERPLASTICITY

Since the first report of superplasticity in a Si_3N_4/SiC nanocomposite, in the early 90' [12], large tensile elongation, over 100%, were obtained in different silicon nitride-based ceramics above 1500°C. Cavitation damage resistant materials lead to flow curves showing a continuous hardening which is mainly due to the alignment of the acicular β-grains toward the tensile axis, and to a lesser extent to the crystallization of secondary phases. In the low ductility range, or for materials exhibiting cavitation, the shape of the curve is rather complex and shows a peak at the end of the primary loading stage, with a subsequent decrease of the stress presumably due to the growth of microcracks induced by cavity coalescence, and an hardening stage leading to ultimate fracture. So far, and in all cases, with regard to our poor state of understanding, the hope for finding a mechanism accounting for the extremely complex nature of the superplastic flow in such materials seems premature.

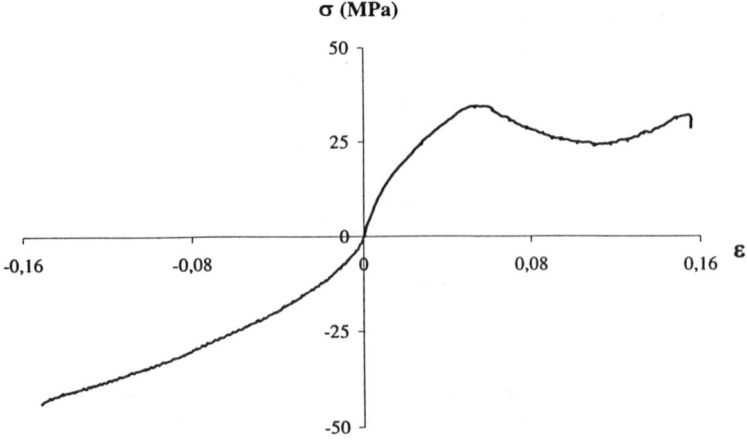

Figure 6- Tension-compression asymmetry - Experiments conducted at 1650°C with a strain rate of $10^{-5}s^{-1}$.

Of interest is the tension-compression asymmetry (Fig.6). We believe that flow in compression is contingent to the problem of packing of a mixture of equiaxed and rodlike particles embedded in a viscous matrix, and data in this area are clearly needed. Also, due to the relatively high flow stresses that are appear in compression, the possibility for dislocation motion, as an accommodating process, should also be considered. The tensile flow has many features in common with the rheology of concentrated suspension. In this latter area, shear thinning due to the flow destroying aggregates and also to aligning the microstructure with the flow is a well-known phenomenon.

REFERENCES

1. X. Milhet, H. Garem, J-L. Demenet, J. Rabier and T. Rouxel, J. Mat. Sci. **31**, p. 3733-3738 (1997).

2. X. Milhet, J-L. Demenet and J. Rabier, Phil. Mag. Lett. **79**, p. 19 (1999).

3. X. Milhet, Doctorate thesis, University of Poitiers (France), January 1999.

4. K. Kawahara, S. Tsurekawa and H. Nakashima, in *Key Engineering Materials* Vols. **171-174**, p. 825-832 (2000).

5. S. Testu, Doctorate thesis, Limoges (France), November 1999.

6. G. Bernard-Granger, J. Crampon, R. Duclos and B. Cales, J. Europ. Ceram. Soc. **17**, 1647-1654 (1997).

7. R. Kohlrausch, Ann. Phys. (Leipzig) **12**, p. 393 (1887). See also: "Fundamentals of inorganic glasses", Ed. A.K. Varshneya, Pub. Academic Press Inc. (Boston, San-Diego, New-York, London, 1994).

8. J. Perez, J.Y. Cavaillé, S. Etienne and C. Jourdan, "Physical interpretation of the rheological behavior of amorphous polymers through the glass transition", Revue Phys. Appl. **23**, 125-135 (1988).

9. T. Rouxel and F. Wakai, Acta metall. mater. **41**, 11, p. 3203-3213 (1993).

10. T. Rouxel, F. Rossignol, J-L. Besson and P. Goursat, J. Mater. Res. **12**, 2, p.480-492 (1997).

11. H. Lemercier., T. Rouxel., D. Fargeot, J-L. Besson and B. Piriou, J. Non-Cryst. Sol. **201**, 128-145 (1996).

12. F. Wakai, Y. Kodama, S. Sakaguchi, N. Murayama, K. Izaki and K. Niihara, Lett. to Nature **344**, p. 421 (1990).

Grain Boundary Structure and Sliding of Alumina Bicrystals

Y.Ikuhara, T.Watanabe, T.Yamamoto, T.Saito*, H.Yoshida** and T.Sakuma**

Engineering Research Institute, School of Engineering, The University of Tokyo, 2-11-16, Yayoi, Bunkyo-ku, Tokyo 113-8656, Japan, ikuhara@ceramic.mm.t.u-tokyo.ac.jp
*Japan Fine Ceramics Center, 2-4-1, Mustuno, Atsuta-ku, Nagoya Sakyo-ku, Kyoto 606-01, Japan.
**Department of Advanced Materials Science, Graduate School of Frontier Sciences, The University of Tokyo, 7-3-1, Hongo, Bunkyo-ku, Tokyo 113-8656, Japan

ABSTRACT

Alumina bicrystals were fabricated by a hot joining technique at 1500°C in air to obtain ten kinds of [0001] symmetric tilt grain boundaries which included small angle, CSL and high angle grain boundaries. Their grain boundary structures were investigated by high-resolution electron microscopy (HREM), and the respective grain boundary energies were systematically measured by a thermal grooving technique. It was found that grain boundary energy strongly depended on the grain boundary characters, e.g., there were large energy cusps at low Σ CSL grain boundaries. But, main part of grain boundary energy is likely to be due to the strain energy around the grain boundary, and the contribution of atomic configuration is not so large. Small angle grain boundaries were consisted of an array of partial dislocation with Burgers vector of $1/3[1\bar{1}00]$ to form the stacking faults between the dislocations. The behavior of grain boundary sliding was also investigated for typical grain boundaries by high-temperature creep test at 1400°C. As the result, the occurrence of grain boundary sliding was found to depend on the grain boundary atomic structure.

INTRODUCTION

High-temperature mechanical properties of structural ceramics strongly depend on their grain boundary structures. For example, superplasticity in fine-grained ceramics is mainly controlled by grain boundary sliding which is influenced by the grain boundary atomic structure [1,2]. On the other hand, grain boundary structure in ceramic materials is expected to be sensitive to the grain boundary characters, which has commonly been observed in metals [3]. Therefore an investigating the relationship between grain boundary structure and its characters is important to understand the effect of grain boundary structure on the mechanical properties in ceramics. For this purpose, bicrystal-experiments have an advantage to be performed because grain boundary characters can be controlled and the fabricated boundaries are easily treated for the subsequent characterization.

In this study, alumina bicrystals were systematically fabricated, and the respective grain boundary structures were characterized by high-resolution electron microscopy (HREM). In addition, grain boundary energies were measured by a thermal grooving technique, and the behavior of grain boundary sliding was investigated by the compression creep test. The relationships of grain boundary atomic structure, grain boundary energy and grain boundary sliding were discussed in detail from the view point of grain boundary characters.

Mat. Res. Soc. Symp. Proc. Vol. 601 © 2000 Materials Research Society

EXPERIMENTAL PROCEDURE

Single crystals of alumina (sapphire) with high purity of 99.99% were used to obtain the bicrystals of the [0001] symmetrical tilt grain boundaries as shown in Fig. 1. The tilt angles 2θ, the corresponding Σ values and grain boundary planes are shown in Table I. The single crystals were cut by a diamond saw with the size of $10 \times 10 \times 5 \text{mm}^3$ so as to have the surface corresponding to the grain boundary plane in the respective bicrystals. The surfaces of the blocks were then mechano-chemically polished using a colloidal silica to a mirror state, and subsequently joined together so as to keep the desirable orientations at 1500°C for 10h in an air to obtain the bicrystals with the size about $10 \times 10 \times 10 \text{mm}^3$. These bicrystals were cut to the several pieces for the subsequent characterization.

Grain boundary structures were observed by high-resolution electron microscopy (HREM). The specimens for the observation were prepared using a standard technique involving mechanical grinding to a thickness of 0.1 mm, dimpling to a thickness of about 20μm and ion beam milling to electron transparency at about 4kV. HREM observations were performed using a Topcon EM-002B transmission electron microscope operating at 200kV with a point to point resolution of 0.18nm.

Grain boundary energies were measured by the thermal grooving technique. The thermal grooving was conducted by annealing the specimens at 1400°C for 2h, and the grooved angles (dihedral angles ϕ) were precisely measured by using an atomic force microscope (AFM).

Grain boundary sliding behavior was investigated by compression creep test under the constant stress of 16MPa at 1400°C in an air. The specimens for creep test were cut into the size of $5 \times 5 \times 7 \text{mm}^3$, and the compression stress was applied to the surface of $5 \times 5 \text{mm}^2$. In the specimen, grain boundary planes were set to incline by 45° with respect to the compression axis so that maximum shear stress can be applied to the grain boundary planes.

Table I   Tilt angles, Σ values and grain boundary planes for the bicrystals fabricated in this study.

Tilt angle 2θ	Σ value	Grain boundary plane
2.0	1	$(1\bar{1}00)(\bar{1}100)$
10.0		$\sim(\bar{5}610)(\bar{6}510)$
21.8	21	$(4\bar{5}10)(\bar{5}410)$
27.8	13	$(3\bar{4}10)(\bar{4}310)$
32.0		$\sim(5\bar{7}20)(\bar{7}520)$
38.2	7	$(2\bar{3}10)(\bar{3}210)$
42.0		$\sim(5\bar{8}30)(\bar{8}530)$
46.8	19	$(3\bar{5}20)(\bar{5}320)$
60.0	3	$(1\bar{2}10)(\bar{2}110)$

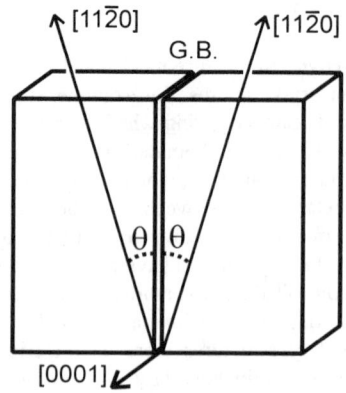

Fig.1 A schematic of a bicrystal with the [0001] symmetrical tilt grain boundary.

RESULTS AND DISCUSSION

In order to estimate the grain boundary energy in the bicrystals, the following equation is generally applied,

$$\gamma_{gb}/\gamma_s = 2\cos(\phi/2) \qquad (1)$$

where γ_{gb}, γ_s and ϕ correspond to grain boundary energy, surface energy and dihedral angle at a grain boundary, respectively. Fig. 2 shows an example of the thermally grooved profiles for the 46.8°(Σ19) grain boundary observed by AFM. Since ϕ can be thus estimated from the grooving, the value of γ_s is needed to calculate the grain boundary energy γ_{gb}. According to Nikolopoulos[4], surface energy of alumina is expressed as a function of temperature T as the following.

$$\gamma_s = 2.559 - 0.784 \times 10^{-3} T \ (J/m^2) \qquad (2)$$

Taking these equations into account, grain boundary energy can be plotted as a function of misorientation angles 2θ as shown in Fig. 3. There are two large energy cusps at the angles corresponding to Σ7 and Σ3. These two CSL grain boundaries have an extremely low energy less than 0.05J/m². From the misorientation angles of 0° to 32°, the grain boundary energy increases with increasing misorientation angle, although Σ21 and Σ13 CSL boundaries exist in this region. The highest grain boundary energy among the examined specimens is about 0.8J/m² for the 42° high angle grain boundary. It is concluded that grain boundary energy strongly depends on the grain boundary characters. It has been reported that the grain boundary energies of the [1$\bar{1}$00] tilt boundaries were about 0.5-0.6J/m² irrespective of misorientation angles [5]. These values are roughly coincided with the average values in this study, however, no CSL grain boundary was

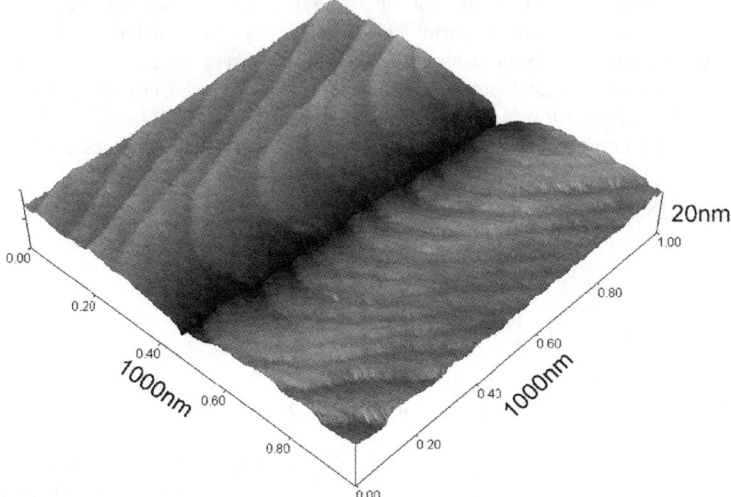

Fig. 2   Profile of the thermarlly grooved grain boundary observed by AFM (46.8°(Σ19) grain boundary).

included in their specimens. Recently, the grain boundary energy for polycrystalline alumina was measured by thermal grooving- AFM technique, and the energy was reported to be 0.85J/m^2 [6]. This value is considered to reflect the average energy, and is approximately equal to the highest energy obtained in this study. This would be because our bicrystals possess symmetrical tilt grain boundaries which has relatively low energy, comparing to random grain boundaries.

Fig. 4 shows high resolution electron micrographs of the grain boundaries in the bicrystals with the misorientation angle of (a) 60° (Σ3), (b) 32° and (c) 46.8° (Σ19). The observed direction for the all boundaries is parallel to the [0001] for both of adjacent grains. The set of {11$\bar{2}$0} planes can be seen in each crystal. As seen in Fig. 4(a), Σ3 boundary is completely coherent boundary. It is obvious that high coherency is the reason why such low Σ CSL boundaries have extremely low grain boundary energy. Grain boundary energy is similar between 32° grain boundary and 46.8° (Σ19) grain boundary as shown in Fig. 3, although the atomic structure is quite different between them as shown in Fig. 4(b) and (c). In the Σ19 boundary, two adjacent crystals are directly contacted without any second phase, and there is some periodic structure which reflects an array of structure unit [7]. On the other hand, grain boundary structure is relaxed to form an amorphous-like film along the grain boundary in 32° grain boundary. In this case, high index planes near (5$\bar{7}$20) and (7$\bar{5}$20) are contacted each other at the boundary, resulting in the formation of high energy boundary. The amorphous-like structure is confirmed to be consisted of aluminum and oxygen atoms by nano-probe TEM-EDS analysis, and considered to be formed by reducing the high grain boundary energy. In other words, high energy grain boundary is extended to form an relaxed structure with some thickness to reduce its energy. We previously proposed the concept of "extended grain boundary" [8], and the present boundary is likely to be the case for the extended grain boundary. It is again noted that grain boundary structure between 32° and 46.8°(Σ19) is quite different although their grain boundary energy is almost the same. This means that main part of grain boundary energy would be due to strain energy around grain boundary, but atomic configuration does not contribute so much to grain boundary energy.

Fig. 5 shows (a) high-resolution electron micrograph of the 10° small angle gain boundary. Dislocation contrast can be observed periodically along the grain boundary as indicated by the arrows. Fig. 5(b) is a Burgers circuit around a dislocation shown in (a). From this circuit, Burgers vector of the dislocation can be identified to be 1/3[1$\bar{1}$0n] since the circuit is just a projected circuit along the [0001] direction. As candidate Burgers vectors, b_1=1/3[1$\bar{1}$02], b_2=1/3[1$\bar{1}$01] and b_3=1/3[1$\bar{1}$00] can be considered, in which b_1 and b_2 are perfect dislocations and b_3 is partial

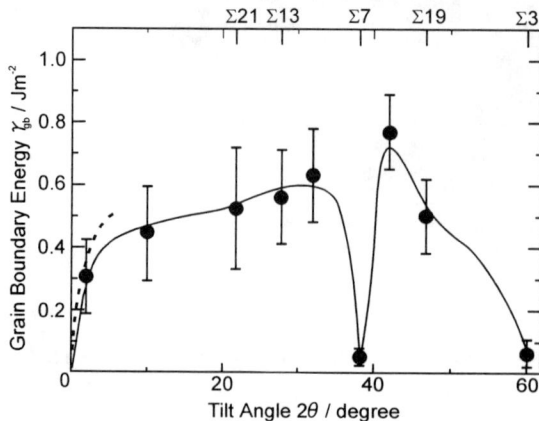

Fig. 3   Grain boundary energy γ_{gb} as a function of misorientation angle 2θ. Dotted line inset is theoretical curve obtained for small angle grain boundaries.

dislocation. The sizes of Burgers vectors are $|b_1|$=0.908nm, $|b_2|$=0.512nm and $|b_3|$=0.274nm, respectively, and, thus, b_3 has the smallest vector among them. In addition, if b_1 and b_2 dislocations are formed along the grain boundary, large twist component is introduced at the boundary. Consequently, b_3 dislocation should be formed periodically at an interval spacing of about 2nm to compensate the misorientation angle of 10° in this case. This can be reasonably explained by Frank's formula [7].

Since the dislocation of $1/3[1\bar{1}00]$ is a partial dislocation, stacking faults are considered to be introduced on the boundary plane between two dislocations for the small angle grain boundaries.

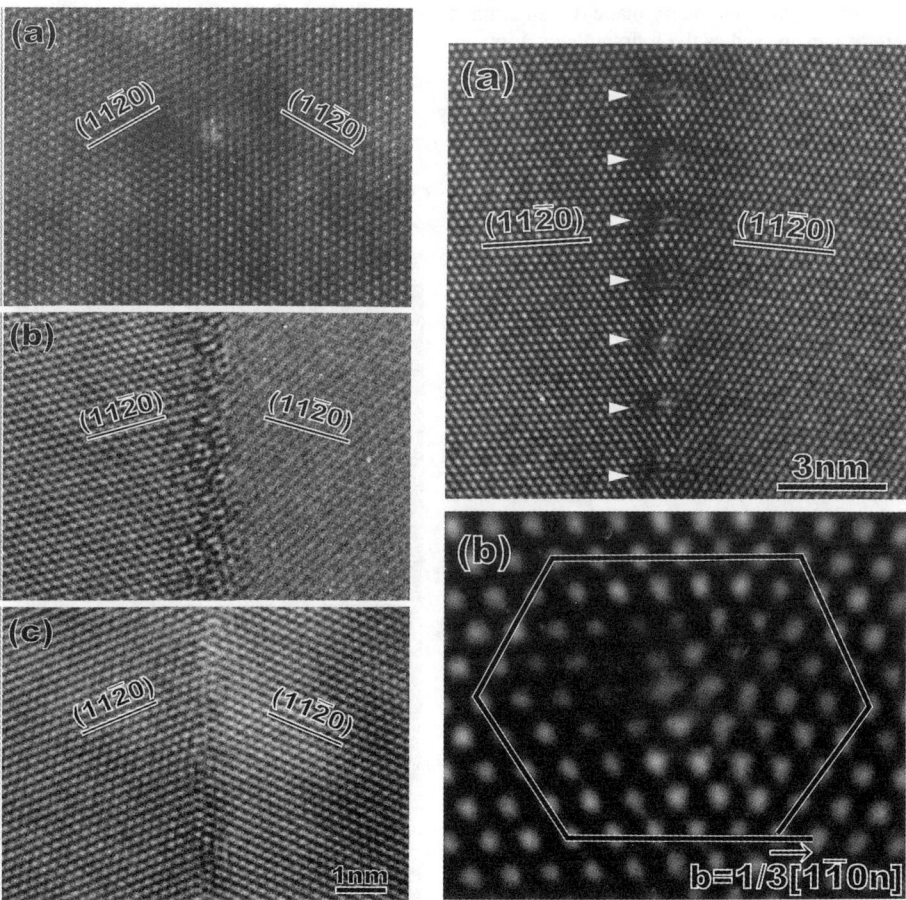

Fig. 4 High-resolution electron micrographs of the grain boundaries in the bicrystlas with the misorientation angle of (a) 60° (Σ3), (b) 32° and (c) 46.8° (Σ19).

Fig. 5 High-resolution electron microgphs of (a) 10° tilt grain boundary and (b) Burgers circuit around the dislocation in (a).

129

Here, grain boundary energy for dislocation boundary can be calculated by the next equation[9].

$$\gamma_{gb} = \frac{Gb\theta}{4\pi(1-\nu)}\left(\ln\left(\frac{b}{2\pi r_0\theta}\right)+1\right) \qquad (3)$$

where G, ν and r_0 are shear modulus, Poisson's ratio and core radius, respectively. In the case of small angle grain boundary in alumina in this study, grain boundary is consisted of partial dislocations with the Burgers vector of $1/3[1\bar{1}00]$ and the stacking fault on the $(1\bar{1}00)$ planes. Since the perfect dislocation with Burgers vector of $[1\bar{1}00]$ is dissociated to three partial dislocations with the Burgers vector of $1/3[1\bar{1}00]$, the stacking faults are formed on the 2/3 area of the whole grain boundary plane. Grain boundary energy of small angle grain boundary can be, therefore, expressed as the following equation.

$$\gamma_{gb} = \frac{Gb_p\theta}{4\pi(1-\nu)}\left(\ln\left(\frac{b_p}{2\pi r_0\theta}\right)+1\right)+\frac{2}{3}\gamma_{sf} \qquad (4)$$

where b_p is a Burgers vector of a partial dislocation and γ_{sf} stacking fault energy. The theoretical curve obtained from this equation is inset as the dotted line in Fig. 3, where 130MPa, 0.24, 0.274nm, $1.5b_p$ and $0.1J/m^2$ are used for G, ν, b_p, r_0 and γ_{sf} [10]. As seen in Fig. 3, the calculated curve agrees well with the experimentally measured dependence.

The behavior of grain boundary sliding was investigated for typical three grain boundaries, i.e., 38.2° (Σ7) grain boundary, 46.8° (Σ19) grain boundary and 42° (high angle) grain boundary. Fig. 6 shows optical micrographs of the specimens after creep test for (a) 38.2° (Σ7) grain boundary and (b) 42° (high angle) grain boundary. Σ7 grain boundary showed fracture along the boundary during creep test, but 42° grain boundary clearly showed the traces of grain boundary sliding. Judging from the trace in the figure, the distance of grain boundary sliding is 17μm which corresponds to 0.24% as strain along the grain boundary. On the other hand, no trace was observed for the 46.8° (Σ19) grain, although fracture did not occur under the present creep condition. This indicates that grain boundary sliding easily takes place in high angle grain boundaries, but not in CSL grain boundaries.

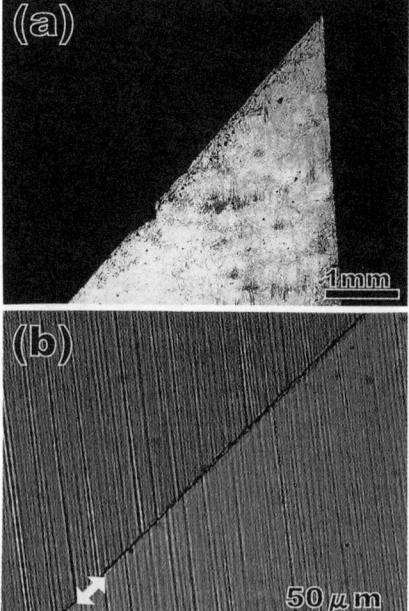

Fig. 6　Optical micrographs of the specimens after creep test for (a) 38.2° (Σ7) and (b) 42° (high angle) grain boundaries.

As discussed in this paper, grain boundary energy is not directly related to the grain boundary atomic structure. The fact that grain boundary sliding takes place in the 42° (high angle) grain boundary, but not in the 46.8° (Σ19) grain boundary implies that the occurrence of grain boundary sliding depends on the grain boundary atomic structure, but not on grain boundary energy.

Fig. 7 shows a high-resolution electron micrograph of the 42° (high angle) grain boundary after creep test for about 350ks under the stress of 16MPa at 1400°C. It can been seen that the shape of the grain boundary is wavy, and grain boundary planes move so as to be parallel to ($\bar{2}$110) or (1 $\bar{2}$10) plane in the two crystals. This would be an evidence that grain boundary sliding accompanies grain boundary migration. Assuming that atomic steps are existed on the grain boundary plane, atom diffusion would occur to accommodate the lattice strain at the step during grain boundary sliding. This diffusion is supposed to make the boundary shape to be wavy so as that the boundary plane moves to the low index planes in the adjacent crystals. The detailed mechanism is under consideration [7].

CONCLUSIONS

Alumina bicrystals were systematically fabricated to obtain ten kinds of [0001] symmetrical tilt grain boundaries. The grain boundary structure and grain boundary energy were investigated for all specimens by HREM and thermal grooving-AFM technique, and the behavior of grain boundary sliding was studied for typical specimens by compression creep test. The results obtained here are summarized as the followings.

1. Grain boundary energy strongly depends on the misorientation angle between two crystals. There are large energy cusps at the angles corresponding to low Σ CSL grain boundaries, but the grain boundary energy increases with increasing misorientation angle up to 32°.

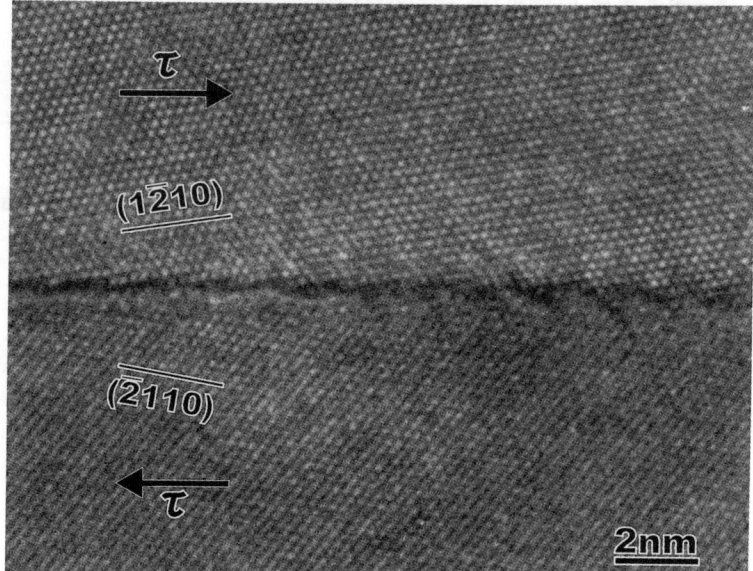

Fig. 7   High-resolution electron micrograph of the 42° (high angle) grain boundary after creep test under the stress of 16MPa at 1400°C.

2. Grain boundary energy is not directly related to the grain boundary atomic structure. Main part of grain boundary energy is due to the strain energy around the grain boundary, and the contribution of atomic configuration is not so large.
3. Small angle grain boundaries are consisted of an periodic array of partial dislocations with Burgers vector of $1/3[01\bar{1}0]$. Theoretical grain boundary energy, which was calculated by combining dislocation strain energy and stacking fault energy, agrees well with the experimentally obtained results.
4. Grain boundary sliding is hard to occur in low Σ CSL grain boundary, but not in high angle grain boundary at the present experimental condition. The occurrence of grain boundary sliding would depend on the grain boundary atomic structure, but not on the grain boundary energy.

ACKNOWLEDGMENTS

This work was supported by the Grant-in-Aid for Scientific Research (11450260) from the Ministry of Education, Science and Culture, Japan. A part of this work was also supposed by Proposal-Based R&D Program (project No. 98Y28019) of New Energy and Industrial Technology Development Organization (NEDO).

REFERENCE

1. S. Primdahl, A. Tholen and T. G. Langdon, Acta Mater., **43**, p.1211 (1995).
2. Y. Ikuhara, P. Thavorniti and T. Sakuma, Acta Mater., **45**, p. 5275 (1997).
3. T.Watanabe, H. Yoshida, T. Saito, T. Yamamoto, Y. Ikuhara and T. Sakuma, Mater. Sci. Forum, **304-306**, p.421 (1999).
4. P. Nikolopoulos, J. Mater. Sci., **20**, p. 3,993 (1985).
5. J. F. Shackelford and W. D. Scott, J. Am. Ceram. Soc., **51**, p688(1968).
6. W. Shin, W-S. Seo and K. Koumoto, J. Eur. Ceram. Soc., 18, p595(1998).
7. T. Watanabe, T. Yamamoto, Y. Ikuhara and T. Sakuma, in preparation.
8. Y. Ikuhara, H. Kurishita and H. Yoshinaga, Proc. 2nd Inter. Conf. Compo. Interfaces, Cleveland, Elsevier Sci. Pub., p.673 (1988).
9. W. T. Read and W. Shockley, Phys. Rev., **78**, p.275 (1950).
10. K. D. P. Lagerlof, T. E. Mitchell, A. H. Heuer, J. P. Riviere, J. Cadoz, J. Castaing and D. S. Phillips, Acta Mater., **32**, p.97 (1984).

SUPERPLASTICITY OF LIQUID-PHASE SINTERED β-SiC

T. Nagano, K. Kaneko, G.D. Zhan*, and M. Mitomo*

Ceramics Superplasticity Project, ICORP, Japan Science and Technology Corporation,
2-4-1, Mutsuno, Atsuta-ku, Nagoya 456-8587, Japan, nagano@ngo.jst.go.jp
*National Institute for Research in Inorganic Materials, 1-1, Namiki, Tsukuba 305-0044, Japan

ABSTRTACT

The compression and the tension tests of liquid-phase sintered β-SiC fabricated by hot-pressing using ultra fine powders were performed at 1973 K ~ 2048 K in N_2 atmosphere. Amorphous phases were observed at the grain boundaries and at multi-grain junctions in the as-sintered material. Strain hardening was observed under all experimental conditions. Stress exponents in the compression test were from 1.7 to 2.1 in the temperature ranging from 1973 K to 2023 K. A maximum tensile elongation of 170 % was achieved at the initial strain rate of 2 X 10^{-5} s^{-1} at 2048 K.

INTRODUCTION

Liquid-phase sintering is an effective way to lower the sintering temperature and obtain the fine-grained material. SiC with the grain size about 0.2 μm could be obtained by liquid-phase sintering [1]. The presence of liquid phase at the grain boundaries promotes the deformation, and contributes the accommodation of stress concentration at the triple points. Authors investigated the high temperature deformation of liquid-phase sintered β-SiC with the additions of oxide additives and obtained about 60 % tensile elongation in Ar [2-4]. The fine initial grain size, low grain growth rate, and stability at elevated temperature are especially important for superplastic deformation. However, the tensile elongation of liquid-phase sintered β-SiC with the additions of oxide additives was prevented by the vaporization of liquid phase and the decomposition of SiC [4].

Authors also investigated the weight loss of liquid-phase sintered β-SiC with oxide additives during the heat treatment, and found that the weight loss in N_2 was suppressed about 1/2 to 1/3 in comparison with the weight loss in Ar [5]. Although we performed the tension test of liquid-phase sintered β-SiC with oxide additives in N_2, maximum tensile elongation reached only about 11 % due to the crystallization of grain-boundary phase [6].

It was reported that the microstructure of SiC with the addition of oxynitride glass was stable at elevated temperatures in N_2, and grain growth and phase transformation were suppressed [7-9]. Accordingly, the deformation of liquid-phase sintered SiC with oxynitride glass in N_2 is thought to cause the suppression of grain growth, the phase transformation, the vaporization of liquid phase and the decomposition of SiC. Moreover, the viscosity of liquid phase with nitrogen is high, and the occurrence of cavitation should be suppressed.

Therefore, we investigated the deformation behavior of liquid-phase sintered β-SiC with the addition of oxynitride glass at elevated temperature in N_2.

EXPERIMENTAL PROCEDURE

Starting material was ultra fine β-SiC powder (T-1 grade, Sumitomo-Osaka Cement Co., Tokyo, Japan) fabricated by chemical vapor deposition (CVD) method. It was oxidized at 873 K for 1 h in air to eliminate free carbon and hydrofluoric acid-treated to remove SiO_2. The particle size was ~ 90 nm as calculated from the specific surface area. A mixture of 55.8 wt% Al_2O_3 (99.9 % pure, Sumitomo Chemical Co., Tokyo, Japan), 36.9 wt% Y_2O_3 (99.9 % pure, Shin-Etsu Chemical Co., Tokyo, Japan), and 7.3 wt% AlN (F Grade, Tokuyuma Soda Co., Tokyo, Japan) was prepared to an oxynitride composition by SiC ball milling in n-hexane. SiC was then mixed with 9 wt% oxynitride powder by SiC ball milling in n-hexane. The mixed powder was hot-pressed at 2073 K for 15 min under a stress of 30 MPa in N_2, and its bulk

133

density was 3.21 g/cm^3.

As-sintered body was cut by diamond cutter and ground by diamond wheel. The size of compressive specimen was 2 X 2 X 3 mm. The specimen surfaces were mirror-polished with diamond pastes. Tensile specimen was dog-bone shape, having rectangular cross section (1 X 0.8 mm) with the gauge length of 4.4 mm. The gauge portion of tensile specimen was also mirror-polished with diamond paste.

Compression and tension tests at constant crosshead speeds were performed by using a universal testing machine at the initial strain rates from 1 X 10^{-4} s^{-1} to 5 X 10^{-6} s^{-1} at the temperature ranging from 1973 to 2048 K in N$_2$.

The specimens before and after the deformation were observed using the scanning electron microscopy (SEM: JSM-6330F, JEOL, Tokyo, Japan) on the polished and plasma etched surface with CF$_4$ + O$_2$ (8%) gas. The average grain size was measured by linear intercept length. X-ray analysis was also conducted (RINT2500, RIGAKU, Tokyo, Japan). The polytypes of SiC were calculated by Tanaka's method [9]. Specimen was cut perpendicular to the compressive axis and prepared by the polishing and the argon-ion-beam thinning for transmission electron microscopy (TEM) observation. High-resolution transmission electron microscopy (HRTEM) observation was performed by 200 kV TEM (JEM-2010MD, JEOL, Tokyo, Japan) and whose point-to-point resolution was 0.18 nm. The grain-boundary composition was analyzed by both electron energy loss spectroscopy (EELS) and energy dispersive X-ray spectroscopy (EDX). The energy loss spectroscopy was performed using an electron energy loss spectrometer attached to a scanning transmission electron microscope (HB601UX, VG, East Grinstead, U.K.), having a minimum probe size of 0.22 nm.

RESULTS

Mechanical behavior

The true stress-true strain curves at 1973 K in compression test are shown in Fig. 1. Crack was not observed from compressed specimens. Strain hardening was observed in all experimental conditions. Flow stress increased with increasing strain rate.

The relationship between flow stress and strain rate in compression test is shown in Fig. 2. Stress exponents were calculated from the slope of lines. Stress exponents were from 1.7 to 2.1 in the temperature ranging from 1973K to 2023 K.

The apparent activation energy in the temperature ranging from 1973 K to 2073 K was 776 kJ/mol. This value is lower than that of lattice diffusion of Si and C (912±5 kJ/mol, 814 ±14 kJ/mol) and higher than that of grain-boundary diffusion of C (569±9 kJ/mol) in high purity CVD β-SiC [11-12].

The true stress-true strain curves at 2048 K in tension test are shown in Fig. 3. Strain hardening was observed in all experimental conditions. Flow stress was higher at lower strain rate in the beginning of deformation. Flow stress at the initial strain rate of 4 X 10^{-5} s^{-1} increased linearly with strain and fractured about 70 % elongation. Strain softening was observed before fracture at the initial strain rates of 1 X 10^{-5} s^{-1} and 2 X 10^{-5} s^{-1}. Tensile elongation of 170 % was achieved at the initial strain rate of 2 X 10^{-5} s^{-1}.

Microstructural change

High-resolution TEM image of grain boundary of as-sintered material is shown in Fig. 4. An amorphous phase about 2 nm thick was clearly observed at the grain boundary.

EELS results of as-sintered material are shown in Figs. 5 and 6. The peaks of Al, O and N which were not detected in SiC grain at the grain boundary. Moreover, the peak of Y was detected at grain boundary in EDX result. Therefore, this amorphous phase is thought to have the structure of Al$_2$O$_3$-SiO$_2$ glass with Y and N.

Microstructural changes before and after compression and tension tests are shown in Fig. 7. As-sintered material was composed of equiaxed fine grains. An amorphous phase existed at the grain boundaries. In the deformed specimens, the anisotropy of grain growth was observed and grains were elongated in the perpendicular direction of the compression axis. The aspect ratio

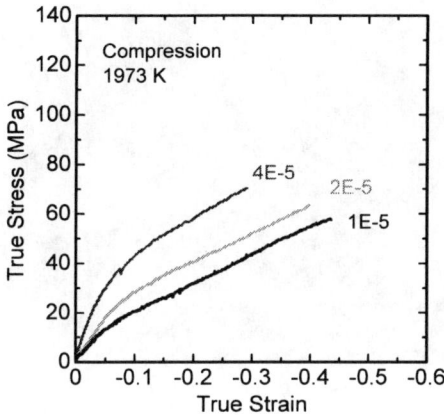

Fig. 1   True stress-true strain curves at 1973 K in N_2.

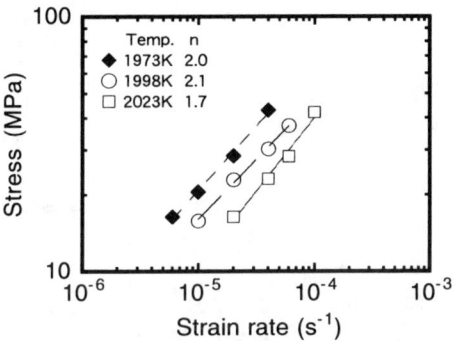

Fig. 2   Relationship between flow stress and strain rate.

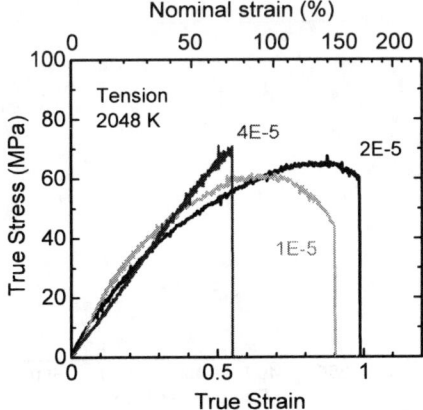

Fig. 3   True stress-true strain curves at 2048 K in N_2.

Fig. 4 High-resolution TEM image at grain boundary in as-sintered material.

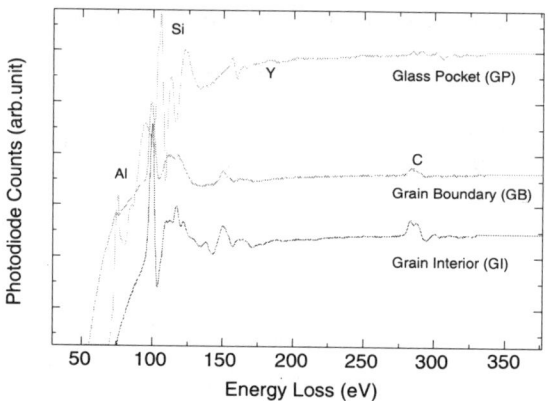

Fig. 5 EELS results for as-sintered material.

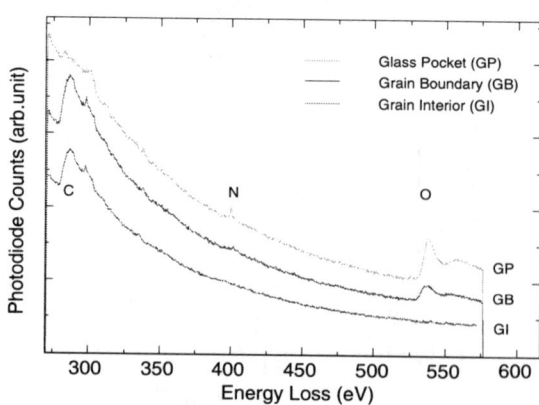

Fig. 6 EELS results for as-sintered material.

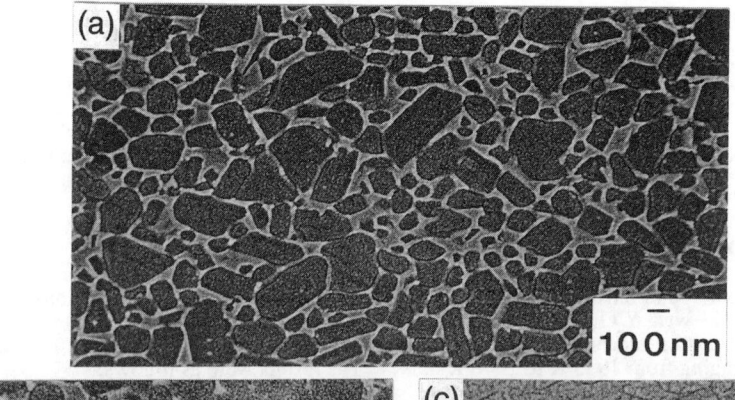

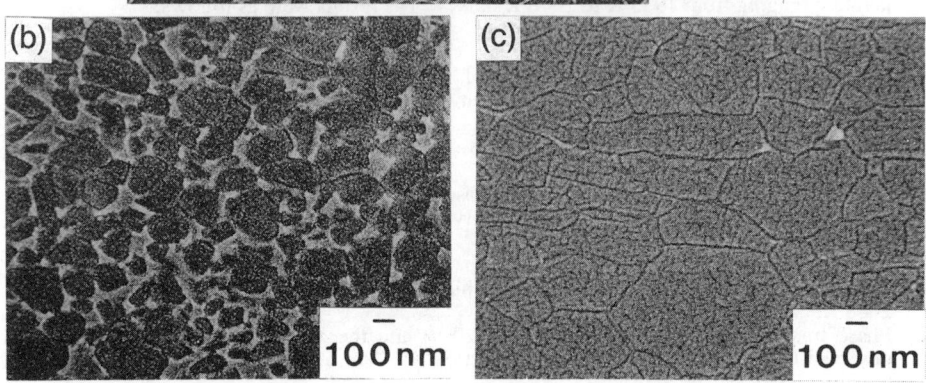

Fig. 7   Microstructure of (a) as-sintered SiC, (b) after compressive deformation (1998 K, 6×10^{-5} s^{-1}, 33 % strain, compressive axis is vertical), (c) after tensile deformation (2023 K, 1×10^{-5} s^{-1}, 153 % elongation, tensile axis is horizontal).

was about 0.82 in the direction of compressive axis.   Directly bonded grain boundaries increased after deformation.

On the other hand, almost amorphous phase was vaporized after the tensile deformation. Most of residual amorphous phase existed at the triple points.   The anisotropy of grain growth was observed and grains were elongated in the direction of tensile axis.   The grain aspect ratio was 1.44 in the direction of tensile axis.

Intergranular strain (ε_g) is defined as follows:

$$\begin{aligned} \varepsilon_g &= \ln (d_{//}/d_\perp{}^2 \cdot d_{//})^{1/3} \\ &= 2/3 \ln (d_{//}/d_\perp) \end{aligned} \qquad (1)$$

where $d_{//}$ is the grain size parallel to the tensile axis and $d_\perp$ is the grain size perpendicular to the tensile axis.   The ratio of intergranular strain to the total strain (ε_{total}) is calculated to be about 24 %.   If we assume that the total strain is the sum of intergranular strain and the strain by grain-boundary sliding, the strain caused by grain-boundary sliding is roughly estimated to be 76 %.

Grain size after the tensile deformation was about 7 times larger than that before the tension test in comparison with the size of short axis.   The cavitation damage at gauge portion was fairly suppressed even after 153 % tensile elongation.

XRD analysis

Crystalline phase of as–sintered material was almost β-SiC(3C) single phase.   In compressed and tensiled specimens, the phase of α-SiC(6H) was detected.   The contents of α-SiC for the compressed specimen and the tensiled specimen were calculated about 1 and 20 %, respectively.   The specimen annealed for the testing time corresponded to 153 % tensile deformation at 2023 K in N_2 was also prepared to investigate the effect of strain for phase transformation from β to α.   As the result, the content of α-SiC(6H) was about 15 % in annealed specimen.

CONCLUSION

Compression and tension tests of liquid-phase sintered β-SiC with the additions of 5.022 wt% Al_2O_3, 3.321 wt% Y_2O_3 and 0.657 wt% AlN at constant crosshead speeds were performed by a universal testing machine at the initial strain rates of 1×10^{-4} s^{-1} to 5×10^{-6} s^{-1} in the temperature ranging from 1973 K to 2048 K in N_2.   The results were as follows:

Compression test
(1) Stress exponents were from 1.7 to 2.1.
(2) Apparent activation energy was 776 KJ/mol.   This value was between activation energy of lattice diffusion of Si and C, and that of grain-boundary diffusion of C.
(3) Phase transformation from β to α was slightly detected.   The anisotropy of grain growth was observed.   The equiaxed grains were elongated in the perpendicular direction of compressive axis.
(4) The crystallization of grain-boundary phase was not observed even after compressive deformation.   Some parts of grain-boundary phase was squeezed out or vaporized during compressive deformation.

Tension test
(5) Superplastic deformation of 170 % was achieved at the initial strain rate of 2×10^{-5} s^{-1} at 2048 K.
(6) Phase transformation from β to α about 20 % and the anisotropy of grain growth were observed in 153 % deformed specimen at 2048 K.   The equiaxed grains were elongated in the direction of tensile axis.
(7) Much of grain-boundary phase was vaporized during tension test.   Most of residual grain-boundary phase existed at the triple points.

The above suggested that critical deformation mechanism was grain-boundary sliding.

REFERENCES

1. M. Mitomo, Y-W. Kim and H. Hirotsuru, J. Mater. Res., **11**, 1601-1604 (1996).
2. T. Nagano, S. Honda, F. Wakai, and M. Mitomo, in Pro. of 6th Int'l Symp. on Ceramic Materials and Components for Engines, Edited by K. Niihara, S. Hirano, S. Kanzaki, K. Komeya, and K. Morinaga, (Japan Fine Ceramics Association, Tokyo, 1998) p.707-712.
3. T. Nagano, H. Gu, Y. Shinoda, M. Mitomo and F. Wakai, Ceramics: Getting into The 2000's, Part D, Edited by P. Vincenzini, (Techna, Faenza, 1999) p. 25-32.
4. T. Nagano, H. Gu, Y. Shinoda, G.D. Zhan, M. Mitomo, and F. Wakai, Mater. Sci. Forum. **304/306**, 507-512 (1999).
5. T. Nagano, K. Kaneko, G.D. Zhan and M. Mitomo, Submitted to J. Am. Ceram. Soc..
6. T. Nagano, H. Gu, G.D. Zhan, and M. Mitomo, Submitted to J. Am. Ceram. Soc..
7. H-W. Jun, H-W. Lee, G-H. Kim, H. Song, and B-H. Kim, Ceram. Eng. Sci. Proc., **18**, 487-504 (1997).
8. Y-W. Kim and M. Mitomo, J. Am. Ceram. Soc., in Press.
9. H. Tanaka and N. Iyi, J. Ceram. Soc. Jpn., **101**, 1313-1314 (1993).
10. A. Mukherjee, J.E. Bird and J.E. Dorn, Trans. ASM **62**, 155-179 (1969).
11. M.H. Hon and R.F. Davis, J. Mater. Sci., **14**, 2411-2421 (1979).
12. M.H. Hon, R.F. Davis and D.E. Newberry, J. Mater. Sci., **15**, 2073-2080 (1980).

Fundamental Aspects
of Superplasticity

NEWTONIAN FLOW IN BULK AMORPHOUS ALLOYS

J. WADSWORTH AND T.G. NIEH
Lawrence Livermore National Laboratory, L-350, P.O. Box 808, Livermore, CA 94551

ABSTRACT

Bulk amorphous alloys have many unique properties, e.g., superior strength and hardness, excellent corrosion resistance, reduced sliding friction and improved wear resistance, and easy formability in a viscous state. These properties, and particularly easy formability, are expected to lead to applications in the fields of near-net-shape fabrication of structural components. Whereas large tensile ductility has generally been observed in the supercooled liquid region in metallic glasses, the exact deformation mechanism, and in particular whether such alloys deform by Newtonian viscous flow, remains a controversial issue. In this paper, existing data are analyzed and an interpretation for the apparent controversy is offered. In addition, new results obtained from an amorphous alloy (composition: Zr–10Al–5Ti–17.9Cu–14.6Ni, in at. %) are presented. Structural evolution during plastic deformation is particularly characterized. It is suggested that the appearance of non-Newtonian behavior is a result of the concurrent crystallization of the amorphous structure during deformation.

INTRODUCTION

Metallic glasses fabricated by rapid quenching from the melt were first discovered in 1960 [1]. Due to the high quench rate requirements ($10^4 - 10^6$ K/s), only thin ribbons and sheets with a thickness less than 0.1 mm could be fabricated. One of the most important recent developments in the synthesis of amorphous materials is the discovery that certain metallic glasses can be fabricated from the liquid state at cooling rates of the order of 10K/s. This enables the production of bulk amorphous alloys with a thickness of ~10 mm. These bulk amorphous alloys have many potential applications resulting from their unique properties, e.g., superior strength and hardness [2], excellent corrosion resistance [3], reduced sliding friction and improved wear resistance [4], and easy forming in a viscous state [5-7]. These properties, and particularly easy forming in a viscous state, should lead to applications in the fields of near-net-shape fabrication of structural/functional components.

The mechanical behavior of metallic glasses is characterized by either inhomogeneous or homogeneous deformation. Inhomogeneous deformation usually occurs when a metallic glass is deformed at low temperature (e.g. room temperature) and is characterized by the formation of localized shear bands, followed by the rapid propagation of these bands, and sudden fracture. Thus, when a metallic glass is deformed under tension it exhibits very limited macroscopic plasticity. It is pointed out that, despite a limited macroscopic plasticity, local strain within these shear bands can be, sometimes quite significant (about 10). These bands are typically 20-30 nm in width and have not yet been microscopically examined [8], although some observations indicated possible crystallization [9]. Whereas there exist many different views on inhomogeneous deformation in metallic glasses (e.g., free-volume model [10, 11] and dislocation theory [12]), there is still no universal agreement. This paper is to address only homogeneous deformation.

Homogeneous deformation in metallic glasses usually takes place at about $0.70T_g$ [11] above which metallic glasses exhibit significant plasticity. It is pointed out that the transition temperature T_{tr} from inhomogeneous to homogeneous deformation (or brittle-to-ductile transition) is strongly dependent upon strain rate. For example, T_{tr} for $Zr_{65}Al_{10}Ni_{10}Cu_{15}$ alloy is about 533K (corresponds to $0.82T_g$) at 5 x 10^{-4} s^{-1}, but it is 652K (corresponds to $1.0T_g$) at 5 x 10^{-2} s^{-1} [6]. The strain rate dependence of T_{tr} (480–525K; 0.61–0.75T_g) has also been demonstrated in $Fe_{40}Ni_{40}B_{20}$ [13]. These results suggest that homogeneous deformation is associated with certain rate (or diffusional relaxation) processes. For the purpose of discussion, we divide the homogeneous deformation of metallic glasses into three regions, according to testing temperatures.

Mat. Res. Soc. Symp. Proc. Vol. 601 © 2000 Materials Research Society

$T < T_g$

Mulder *et al* [13], in a study of the deformation of $Fe_{40}Ni_{40}B_{20}$ metallic glass in tension at elevated temperatures, found that the transition temperature from inhomogeneous to homogeneous deformation is about 480-525K (0.68–0.75 T_g), as predicted by a free volume model [11]. The also conducted creep experiments at temperatures (523–548K) below the glass transition temperature with a relatively high stress (> 1.0 GPa). Experimental results showed that the stress exponent was rather high (n=8.5). The activation energy was determined to be between 250 and 280 kJ/mol, which is similar to that for eutectic crystallization below the glass transition temperature.

Taub and Luborsky [14] also conducted tensile creep experiments on amorphous $Fe_{40}Ni_{40}P_{14}B_6$ ribbons. They found that within the temperature range of 383–582K (T_g=663K and T_x=673K, where T_x is the crystallization temperature) and at a constant tensile stress of 312 MPa the strain rate varies inversely with time but only after an initial transition. They also performed stress relaxation experiments and showed that the stress dependence of the strain rate obey a hyperbolic sine relationship (nonlinear), consistent with transition state theory [15]. To further reconcile all data, including those from the initial transition, a threshold stress (=39±4MPa) was introduced into the formulation, although the physical meaning of the threshold stress was unclear.

Most recently, Kawamura *et al* [6] studied the high-temperature deformation properties of a $Zr_{65}Al_{10}Ni_{10}Cu_{15}$ metallic glass (T_g=652K, T_x=757K) with a wide range of ΔT (=T_g-T_x) produced by a melt spinning method. They found that, within $T_{tr}<T<T_g$, the alloy has a low strain rate sensitivity value (m<0.25) and the tensile elongation is also low (<100%). Similar observations were also made in $Pd_{40}Ni_{40}P_{20}$ metallic glasses, and results showed that at a test temperature of 560K (T_g=578-597K), the alloy exhibited a low strain rate sensitivity of only 0.20 and tensile elongation < 50%.

From the above results, we may conclude that homogeneous deformation of metallic glasses at temperatures below T_g are characterized by a low strain rate sensitivity (i.e. high stress exponent) and ductility (<100%). This is attributable to the fact that structural relaxation and recovery are still difficult as a result of sluggish diffusion in this temperature range.

$T_g < T < T_x$ (supercooled liquid region)

As early as 1980, Homer and Eberhardt [16] reported the observation of superplasticity in amorphous $Pd_{78.1}Fe_{5.1}Si_{16.8}$ ribbons (T_g=668K, T_x=683K) during non-isothermal creep experiments. In the experiments, test samples were rapidly heated to the maximum temperature of 698K under a constant load (range: 25–150 MPa). The resulted creep rate was rather high; for example, an applied stress of 150 MPa produced a creep rate of 0.5 s^{-1}. The strain rate sensitivity value was estimated to be about one, suggesting possible Newtonian flow. Since 698K is higher than T_x, a dispersion of 0.4 μm grains in the amorphous matrix was observed in the test samples after superplastic deformation. It is noted that slow heating during creep test resulted in the disappearance of superplasticity. This is apparently caused by easy crystallization in the alloy, as indicated by a narrow ΔT (=15K).

Zelenskiy *et al.* [17] studied the formability of amorphous $Co_{68}Fe_7Ni_{13}Si_7B_5$ (T_g=836K, T_x=856K) at temperatures between 773 and 913K. They observed large tensile ductility at a relatively fast strain rate of 10^{-2} s^{-1} within 823-853K (in the supercooled liquid region). Specifically, the maximum elongation of 180% was recorded at a corresponding minimum stress of about 150 MPa. However, strain rate sensitivity was not measured. TEM microstructural examinations indicated that 853K annealing produces nanometer grains (~50–70 nm) in the alloy. From these results, the authors argued that the presence of a large amount of grain/amorphous matrix interfacial area is necessary for the observed superplasticity. However, we want to point out that this may not be true. In fact, several experiments indicated a reduced ductility in the presence of nanograins [6, 16]. A recent study of Busch *et al* [18] also showed that the viscosity of a metallic glass increases sharply once the temperature is above the crystallization temperature.

As expected, an increase in viscosity leads to an increasing resistance to plastic flow and, thus, a decrease in ductility.

To further understand superplasticity and extended plasticity in metallic glasses, Khonik and Zelenskiy [19] analyzed available mechanical data from fifteen different metallic glasses, including both metal-metal and metal-metalloid systems. They made several important observations. First, superplasticity occurs in alloys with a large ΔT, typically about several tens degrees. The larger is ΔT, the larger is the tensile elongation, provided tests were conducted in the supercooled liquid region. This indicates the importance of thermal stability of a metallic glass during superplastic deformation. They noticed also that a faster heating rate usually produced a larger elongation. Apparently, this is associated with structural stability since slower heating results in an earlier crystallization.

In studying the formability of a $La_{55}Al_{25}Ni_{20}$ alloy, Kawamura et al [7] reported that the alloy in the supercooled liquid range (480-520K) behaves like a Newtonian fluid, i.e., $m=1$. A tensile elongation of over 1,800% was recorded at 503K at a strain rate of 2×10^{-1} s^{-1}. The glassy solid below the glass transition temperature exhibited non-Newtonian viscosity, and the supercooled liquid revealed a Newtonian viscosity but changed to non-Newtonian with increasing strain rate. The elongation was reduced by the transition to non-Newtonian viscosity and crystallization. However, a careful examination of their stress-strain rate data indicated that the strain rate sensitivity tends to decrease to less than one when testing temperature (e.g. 510, 520K) approaches T_x. Again, this is probably associated with a partial crystallization in amorphous structure during testing.

Kawamura et al [6] recently studied the high-temperature deformation of a $Zr_{65}Al_{10}Ni_{10}Cu_{15}$ metallic glass with a wide range of ΔT ($T_g=652K$, $T_x=757K$). In the supercooled liquid region, they found that plastic flow were strongly dependent on strain rate and the strain rate sensitivity value exceeded 0.8, but less than one. The high strain rate sensitivity produces a corresponding high tensile elongation. For example, a tensile elongation of 340% was obtained at a strain rate of 5×10^{-2} s^{-1} and at 673K. However, a true Newtonian behavior ($m=1$) was not observed in the alloy.

To further investigate superplasticity in metallic glass systems, Kawamura et al [6] tested a $Pd_{40}Ni_{40}P_{20}$ alloy prepared by rapid solidification. Within 560-620K, the alloy exhibits a similar deformation behavior to that of $Zr_{65}Al_{10}Ni_{10}Cu_{15}$, namely, a high-strain-rate-sensitivity value accompanied by extended tensile ductility in the supercooled liquid region ($T_g=578–597K$, $T_x=651K$). In contrast to $Zr_{65}Al_{10}Ni_{10}Cu_{15}$ which is non-Newtonian, $Pd_{40}Ni_{40}P_{20}$ can behave like a true Newtonian fluid (i.e. $m=1$) under appropriate testing conditions. The difference may be associated with the fact that $Pd_{40}Ni_{40}P_{20}$ is thermally more stable than $Zr_{65}Al_{10}Ni_{10}Cu_{15}$ in the supercooled liquid state, as pointed out by Kawamura et al [6]. ($\Delta T= 72$ and 100K for $Zr_{65}Al_{10}Ni_{10}Cu_{15}$ and $Pd_{40}Ni_{40}P_{20}$, respectively.) Therefore, during high-temperature deformation, $Pd_{40}Ni_{40}P_{20}$ can still retain its amorphous state, whereas crystallization may have already taken place in $Zr_{65}Al_{10}Ni_{10}Cu_{15}$. This is indirectly indicated by the fact that the viscosity of $Pd_{40}Ni_{40}P_{20}$ is about one order of magnitude lower than that of $Zr_{65}Al_{10}Ni_{10}Cu_{15}$.

Thus, at $T_g < T < T_x$, large tensile ductility can be obtained from metallic glasses with large ΔT. The maximum ductility is expected to occur at a temperature near T_x, where the flow stress (or viscosity) is low, and high strain rates at which the alloy can retain its amorphous structure during deformation. For convenience, the above data are summarized in Table 1.

Table 1 Summary of the deformation data of some metallic glasses in the supercooled liquid region

Alloys	T_g	T_x	m	Elongation	Ref.
$Pd_{78.1}Fe_{5.1}Si_{16}$	668K	683K	~1.0	N/A	[16]
$Co_{68}Fe_7Ni_{13}Si_7B_5$	836K	856K	N/A	180	[17]
$La_{55}Al_{25}Ni_{20}$	480K	520K	1	1,800	[7]
$Zr_{65}Al_{10}Ni_{10}Cu_{15}$	652K	757K	>0.8	340	[6]
$Pd_{40}Ni_{40}P_{20}$	578–597K	651K	1.0	N/A	[6]
$Ni_{77.5}Si_{7.5}B_{15}$	N/A	N/A	1.09	N/A	[20]

$T_x < T$

At a temperature higher than T_x, metallic glass alloys are readily crystallized and form nanocrystalline structures. Whereas there are some data on plasticity of nanocrystalline solids at temperatures much greater than T_x, there exist only limited information on the mechanical behavior of metallic glasses at temperatures slightly above T_x.

Ashdown et al [21] studied the superplastic behavior of a crystallized Fe-Cr-Ni-B glassy alloy. By controlling crystallization, a material with a grain size of 0.2 μm was produced. Tensile elongation of over 200% was readily obtained from the material tested at strain rates of over 10^{-2} s^{-1} and at a temperature as low as 1073K; the maximum elongation of 450% was recorded at 1273K and a strain rate of 10^{-2} s^{-1}. These results are in accordance with the conventional fine-grained superplasticity.

Brandt et al [22] also studied the superplastic behavior of a microcrystalline (0.5 μm) $Ni_{78}Si_8B_{14}$ produced by annealing amorphous samples at 1073K. The material contains equiaxed Ni_3B and Ni_3Si grains and showed a tensile elongation of over 120% at 823–1023K. The strain rate sensitivity value was essentially constant ($m = 0.85$) over a wide range of strain rate (10^{-6} to 10^{-2} s^{-1}) and the activation energy was 72.4 kJ/mol. These results indicated the occurrence of conventional fine-grained superplasticity.

Using a similar technique, Wang and coworkers [23, 24] crystallized an amorphous $Ni_{80}P_{20}$ alloy at 603K. The crystallized alloy was nanocrystalline, consisting of 80vol% Ni_3P and 20vol% Ni. Creep experiments were subsequently conducted at 543-583K on the crystallized samples and the data showed a stress exponent of 1.2 ($m = 0.8$) and activation energy 68 kJ/mol. From these data, the authors argued that Coble creep was responsible for the deformation.

Most recently, two submicrocrystalline bulk alloys, Al–14mass% Ni–14mass% Misch Metal [25] and Mg-8.3wt%Al-8.1wt%Ga [26], were produced by the extrusion of rapidly solidified amorphous powders. The grain sizes of the extruded Al and Mg alloys were 0.1 and 0.6 μm, respectively. Resulting from fine grain sizes, both alloys were highly superplastic (elongation = 600% for Al and >1000% for Mg) and the strain rates at which superplasticity took place were also high (>10^{-2} s^{-1}).

The above experiments were all carried out at temperature much higher than T_x. In fact, the testing temperatures were sometimes close to T_m, the melting point of the alloys. At these temperatures material are no longer amorphous, but rather nanocrystalline or even microcrystalline, depending upon the thermal stability of the alloys. Conventional mechanisms for high temperature deformation, and particularly superplastic deformation, are expected to prevail.

In summary, for homogeneous deformation in metallic glasses, large tensile ductility can generally be obtained in the supercooled liquid. The exact deformation mechanism, however, and in particular whether an alloy deforms by Newtonian viscous flow or not remains a controversial issue. The purpose of this paper is to report an example of non-Newtonian behavior in a cast amorphous Zr-10Al-5Ti-17.9Cu-14.6Ni alloy in the supercooled liquid region.

EXPERIMENTS

The material used in the present study has a composition of Zr-10Al-5Ti-17.9Cu-14.6Ni. Zone-purified Zr bars (containing 12.3 appm O and 10 appm Hf), together with pure metal elements, were used as charge materials. The alloys were prepared by arc melting in inert gas, followed by drop casting into 7.0-mm-diameter by 7.2cm-long Cu molds at Oak Ridge National Laboratory. The details of fabrication of the alloy have been described previously [2]. Differential scanning calorimetry was used to characterize the thermal properties of the alloy. The temperatures for the onset and end of glass transition, and the crystallization temperature (T_x), have been previously measured using differential scanning calorimetry (DSC) [27]; specifically, these temperatures are 631, 705, and 729K, respectively, at a heating rate of 20K/min.

Tensile sheet specimens were fabricated from the as-cast material by means of electrical discharge machining. They had a gage length of 4.76 mm, a thickness of 1.27 mm and a width of 1.59 mm, as shown in Fig. 1. Tensile tests were conducted using an Instron machine equipped with an air furnace. Because of structural instability during testing of samples at high temperatures, the heating rate must be rapid to minimize crystallization. Typically, the heating-plus-holding time prior to testing was about 25 minutes. For example, for a test at 683K at a constant strain rate of 10^{-2} s^{-1}, the temperature profile was: 578K (5 min), 644K (10 min), 670K (15 min), 680K (20 min), and 683 (47 min). Constant strain rate tests were performed at a constant strain rate of 10^{-2} s^{-1} with a computer-controlled machine within a temperature range of 663-743K. To measure strain rate sensitivity exponents, both strain rate cycling (i.e. cycling between 10^{-3} and 10^{-2} s^{-1}) and strain rate increase tests were performed. The amorphous nature of the alloy was confirmed using transmission electron microscopy (TEM), as shown in Fig. 2. The TEM sample was prepared by chemical milling.

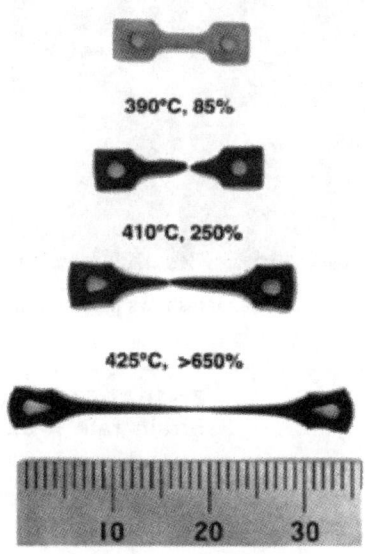

Fig. 1 Amorphous samples fractured at different temperatures. Sample necking is apparent. An untested sample is included for comparison.

RESULTS AND DISCUSSION

The stress-strain curves for the alloy at different temperatures at a strain rate of 10^{-2} s^{-1} is shown in Fig. 3. A yield drop phenomenon is readily observed at low temperatures, and in particular at 663 and 683K. In fact, at 663K the yield drop is 750 MPa (i.e. from 1600MPa to 850MPa), which is about the same magnitude as its 'normal' yield strength (~850 MPa). The yield drop phenomenon has also been observed and studied by Kawamura et al [5] during testing of a $Zr_{65}Al_{10}Ni_{10}Cu_{15}$ metallic glass in the supercooled liquid region. They attributed the yield drop to a 'transient phenomenon', but the associated concurrent change of atomic structure associated with the phenomenon was not given.

It is noted that similar phenomena have also been observed in the homogeneous flow of glassy polymers and in the plastic deformation of crystalline metal alloys. In the case of glassy polymers, this behavior is associated with the stress-effected, segmental chain displacements and the preferential alignment of the long axis of the molecules along the tensile axis. Both chain displacements and molecular alignments are achievable by the nucleation and propagation of 'double kinks' along the chain axis. By contrast, in crystalline metal alloys, it results from the locking of dislocations by solute atoms, e.g. Cottrell locking [28], or the shearing of coherent precipitates by dislocations [29]. For metallic glasses, however, neither the theory for glassy polymers or for crystalline metals appears to be applicable. Kawamura et al [30] argued that the yield drop was caused by an initial increase in atomic mobility at high strain rates, but, upon yielding, the atomic mobility decreases and structure relaxes. However, the exact physical process that leads to the observed yield drop is still unclear [30]. Structural clustering and chemical short range ordering are expected to impede the propagation of shear bands and may be responsible, in part, for the yield drop.

Fig. 2 Transmission electron micrograph shows the amorphous nature of the alloy.

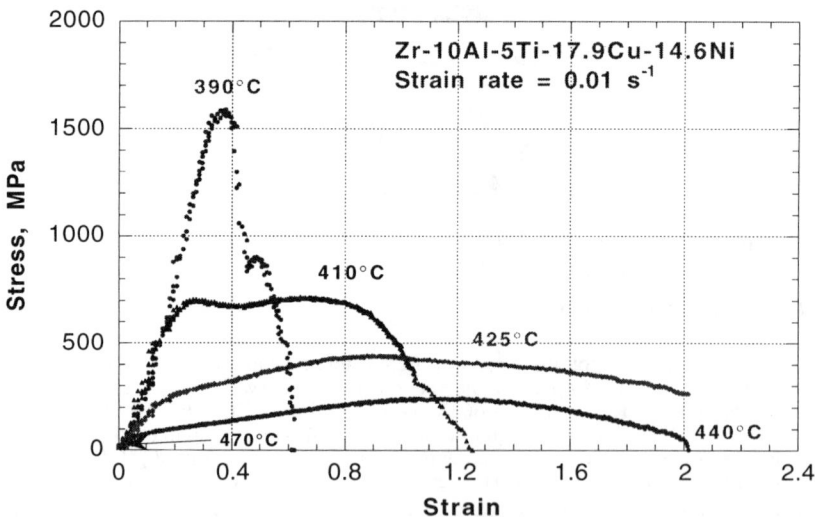

Fig. 3 Stress-strain curves of Zr-10Al-5Ti-17.9Cu-14.6Ni obtained at temperatures near the supercooled liquid region. The yield drop phenomenon is readily seen, especially at low temperatures.

The yield drop phenomenon disappears at a strain rate of 10^{-2} s^{-1} and at temperatures higher than 683K. In these temperatures, there is an initial hardening, followed by a gradual decrease in flow stress until final fracture. The fracture strain increases with increasing test temperature and reaches a maximum value of 2.0 (~650% elongation) at 698 and 713K. At 743K, which is above the crystallization temperature (729K), the alloy becomes extremely brittle; in fact, the test sample failed at the location of the loading pin. It is noted that samples, which were deformed in the supercooled liquid region, exhibit gradual necking, as shown in Fig. 2. In fact, some samples necked down nearly to a point. The final decrease in flow stress is, therefore, not a result of softening, but reduction in load bearing. This fracture appearance is different from that observed in a $Zr_{65}Al_{10}Ni_{10}Cu_{15}$ metallic glass [5], in which uniform deformation was observed. The difference may be caused by the fact that the samples used by Kawamura *et al* were very thin (0.02 mm). As a result, the samples were subject to a plane stress condition.

It is evident in Fig. 3 that both the flow stress and fracture strain are extremely sensitive to testing temperature. For example, with only a 15K difference in testing temperature, the flow stress drops from 700MPa at 683K to about 400MPa at 698K. The tensile elongation is almost tripled (230% to 630%). The flow stresses are generally quite high; for example, at even 713K the flow stress is about 200MPa. A high flow stress was also observed in another superplastic $Zr_{65}Al_{10}Ni_{10}Cu_{15}$ metallic glass, which was over 100MPa [6]. These values are considerably higher than the flow stresses generally observed in metals or ceramics exhibiting superplasticity or extended ductility [31]. Flow stresses for a superplastic metal or ceramic are typically lower than 35MPa. Technologically, a high flow stress can cause fast wear of the forming dies.

To characterize the deformation behavior, strain rate cycling tests were carried out at 683K to measure the strain rate sensitivity value. The result is shown in Fig. 4; the values of strain rate sensitivity m in equation $\dot{\varepsilon} = K \cdot \sigma^m$, where $\dot{\varepsilon}$ is the strain rate, σ is the flow stress, and K is a constant, were measured by strain rate cycling between 10^{-2} and 7×10^{-3} s^{-1}. There is no steady state region after each cycle, making it difficult to determine accurately the strain rate sensitivity value. This difficulty may be associated with structural instabilities during testing. It is noted that an external applied stress can promote crystallization in amorphous alloys [32]. Thus, despite the fact that 683K is below the crystallization temperature, it is believed that some nano-scale, crystallized phase already evolved during the course of the test. The presence of nanocrystalline phases can significantly affect the mechanical properties of a metallic glass. For example, Busch *et al* [18] recently showed that the presence of crystalline phases increases the viscosity of a $Zr_{46.75}Ti_{8.25}Cu_{7.5}Ni_{10}Be_{27.5}$ metallic glass. This observation is also consistent with the results of Kim *et al* [33, 34] who reported that the fracture strength of an amorphous $Al_{88}Ni_{10}Y_2$ was doubled when the alloy was crystallized and contained 5-12 nm-size Al particles. Thus, in the present strain rate cycle test, continuous strengthening is proposed to be a result of the continuous precipitation of nanocrystals in the amorphous matrix. In fact, this is also reflected by a slight increase in stress after the initial yield drop (strain >0.4) shown in Fig. 4.

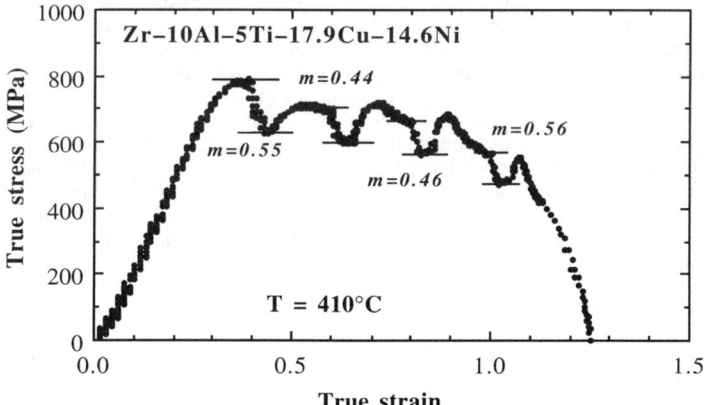

Fig. 4 Strain rate cycling test at 683K showing that the strain rate sensitivity value is only 0.45-0.55.

As shown in Fig. 4, after each strain rate decrease, except for the first one, there is no steady-state flow region. The gradual decrease in flow stress after decreasing the strain rate results from sample necking. Data from Fig. 4 indicate that the onset of sample necking occurs at a strain of approximately 0.8, which is consistent with the results observed in single-strain-rate tests (Fig. 3). The fracture obtained in the strain rate cycling test is also similar to that obtained in a constant strain rate test.

It is worth noting that from Fig. 4 the "apparent" strain rate sensitivity for the present Zr-10Al-5Ti-17.9Cu-14.6Ni alloy is computed to be about 0.5. Although structural instability can contribute to some variations in determining the "true" strain rate sensitivity value, its influence is not expected to be sufficiently great to imply a "true" strain rate sensitivity value of as high as one. In other words, the present alloy does *not* behave like a Newtonian fluid. The non-Newtonian behavior may be caused by the fact that the structure of the alloy in the supercooled liquid region is thermally unstable. Upon thermal exposure, and particularly under an external applied stress, the amorphous structure tends to crystallize and results in a mixture of crystalline and amorphous structure. Experimentally, it is challenging to examine the structure of a superplastically deformed specimen, primarily because tested samples necked down to nearly a point. Sample preparation for either x-ray diffraction or TEM study is, therefore, difficult. Moreover, the present experiments were performed in air, which resulted serious oxidation on test samples (see next section). The presence of a diffraction pattern may be simply indicative of oxide formation.

However, if an alloy indeed has a mixed crystalline-plus-amorphous structure, to a first approximation, the total deformation rate can be expressed by:

$$\dot{\varepsilon}_{total} = (1 - f_v) \cdot \dot{\varepsilon}_{am} + f_v \cdot \dot{\varepsilon}_{cry} \tag{1}$$

where $\dot{\varepsilon}_{total}$ is the total strain rate, $\dot{\varepsilon}_{am}$ and $\dot{\varepsilon}_{cry}$ are the strain rates caused by the amorphous and crystalline phases, respectively, and f_v is the volume fraction of the crystalline phase. Since the plastic flow of an amorphous alloy can be described by $\dot{\varepsilon}_{am} = A\sigma$, and the plastic flow of a nanocrystalline, superplastic alloy can be described by $\dot{\varepsilon}_{cry} = B\sigma^2$, where σ is the flow stress, and A and B are material constants, Equation (1) can be rewritten as:

$$\dot{\varepsilon}_{total} = (1-f_v) \cdot A\sigma + f_v \cdot B\sigma^2 \tag{2}$$

It is obvious that the strain rate sensitivity, which is the reciprocal of the stress exponent, would be between 0.5, the value for grain boundary sliding mechanism in fine-grained crystalline material, and unity, the value for Newtonian viscous flow.

It is interesting to comment on the drastic reduction in tensile elongation from 650% at 713K to virtually zero at 743K. A temperature of 743K is above the crystallization temperature of the alloy (~729K). At this temperature, the alloy has a nanocrystalline structure consisting of many intermetallic phases. Conventional wisdom suggests that an ultrafine grain size alloy should have a large tensile elongation, presumably resulting from extensive grain boundary sliding. However, it must be pointed out that, in the case of grain boundary sliding, sliding strain must be properly accommodated either by diffusional flow or by dislocation slip (e.g., climb or glide) across neighboring grains, in order to prevent cavitation and, thus, fracture. However, dislocation slip in an ordered, multicomponent intermetallic compound is difficult at temperatures near T_x. Also, diffusional processes are not expected to be sufficiently fast to accommodate sliding strains at strain rates of ~10^{-2} s^{-1}. This offers an explanation for a high *m* value but low tensile elongation in metallic glasses at temperatures near T_x. It is worth noting that, no accommodation is needed for pure Newtonian flow.

Microstructure and Fracture surface

Zr-based metallic glasses have poor oxidation resistance. For example, the cross-section microstructure in the vicinity of the fracture surface of the sample tested at 713K and 10^{-3} s^{-1} is shown in Fig. 5. As expected, the internal structure is featureless. However, an oxide layer, estimated to be about 4 μm thick, is present. The layer was extremely brittle and non-uniform in thickness. The uneven thickness suggests that oxide layer was fractured and new surface was continuously created, but immediately oxidized, as a result of sample stretching. The formation and fracturing of oxide layer is not expected to affect significantly the flow stress since the layer thickness is relatively low. It has also little effect on the ductility, since the material is not notch sensitive in the supercooled liquid region, except in the final stage of deformation.

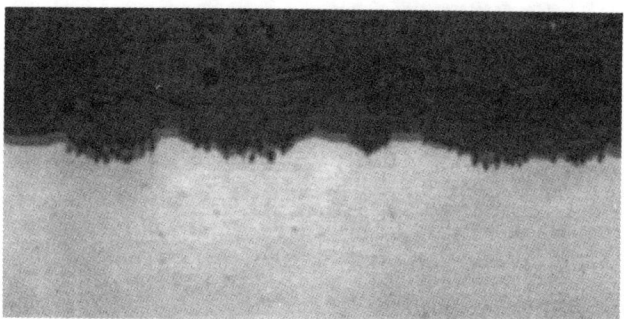

Fig. 5 Cross sectional view of the sample fractured at 713K and 10^{-3} s^{-1}. A non-uniform oxide layer is readily seen. The maximum thickness is about 4 μm.

It is well known that, at room temperature, the fracture surface of a metallic glass exhibits a vein pattern (Fig. 6), as a result of the sudden release of elastic energy at fracture [2]. Also, sample does not show any necking after fracturing. These are typical fracture characteristics for metallic glasses deformed inhomogeneously. In contrast, metallic glasses show strong resistance to necking in the supercooled liquid region; this is readily observed in Fig. 1. The strong necking resistance is apparently a result of high strain rate sensitivity.

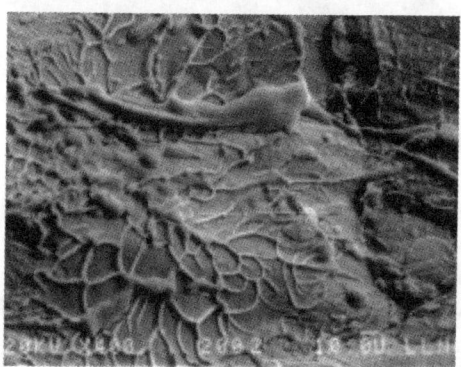

Fig. 6 Vein pattern formed on the fracture surface of amorphous of Zr-10Al-5Ti-17.9Cu-14.6Ni tested in tension at room temperature. Melted droplets are readily seen.

The fracture surface of the sample tested at 663K and at a strain rate of 10^{-2} s^{-1} is shown in Fig. 7. It appears that vein pattern persists even at 663K, where homogeneous deformation prevails. However, a close examination of Fig. 7 indicates that the pattern is actually different from that observed at room temperature. Specifically, the ridges between voids are much higher and there is no indication of melting. In addition, large voids were present. These voids were obviously developed prior to the final fracture of the sample; they were formed under a triaxial stress. As shown in Fig. 1, the sample exhibits considerable necking. The reduction in area for the 663K sample was about 96%. At 683K sample fractured nearly to a chisel point (reduction in area>99%). The fracture surface, as shown in Fig. 8, reveals a ductile dimple fracture with the absence of vein pattern. The brittle-to-ductile transition from inhomogeneous to homogeneous deformation is clearly revealed by the fracture surface appearance.

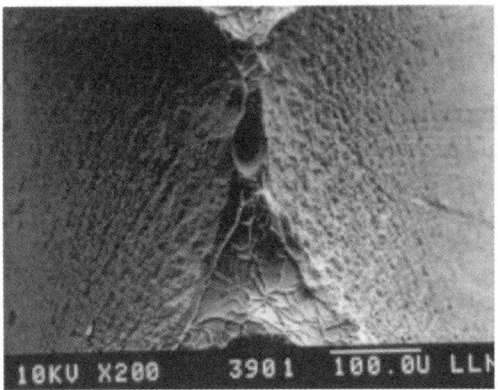

Fig. 7 Fracture surface of the sample tested at 663K and at a strain rate of 10^{-2} s^{-1}.

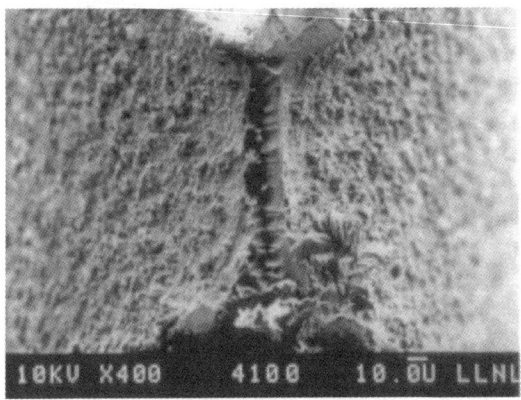

Fig. 8 Fracture surface of the sample tested at 683K and at a strain rate of 10^{-2} s^{-1} reveals a ductile dimple appearance with the absence of vein pattern.

SUMMARY

The deformation behavior of a Zr–10Al–5Ti–17.9Cu–14.6Ni metallic glass was characterized in the supercooled liquid region. The alloy was observed to exhibit a large tensile elongation in this region; a maximum tensile elongation of over 600% was recorded at 698-713K at a high strain rate of 10^{-2} s^{-1}. The superplastic properties, e.g. flow stress and elongation, are found to be very sensitive to testing temperature. As a result of structural instabilities it is difficult to determine the "true" strain rate sensitivity value. Despite this difficulty, experimental results indicated that the alloy does not behave like a Newtonian fluid (m=1). It is suggested that the non-Newtonian behavior is caused by the concurrent crystallization of the amorphous structure during deformation; a mixed crystalline-plus-amorphous structure was actually tested. At temperatures above the crystallization temperature, in spite of having a nanocrystalline structure, the alloy exhibit limited ductility. This is a result of poor strain accommodation at grain triple junctions. Microstructural examination of the fracture sample is now underway.

ACKNOWLEDGMENT

This work was performed under the auspices of the U.S. Department of Energy by Lawrence Livermore National Laboratory under contract No. W-7405-Eng-48. The authors would like to thank Dr. Luke Hsiung for his contribution of Fig. 2.

REFERENCES

1. W. Klement, R. Willens, and P. Duwez, *Nature*, **187** (1960) 869.
2. C.T. Liu, *et al.*, *Mater. Trans. A*, **29A** (1998) 1811.
3. K. Hashimoto, N. Kumagai, H. Yoshioka, H. Habazaki, A. Kawashima, K. Asami, and B.-P. Zhang, *Mater. Sci. Eng.*, **A133** (1991) 22.
4. D.G. Morris, in *Proc. 5th Int'l Conf. on Rapidly Quenched Metals*, p. 1775, edited by S. Steeb and H. Warlimont, Elsevier Science Publishers B.V., 1985.
5. Y. Kawamura, T. Shibata, A. Inoue, and T. Masumoto, *Scr. Mater.*, **37** (1997) 431.
6. Y. Kawamura, T. Nakamura, and A. Inoue, *Scr. Mater.*, **39(3)** (1998) 301.
7. Y. Kawamura, T. Nakamura, A. Inoue, and T. Masumoto, *Mater. Trans. JIM*, **40(8)** (1999) 794.
8. T. Masumoto and R. Maddin, *Mater. Sci. Eng.*, **19(1)** (1975) 1.
9. H. Chen, Y. He, G.J. Shiflet, and S.J. Poon, *Nature*, **367(6463)** (1994) 541.
10. F. Spaepen, *Acta Metall.*, **25** (1977) 407.
11. A.S. Argon, *Acta Metall.*, **27** (1979) 47.
12. J.C.M. Li, in *Proc. 4th Int'l Conf. on Rapidly Quenched Metals*, p. 1335, edited by T. Masumoto and K. Suzuki, Japan Institute of Metals, Sendai, Japan, 1982.
13. A.L. Mulder, R.J.A. Derksen, J.W. Drijver, and S. Radelaar, in *Proc. 4th International Conf. on Rapidly Quenched Metals*, p. 1345, edited by T. Masumoto and K. Suzuki, Japan Institute of Metals, Sendai, Japan, 1982.
14. A.I. Taub and F.E. Luborsky, *Acta Metall.*, **29** (1981) 1939.
15. H. Eyring, *J. Chem. Phys.*, **4** (1936) 283.
16. C. Homer and A. Eberhardt, *Scr. Metall.*, **14** (1980) 1331.
17. V.A. Zelenskiy, A.S. Tikhonov, and A.N. Kobylkin, *Russian Metallurgy*, **4** (1985) 152.
18. R. Busch, E. Bakke, and W.L. Johnson, *Acta Mater.*, **46(13)** (1998) 4725.
19. V.A. Khonik and V.A. Zelenskiy, *Phys. Met. Metall.*, **67(1)** (1989) 196.
20. K. Csach, Y.V. Fursova, V.A. Khonik, and V. Ocelik, *Scr. Mater.*, **39(10)** (1998) 1377.
21. C.P. Ashdown, Y. Zhang, and N.J. Grant, *Int'l J. Powder Metall.*, **23(1)** (1987) 33.
22. H. Brandt, J. Gossing, G. Mathiak, and H. Neuhauser, *Z. Metallkd.*, **84(4)** (1992) 273.
23. D.L. Wang, Q.P. Kong, and J.P. Shui, *Scr. Metall. Mater.*, **31(1)** (1994) 47.
24. J. Deng, D.L. Wang, Q.P. Kong, and J.P. Shui, *Scr. Metall. Mater.*, **32(3)** (1995) 349.
25. K. Higashi, T. Mukai, A. Uoya, A. Inoue, and T. Masumoto, *Mater. Trans. JIM*, **36(12)** (1995) 1467.
26. A. Uoya, T. Shibata, K. Higashi, A. Inoue, and T. Masumoto, *J. Mater. Res.*, **11(11)** (1996) 2731.
27. T.G. Nieh, J. Wang, J. Wadsworth, T. Mukai, and C.T. Liu, in *Symposium on Bulk Metallic Glasses*, edited by W.L. Johnson, C.T. Liu, and A. Inoue, Materials Research Soc., Pittsburgh, PA, 1999. -in press
28. A.H. Cottrell and B.A. Bilby, *Proc. Phys. Soc. (London)*, **62A** (1949) 49.
29. T.G. Nieh and W.D. Nix, *Metall. Trans.*, **17A** (1986) 121.
30. Y. Kawamura, T. Shibata, A. Inoue, and T. Masumoto, *Mater. Trans. JIM*, **40(4)** (1999) 335.
31. T.G. Nieh, O.D. Sherby, and J. Wadsworth, *Superplasticity in Metals and Ceramics*, Cambridge University Press, Cambridge, UK, 1997.
32. R. Maddin and T. Masumoto, *Mater. Sci. Eng.*, **9** (1972) 153.
33. Y.H. Kim, A. Inoue, and T. Masumoto, *Mater. Trans. JIM*, **31** (1990) 747.
34. Y.H. Kim, K. Hiraga, A. Inoue, T. Masumoto, and H.H. Jo, *Mater. Trans. JIM*, **35(5)** (1994) 293.

RECENT DEVELOPMENTS IN SUPERPLASTICITY

S. X. McFADDEN[*], R. S. MISHRA[*+], and A. K. MUKHERJEE[*]
[*]Dept. of Chemical Eng. & Materials Science, University of California, Davis, CA 95616
[+]Now at the Dept. of Metallurgical Engineering, University of Missouri, Rolla, MO 65409

ABSTRACT

The phenomenon of superplasticity is explored in the range of high strain rate for both nanocrystalline and microcrystalline materials. True tensile superplasticity has been demonstrated in nanocrystalline grain size range. The difference in the details of such superplasticity between nanocrystalline and microcrystalline state is emphasized.

INTRODUCTION

Nanocrystalline materials contain a very large density of grain boundaries [1]. Therefore, it has attracted considerable discussion on the extension of grain boundary related deformation mechanisms. Superplasticity is a well-established grain boundary dependent phenomenon. Bulk nanocrystalline materials provide an opportunity to investigate the scalability of grain size dependent phenomenon to a much finer scale than was previously deemed to be possible.

The elevated temperature crystalline plasticity is given by the Mukhejee-Bird-Dorn (MBD) correlation [2]. This correlation has been validated in the context of superplasticity for metals, intermetallics, and ceramics materials [3]. A common issue for investigation of superplasticity is: "does the formalism for superplastic mechanism that has been established for microcrystalline materials scale with grain size to the nanocrystalline range or is there a transition in superplastic behavior at such fine grain size?"

We briefly review some of the experimental observations of superplasticity in nanocrystalline materials. A comparison with microcrystalline materials leads to some general understanding. There are significant differences in the details of superplasticity in nanocrystalline and microcrystalline structures.

The commercially attractive area of high strain rate superplasticity (HSRS) for various microcrystalline powder metallurgy and mechanically alloyed aluminum or nickel-based alloys is briefly reviewed next and the results are compared with observations on nanocrystalline structures.

Nanocrystalline Superplasticity

We have observed significant levels of superplasticity in nanocrystalline Ni, Ni_3Al, and 1420 Al alloy [4]. Figure 1 shows deformed tensile samples of these materials. The Ni_3Al and 1420-Al alloy were processed from ingot material by severe plastic deformation (SePD), namely torsion straining (true strain of 6 to 7) under high pressure of 1.2 GPa. Details are given in ref [8]. The nanocrystalline Ni was electrodeposited. All samples were deformed at constant strain rates while heated in air. Our experimental observations include both low temperature and high strain rate superplasticity.

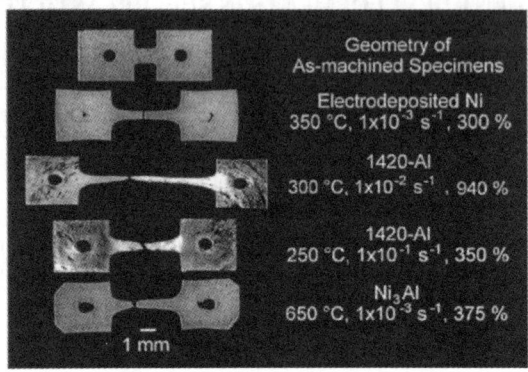

Figure 1. Tensile specimens of nanocrystalline Ni, 1420-Al, and Ni$_3$Al.

High Strain Rate Superplasticity

High strain rate superplasticity (HSRS) is defined [5] by strain rates greater than 10^{-2} s^{-1} and a minimum elongation of 200%. The nanocrystalline 1420-Al alloy was found to be superplastic at a strain rate of 10^{-1} s^{-1}. Table I shows a comparison of elongation, deformation temperature, strain rate, and flow stress for nanocrystalline and microcrystalline 1420-Al. The data shown in Table I was selected using an elongation of around 390% as a frame of reference from which to compare the effect of grain size on strain rate and flow stress. Clearly, for an equivalent elongation, the superplastic strain rate increased with a reduction in grain size. However, nanocrystalline grain size alone is not sufficient to produce HSRS, as evidenced (Table II) by the results with Ni$_3$Al having a grain size of around 50 nm. Although Table II shows that the superplastic strain rate increased with a reduction in grain size from 6 μm to 50 nm, HSRS was not obtained with nanocrystalline Ni$_3$Al.

Table I: Comparison of superplastic results for 1420-Al

	100 nm grain size	1.2 μm grain size (Berbon et al.[6])	6 μm grain size (Kaibyshev [7])
% Elongation	390	380	400
Temperature (°C)	300	300	450
Strain rate (s^{-1})	5×10^{-1}	1×10^{-1}	1×10^{-3}

Table II: Comparison of superplastic results for Ni$_3$Al [8]

	50 nm grain size	6 μm grain size
% Elongation	560	560
Temperature (°C)	725	1050
Strain rate (s^{-1})	1×10^{-3}	6×10^{-4}

Low Temperature Superplasticity

Nanocrystalline materials have consistently shown a reduction in the superplastic temperature. A comparison of the results for nanocrystalline materials shows that the superplastic temperature decreased (as compared to microcrystalline structure) by ~200°C for 1420-Al, ~325°C for Ni_3Al, and ~400°C for Ni. Salischev et al. has shown the same trend for nanocrystalline Ti-6Al-3.5Mo [9,10]. These results show a significant drop in terms of homologous temperature for superplasticity when the grain size changes from microcrystalline to nanocrystalline state. The reduction of superplastic temperature for Ni_3Al has direct technological relevance. For example, a Ni_3Al alloy with 6 μm grain size shows superplasticity at 1050°C. This temperature is higher than the current industrial practice of ~900°C as an upper temperature limit for conventional tooling for forming operations. The decrease of superplastic temperature in nanocrystalline Ni_3Al brings it within the conventional superplastic forming range of titanium alloys. As a result, the current die and tooling practices could be used.

The most dramatic reduction in superplastic temperature was observed with nanocrystalline Ni. Superplastic elongation was obtained at temperatures as low as 350°C, or $0.36T_m$, where T_m is the melting temperature of nickel in absolute degrees. A maximum elongation of 895% was achieved at 420°C, and a strain rate of 1×10^{-3} s^{-1}. Previous work on microcrystalline Ni [13,14] resulted in a maximum elongation of 250% and 180% at 820°C and 800°C, respectively. The results on nanocrystalline Ni come closest to the early predictions of enhanced superplasticity in nanocrystalline materials. However, we are still in the process of evaluating the possible effects of impurities, particularly sulphur, on the observed behavior.

Grain Growth

The large volume fraction of grain boundary area in nanocrystalline materials results in a strong driving force for grain growth. We have studied grain growth in nanocrystalline Ni, 1420-Al, and Ni_3Al, using differential scanning calorimetry (DSC), *in situ* TEM heating, and TEM investigation of annealed specimens. Figure 2(a,b) shows the DSC curves obtained for Ni_3Al and 1420-Al, respectively. In the as-processed condition (SePD) both of these materials have microstructures containing significant levels of internal strain, evidenced by diffuse contrast and indistinct grain boundaries as shown in Figure 3(a,b). With heating, recovery results in a reduction of the internal strain followed by grain growth. In materials processed by SePD, normal grain growth has been observed, while in the case of Ni, a period of abnormal growth has been shown, followed by normal growth after a grain size of around 0.5 μm when a lognormal distribution has been reached.

From Figure 2, it is clear that nanocrystalline superplasticity is associated with grain growth. In the case of nanocrystalline Ni, the majority of superplastic deformation at 350°C occurred with a grain size of 500 nm. The final grain size after deformation was 1 μm. Historically, the lack of superplastic observations in pure metals has been attributed to rapid grain growth, which can increase the grain size beyond the general range for microcrystalline superplasticity i.e., a grain size less than 10μm [14]. While starting with nanocrystalline material can lower the superplastic temperature and thereby reduce the effects of grain growth, in the most rigorous sense, superplasticity in pure metals and nanocrystalline grain size appear to be inherently incompatible.

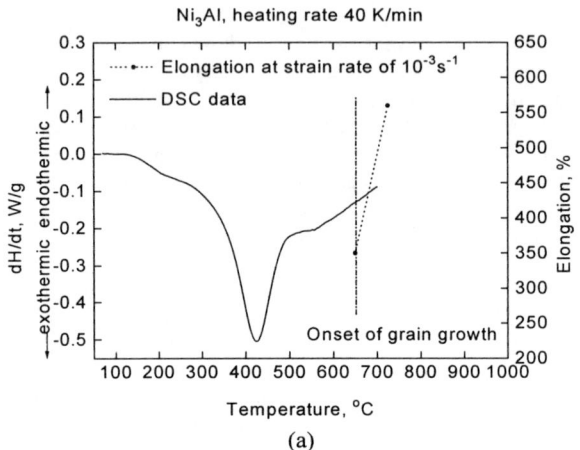

(a)

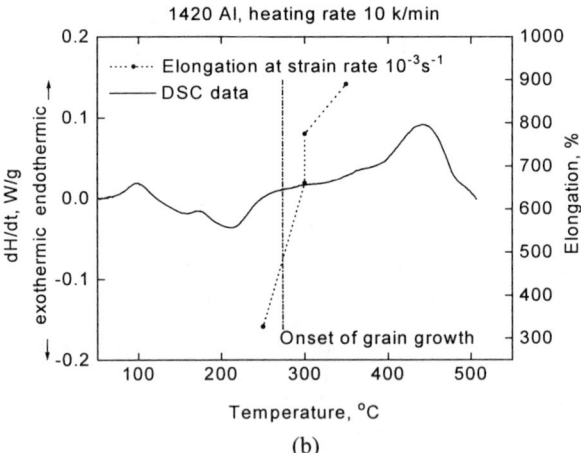

(b)

Figure 2. Differential scanning calorimetry curves for (a) nanocrystalline Ni₃Al and (b) nanocrystalline 1420-Al. The large peak in (a) is attributed to recovery and reordering of the severely deformed structure. The first three peaks in (b) are attributed to precipitate coarsening (exothermic) and precipitate dissolution (endothermic). The onset of grain growth is indicated on each plot. Tensile elongation values are also plotted to show that the onset of superplastic behavior coincided with the onset of microstructural instability.

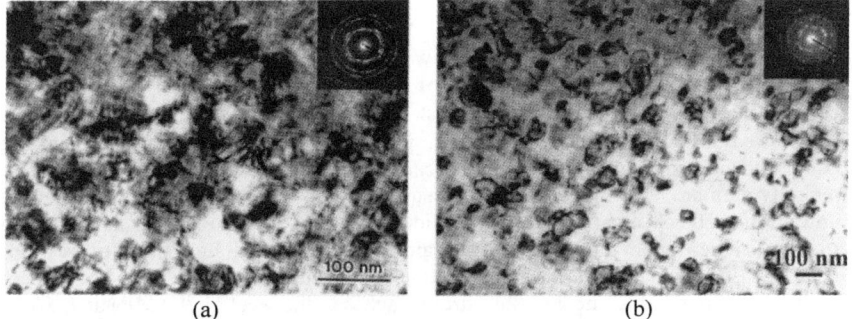

(a)   (b)

Figure 3. Bright-field TEM of (a) nanocrystalline Ni₃Al and (b) nanocrystalline 1420-Al. The diffuse contrast in (a) is attributed to large residual strain often observed in materials processed by severe plastic deformation. The comparatively good contrast in (b) is attributed to room temperature recovery from the strains introduced by severe plastic deformation in this aluminum alloy.

At the lowest superplastic temperature, grain growth in 1420-Al was not as extensive as in Ni. After deformation at 250°C, the grain size was 300 nm. At 300°C, the final grain size was 1 μm. Difference between grain growth in Ni and 1420-Al can be attributed to the effects of solute atoms and precipitates on inhibiting grain growth. The ordered structure of Ni₃Al presented an even more effective barrier to grain growth. Following deformation at 650°C, TEM investigation showed a grain size of 100 nm, which is within the currently accepted notion of nanocrystalline range.

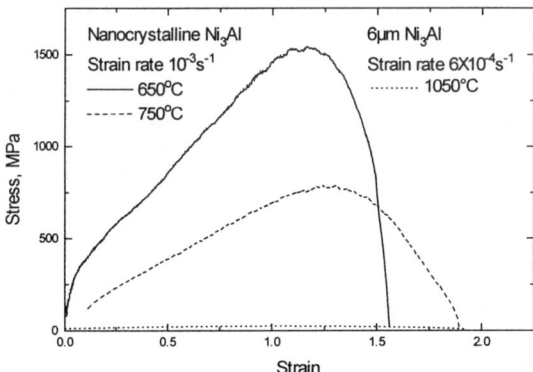

Figure 4. Stress-strain curves for nanocrystalline Ni₃Al. A stress-strain curve for microcrystalline Ni₃Al was included on the plot to demonstrate the very highflow stresses observed from nanocrystalline Ni₃Al.

High Flow Stresses

Figure 4 shows a comparison of the flow curves during superplasticity in Ni$_3$Al alloy. The high flow stresses during superplasticity of nanocrystalline Ni$_3$Al alloy are apparent. If we compare the 6μm grain size specimen tested at 1050°C and the 50nm nanocrystalline specimen tested at 650°C, although the test temperature and grain size was different for the two curves, the comparison is interesting because the overall ductility in both cases was inadvertently similar. One of the features of superplasticity in microcrystalline materials is low flow stress. The observation of superplasticity with high flow stresses in nanocrystalline material needs a new approach in order to explain the origin of such high flow stresses. First, the flow stress near yielding (for nanocrystalline specimens) is around 170 MPa. Second, after significant strain hardening, the flow stress reaches to the level of 780 MPa at 725°C. At 650 °C, the flow stress values are even higher [8], i.e., 1.5 GPa. Although the well known temperature dependent strength anomaly and strong strain hardening in Ni$_3$Al occurs near the test temperature of 725°C reported for this work, a ductility minimum on the order of 20 % elongation also occurs at this temperature for Ni$_3$Al [11]. Furthermore, fine-grained materials (2.9 to 9.5 μm) have not shown the strength anomaly [12]. Therefore, the temperature dependent stress anomaly alone is not sufficient to account for the high flow stresses observed, particularly in light of the large elongation achieved.

HSRS in Microcrystalline Matrix

During the early years of development of this field, the grain sizes used to be approximately 15 μm and the optimum strain rate used to be about 10^{-4} s^{-1} (especially for aluminum alloys). With the development of processing methods to produce ultra-fine grain size, it has been possible to move up the optimum forming rates to approximately 10^{-1} to 10^{-2} s^{-1}. In this range, High Strain Rate Superplasticity (HSRS) is being explored aggressively in recent times.

The general features [21] of HSRS are the following: (a) The parametric dependencies tend to depend on the reinforcement size and grain size, (b) the optimum superplastic temperatures change with the matrix material, and (c) the optimum superplastic conditions can be significantly altered by prior thermomechanical processing.

The natural question that arises can be phrased as: "Is the rate controlling mechanism the same for HSRS in all of the materials?" It is shown here that the size of second phase particles influences the parametric dependencies, particularly the high activation energy for HSRS for large reinforcing particle sizes. The results are explained on the basis of a change in the rate controlling mechanism for deformation with particle size.

ANALYSIS AND DISCUSSION

The present analysis is based on the existence of a threshold stress for HSRS of dispersion strengthened materials. The variation of true activation energy (calculated after taking into account the presence of a temperature dependent threshold stress) with particle size is plotted in Fig. 5 for a number of mechanically alloyed aluminum alloys and aluminum matrix composites. The important observation that is apparent from this figure is the change in activation energy for alloys with bigger particle sizes. It will be appropriate to examine the role of particle size on the accommodation process during superplasticity.

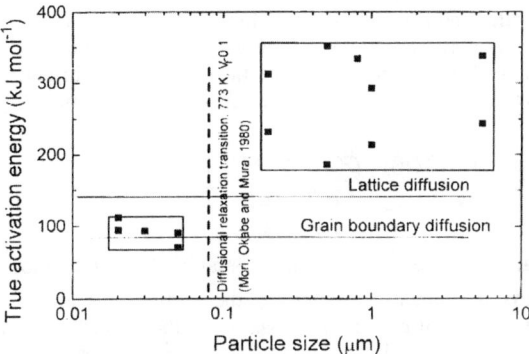

Figure 5. The variation of true activation energy with particle size for a number of mechanically alloyed aluminum alloys and aluminum matrix composites. Note that the change can be predicted by diffusional relaxation models.

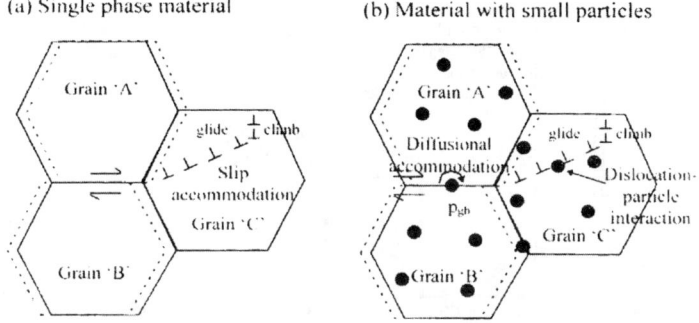

Figure 6. The effect of second phase particles on slip accommodation during grain boundary sliding. (a) Typical micrograin superplasticity in single-phase material or when all phases are deformable. (b) The presence of small particles will modify the grain boundary sliding and dislocation movement during slip accommodation.

Figure 6(a) shows the classical concept of slip accommodation during grain boundary sliding. This has been shown to be applicable for a number of materials [15]. We now examine the influence of second phase particle on the process. The particles at grain boundaries (e.g., particle marked P_{gb} in Fig. 6(b)) impede the grain boundary sliding and lead to stress concentration. This stress concentration must be lowered for continuous sliding to take place and to avoid cavity nucleation. The stress relaxation can occur by diffusional flow of atoms around the particles as depicted in Fig. 6(b). If the rate of diffusional relaxation is fast enough to remove the stress build up, the overall grain boundary sliding would not be influenced parametrically. The rate of such diffusional relaxation of stress concentration has been calculated by Koeller and Raj [16] and by Mori et al. [17]. In addition, Koeller and Raj have pointed out that the increase in stress at the particle matrix interface at above a critical strain rate may result in decohesion at that interface leading to formation of voids, and a consequent loss of ductility. The critical strain rate for cavity nucleation is given by [16]

$$\dot{\varepsilon}_c \geq C \frac{(1-\nu)[1-2\nu+(2/\nu)]}{[5/6-\nu]^2} \left(\frac{G\Omega}{KT}\right)\left(\frac{V_f \delta D_g}{d_p^3}\right) \tag{1}$$

where $\dot{\varepsilon}_c$ is the critical strain rate, ν is the Poisson's ratio, G the shear modulus, Ω the atomic volume, V_f the volume fraction of particles, d_p the particle diameter, and δ is the grain boundary width. The numerical constraint C has the value of 118 in Koeller and Raj [16]. Subsequently, Mori et al. [17] found that a more rigorous analysis results in a lower value for the constant C equal to 9. This equation is derived on the basis of interfacial diffusional relaxation.

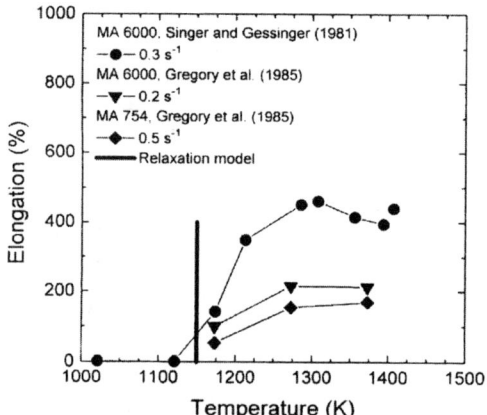

Figure 7. The onset of high strain rate superplasticity in mechanically alloyed nickel based alloys as a function of temperature.

It will be instructive to evaluate the applicability of the diffusional relaxation on the onset of HSRS in MA6000 (an ultra-fine oxide dispersion strengthened nickel-based alloy). Taking values of $\nu = 0.33$, $V_f = 0.025$, $d_p = 10$ nm, D_v, and G and Ω from ref. [18], $\dot{\varepsilon}_c = 0.3$ s^{-1} and the constant C = 9 (according to Mori et al.), the critical temperature at which the diffusional relaxation rate is fast enough to relax the stress build up comes out to be ~ 1150K. The prediction of Equation (1) is compared with the data of Singer and Gessinger [19] for MA6000 alloy in Fig 7. It can be noted that the increase in observed ductility agrees well with the prediction of diffusional relaxation model. Another way to verify the applicability of this concept is to calculate the critical particle size for a given temperature and strain rate below which diffusional relaxation will be adequate. The result of such a calculation for aluminum alloys is incorporated in Fig. 5 (at 773K and $V_f = 0.1$ and for $\dot{\varepsilon}_c = 100$ s^{-1}, see ref.

20), the critical particle size comes out to be 0.08 μm (using the constant of $C = 9$ from Mori et al. in Equation (1)). The change in activation energy agrees well with the critical particle size for diffusional relaxation by grain boundary diffusion, according to Equation (1).

In the case of metal matrix composites, the microstructure is quite different. This situation has been treated in detail by Mukherjee et al. [21] and will not be treated here any further. However, that analysis does explain the experimentally observed activation energy that often is much higher than that for volume division of the matrix.

CONCLUSIONS

1. Tensile superplasticity has been demonstrated in nanocrystalline materials.
2. The superplastic deformation kinetics are slower in nanocrystalline materials after grain size and temperature compensation.
3. The micromechanisms and constitutive relationships developed for superplasticity in microcrystalline materials are not simply scaleable to the nanocrystalline range.
4. The activation energy for high strain rate superplasticity in microcrystalline materials changes with particle size. Diffusional relaxation models can be used to explain this change as well as for predicting the temperature for onset of high strain rate superplasticity in mechanically alloyed nickel based or aluminum based alloys.

ACKNOWLEDGEMENTS

The authors acknowledge the support from the U. S. National Science Foundation under grant NSF-DMR-9903321. We are thankful to Professor Ruslan Valiev for providing us with the severe plastically deformed samples under the U. S. NSF-International Program grant NSF-DMR-9630881.

REFERENCES

1. H. Gleiter, Prog. Mater. Sci., **33**, p. 223, (1990).
2. A. K. Mukherjee, J. E. Bird and J. E. Dorn, Trans. ASM, v. **62**, p. 155 (1969)
3. A. K. Mukherjee in: *Plastic Deformation and Fracture of Materials*, vol. **6**, edited by H Mughrabi, VCH Publishers, Weinheim, Germany, 1993, pp. 407-460.
4. S. X. McFadden, R. S. Mishra, R. Z. Valiev, A. P. Zhilyaev, A. K. Mukherjee, Nature, **398**, p. 684 (1999).
5. Glossary of Terms Used in Metallic Superplastic Materials, JIS-H-7007, Japanese Standards Association, Tokyo, Japan, p. 3, (1995).
6. P. B. Berbon, N. K. Tsenev, R. Z. Valiev, M. Furukawa, Z. Horita, M. Nemoto, and T. G. Langdon, Metall. Mater. Trans., **29A**, 2237 (1998).
7. O. A. Kaibyshev, *Superplasticity of Alloys, Intermetallics, and Ceramics*, Springer-Verlag, Berlin, 1992, p. 151.
8. R. S. Mishra, R.Z. Valiev, S.X. McFadden, A. K. Mukherjee, Mat. Sci. & Eng., **A252**, p. 174, (1998).
9. G. A. Salischev, O. R. Valiakhmatov, V. A. Valitov and S. K. Mukhtorov, Materials Science Forum, **170-172**, p. 121 (1994).
10. G. A. Salischev, R. M. Galeyev, S. P. Malisheva and O. R. Valiakhmetov, Materials Science Forum, **243-245**, p. 585 (1997).
11. C. T. Liu, V. Sikka, J. Met., **38**, p. 19 (1996).

12. E. M. Schulson, I. Baker, H. J. Frost: in *High Temperature Ordered Intermetallic Alloys II*, edited by N. S. Stoloff, (Mater. Res. Soc. Symp. Proc., **81**, 1987) pp. 195-205.
13. S. Floreen, Scripta Metallurgica, **1**, p. 19 (1967).
14. O. A., Kaybyshev, A. A. Markelov, Fiz. Metal. Metalloved. **41**, No 1, p. 190 (1976).
15. A. K. Mukherjee, T. Bieler and A. Chokshi, *Materials Architecture*, edited by J. B. Bilde-Sorenson et al. Riso, Denmark, 1989, p. 207.
16. R. C. Koeller, R. Raj, Acta Met., **26,** p. 1551 (1978).
17. T. Mori, M. Okabe, T. Mura, Acta Met., **28**, p. 319 (1980).
18. H. J. Frost, M. F. Ashby, *Deformation Mechanism Map*, Pergamon, London, 1982.
19. R. F. Singer, G. H. Gessinger, in *Deformation of Polycrystals: Mechanism and Microstructures*, 2nd, edited by N. Hansen et al., (RISO International Symposium, RISO National Laboratory, Roskilde, Denmark, 1981), p. 385.
20. R. S. Mishra, T. R. Bieler, A. K. Mukherjee, Acta Mater., **45**, p. 561 (1997).
21. A. K. Mukherjee, R. S. Mishra, T. R. Bieler, Mater. Sci. Forum, **233-234**, p. 217 (1997)

CERAMICS SUPERPLASTICITY
- New Properties from Atomic Level Processing -

Fumihiro WAKAI*,**
*Ceramics Superplasticity Project, ICORP, Japan Science and Technology Corporation
JFCC 2F, 2-4-1 Mutsuno, Atsuta, Nagoya 456-8587, Japan
**Center for Materials Design, Materials and Structures Laboratory
Tokyo Institute of Technology, 4259 Nagatsuta, Midori, Yokohama 226-8503, Japan,
wakai@rlem.titech.ac.jp

ABSTRACT

The international joint project "Ceramics Superplasticity" has been conducted from 1995 till 1999 by a cooperation between the Japan Science and Technology Corporation (JST) and Max-Planck-Institute for Metals Research (MPI-MF). The new type of covalent ceramics was developed through atomic level processing, and novel properties such as superplasticity were investigated. The development of local analysis technique of disordered structure combined the two research fields, grain boundary of crystalline solids and structure of amorphous.

INTRODUCTION

Materials and the tools made from them play a crucial role in the history of human development. The search for new materials with outstanding properties through new synthesizing route is a key to the next millennium. Si_3N_4 and SiC are the ceramic compounds that adopt network structures featuring vertex sharing of, respectively, $[SiN_4]$, and $[SiC_4]$ tetrahedra. (Fig. 1) Silicon nitride is an important structural ceramic and fiber-reinforced composite matrix, while silicon carbide is a principal reinforcing fiber, a potential high-temperature device semiconductor, and a leading candidate (in SiC-SiC fiber-reinforced ceramic matrix composite form) for aerospace applications. Networks are not constrained to be crystalline arrangements of connected polyhedra; highly

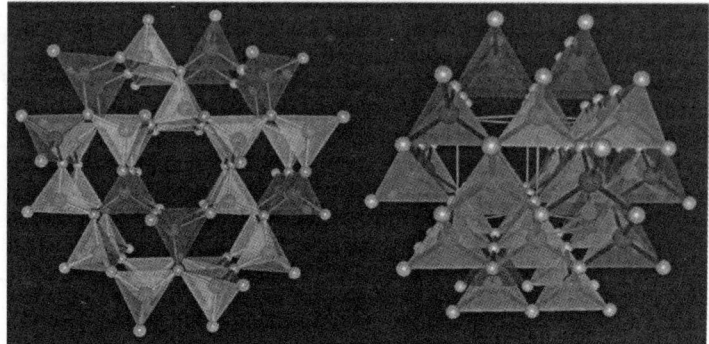

Fig. 1 Network structure of tetrahedra in Si_3N_4 and SiC crystals. (MPI-MF)

Mat. Res. Soc. Symp. Proc. Vol. 601 © 2000 Materials Research Society

covalent networks, especially, are more likely to be found in non-crystalline forms than are less covalent ceramics. Silica, for example, is easily amorphized by cooling from its liquid. The preparation of inorganic materials by thermolysis of preceramic compounds, the precursor-derived ceramics, brought about new class of covalent amorphous materials (Si-C-N, Si-B-C-N, Si-C-O) and also nanocrystalline materials (Si_3N_4 / SiC).[1-3]

Although it is well known that amorphous materials such as silicate glass flows at elevated temperatures, polycrystalline ceramics had been regarded as brittle materials which are broken without any significant plastic deformation. The finding of "Ceramics Superplasticity" has overthrown this old concept.[4,5] Superplasticity refers to an ability of polycrystalline solids to exhibit exceptionally large elongation in tension. The application of superplasticity make it possible to fabricate ceramic components by superplastic forming (SPF), SPF concurrent with diffusion bonding, and superplastic sinter-forging just like superplastic metals. Furthermore the superplastic deformation plays an important role in stress-assisted densification processes such as hot isostatic pressing (HIP) and hot pressing (HP). The ceramics superplasticity has been one of intensive research fields in the last decade. Although most of reports are still limited to those of zirconia, new developments have been achieved in superplasticity of Si_3N_4 and SiC in recent years.[6,7] It is clearly demonstrated that the superplasticity is one of the common natures of fine-grained ceramics and nanocrystalline ceramics at elevated temperatures.

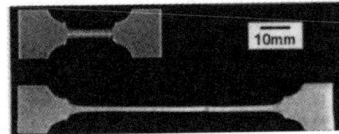

Fig. 2 Superplastic elongation of Si-Al-O-N (N. Kondo, NIRI, Nagoya)[8]

The objectives of our joint project are to develop new type of covalent materials through atomic level processing and to explore the novel properties at elevated temperatures, superplasticity in nonocrystalline materials and viscous flow of amorphous materials. The principal motivation was a simple question, "How polycrystalline materials will behave when the grain size becomes smaller and smaller? Will it approach to the nature of amorphous solids ultimately?" The bridge to combine two research fields, nanocrystalline ceramics and amorphous solids, is the local analysis of disordered region in amorphous structure and in grain boundaries. The new method to obtain quantitative information on interface was developed in this project.

SELECTED TOPICS

Topological Change in Superplasticity

The grain boundary of polycrystalline solids forms complex three-dimensional network structure. The structure is neither periodic nor random. If the distribution of grain size maintains the self-similar shape which is independent of time, it is known as the steady structure in normal grain growth. The topology of the network changes by movement of grains (superplasticity)[9] and by boundary motion of curvature (grain growth)[10]. The elemental process of topological change in superplasticity is grain switching (T1 process). The sliding of rigid grains generates cavities and cracks inevitably. Then the essential mechanism of superplasticity is the accommodation process of grain boundary sliding so that the polycrystalline materials can be stretched extensively without fracture. The three-dimensional simulation indicated that how the dynamical change in shape of grain occurs in compatible deformation where individual grain contacts and separate from other grain. (Fig. 3) A vector in Fig.4 represents the motion of individual grain in superplastic flow. The length of arrow is proportional to the velocity of grain. The displacement of the centers of mass of the grain is similar to the movement of molecules in viscous flow of liquid (or amorphous).

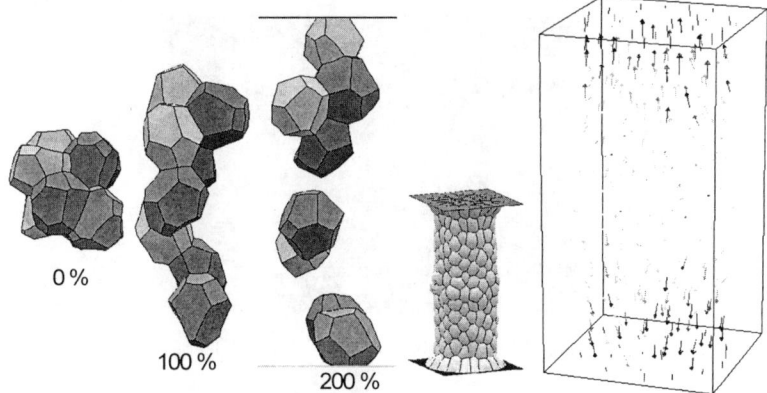

Fig. 3 Rearrangement of grains in superplasticity.   Fig.4 Motion of grains in deformation.

The idealized grain growth in three-dimension (3D) was studied by a program that models the process of boundary motion by curvature to minimize the boundary energy. Even starting from arbitrary packing of uniform grains, the boundary network reached to a steady structure in time after incubation period and transient period. The parabolic law in grain growth was observed only in a region where the steady structure was maintained. The more general von Neumann-Mullins law on kinetics of grain growth held in both transient period and normal grain growth period. The grain size distribution function and the distribution of number of faces in steady structure were analyzed in 3D, and compared with the microstructure in cross section. The perimeter law and Aboav-Weaire law in 3D on topological nature of boundary network structure held not only in the steady structure but also in transient structures.[10]

Superplasticity Enhanced By Intergranular Liquid – Si_3N_4 and SiC

The structure and the nature of grain boundary affect accommodation processes in grain boundary sliding significantly. In many metals and ceramics impurity atoms segregate at grain boundaries, but no glass film has been detected by high-resolution transmission electron microscopy (HRTEM). On the other hand, glass phase pockets remain frequently at grain boundaries of liquid-phase sintered ceramics. The glass phase behaves as a viscous fluid (super-cooled liquid) at temperatures higher than those of the glass transition. When the glass wets the crystal completely, thin glass film with the thickness from 0.5 to a few nm is observed at two-grain junctions.

The glass phase acts as a lubricant for grain boundary sliding, and also as a path for matter transport through solution-precipitation process.[11] Then the mechanism of superplasticity for glass-containing ceramics will be different from the intrinsic superplasticity of metals and ceramics that do not contain glass film. The Si-Y-Al-O-N glass phase exists at

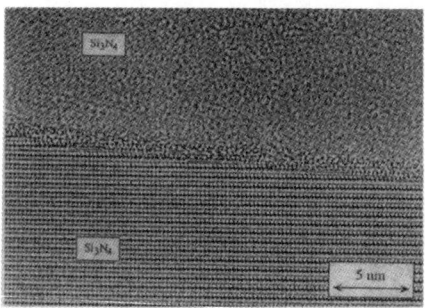

Fig. 5 Intergranular film at Si_3N_4 (J. Kleebe)

grain boundaries of the Si_3N_4 which were sintered with Y_2O_3 and Al_2O_3 as additives. It has been already known that even the materials which contain rod-shaped β- Si_3N_4 grains can be superplastically deformed assisted by glass phase. In this "Non-Classical" superplasticity the anisotropic β-grains tend to align with strain, producing a fiber-strengthening effect. These phenomena lead a strengthening and toughening of Si_3N_4 by compressive superplastic deformation.[8,12] The enhancement of superplastic deformation by intergranular glass phase was also applicable to liquid-phase sintered SiC very recently.[13] The oxide glass film at grain boundary of SiC is a Si-C-O based amorphous. Therefore the understanding on structure and properties of bulk Si-C-O [14] gives valuable information to understand the nature of intergranular phase.

Intrinsic Superplasticity in Covalent Ceramics - B, C doped SiC

Is the superplasticity of covalent polycrystalline solids possible where no help of intergranular glass phase is expected? The superplastic elongation was achieved for B, C-doped SiC which was fabricated by hot isostatic pressing.[15,16] The intergranular glass film was not detected under TEM observation. This result illustrates that the phenomena of superplasticity can be observed in all fine-grained polycrystalline solids regardless of the difference in atomic bond and structure of grain boundaries.

Characterization of Grain Boundary and Amorphous

The characterization of grain boundary structrure and its chemistry are essential to understand sintering, grain growth and deformation at elevated temperatures. In order to obtain the information on the nature of chemical bond at grain boundary, the method to separate the electron energy loss signal (EELS) from grain boundary has been developed. The method provides quantitative information on segregation of impurity atoms and "chemical width" of grain boundary.[17,18] For example, boron segregates at grain boundaries of B, C-doped SiC.[19] The boron takes the place of silicon but forms bonds in a local environment that are similar to those in B_4C structure. The presence of B-C and B-B bond at the grain boundary is beneficial to increase the self-diffusion rate, because the local structure and bonding have more varieties. This is why boron enhanced the sintering and superplasticity of SiC.

The local structure and chemistry of the amorphous materials hold the key for the fundamental understanding of the new meta-stable states and for the evaluation of the ultimate potential of the new material processing method. The nano characterization of precursor-derived ceramics (Si-B-C, Si-B-C-N, Si-C-O) is undertaken by using analytical electron microscopy and EELS which is sensitive to chemical environmental change surrounding the probed elements. The nano-scale phase separation in the amorphous structure was detected for the first time.

Deformation of Precursor-Derived Ceramics

The extensive investigations of viscosity of silicate melts, relaxation, rheology, and the glass transition have yielded models for these physical properties as a function of composition. These properties are controlled by the atomic coordination and Si-O bond strengths of the melt. The precursor-derived covalent amorphous (Si-B-C, Si-B-C-N) are novel materials. The research field on mechanical properties of these materials is quite a frontier.[21] These materials maintain the excellent mechanical strength at temperatures higher than 1800 K. What will these amorphous solid deforms at elevated temperatures? Is it similar to silicate glass? It is revealed that the deformation can be described as viscous flow. The viscosity was in good accordance with the prediction of deformation model of **metallic glass**.[22] The ability for deformation was applied to fabricate completely dense precursor-derived ceramics by hot isostatic pressing.[23]

CONCLUSION

The superplasticity of covalent ceramics and viscosity of covalent amorphous solids may be regarded as extremity of deformation and fracture of materials. However, I believe that these knowledge give us a new insight on the subject, how the materials behave at elevated temperatures in which atomic bonds exchange continuously. The new technologies will be developed from the combination of these knowledge and imagination in future, for example, developments of new composite materials by using superplasticity or precursor-derived matrix.

REFERENCES

1) Edited by J. Bill, F. Wakai, F. Aldinger: Precursor-Derived Ceramics, Wiley-VCH, Weinheim 1999.
2) J. Bill and F. Aldinger, Adv. Mater., 7, p.775-787 (1995).
3) R. Riedel H.J. Kleebe, H Schönfelder and F. Aldinger, Nature, 374, p.526-528 (1992).
4) F. Wakai, S. Sakaguchi, and Y. Matsuno, Advanced Ceramic Materials, 1 [3] p.259-63 (1986).

5) F. Wakai, Y. Kodama, S. Sakaguchi, N. Murayama, K. Izaki and K. Niihara, Nature, **344** [6265] p.421-423 (1990).
6) F. Wakai, N. Kondo, H. Ogawa, T. Nagano and S. Tsurekawa, Materials Characterization, **37** [5] p.331-341(1996).
7) F. Wakai, N. Kondo and Y. Shinoda, Current Opinion in Solid State & Materials Science, in press.
8) N. Kondo, T. Ohji and F. Wakai, J. Ceram. Soc. Japan, **106** [10] p.1040-1042 (1998).
9) Edited by E. Yasuda, F. Wakai, L.M. Manocha, Y. Tanabe: Time Dependent Mechanical Response of Engineering Ceramics, Trans Tech Publications, Zuerich, 1999.
10) F. Wakai, N. Enomoto, and H. Ogawa, Acta mater., in press.
11) F. Wakai, Acta Metall. Mater. **42** [4] p.1163-1172 (1994).
12) N.Kondo, T. Ohji and F. Wakai, J. Am. Ceram. Soc., **81** [3] p.713-16 (1998).
13) T. Nagano, H. Gu, Y. Shinoda, G.D. Zhan, M. Mitomo and F. Wakai, Mater. Sci. Forum, **304-306**, p.507-512 (1999).
14) Y. Shinoda, T. Nagano, and F. Wakai, J. Am. Ceram. Soc., **82** [3] p.771-73 (1999).
15) K. Kakimoto, F. Wakai, J. Bill, and F. Aldinger, J. Am. Ceram. Soc., **82** [9] p.2337-41 (1999).
16) Y. Shinoda, T. Nagano, H. Gu, and F. Wakai, J. Am. Ceram. Soc., **82**[10] p.2916-2918 (1999).
17) H. Gu, Ultramicroscopy **76,** p.159-172 (1999).
18) H. Gu, Ultramicroscopy **76,** p.173-185 (1999).
19) H. Gu, Y. Shinoda and F. Wakai, J. Am. Ceram. Soc., **82** [2] p.469-72 (1999).
20) H. Gu and F. WakaiJ. Mater. Synthesis and Processing, **6** [6] p.393-399 (1998).
21) B. Baufeld, H. Gu, J. Bill, F. Wakai, and F. Aldinger, J. Euro. Ceram. Soc., **19** [16], p.2797-2814 (1999).
22) G. Thurn, M. Christ, J. Canel and F. Aldinger, private communication.
23) S. Ishihara, F. Wakai, F. Aldinger, and J. Bill, Mater. Sci. Forum, **304-306**, p.501-506 (1999).

ROLE OF VACANCIES AND SOLUTE ATOMS ON GRAIN BOUNDARY SLIDING

J.S. VETRANO, C.H. HENAGER, JR. AND E.P. SIMONEN
Pacific Northwest National Laboratory, Richland, WA 99352, john.vetrano@pnl.gov

ABSTRACT

It is necessary for grain boundary dislocations to slide and climb during the grain boundary sliding process that dominates fine-grained superplastic deformation. The process of climb requires either an influx of vacancies to the grain boundary plane or a local generation of vacancies. Transmission electron microscopy (TEM) observations of grain boundaries in superplastically deformed Al-Mg-Mn alloys quenched under load from the deformation temperature have revealed the presence of nano-scale cavities resulting from a localized supersaturation of vacancies at the grain boundary. Compositional measurements along interfaces have also shown an effect of solute atoms on the local structure. This is shown to result from a coupling of vacancy and solute atom flows during deformation and quenching. Calculations of the localized vacancy concentration indicate that the supersaturation along the grain boundary can be as much as a factor of ten. The effects of the local supersaturation and solute atom movement on deformation rates and cavity nucleation and growth will be discussed.

INTRODUCTION

Fine-grained superplasticity is a process by which materials with fine (<10 μm), stable grain structures can deform to high elongations. This process is generally thought to occur by the sliding of grains past one another with the motion of matrix dislocations assisting in accommodation processes at structural asperities [1-3]. In many materials, the final elongation is determined by the breakdown of the accommodation processes and the resulting cavitation. Microstructural studies in aluminum alloys have shown that the majority of these cavities form at large intermetallic particles called eutectic constituents (EC) that form during casting and are broken up during subsequent thermomechanical processing [4,5]. These particles also serve as potent nuclei during recrystallization so they typically lie along grain boundaries [6].

Models of grain boundary sliding (GBS) and grain migration have showed that vacancies play an important role for the motion and climb of grain boundary dislocations [7,8]. This idea is strengthened by the observation that the activation energy for superplasticity in many metals is equal or close to that for grain boundary diffusion [1]. It is unclear whether or not grain boundary dislocations as singularities exist at the high homologous temperatures typical for superplastic deformation, in particular at the high-energy grain boundaries that are most likely to slide, but computer models show that substantial vacancy motion is required to allow these dislocations to glide and climb in the grain boundary plane [9].

Compositional variations along the grain boundary and at particle-matrix interfaces have been shown to result from the process of superplastic deformation [10,11]. Segregation has also been measured at reinforcement-matrix interfaces in aluminum-based metal matrix composites [12]. These regions of solute-rich material at particle-matrix interfaces are thought to act as "accommodation helpers" during high-rate superplasticity [13]. In some cases, however, impurity segregation has been shown to increase cavitation during superplastic deformation [14]. These observations indicate that the localized solute and impurity concentration can play a large role in GBS as well as the subsequent cavitation and failure of the material.

169

In this paper we characterize the structure and composition of grain boundaries and triple junctions of superplastically deformed Al-Mg based alloys to gain a better understanding of the mechanisms for GBS and cavity formation. In particular, we have utilized high-resolution analytical microscopy to measure localized compositional inhomogeneities following superplastic deformation.

EXPERIMENTAL PROCEDURES

Materials used in this study were purchased from Kaiser Center for Technology as 50 mm x 250 mm x 200 mm book mold castings. The two alloys had the following compositions in wt.%: 4.2 (4.0) Mg, 1.0 (1.0) Mn, 0.52 (0.52) Sc, 0.29 (0.27) Si, 0.39 (0.27) Fe and (0.034) Sn—bal. Al. The ingots were scalped, warm rolled at 300°C to 6.25 mm thickness, given a recrystallization/precipitation heat treatment of 2 hrs. @ 500°C, then cold rolled 80% to 1.25 mm. Dog-bone shaped tensile samples with a 25 mm x 6.25 mm gage were machined and the samples given a recrystallization heat treatment of 10 min. @ 400°C prior to testing.

Mechanical testing was carried out on an MTS servo-hydraulic test frame with a vertical clamshell-type furnace (T held to ± 2°C along the gage length) at constant true strain rates of $1x10^{-4}$ s^{-1} to $1x10^{-2}$ s^{-1} and temperatures from 450°C to 550°C. Some samples were pulled to failure and others were quenched to retain as much of the deformation structure and compositional distribution as possible. Upon reaching the desired strain, samples were quenched by spraying with liquid nitrogen while held under constant load. Thermocoupled samples revealed a quench rate of approximately 75°C s^{-1}.

Samples for electron microscopy were prepared by mechanically thinning the mechanical test samples, punching out 3 mm disks, then electropolishing to electron transparency in a Struers Tenupol double-jet polisher using a 5% $HClO_4$-methanol solution (-40°C, 0.12 mA). The materials were examined in a JEOL 2010F field-emission gun transmission electron microscope (FEG-TEM). Compositional analysis at grain boundaries and triple junctions was performed with a 0.7-nm diameter probe utilizing an Oxford ATW x-ray detector and the Link ISIS software.

RESULTS

One of the advantages of quenching the samples following testing is that much of the deformation microstructure is retained. In particular, the grain boundaries are shown to be quite curved when the material is quenched. Ge [15] showed that there are substantial grain shape changes (manifested in a contraction of the grains in the tensile direction) if the sample sits at temperature for even a short time following the completion of deformation. Figure 1 shows a typical section of grain boundary that is intersected by an Al_6Mn particle (added to prevent grain growth). The boundary is strongly curved in the vicinity of the particle, but is not perfectly straight even some distance from the particle. In general, most boundaries are curved over a substantial portion of their length between triple points. This means that the idealized computer models of grain boundary sliding where the boundary is flat will not be entirely correct. In addition to the view of extrinsic grain boundary dislocation motion causing boundary sliding, substantial movement of material along these curved boundaries needs to occur via diffusion or dislocation motion.

The presence of excess vacancies under these conditions is revealed by the presence of nano-scale cavities along the grain boundaries and at triple points. Figure 2 shows a triple point where there are nanocavities along the boundaries in addition to decorating the triple-junction line.

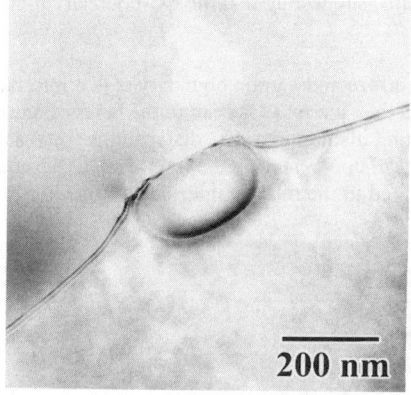

Figure 1. TEM micrograph of a grain boundary curved around an Al$_6$Mn particle following superplastic deformation.

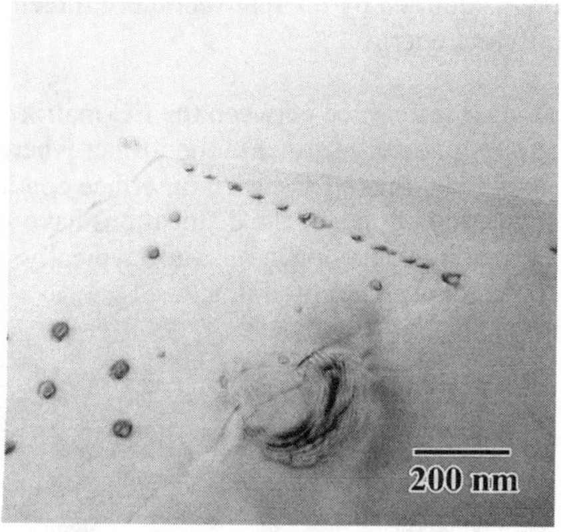

Figure 2. Micrograph showing nanocavities along grain boundaries and the triple point. Strain = 1.1

Under the conditions of this micrograph the nanocavities appear as small dark strain fields on a light background. They are also present at particle-boundary interfaces where particles intersect grain boundaries. A calculation was performed to assess the concentration of vacancies in the near-boundary vicinity. Using typical measured values of 5 nm cavities with a 100 nm spacing, the atomic fraction of vacancies in a 4-plane-thick grain boundary was calculated to be about 0.005. A vacancy formation energy of 0.65 eV was used. This is about 10 times higher than the equilibrium concentration at 450°C for the samples examined and five times higher than the equilibrium concentration at the melting temperature. This number can obviously change depending on the assumed width of the grain boundary region and we are currently pursuing a more quantitative method of determining this value. It is noted that the nanocavity density is also higher on the particle-matrix interfaces than the grain boundaries

Optical micrographs of samples strained to 200% reveal that in these alloys cavity nucleation occurs at EC-matrix interfaces, similar to other aluminum alloys. Figure 3 shows such a sample with arrows indicating examples of EC particles that are connected with large cavities. This behavior is generally attributed to the build-up of stress in the vicinity of hard particles but the effect of local composition changes and boundary cohesive energy have not been generally considered.

Compositional inhomogeneities were measured along the grain boundaries and triple junctions following superplastic deformation. As previously reported [10, 11] the Mg level is depleted along grain boundaries but enriched at the nanocavities. Another interface that often contained impurities was a triple junction resulting from the intersection of an EC and two grains (i.e. where an EC was on a grain boundary). Figure 4 shows such an intersection and the impurity-rich region is revealed as a slightly brighter region as indicated by the arrow. Figure 5 shows an energy dispersive x-ray spectroscopy (EDS) spectrum of this region and indicates that it is rich in silicon and oxygen. Other similar regions show magnesium enrichment as well, and in the tin-containing alloys that is the primary location where tin is detected. The fact that it shows up as a light region in the TEM micrographs indicates that the density is lower there and the material is probably porous. It is interesting that these impurity pockets have not been seen at standard triple points defined by the intersection of three grains, suggesting that the EC-boundary junction has a different energy.

An additional difference between the EC-matrix interface and a grain boundary is that interfacial dislocations appeared distinct in the former whereas they were had spread in the latter. Figure 6 shows a region with an EC-matrix interface containing distinct extrinsic dislocations whereas in the nearby grain boundary the dislocations have spread upon entering the interface. Differences in core spreading behavior in metals is typically related to the relative interface diffusivity [8] but it could also be structural in nature.

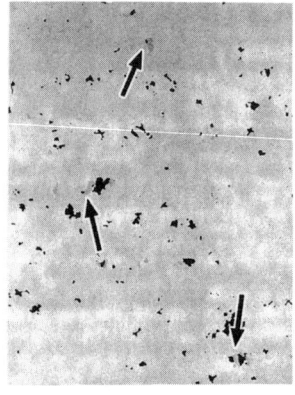

Figure 3. Cavity formation on EC particles (arrowed)

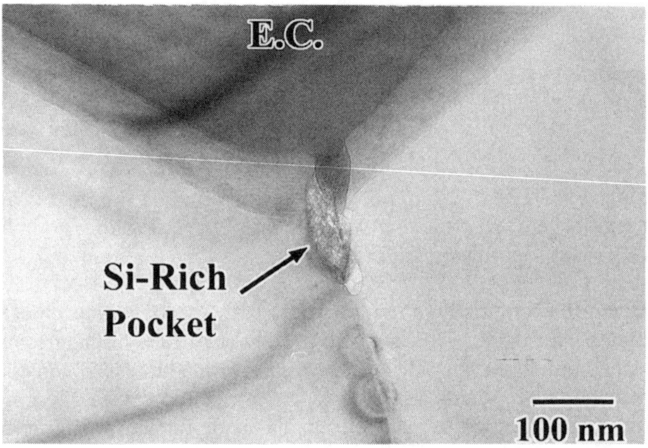

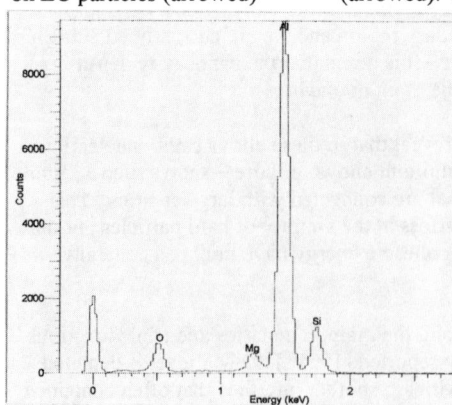

Figure 4. Si-rich pocket at EC-grain boundary interface (arrowed). Note porous appearance.

Figure 5. EDS spectrum of Si- and O-rich pocket at EC-grain boundary interface.

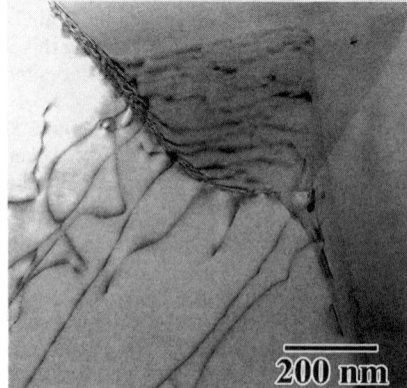

Figure 6. EC-matrix boundary showing distinct interfacial dislocations and adjacent grain boundary with none.

DISCUSSION

Superplasticity in structural aluminum materials such as Al-Mg-based 5xxx alloys is made complex by a heterogeneous microstructure consisting of curved grain boundaries, second-phase particles in the matrix and at grain boundaries, and solute additions and impurity atoms that are mobile at the deformation temperature. All of these microstructural features affect the ability of grains to slide and rotate around each other and the accommodation processes at junctions.

In random high-angle, high-energy grain boundaries such as are present in these alloys it is likely that most grain boundary dislocations do not have the grain boundary plane as their glide plane. Therefore, substantial climb must take place to move the grains past one another. Curvature and hard particles that intersect the boundaries will further hinder this process. The presence of nanocavities in quenched samples is evidence for the locally high vacancy concentration. The formation of grain boundary nanocavities and their associated compositional inhomogeneities has been shown to result from directed vacancy and solute motion from the vacancy-supersaturated region during the quench [11].

In calculating the exact concentration it is important to use a reasonable value of grain boundary "width" that the vacancies are confined to. We have used a relatively standard grain-boundary width of four lattice planes but it is quite likely that the vacancy concentration is influenced outside of this region as well since the concentration five-times higher than at the melting point seems unrealistic. Though particle-matrix interfaces in the middle of the grains do not contain nanocavities, some particles very near the boundary (but evidently not touching) do have nanocavities at the interface. This is an indication of the effective width for the excess concentration of vacancies and we are currently collecting statistical information on that distance.

Cavity nucleation at EC particles has generally been ignored from an interface point of view and has generally been looked upon from a mechanical standpoint. However, results from high strain rate superplastic deformation in metal matrix composites [13] and Sn-doped Al-Mg alloys [16] indicate that localized compositional changes can influence the accommodation processes at the EC-boundary interfaces. Though at this time it is not clear what effect the Si- and Mg-rich regions are having on the cavity nucleation but computer simulations have shown a reduction in boundary cohesive energy with increasing Mg content[17] and experiments in similar alloys have shown a detrimental effect on cavitation with Si additions [18]. These elements will most likely also influence the local diffusivity and modulus which are known to affect the nucleation and growth of cavities in superplastic materials.

CONCLUSIONS

It has been shown that during superplastic deformation there are substantial changes in local vacancy concentration, perhaps as much as a ten-fold increase, and solute segregation. These changes have been revealed by quickly quenching the samples under load from the deformation temperature. Pockets of impurities have been located near EC-grain boundary triple junctions and these are thought to contribute to cavitation during superplasticity, which typically nucleates at these boundaries. Cavity nucleation may be exacerbated by the apparent structural and/or diffusivity differences at particle-matrix interfaces as revealed by the retention of distinct extrinsic dislocations.

ACKNOWLEDGEMENTS

The authors would like to thank Dr. Val Gertsman and Konrad Lasota for assistance with mechanical testing and electron microscopy. This work was supported by the Materials Science Division, Office of Basic Energy Sciences, U.S. Department of Energy (DOE) under Contract DE-AC06-76RLO 1830.

REFERENCES

1. O. Sherby and J. Wadsworth, Progress in Materials Science, **33**, p. 169 (1989).
2. R.Z. Valiev and T.G. Langdon, Acta metall. mater., **41**, p. 949 (1993).
3. T.G. Langdon, Acta metall. mater., **42**, p. 2437 (1994).
4. A.H. Chokshi and A.K. Mukherjee, Acta metall., **37**, p. 3007 (1989).
5. D.H. Shin and K.-T. Park, Mat. Sci and Engineering, **A268**, p. 55 (1999).
6. J.S. Vetrano, C.H. Henager, Jr., M.T. Smith and S.M. Bruemmer in *Hot Deformation of Aluminum Alloys II*, (TMS, Warrendale, PA 1998) p. 407.
7. D.A. Smith, Ultramicroscopy, **29**, p. 1 (1989).
8. W. Lojkowski, J.W. Wyrzykowski, J. Kwiecinski, D.L. Beke and I. Godeny, Defect and Diffusion Forum, **66-69**, p. 701 (1989).
9. R.C. Pond, A. Serra and D.J. Bacon, Acta mater., **47**, p. 1441 (1999).
10. J.S. Vetrano, E.P. Simonen and S.M.. Bruemmer, Mat. Science Forum, **243-245**, p. 493 (1997).
11. J.S. Vetrano, E.P. Simonen and S.M. Bruemmer, Acta mater., in press.
12. M. Strangwood, C.A. Hippsley and J.J. Lewandowski, Scripta Met. et Mat., **24**, p. 1483 (1990).
13. M. Mabuchi and K. Higashi, Acta mater., **47**, p. 1915 (1999).
14. K.-T. Park, S.T. Yang, J.C. Earthman and F.A. Mohamed, Mat. Sci, and Engineering, **A188**, p. 59 (1994).
15. Y. Ge, Master's Thesis, Washington State University (1997).
16. C.H. Henager, Jr., J.S. Vetrano, V. Gertsman and S.M. Bruemmer, This proceedings.
17. X.-Y. Liu and J.B. Adams, Acta mater., **46**, p. 3467 (1998).
18. H. Hosakawa, H. Iwasaki, T. Mori, M. Mabuchi, T. Tagata and K. Higashi, Acta mater., **47**, p. 1859 (1999).

MODIFIED CONSTITUTIVE EQUATION OF SUPERPLASTICITY INCLUDING GRAIN GROWTH RATE

H. MIYAZAKI*, T. ISEKI*, T. YANO**
*Department of Inorganic Materials, Tokyo Institute of Technology, Meguro-ku, Tokyo 152-8552, Japan, hmiyazak@o.cc.titech.ac.jp
**Research Laboratory for Nuclear Reactors, Tokyo Institute of Technology, Meguro-ku, Tokyo 152-8550, Japan

ABSTRACT

Superplastic behavior associated with grain growth in ceramics at high temperature is predicted by using a modified constitutive equation for superplasticity that incorporates the rate of grain growth. The equation reveals that there is an optimum deformation conditions – i.e., the strain rate and deformation temperature - so that ceramics have low flow stress when the grain growth is significant. The equation provides an explanation of increased flow stress for liquid-phase-sintered SiC in the slow-strain-rate region or higher-temperature region in which grain growth plays an important role.

INTRODUCTION

The well-known constitutive equation suggests that increasing the deformation temperature decreases the flow stress. The flow stress of TiO_2-doped TZP decreases with increasing the deformation temperature, although grains grew more than three times larger after the tensile deformation at 1500°C [1]. However, the decrease in the flow stress with increasing the deformation temperature may be canceled if grains grow significantly during deformation. Nagano et al. reported that the flow stress of liquid-phase-sintered SiC at strain $\varepsilon > 0.3$ increased when temperature was increased from 2023K to 2073K [2]. The conventional constitutive equation for superplasticity does not take into account this dynamic deformation observed with grain coarsening in higher-temperature region and/or very slow-strain-rate regions. Few studies have centered on the condition under which the significant effect of grain coarsening upon the flow stress appears.

In this paper, we present a new equation for the superplasticity of ceramics that includes the grain-coarsening effect. The effect of the value of each parameters in the equation upon the flow stress dependence on the test condition was estimated. Superplastic behavior of SiC in high-temperature region or slow-strain-rate region in which grain growth occurred is discussed using the equation.

Mat. Res. Soc. Symp. Proc. Vol. 601 © 2000 Materials Research Society

THEORY

Kondo et al. presented an equation that predicted the grain size change during high temperature deformation [1]. Application of their equation is limited to the case where static grain growth follows parabolic law. To generalize the equation, we used the next form for static grain growth at high temperature.

$$r_{\mathrm{a}}^{c} - r_{0}^{c} = kt \tag{1}$$

where r_{a} is the grain size at time, t ; r_0 the initial grain size; k the growth constant; and c the grain-growth exponent. Grain growth increases during deformation at high temperature [1, 3]. The grain size of the deformed sample, r is estimated by [3].

$$r = \left(r_{0}^{c} + kt\right)^{\frac{1}{c}} \exp(\alpha\varepsilon) \tag{2}$$

where ε is the true strain and α constant. The deformation time t is replaced by $\varepsilon/\dot{\varepsilon}$. The parameter k usually is written in the form $k=k_0\exp(-Q_{gg}/RT)$, where Q_{gg} is the activation energy for grain growth; T deformation temperature; R the gas constant; and k_0 is constant. The following equation results.

$$r = \left(r_{0}^{c} + k_{0} \exp\left(-\frac{Q_{gg}}{RT}\right) \times \frac{\varepsilon}{\dot{\varepsilon}}\right)^{\frac{1}{c}} \exp(\alpha\varepsilon) \tag{3}$$

The well known constitutive equation of superplastic behavior is

$$\dot{\varepsilon} = A \frac{\sigma^{n}}{r^{p}} \exp\left(-\frac{Q_{\mathrm{sup}}}{RT}\right) \tag{4}$$

Here, σ is the true stress; n and p the stress and grain-size exponents, respectively; Q_{sup} the apparent activation energy for superplastic deformation; and A a constant depending on the material. Substituting Eq. (3) into Eq. (4) yields

$$\sigma^{n} = \frac{\dot{\varepsilon}}{A} \left[\left\{r_{0}^{c} + k_{0} \exp\left(-\frac{Q_{gg}}{RT}\right) \times \frac{\varepsilon}{\dot{\varepsilon}}\right\}^{\frac{1}{c}} \exp(\alpha\varepsilon)\right]^{p} \exp\left(\frac{Q_{\mathrm{sup}}}{RT}\right) \tag{5}$$

SIMULATION OF FLOW STRESS DEPENDENCE UPON BOTH TEMPERATURE AND STRAIN RATE

When the deformation temperature (T) is low and the strain rate ($\dot{\varepsilon}$) high, the grains grow little during deformation, that is, $r_0^c \gg k_0 \exp[-(Q_{gg}/RT)](\varepsilon/\dot{\varepsilon})$, then Eq. (5) simplifies to Eq. (6), almost the same as the conventional equation except for a term denoting strain hardening.

$$\sigma'' = \frac{\dot{\varepsilon}}{A} r_0^{\ p} \exp(p\alpha\varepsilon) \times \exp\left(\frac{Q_{sup}}{RT}\right) \tag{6}$$

Equation (6) indicates that σ at a given strain decreases with increasing T at a constant $\dot{\varepsilon}$, whereas σ increases with increasing $\dot{\varepsilon}$ at constant T.

In contrast, when T is high and/or $\dot{\varepsilon}$ low, the grains grow severely during deformation, that is, $r_0^{\ c} \ll k_0 \exp[-(Q_{gg}/RT)](\varepsilon/\dot{\varepsilon})$, then Eq. (5) approximates as

$$\sigma'' = \frac{(k_0\varepsilon)^{\frac{p}{c}}}{A} \dot{\varepsilon}^{1-\frac{p}{c}} \exp(p\alpha\varepsilon) \times \exp\left(\frac{Q_{sup} - \frac{p}{c}Q_{gg}}{RT}\right) \tag{7}$$

Equation (7) reveals that the dependence of σ at a given strain value can be classified into four categories, according to the values of $Q_{sup} - (p/c)Q_{gg}$ and $1 - p/c$. The patterns of each type are shown in Fig.1 (a)-(d), with stress presented in contour. The x-axis in Fig.1 represents the reciprocal of temperature, $1/T$ and the y-axis is $\log\dot{\varepsilon}$. These patterns are calculated from the values listed in Table I. The four patterns of flow stress illustrated in Fig. 1 are as follows.

(a) When $Q_{sup} > (p/c) Q_{gg}$ and $p < c$: Fig.1 (a) is calculated with the reported value of 3Y-TZP [4,5]. The σ value decreases with increasing T at a constant $\dot{\varepsilon}$, whereas σ increases with increasing $\dot{\varepsilon}$ at constant T. Effect of grain coarsening on flow stress appears slightly at high T and low $\dot{\varepsilon}$. The spacing of the contour lines become wider when $\sigma < 10\text{MPa}$. That wider spacing means a lower dependence of flow stress on strain rate; i.e., a high n value. Thus, the n value increases when the strain rate decreases significantly at constant T. These features are consistent with the reported deformation behavior of 3Y-TZP.

(b) When $Q_{sup} < (p/c) Q_{gg}$ and $p < c$: In the lower range of $\dot{\varepsilon}$, σ decreases with increasing T in the range of lower T, whereas σ increases slightly with increasing T in a higher range of T because of grain coarsening. The lowest σ value appears at low $\dot{\varepsilon}$ and medium T.

(c) When $Q_{sup} > (p/c) Q_{gg}$ and $p > c$: In the lower range of $\dot{\varepsilon}$, σ increases with decreasing $\dot{\varepsilon}$ because of grain coarsening. The lowest σ value appears at high T and medium $\dot{\varepsilon}$.

(d) When $Q_{sup} < (p/c) Q_{gg}$ and $p > c$: Grain growth has a significant effect on σ at high T and low $\dot{\varepsilon}$. In the region where $\dot{\varepsilon}$ is low and T high, σ increases with decreasing $\dot{\varepsilon}$ at constant T because of grain coarsening. The σ value increases with increasing T at constant $\dot{\varepsilon}$ because of grain coarsening. The other peak in σ appears when T is high and $\dot{\varepsilon}$ low. The region of minimum σ lies between the peaks.

It is clear that there is an optimum deformation condition so that the ceramics have low flow stress when the effect of grain growth is large.

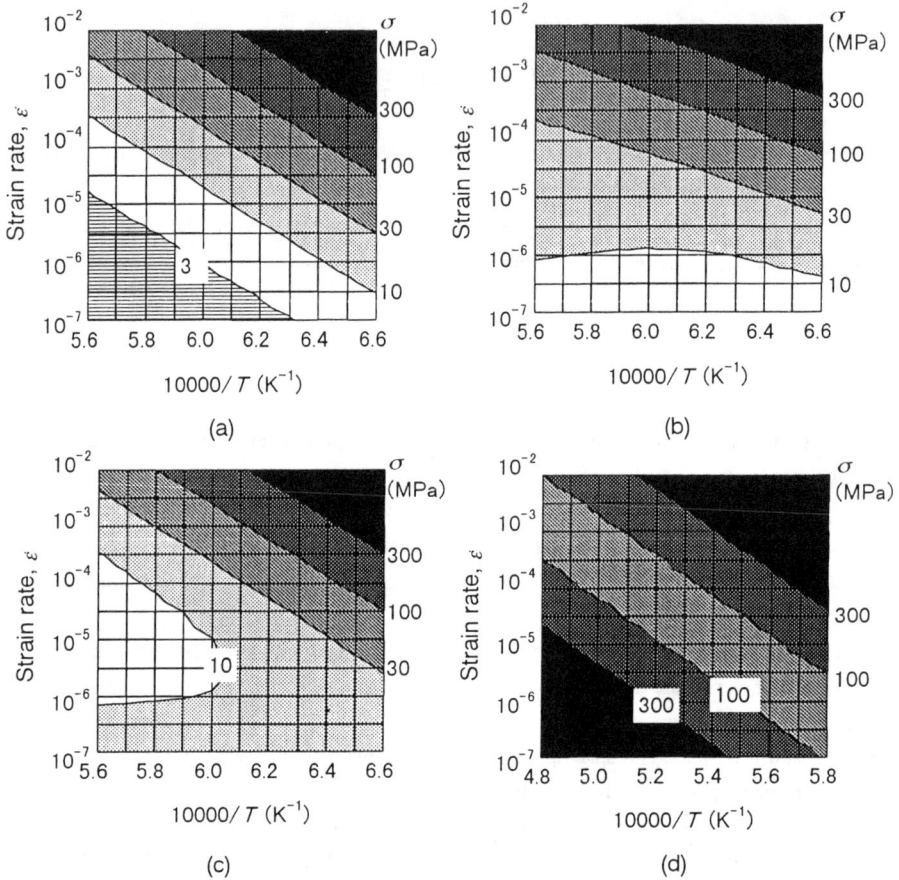

Fig. 1. Dependence of flow stress upon both strain rate ($\dot{\varepsilon}$) and deformation temperature (T) calculated form Eq. (5) with the values in Table I.   The flow stress is represented in contour. (a): $Q_{sup} > (p/c)Q_{gg}$ and $p<c$, (b): $Q_{sup} < (p/c)Q_{gg}$ and $p<c$, (c): $Q_{sup} > (p/c)Q_{gg}$ and $p>c$, (d): $Q_{sup} < (p/c)Q_{gg}$ and $p>c$.

SIMULATION OF DEFORMATION BEHAVIOR OF LIQUID-PHASE-SINTERED SiC AT ELEVATED TEMPERATURE

Nagano et al. [2] studied deformation of liquid-phase-sintered silicon carbide at elevated temperature and reported an apparent activation energy for deformation, Q_{sup}=680 kJmol^{-1} and a grain size exponent, $p \doteq 2$.   Mitomo's [6] study of the coarsening of SiC at high

Table I. Value of each parameters used in calculation of Fig. 1 with Eq.(5)

Type	Q_{sup} (kJmol^{-1})	Q_{gg} (kJmol^{-1})	p	c	r_0 (μm)	α	ε	A	Ref.
a	590	584	2	3	0.3	0.6	0.05	6×10^{-14}	4,5
b	350	590	2	3	0.3	0.6	0.5	1×10^{-21}	—
c	600	390	3	2	0.3	0.6	0.5	1×10^{-19}	—
d	680	680	2	1	0.1	0	0.5	2.5×10^{-12}	2,6

Stress exponent, n =2 and k_0=2 $\times 10^{-6}$ for (a),(b),(c).   For (d), n=1.6 and k_0=2 $\times 10^7$.   The values of k_0 and A for type (d) are estimated from the data in Nagano's [2] study provided that Q_{gg}=Q_{sup} and c =1. The values for type (b) and (c) are imaginary.

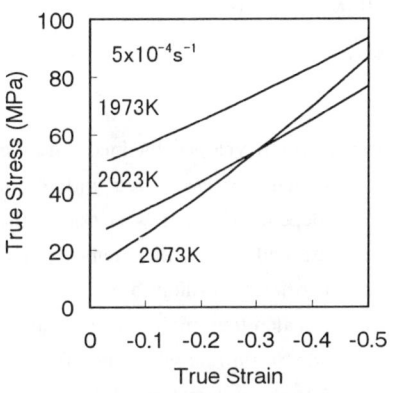

Fig. 2 Calculated true stress vs. true strain curves for liquid-phase-sintered SiC at the strain rate of 5x10^{-4}s^{-1}.

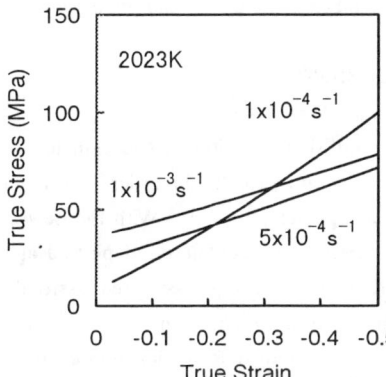

Fig. 3 Calculated true stress vs. true strain curves for liquid-phase-sintered SiC at 2023K.

temperature, gave c < 1 although the law of anisotropic grain growth of liquid-phase-sintered SiC was not clarified.   The small value of c may be attributable to anisotropic grain growth. It is not reasonable to think that Q_{sup}> 2Q_{gg}, since both accommodation process for superplastic deformation and grain growth are controlled by the diffusion process.   Thus, the flow-stress dependence of the SiC would be categorized as type (d).   Fig 1(d) shows the plot calculated with those values, provided that $Q_{sup} \doteqdot Q_{gg}$.

Fig. 2 shows true stress versus true strain carves calculated with those values when the initial strain rate is fixed to 5×10^{-4}s^{-1} and testing temperature is changed from 1973 to 2073K.   The flow stress at 2073K increased lineally with increasing strain and surpassed that of the sample deformed at 2023K in the strain range > 0.3.

Fig. 3 shows true stress versus true strain carves calculated with those values when the testing temperature is fixed to 2023K and initial strain rate is changed from 1×10^{-4}s^{-1} to 1

$\times 10^{-3} \text{s}^{-1}$. The flow stress at $1 \times 10^{-4} \text{s}^{-1}$ increased lineally with increasing strain and surpassed that of the sample deformed at higher strain rate in the strain range > 0.2-0.3.

Nagano et al. [2] showed that the flow stress at a constant strain rate of $5 \times 10^{-4} \text{s}^{-1}$ increased when the temperature was increased from 2023K to 2073K and the strain >0.35. Those researchers also reported that flow stress of liquid-phase-sintered SiC at strain $\varepsilon > 0.3$ decreased when the strain rate was increased from $1 \times 10^{-4} \text{s}^{-1}$ to $5 \times 10^{-4} \text{s}^{-1}$ at 2023K and that the flow stress recovered at $1 \times 10^{-3} \text{s}^{-1}$. These observed tendencies of the flow stress are consistent with the estimated variation of the flow stress in Fig. 2, Fig. 3 and Fig. 1(d). It is clear that the deformation behavior of ceramics can be calculated with Eq. (5). These results suggest that the optimum deformation condition so that the ceramics have low flow stress can be calculated by using the parameters of the flow-stress plot.

CONCLUSIONS

A modified constitutive equation for superplasticity was developed by incorporating the rate of grain growth, in order to predict superplastic behavior associated with grain growth at high temperature. With the new equation, the dependence of flow stress on the test condition was calculated and plotted against strain rate and deformation temperature. The patterns of the flow stress were classified into four categories, according to the values of parameter used in the equation. The plots of flow stress revealed that minimum flow stress appeared at medium deformation temperature and/or medium strain rate when the effect of grain growth is significant. The plot of flow stress for liquid-phase-sintered SiC showed good agreement between calculated and observed results. Thus, it is expected that deformation condition can be optimized to minimize the flow stress by calculating the deformation behavior with the equation.

REFERENCES

1. K. Kondo, Y. Takigawa and T. Sakuma, Mater. Sci. Eng., **A231**, 163 (1997).
2. T. Nagano, S. Honda, F. Wakai and M. Mitomo in *Superplasticity and Superplastic Forming*, edited by A. K. Ghosh and T. R. Bieler (The Minerals, Metals and Materials Society, 1998), pp. 247-256.
3. E. Sto, K. Kuribayashi, T. Horiuchi, in *Superplasticity and Superplastic Forming*, edited by C.H. Hamilton and N. E. Paton (The Minerals, Metals and Materials Society, Warendale, PA, 1988), pp.115-119.
4. F. Wakai, S. Sakaguchi and Y. Matsuno, Adv. Ceram. Mater., **1**, 259 (1986).
5. J. Zhao, Y. Ikuhara and T. Sakuma, J. Am. Ceram. Soc., **81**, 2087 (1998).
6. Y-W. Kim, M. Mitomo and H. Hirotsuru, J. Am. Ceram. Soc. **78**, 3145 (1995).

A SUPERPLASTICITY THEORY BASED ON DYNAMIC GRAIN GROWTH

M.J. MAYO and J.R. SEIDENSTICKER
Dept. Materials Science & Engineering
The Pennsylvania State University
University Park, PA   16802
mayo@ems.psu.edu; Joseiden@aol.com

ABSTRACT

The linear relation between dynamic grain growth rate and strain rate appears to be constant across all superplastic materials—regardless of whether the system examined is metallic or ceramic, and regardless of the stress exponent exhibited during deformation. The simplicity and universality of the dynamic grain growth law suggest it might be useful as a foundation for a theory for superplasticity. One attempt at such a theory is presented. It is argued that stress leads to the development of anisotropic grain shapes that then require recovery through directionally biased grain growth events. Once this mathematical relationship between stress and grain growth rate is developed, it is inserted into the existing dynamic grain growth - strain rate law to arrive at a phenomenological law for superplasticity.

INTRODUCTION

The phenomenon of superplasticity has enjoyed widespread attention since Pearson published his dramatic photograph showing a Bi-44%Sn sample elongated to such extremes (1,950%) that the sample had to be coiled around itself multiple times in order to fit within the frame of the photograph.[1] Since that time, many theoretical explanations for superplasticity have emerged to explain the dramatic ductility. Many, if not most, such theories focus on the phenomenological law obeyed by superplastic materials, viz.

$$\dot{\varepsilon} = A \frac{\sigma^n}{d^p} D_0 \exp(-Q/RT),\qquad\qquad(1)$$

where $\dot{\varepsilon}$ is the strain rate of the material, A is a constant, σ is the applied stress, n is the stress exponent, d is the grain size exponent, and the term $D_0 \exp(-Q/RT)$ represents the diffusivity, typically taken to be the grain boundary diffusivity. To arrive at a theory for superplasticity using Eq. 1 as the foundation, it is necessary to first assume there is a value (or limited range of values) for n that is unique to superplasticity. For many years, Equation 1 with n=2 was taken to be the characteristic phenomenological law for superplasticity.[2-5] Unfortunately, as more and more data emerge, it is becoming apparent that superplasticity is not easily associated with a unique value of n. Even for a single material, such as tetragonal yttria-stabilized zirconia (Y-TZP), macroscopically superplastic behavior (i.e., large deformation to failure) has been experimentally observed under conditions when n=1 (Ref. 6), n=2 (Refs. 7,8), and n=3 (Refs. 9,10), or sometimes, within the same laboratory, all three (Ref. 11).   The problem of uncertain n value is further compounded when attempting to explain superplastic behavior across a range of materials. In the end, it becomes difficult to arrive at a truly universal theory for superplasticity, when the starting point for the derivation is not itself universally obeyed.

A UNIVERSAL LAW FOR DYNAMIC GRAIN GROWTH

It seems logical that a starting point for a theory for superplasticity should be a law that is universal across *all* superplastic materials, even those with widely varying n. Interestingly, there is such a law: the dynamic grain growth law. As Figure 1 shows, the relationship between (normalized) dynamic grain growth rate and strain rate is constant for all superplastic systems for which data can easily be found. In Figure 1, it can be seen that the dynamic grain growth rate, Γ,

181

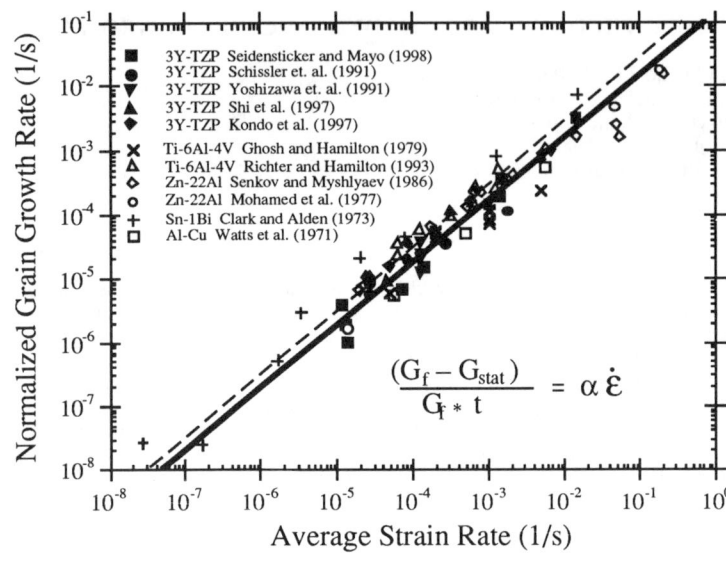

Figure 1. Dynamic grain growth rate vs. strain rate. Solid symbols are from ceramic (3Y-TZP) studies; open symbols are from studies on metallic systems.[12] Dashed line represents the theoretically derived proportionality constant, α, of 0.28 (Ref.13), while the solid line represents the best fit line to the 3Y-TZP data.[12]

Legend in figure:
- ■ 3Y-TZP Seidensticker and Mayo (1998)
- ● 3Y-TZP Schissler et. al. (1991)
- ▼ 3Y-TZP Yoshizawa et. al. (1991)
- ▲ 3Y-TZP Shi et al. (1997)
- ◆ 3Y-TZP Kondo et al. (1997)
- ✕ Ti-6Al-4V Ghosh and Hamilton (1979)
- △ Ti-6Al-4V Richter and Hamilton (1993)
- ◇ Zn-22Al Senkov and Myshlyaev (1986)
- ○ Zn-22Al Mohamed et al. (1977)
- + Sn-1Bi Clark and Alden (1973)
- ☐ Al-Cu Watts et al. (1971)

$$\frac{(G_f - G_{stat})}{G_f * t} = \alpha\,\dot{\varepsilon}$$

is defined as

$$\dot{\Gamma} = \frac{G_f - G_s}{G_f \cdot t}, \qquad (2)$$

where G_f is the final, post-test grain size, G_{stat} is the static grain size (the grain size which would have been obtained by a furnace exposure alone), and t is time. Figure 1 is very similar to the dynamic grain growth plots put forth by Wilkinson and Caceres[14] and Yoshizawa and Sakuma[15], with one crucial difference: in the present work, the grain growth rate is normalized by the final grain size, G_f, rather than the initial grain size, G_0. This difference allows data from both metals and ceramics to fall on a single curve. The more mathematically correct normalization factor would be G, the time-averaged grain size during dynamic grain growth. However, both G_f and G_s turn out to be better approximations to G than G_0, given the fact that grain growth is non-linear in time. The microstructural origin of Eq. 2 has been described in terms of a model in which grain growth opens up geometrically necessary "gaps" in the microstructure into which adjacent grains can slide, thereby leading to sample shortening in compression or elongation in tension.[23] Whether one accepts this explanation or not, the linear relationship between dynamic grain growth rate and strain rate serves as a simple, unambiguous starting point for a theory for superplasticity:

$$\dot{\Gamma} = 0.28\dot{\varepsilon} \qquad (3)$$

STRESS-ASSISTED GRAIN GROWTH IN FINE-GRAINED MATERIALS

To develop a stress-strain rate relationship for superplasticity from Eq. 3, the dependence of dynamic grain growth rate on stress first needs to be known. Such a relationship is derived below.

Grain Enlargement and Shape Recovery Due to Grain Growth

A key aspect of the theory is that grain elongation due to creep is followed by a series of grain growth steps that serve to both enlarge the grain and return it to an equiaxed shape. The mechanics of the "relaxation-by-grain-growth" process do not enter into the derivation below, as it is assumed these steps are fast relative to the process by which the grains elongate in the first place. For the curious, however, a schematic diagram of such events is shown in Figure 2. Grain boundaries perpendicular to the tensile axis become more highly curved (and those parallel to the tensile axis, much less curved), as a result of sample elongation. The change in curvature is greatest for small grains, greatly increasing their disappearance rate during grain growth events, and accelerating the growth rate of the remaining grains. An interesting effect arises from the fact that grain elongation is not random, but heavily (100%) biased in the direction of the applied stress. The flatter sides of large elongated grains, which have the lowest curvature, will tend to move more rapidly to consume small grains adjacent to them, which have the highest curvature. The net effect is to grow the skinny dimension of an elongated grain to a much larger dimension, i.e. , to encourage boundary migration in the direction *opposite* to the tensile stress direction. Grain growth thus also turns out to be a grain shape recovery mechanism for crept samples.

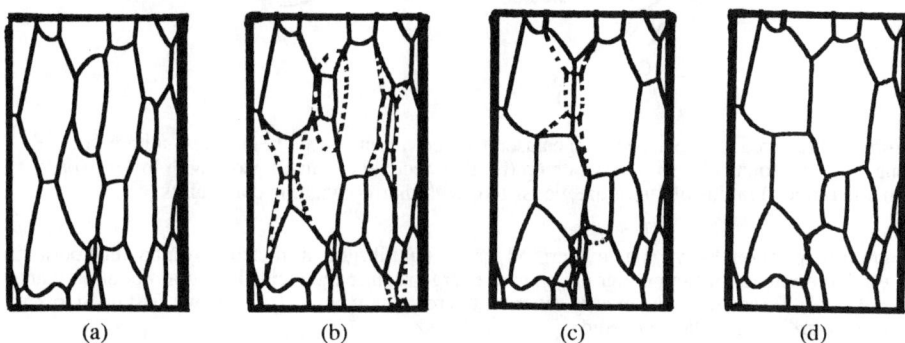

(a)　　　　　　　(b)　　　　　　　(c)　　　　　　　(d)

Figure 2.　Sequence of grain growth events leading to enlargement of grain size and recovery of grain shape after creep.

Mathematical Derivation

The following mathematical derivation simplifies the grain shape change sequence as shown schematically in Figure 3. As per Figure 3, we first assume that grain elongation (Figure 3b) takes place by creep, specifically Coble creep, using the equation derived by Nix.[16] The creep rate is written in terms of the elongation of a single grain in the a_2 direction:

$$\frac{1}{a_2}\frac{da_2}{dt} = \frac{24\delta D_{gb}\Omega}{G^3 kT}\left[\left(\frac{a_2}{a_1}\right)^2 + \left(\frac{a_2}{a_1}\right) + 1\right]^{-1}\sigma \qquad (4)$$

where δD_{gb} is the grain boundary diffusivity, Ω is the atomic volume, G is the grain size, k is Boltzman's constant, T is absolute temperature, σ is the applied stress, and a_2 and a_1 are the long and short axes of the ellipsoid, respectively. The forward creep relation, however, needs to be modified to include the backstress due to boundary curvature at the grain ends.

Since the stress associated with a curved grain boundary is $2\gamma/R$ (γ is the grain boundary energy, and R is the radius of curvature of the grain boundary), and the radius of curvature at the a_2 terminus of the ellipsoid is $R = a_1^2/a_2$, the backstress due to curvature is just $2\gamma a_2/a_1^2$. Inserting the curvature backstress as an offset to the applied stress in the Coble creep relation then results in an equation for elongation of the ellipsoidal grain in the a_2 direction:

$$\frac{1}{a_2}\frac{da_2}{dt} = \frac{24\delta D_{gb}\Omega}{G^3 kT}\left[\left(\frac{a_2}{a_1}\right)^2 + \left(\frac{a_2}{a_1}\right) + 1\right]^{-1}\left(\sigma - \frac{2\gamma a_2}{a_1^2}\right) \qquad (5)$$

For the subsequent lateral grain expansion (Figure 3c), the only constraint applied is that

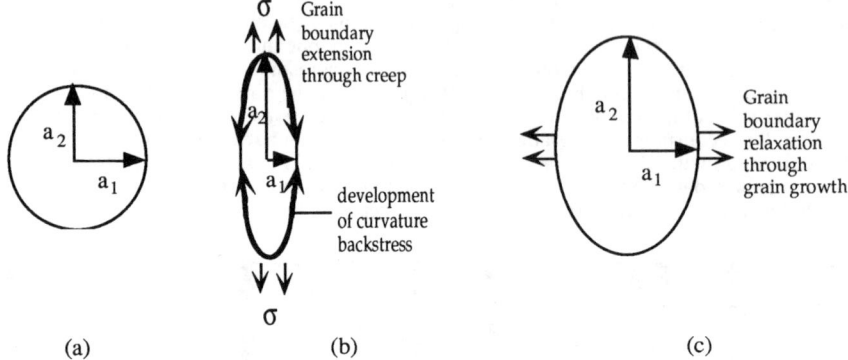

Figure 3: Idealized grain shape changes used for derivation. Grain begins as a sphere (a), elongates to a prolate ellipsoid under creep (b), then recovers to an ellipsoid with lower aspect ratio and reduced radius of curvature, consistent with thermodynamic constraints (c).

the grain shape should always be in thermodynamic equilibrium, as dictated by a balance between the work put into deforming the grain, and the energy required to create new grain boundary area, i.e., dW=dE. To calculate the energy required to create new boundary area, we first need to know the surface area of a prolate ellipsoid:

$$A = 2\pi a_1^2 + \frac{2\pi a_1 a_2}{e}\sin^{-1}e \approx 2\pi\left(a_1^2 + a_1 a_2\right) \text{ (for small } e\text{)} \tag{6}$$

In Eq. 6, e is the eccentricity of the ellipsoid. The energy change (per unit volume) associated with a boundary area change is then

$$dE = \frac{\gamma dA}{V} = \frac{\gamma\pi\left(2a_1\, da_1 + a_1\, da_2 + a_2 da_1\right)}{4/3\,\pi a_1^2 a_2} \tag{7}$$

The work put into the ellipsoid can be calculated from the stress and strain applied, namely,

$$dW = \sigma\, d\varepsilon = \sigma\,\frac{1}{a_2}da_2 \tag{8}$$

Equating the increment in the energy of the system to the increment in the work applied leads to a relationship between growth in the long and short axes of the ellipse, as shown below:

$$dE = dW \tag{9}$$

$$\sigma\frac{1}{a_2}\,da_2 = \frac{\gamma\pi\left(2a_1\, da_1 + a_1\, da_2 + a_2 da_1\right)}{4/3\,\pi a_1^2 a_2} \tag{10}$$

$$\frac{\left(0.75\,\sigma a_1^2 - \gamma a_1\right)}{\gamma\left(2a_1 + a_2\right)}da_2 = da_1 \tag{11}$$

Note that at this point, we now have both an expression for da_2/dt (Eq. 5) and a way to convert da_1 to da_2 (Eq. 12). All that is left is to define the grain growth rate in terms of da_1/dt and da_2/dt, and insert the appropriate terms:

184

$$\dot{\Gamma} = \frac{1}{G}\frac{dG}{dt} = \frac{1}{3V}\frac{dV}{dt} = \frac{1}{3\left(\frac{4}{3}\pi a_1^2 a_2\right)}\frac{d}{dt}\left(\frac{4}{3}\pi a_1^2 a_2\right) = \frac{1}{3a_2}\frac{da_2}{dt} + \frac{2}{3a_1}\frac{da_1}{dt} \qquad (12)$$

$$\dot{\Gamma} = \frac{1}{3a_2}\left(\frac{24\delta D_{gb}\Omega}{G^3 kT}\right)a_2\left[\left(\frac{a_2}{a_1}\right)^2 + \left(\frac{a_2}{a_1}\right) + 1\right]^{-1}\left(\sigma_{app} - \frac{2\gamma a_2}{a_1^2}\right) +$$

$$\frac{2}{3a_1}\left(\frac{0.75\sigma\, a_1^2 - \gamma a_1}{\gamma(2a_2 + a_1)}\right)\left(\frac{24\delta D_{gb}\Omega}{G^3 kT}\right)a_2\left[\left(\frac{a_2}{a_1}\right)^2 + \left(\frac{a_2}{a_1}\right) + 1\right]^{-1}\left(\sigma_{app} - \frac{2\gamma a_2}{a_1^2}\right) \qquad (13)$$

After much algebra, the final solution is obtained (Eq. 14, below):

$$\dot{\Gamma} = \left(\frac{8\delta D_{gb}\Omega}{G^3 kT}\right)\left[\left(\frac{a_2}{a_1}\right)^2 + \left(\frac{a_2}{a_1}\right) + 1\right]^{-1}\left[\sigma^2\left(\frac{3a_1 a_2}{2\gamma[2a_1 + a_2]}\right) - \sigma\left(\frac{3a_2^2 + 2a_1 a_2}{a_1[2a_1 + a_2]} - 1\right) - \left(\frac{4\gamma a_{1a_2} - 2\gamma a_2^2}{a_1^2[2a_1 + a_2]}\right)\right]$$

Note that if the grains are indeed able to maintain a fairly equiaxed shape throughout deformation by rapid recovery, $a_1 \approx a_2 \approx G/2$, and Eq. 14 simplifies to

$$\dot{\Gamma} = \frac{8\delta D_{gb}\Omega}{G^3 kT}\left[\frac{G}{12\gamma}\sigma^2 - \frac{2}{9}\sigma - \frac{4\gamma}{9G}\right] \qquad (15)$$

Eq. 15 can be plotted as a stress-strain rate curve, knowing that $\dot{\Gamma}$ and $\dot{\varepsilon}$ are directly comparable, as per the dynamic grain growth law (Eq. 3). The result is shown in Figure 4.

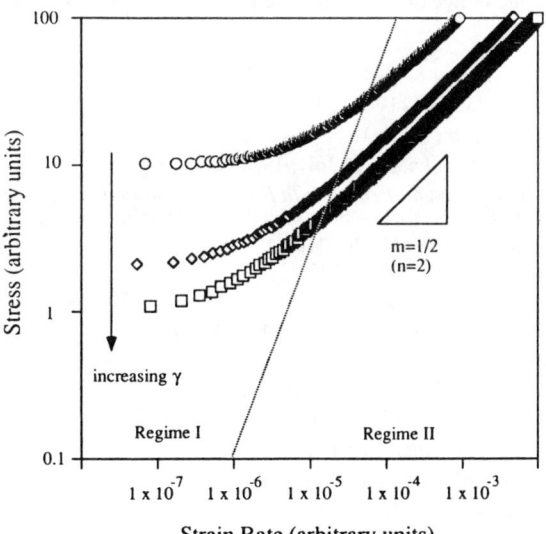

Figure 4. Schematic stress-strain rate curve resulting from Eq. 15.

DISCUSSION

As can be seen from Figure 4, the stress exponent naturally resulting from the present dynamic grain growth-based approach is n=2; however, this stress exponent will vary according to the initial creep mechanism chosen for grain elongation. Generally, the stress exponent will be n=n'+1 where n' is the stress exponent for the underlying creep mechanism. In this manner, all superplastic materials can obey the dynamic grain growth law and experience fundamentally the same microstructural chain of events, but still manifest different stress exponents. The ability to obtain a superplastic relation starting from diffusional creep mechanisms, as is done here, explains why ceramics can be superplastic: no dislocation activity is fundamentally required.

Regime I falls out naturally from the present derivation. It is a direct result of the curvature backstress, which becomes increasingly important at smaller applied stresses, until finally the backstress exceeds the applied stress, and deformation stops (i.e., a threshold stress is reached.) Changes in grain boundary energy affect the magnitude of the backstress and hence the location of Regime I. This result has also been observed experimentally: the work of Jiang et al.[17] found that the transition strain rate between Regime II and Regime I in Fe-doped Zn-22wt%Al decreased as sample purity increased (which, for grain boundary segregating impurities, is equivalent to increasing the grain boundary energy).

A high degree of anelastic backstrain is also predicted by the present model, because the curvature backstress developed through creep will drive reverse creep in the absence of a forward applied stress. Notably, anelastic backstrain has long been recognized as a major feature in superplastic materials.[18]

The present approach also suggests a reason why superplasticity is observed only in fine-grained materials: for grain curvature to play the significant role that is required, it needs to be on the same order as the applied stress ($2\gamma/R \sim \sigma$); in other words, a grain size on the order of microns is needed to observe superplastic effects in the stress range of MPa. Curvature also drives grain shape recovery during grain growth. Finally, there may also be an explanation for the large elongations to failure: since the microstructure is continually being recovered (returned to equiaxed form) by grain growth even as it is being strained, the process of straining can go on ad infinitum.

ACKNOWLEDGMENT

This work was sponsored by the Office of Naval Research, grant #N00014-98-1-0637.

REFERENCES

[1]C.E. Pearson, *J. Inst. Met.* **54** (1934) 111.

[2]C.M. Packer and O.D. Sherby, *Trans. ASM* **60** (1967) 21.

[3]A. Ball and M.M. Hutchison, *Met. Sci. J.* **3** (1969) 1.

[4]J.H. Gittus, *J. Engineering Materials and Technology* (July, 1977) 244.

[5]A.K. Mukherjee, T.R. Bieler, and A.H. Chokshi, *Proc. Tenth Risø International Symposium on Metallurgy and Materials Science: Materials Architecture*, edited by N. Hansen et al. Roskilde, Denmark: Risø National Laboratory, Roskilde, Denmark, 1989. p. 207.

[6]S.P.S. Badwal et al., *J. Am. Ceram. Soc.* **73** (1990) 2305.

[7]M. Gust, G. Goo, J. Wolfenstine, and M.L. Mecartney, *J. Am. Ceram. Soc.* **76** (1993) 1681.

[8]F. Wakai, S. Sakaguchi, and H. Kato, *J. Ceram. Soc. Japan* **94** (1986) 721.

[9]Y. Ma and T. G. Langdon, *Acta Metall. Mater.* **42** (1994) 2753.

[10]T.G. Nieh and J. Wadsworth, *Acta Metall. Mater.* **38** (1990) 1121.

[11]M. Nauer and C. Carry, *Scripta Metall. Mater.* **24** (1990) 1459.

[12]J.R. Seidensticker and M.J. Mayo, *Scripta Materialia* **38** (1998) 1091.

[13]J.R. Seidensticker and M.J. Mayo, *Acta Materialia* **46** (1998) 4883.

[14]D.S. Wilkinson and C.H. Caceres, *Acta Metall* **32** (1984) 1335.

[15]Y.-I. Yoshizawa and T. Sakuma, *Superplasticity in Advanced Materials: ICSAM '91*, edited by S. Hori et al. Osaka, Japan: Japan Society for Research on Superplasticity, 1991. p. 251.

[16]W.D. Nix, *Metals Forum* **4** (1981) 38-43.

[17]X-G Jiang et al., *Metall. and Mater. Trans. A* **27A** (1996) 863-872.

[18]J.H. Schneibel and P.M. Hazzledine, *Acta Metall.* **30** (1982) 1223-1230.

SIMULATION OF MICROSTRUCTURAL EVOLUTION DURING SUPERPLASTIC DEFORMATION

B.-N. KIM, K. HIRAGA
National Research institute for Metals, Tsukuba, Ibaraki 305-0047, Japan

ABSTRACT

Superplastic tensile deformation is simulated in 2 dimensions by incorporating grain boundary diffusion and concurrent grain growth derived from static and dynamic growth mechanisms. The following relationship is found between microstructural changes and deformation behavior for constant stress conditions. Grain boundary diffusion produces an increase in the aspect ratio of the matrix grains during deformation and the increased aspect ratio causes a change in creep rate parameters: the stress exponent is decreased from the initial value of 1.0 for equiaxed grains and the grain size exponent is increased from the initial value of 3.0. Accelerated grain growth is also found by the present simulation.

INTRODUCTION

When a polycrystalline material is subjected to stress at elevated temperatures, it may deform by dislocation processes or by stress-directed diffusion. In diffusional deformation, the individual grains become elongated, and the grains are displaced to each other in order to maintain microstructural coherency. This process is called Lifshitz grain boundary sliding [1] and is known as one of the superplastic deformation mechanisms. Lifshitz grain boundary sliding may become a primary mechanism when the deformation is dominated by the diffusion of vacancies. The vacancies may flow either through the polycrystalline lattice or along the grain boundaries. However, since grain boundary diffusion is usually dominant in polycrystals with small grain sizes, Lifshitz grain boundary sliding associated with grain boundary diffusion is adopted as the governing deformation mechanism in the present simulation.

In addition to the change in the shape of grains, the size of grains increases during high temperature deformation by both static and dynamic grain growth, particularly in superplastic materials with small grain sizes. In this study, microstructural evolution is simulated in 2 dimensions by incorporating the mechanisms of Lifshitz grain boundary sliding and concurrent grain growth. The related mechanical properties are also examined.

SIMULATION

Basing on an atomic jump model of grain boundary migration, one of the authors simulated the static grain growth behavior in 2 dimensions, where the microstructure was composed of straight grain boundaries and the migration of triple points were determined numerically [2]. The simulation resulted in the 1/2-power growth law for the average grain size, and the grain size distribution consisted well with the Louat distribution function. The same algorithm is incorporated in the present simulation. Also, the authors recently proposed a model for dynamic grain growth based on grain boundary diffusion, in which model the diffused matter is assumed to contribute to grain growth during deformation [3]. For the consideration of dynamic grain growth, the same basic concept is introduced to the following Lifshitz grain boundary sliding associated with grain boundary diffusion.

Mat. Res. Soc. Symp. Proc. Vol. 601 © 2000 Materials Research Society

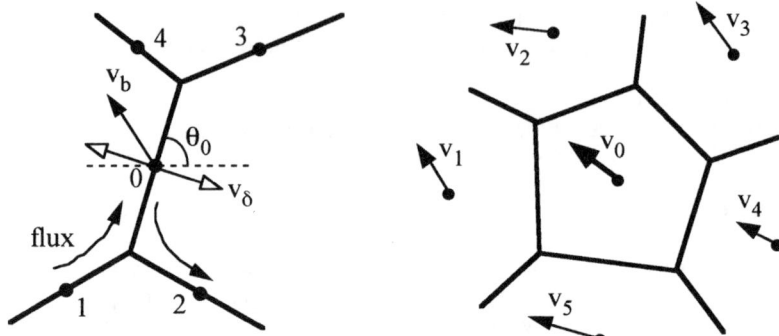

Fig. 1 Grain boundary diffusion model.　　Fig. 2 Velocities of diffusion centers.

In order to enable large deformation in microstructure, the grain boundaries are treated as straight lines. The potential on the grain boundary facet is assumed to be constant, and the stress concentration at triple point is neglected. It is also assumed that shear stresses on the grain boundary facet are fully released during deformation and the potential generated by the applied stress σ on the grain boundary facet of the angle θ with respect to the stress axis is represented by $\sigma\cos\theta^2$. Then, the flux J' in the grain boundary facet 0 shown in Fig. 1 is approximately represented by

$$J' = - \frac{D_b\delta}{\Omega k_B T} \sum_i^4 \frac{\Delta\mu_i}{\Delta s_i} \tag{1}$$

where D_b is the grain boundary diffusion coefficient, δ is the grain boundary thickness, Ω is the atomic volume, k_B is Boltzman's constant, T is the absolute temperature, and Δs_i and $\Delta\mu_i$ are the average distance and the potential difference between the neighboring grain boundary facets 0 and i, respectively. From this equation, the relative velocity v_δ of two boundary surfaces of the grain boundary facet is obtained by

$$v_\delta = - \Omega\, \text{div} J' = \frac{\Omega D_b\delta}{l_0 k_B T} \sum_i^4 \frac{(\cos^2\theta_0 - \cos^2\theta_i)}{\Delta s_i} \tag{2}$$

where l_0 is the length of the grain boundary facet. $+v_\delta$ means the deposition of matter on the two grain boundary surfaces.

The macroscopic strain rate can be calculated by using the relative velocities at the boundary facets within the specimen. When the specimen with a width of L_x and a length of L_y is subjected to the uniaxial stress σ in the longitudinal direction (y-axis), the contribution $d\dot{\varepsilon}$ of the grain boundary facet of the length ds and the angle θ to the strain rate is $v_\delta \cos\theta^2 ds/L_x L_y$. The macroscopic strain rate $\dot{\varepsilon}$ is obtained by integrating all the contributions as

$$\dot{\varepsilon} = \frac{1}{L_x L_y} \int v_\delta \cos^2\theta ds . \tag{3}$$

The macroscopic strain roughly determines the position of each grain after deformation.

Let us consider a point called diffusion center where the vacancy concentration gradient is zero [4]. The velocity of the point will be equal to that of the grain, as schematically shown in Fig. 2, and the velocities of two neighboring grains, v_0 and v_i, have a relationship of $v_0 - v_i = v_{\delta i}$, where $v_{\delta i}$ is the relative velocity of the grain boundary facet between grains 0 and i. For grain 0 which contacts with N grains, the velocity v_0 is represented by

$$v_0 = \frac{1}{N} \sum_{i}^{N} v_{\delta i} + \frac{1}{N} \sum_{i}^{N} v_i \ . \tag{4}$$

The right second term of Eq. (4) is the average velocity of the neighboring grains, so that it can be obtained approximately from $\dot{\varepsilon}$ under the assumption of uniform deformation. In this study, the diffusion center is assumed to correspond to the gravitational center of each grain.

The absolute velocity of the grain boundary facet is determined by the grain boundary diffusion model [3]. For example, let the grain boundary facet between grains 0 and 1 move into the side of grain 0 by static grain growth. It means that the potential of the boundary surface of grain 0 is higher than that of grain 1. If the matter entered the facet, it would be deposited on the boundary surface of grain 1 of lower potential. Then, the velocity of the grain boundary facet v_b is equal to v_0. This is the main concept of the grain boundary diffusion model for dynamic grain growth. If the potential difference is not considered, v_b would be obtained simply as $(v_0 + v_1)/2$, which results in no dynamic effect on grain growth.

Finally, the velocity of triple point is obtained by averaging the three velocities of the connected grain boundary facets. The deformed microstructure is constructed by connecting the moved triple points. Although the grain boundary sliding is not considered directly in this algorithm, it is incorporated implicitly.

For an infinitesimal time, it is assumed that the above three mechanisms (Lifshitz grain boundary sliding, static grain growth and dynamic grain growth) work simultaneously and independently. The initial microstructure is composed of about 3300 equiaxed grains, as shown in Fig. 3(a). The simulation is carried out under constant stress conditions.

RESULTS AND DISCUSSION

<u>Microstructural Evolution</u>

The simulated microstructural changes during superplastic deformation are shown in Fig. 3

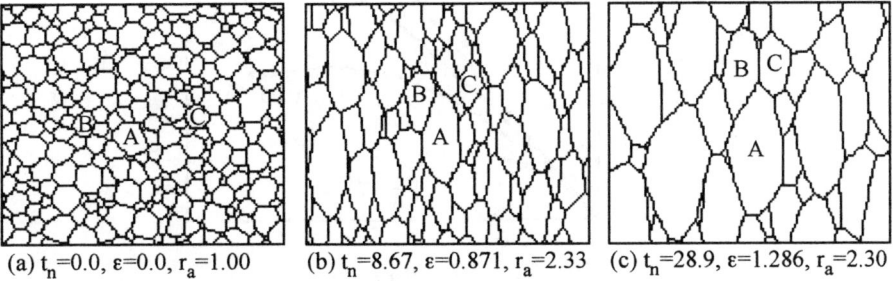

(a) t_n=0.0, ε=0.0, r_a=1.00   (b) t_n=8.67, ε=0.871, r_a=2.33   (c) t_n=28.9, ε=1.286, r_a=2.30

Fig. 3 Microstructural evolution at a constant stress of $\Omega\sigma/k_B T$=0.15, where t_n is the normalized time of $tD_b\delta R_0^{-3}$.

for $\Omega\sigma/k_BT=0.15$ and $M\gamma R_0^{-2}t_{1.5}=3.08$, where M is the grain boundary mobility, γ is the grain boundary energy, R_0 is the average initial radius of grains, and $t_{1.5}$ is the time required for static grain growth from R_0 to $1.5R_0$. By grain boundary diffusion, the grains are elongated along the stress axis and shrank in the direction normal to the stress axis. As a result, grains B and C approach grain A, leading to the impingement of these grains. It should be noted that the impingement is followed by the displacement of the relative positions of these grains.

In the Lifshitz sliding mechanism, grain boundary sliding and diffusion occur simultaneously. However, close inspection of Fig. 3 reveals that grain boundary sliding is deeply related with grain growth or grain boundary migration in a phenomenology. As seen in Fig. 3(b) and (c), a shrinking grain exists between grains B and C, and the shrinkage and annihilation of the intervening grain contribute to the sliding of grains B and C. Apparently, the microstructural incoherency produced by sliding is relaxed by grain boundary migration, that is, grain growth.

Grain boundary migration is also found to control the aspect ratio. The elongation of grains due to grain boundary diffusion is restricted by the recombination of triple points during the grain growth process, and hence the rate of grain elongation depends both on the grain boundary mobility and on the applied stress. Figure 4 shows the variation of the aspect ratio as a function of strain for various stress levels. At high stresses, the rate of grain elongation is quite higher than the relaxation rate by grain boundary migration, so that the macroscopic strain approaches the strain of the grains. At low stresses, however, the elongation is restricted by grain boundary migration. Under constant stress conditions, since the strain rate decreases with increasing strain owing to grain growth, and since the migration velocity of grain boundaries is invariant, the aspect ratio r_a decreases with increasing strain, as shown in Fig. 4.

Figure 5 represents the simulated variation of the grain size at different stress levels, showing that grain growth is accelerated during superplastic deformation. The dynamic effect on grain growth increases with increasing stress levels. At $\Omega\sigma/k_BT=0.2$, the dynamic effect becomes nearly equal to the static contribution to the grain growth. This is because higher applied stress results in larger amount of the diffused matter, that is, larger contribution to the dynamic grain growth. Provided that the diffused matter contributes to dynamic grain growth, the grain boundary diffusion model is independent of the rate-controlling mechanism of superplastic deformation, because the amount of the diffused matter is solely determined by the applied stress [3].

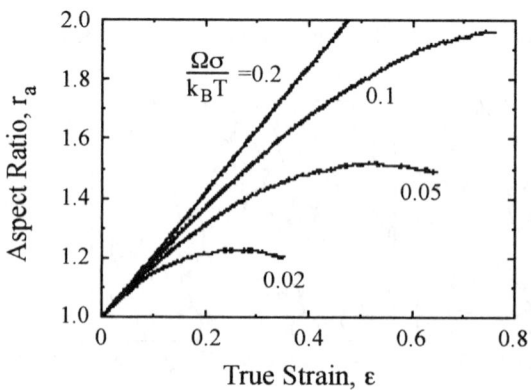

Fig. 4 Variation of the aspect ratio for $M\gamma R_0^{-2}t_{1.5}=3.08$.

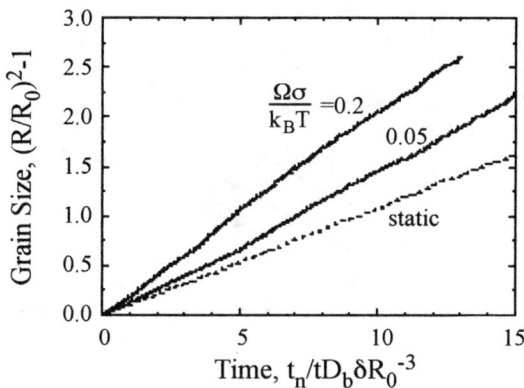

Fig. 5 Variation of the grain size for $M\gamma R_0^{-2}t_{1.5}=3.08$.

Mechanical Properties

The strain rate by grain boundary diffusion is represented by [5]

$$\dot{\varepsilon} = \beta \frac{D_b\delta\Omega\sigma}{k_BTR^3} \qquad (5)$$

where β is a proportional constant. Several models were proposed to give 1.15~41.6 as β for 2-dimensional hexagonal polycrystals [5-8]. The present value of β for equiaxed grains is 1.94, which is within the proposed range. However, grain elongation reduces the strain rate owing to the elongated diffusion distance. The influence of the grain elongation on the strain rate is shown in Fig. 6. While the grain size exponent is 3 for equiaxed grains, it becomes larger for larger aspect ratios. Owing to the grain elongation, the simulated relationship between $\dot{\varepsilon}$ and R deviates from

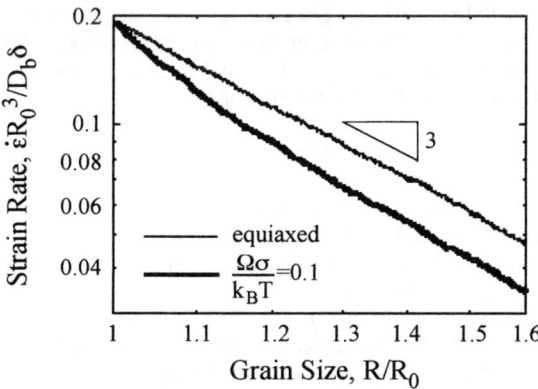

Fig. 6 Influence of grain elongation on the relationship between strain rate and grain size.

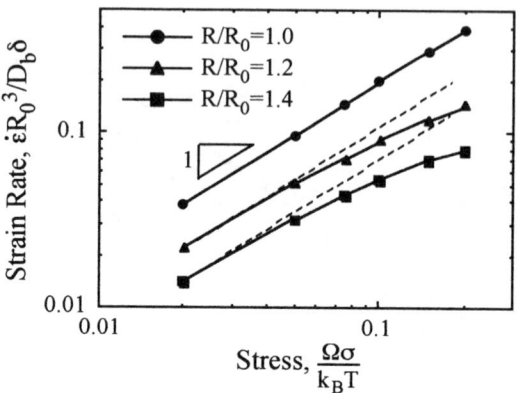

Fig. 7 Influence of grain elongation on the relationship between strain rate and stress.

Eq. (5). When the aspect ratio is nearly constant, the grain size exponent approaches 3.

The influence of grain elongation on the stress exponent is shown in Fig. 7. From simulations at different stress levels, the strain rate was plotted with respect to the stress at the same grain size. The stress exponent of 1.0 in the initial state for equiaxed grains decreases with increasing grain size, that is, increasing aspect ratio. The degree of the deviation increases with increasing stress, as seen in Fig. 7. This result indicates that the consideration of the grain aspect ratio is necessary for the exact examination of the stress exponent.

CONCLUSIONS

Superplastic deformation behavior was simulated in 2 dimensions by incorporating the mechanisms of Lifshitz grain boundary sliding, static and dynamic grain growth. The microstructural changes during deformation were examined along with the related mechanical properties. From the observation of the simulated microstructural changes, grain boundary sliding is found to relate deeply to grain boundary migration in a phenomenology: the shrinkage and annihilation of grains contribute to the apparent sliding behavior. The grain boundary migration is also found to control the grain aspect ratio through the recombination process of triple points. With an increase in the aspect ratio, the stress exponent decreases from the initial value of 1.0 for equiaxed grains and the grain size exponent increases from the inititial value of 3.

REFERENCES

1. I.M. Lihshitz, Soviet Phys. JETP **17**, 909 (1963).
2. B.-N. Kim and T. Kishi, J. Japan Inst. Metals **61**, 1341 (1997).
3. B.-N. Kim, K. Hiraga, Y. Sakka and B.-W. Ahn, Acta Mater. **41**, 3433 (1999).
4. R.N. Stevens, Phil. Mag. **23**, 265 (1971).
5. R.L. Coble, J. Appl. Phys. **34**, 1679 (1963).
6. M.F. Ashby, G.H. Edward, J. Davenport and R.A. Verrall, Acta Metall. **26**, 1379 (1978).
7. J.R. Spingarn and W.D. Nix, Acta Metall. **26**, 1389 (1978).
8. M.F. Ashby and R.A. Verrall, Acta Metall. **21**, 149 (1973).

ON THE RELATIONSHIP BETWEEN GRAIN BOUNDARY SLIDING AND INTRAGRANULAR SLIP DURING SUPERPLASTIC DEFORMATION

A. D. SHEIKH-ALI[1, 2]§, J. A. SZPUNAR[2] and H. GARMESTANI[1]
[1]Laboratory for Micromechanics of Materials, National High Magnetic Field Laboratory,
1800 E. Paul Dirac Drive, Tallahassee, Florida 32310, USA
[2]Department of Metallurgical Engineering, McGill University,
3610 University Street, Montreal, PQ, Canada, H3A 2B2

ABSTRACT

This paper examines grain boundary sliding under the conditions of plastic strain incompatibility that is the most frequent case in polycrystalline materials. Two components of grain boundary sliding: dependent and independent on intragranular slip are distinguished. Theoretical estimate of a ratio between slip induced sliding and intragranular slip is obtained. It is concluded that slip and sliding are rather independent than interrelated processes.

INTRODUCTION

Usual high-temperature deformation is associated with accumulation of lattice dislocations in grains that often induces recrystallization processes. Structural superplasticity is a specific type of high-temperature deformation accompanied by insignificant accumulation of dislocations thus produces minor changes in microstructure. Such behavior is primarily attributed to the process of grain boundary sliding (GBS) that operates with intragranular (crystallographic) slip and diffusional creep in a favorable combination. Sliding and slip make their maximum contributions to total strain at the optimum and high strain rates respectively. There is a link between these processes. Under deformation with a constant strain rate the accumulation of lattice dislocations in grains is decreased by operation of GBS. Lattice dislocations entered grain boundaries are dissociated into extrinsic grain boundary dislocations (GBDs) that participating in the process of cooperative GBS can be carried out to the specimen surface. This makes possible further development of slip. At the same time giving a significant contribution to overall strain, GBS decreases the contribution of slip. Such complex interconnection between sliding and slip results in appearance of two different concepts of structural superplasticity. According to the first one [1] slip and sliding are closely related to each other in the optimum region of superplasticity: facilitation of intragranular slip increases the rate grain boundary sliding. Following the other concept [2] sliding and slip are independent and competing processes. In this paper, results that we believe can really support these different concepts are analyzed and relation between slip and sliding is determined. To elucidate the problem clearly we consider experimental results obtained by different researchers on bicrystals and polycrystalline materials.

INTERACTION BETWEEN SLIDING AND SLIP IN BICRYSTALS

A number of experiments with bicrystals has demonstrated that GBS is strongly related to the development of intragranular slip and that the two processes can not be viewed as non-interacting [3-8]. At the same time it has been shown that crystallographic slip is not a prerequisite for sliding and the latter called pure GBS can develop without slip [4, 5]. However,

---

§ On leave from the Institute for Metals Superplasticity, Russian Academy of Sciences, Ufa, Bashkortostan, Russian Federation.
E-mail: sheikh-ali@usa.net

Mat. Res. Soc. Symp. Proc. Vol. 601 © 2000 Materials Research Society

some coincidence boundaries that are quite rare in polycrystals can slide only in the presence of intragranular slip [6]. Slip activation can result in a marked increase of GBS rate [3-5, 7]. Two mechanisms can be responsible for the increase in GBS rate: interaction between lattice dislocations and boundary, which produces additional glissile grain boundary dislocations (GBDs) [8] and accommodation of sliding at boundary irregularities by the emission of lattice dislocations into the grain interior [3, 8, 9]. Grain boundary dislocations that can contribute to GBS appear when neighboring grains deform non-similarly relative to boundary plane or incompatibly. From geometrical viewpoint this is the most frequent case in polycrystals. Similar deformation of grains or deformation at plastic compatibility results in generation of GBDs that can make a contribution only to local sliding and are not responsible for the increase in the rate of macroscopic sliding [7, 10]. The increase in GBS rate at incompatible conditions, apparently, is a result of accommodation of sliding by slip at grain boundary irregularities. It is worth noting that significant increase in sliding rate occurs only in the case of compatible deformation [4, 5, 7] or some coincidence boundaries [11].

CONCEPT OF SLIDING OPERATING UNDER CONDITIONS OF INCOMPATIBILITY

Let us consider a hypothetical two-dimensional bicrystal containing a finite boundary with the ends designated as O and C (Fig. 1 (a)). If the applied shear stress is sufficient to initiate GBS but not enough to deform grains this is the case of pure sliding (Fig. 1(b)). It is worth noting that Horton et al. [9] used the term of pure sliding when it was accommodated by a small slip. At microscopic scale pure sliding can be considered as the motion of glissile GBDs of opposite signs generated in the boundary plane by numerous grain boundary sources distributed along the whole boundary including points O and C (Fig. 1(b)). When initiation of grain boundary shear or generation of GBDs by grain boundary sources is inhibited, sliding can be produced purely by

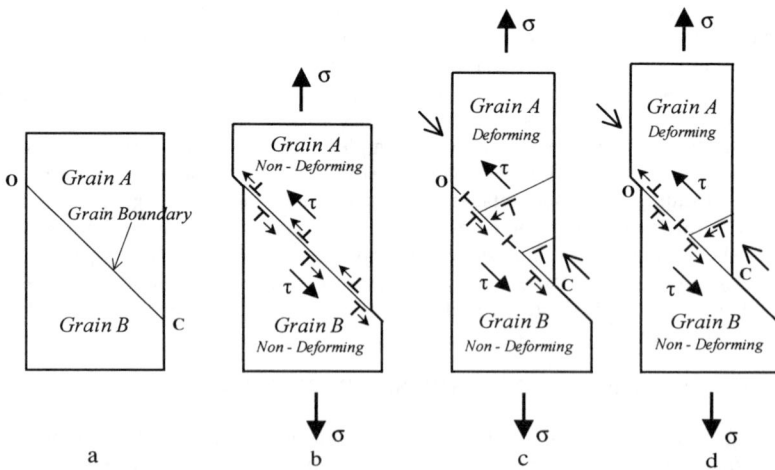

Fig. 1. (a) Original hypothetical bicrystal; (b) pure GBS provided by GBDs generated from boundary sources; (c) slip induced sliding provided entirely by GBDs generated as a result of interaction of lattice dislocations with boundary; (d) combination of pure sliding and slip induced sliding.

the difference of deformation between neighboring grains [6]. In the case of incompatible deformation shown in Figure 1(c) the upper grain A deforms and the lower grain B does not. Only GBDs of one sign are produced as a result of interaction of lattice dislocations with the boundary. If motion of GBDs is governed mainly by applied stresses the amount of sliding increases from zero at point O to some value at point C. Often, sliding is a combination of pure sliding and sliding induced by intragranular deformation (Fig. 1(d)). Assuming that these processes are independent the total amount of sliding is a sum of two kinds of sliding:

$$S = S_i + S_{is},$$ (1)

where S is total amount of sliding, S_i is a value of pure sliding and S_{is} is a value of slip induced sliding. Due to non-uniform distribution of sliding induced by slip the overall sliding is also non-uniform. However in this case, some amount of sliding is reached at point O and this is a pure sliding whereas sliding at point C is a sum of the amounts of pure sliding and sliding induced by slip. At dislocation level the independence of different types of sliding means the non-interacting and independent operation of two different kinds of GBD sources: lattice dislocations impinged into the boundary and grain boundary sources of GBDs. The interaction between GBDs of different origin results in remaining of glissile GBDs of one sign (Fig. 1(d)).

RELATION BETWEEN SLIP AND SLIDING UNDER SUPERPLASTIC CONDITIONS

Experimental Results

The study of anisotropic behavior of Zn-0.4%Al sheet alloy during superplastic flow have shown that straining in the direction favorable for basal slip reduces the strain-rate sensitivity exponent m [12] (Fig. 2). The analogous results have been obtained for fine-grained cadmium and magnesium alloys [13, 14]. The value of m correlates with contribution of sliding to overall strain [15, 16]. Therefore, these results may indicate on concurrent and independent character of sliding and slip during superplastic deformation: the greater contribution of GBS to total strain, the lower contribution of slip and vice versa.

The notion of the two independent processes contradicts to the results obtained by Matsuki

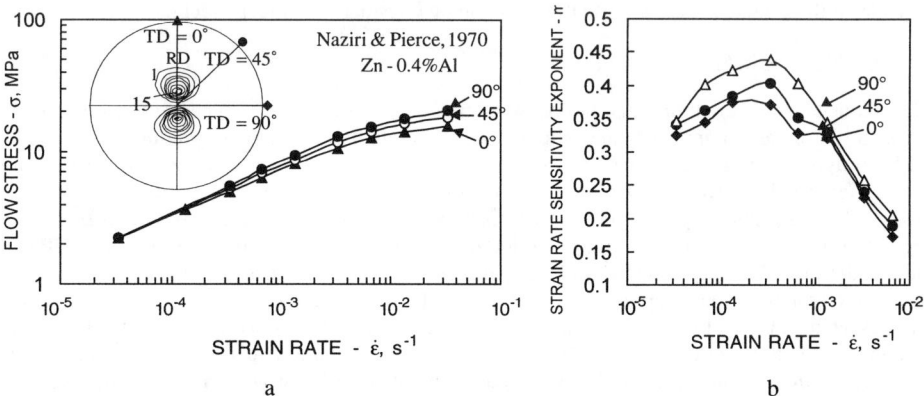

Fig. 2. Anisotropic superplasticity in Zn-0.4%Al [12]. (a) Flow stress-strain rate relations for specimens cut at 0, 45 and 90° to the rolling direction and (b) variation of m with strain-rate and with direction of straining.

195

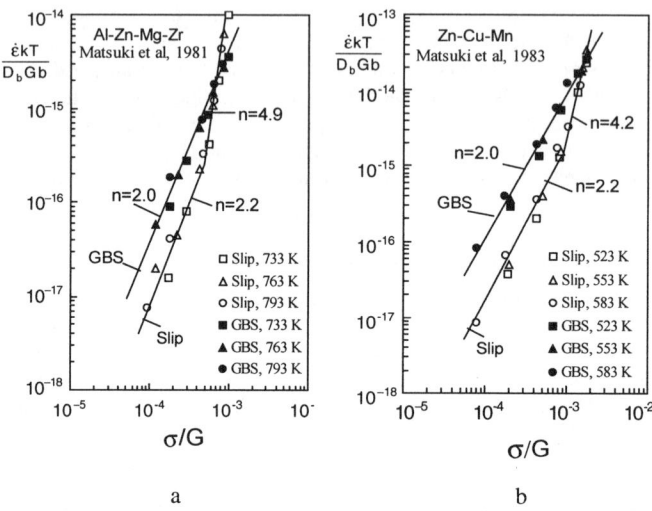

Fig. 3. Relation between individual strain rates of grain boundary sliding and intragranular slip and flow stress in dimensionless plots: $D_b=D_o\exp(-Q/RT)$ for Al (a) and Zn (b) alloys [17, 18].

et al. [17, 18] on Al-10.8%Zn-0.85%Mg-0.29%Zr and Zn-0.92%Cu-0.60%Mn alloys. For these alloys the stress exponents for GBS and slip were measured separately. Figure 3 demonstrates that the stress exponents for sliding and slip are similar in the optimum region of superplasticity ($n_{GBS}=2.0$ and $n_{IS}=2.2$ for both alloys) and dissimilar in the region of the higher strain rates ($n_{GBS}=2.0$ and $n_{IS}=4.9$ for aluminum alloy, $n_{GBS}=2.0$ and $n_{IS}=4.2$ for zinc alloy). These results show that in the optimum region of superplasticity GBS and slip are interdependent processes whereas in the region of high strain rates they can be considered as independent.

Theoretical Consideration

Consideration of sliding in a hypothetical bicrystal shows that two types of GBS coexist in the same boundary: pure GBS and slip induced sliding. Pure sliding is produced by glissile GBDs originated from grain boundary sources. In polycrystalline materials pure sliding is a rare event. Nevertheless, we use the term "pure" to designate the component of sliding that is produced from grain boundary dislocation sources. Slip induced sliding is provided by GBDs generated as a result of interaction of lattice dislocations with grain boundaries. Pure GBS is independent on slip. Slip induced sliding is directly connected with intragranular deformation. The latter makes a contribution to overall strain if motion of GBDs is governed by applied stresses. Hence, GBS along the same boundary can be separated into two components: dependent and independent on slip, each of them can make a contribution to total strain. We denote processes which are interdependent by including the appropriate strains in square brackets and processes which are independent, including the resultants of sequential pairs, just by addition signs. Thus, the total strain ε_{TOT} can be expressed in terms of strains due to pure GBS ε_{GBS}, intragranular slip ε_{IS} and slip induced sliding ε_{SIS}:

196

$$\varepsilon_{TOT} = \varepsilon_{GBS} + [\varepsilon_{IS} + \varepsilon_{SIS}] \qquad (2)$$

Let us consider the optimum region of superplasticity. If the intragranular slip is inhibited, the contribution of sliding to overall strain is enhanced. This type of sliding is mostly provided by pure GBS. Sliding rate can increase as a result of increase in the rate of motion of GBDs, number of boundary dislocation sources and/or the rate of generation of GBDs. Facilitation of intragranular slip (e.g., due to switching over to favorable crystallographic orientation) increases its contribution to overall strain via grain elongation. At the same time intragranular slip can increase sliding rate by making a direct contribution to grain boundary shear. Hence, it is still unclear why the total contribution of GBS to overall strain decreases in this case.

Contribution of intragranular deformation and slip induced sliding to total strain can be assessed assuming that all sliding is induced by slip and there is no pure GBS. Figure 4 (a) illustrates two hexagonal grains having a common boundary. The upper grain is subjected to uniform elongation, the other is non-deforming (Fig. 4 (b)). The upper grain is allowed to slide when it deforms. In general, there are two different cases of sliding induced by slip. In the first case illustrated in Fig. 4 (c) GBS makes a direct contribution to strain and simultaneously accommodates grain deformation. Apparently, this type of sliding occurs in the optimal superplastic region being a part of cooperative GBS. In the second case slip induced sliding accommodates grain deformation without contribution to total strain (Fig. 4 (d)). The latter seems to be frequent during high-temperature non-superplastic deformation and used for separation of slip induced sliding from total strain. Strain produced purely by slip induced sliding is the difference between total strain and intragranular strain:

$$\varepsilon_{SIS} = \varepsilon_{TOT} - \varepsilon_{IS} = \frac{l_1 - l_2}{l_0} \qquad (3)$$

Here l_0 - distance between points C and M before deformation, l_1, l_2 - distance between the

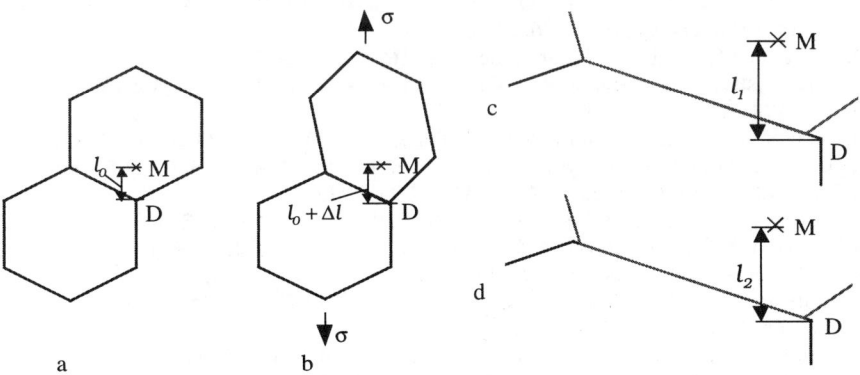

Fig. 4. Geometrical model for slip induced sliding. (a) Initial hexagonal grains. (b) Two-dimensional deformation of the upper grain. Slip induced sliding (c) with and (d) without contribution to total elongation. M is a fiducial marker point.

same points after deformation. At intragranular strain of 0.1 in the upper grain the contribution of slip induced sliding to total strain reaches ~15%. In reality the contribution of slip induced sliding may vary in a wide range. However, it is difficult to expect that maximum contribution of strain induced sliding can significantly exceed the value obtained from two-grain model in Figure 4. Thus, slip induced sliding makes a small contribution to total strain, which can explain independent relation between slip and sliding in the optimum superplastic region. Similarity of stress exponents for slip and sliding in the optimum superplastic region just shows that the rate of slip is limited by sliding. Slip induced sliding and slip produced from grain boundary dislocation sources are governed by applied stresses and have the same stress exponents. In the region of high strain rates the contribution of slip increases that may decrease the contribution of slip induced sliding to the total strain. The difference in stress sensitivity exponents for slip and GBS may indicate that slip induced sliding is governed by internal stresses induced by intragranular slip and its contribution to total strain can decrease. This type of slip induced sliding may also destroy the cooperative character of GBS.

CONCLUDING REMARKS

Grain boundary sliding can be separated into two components that are dependent and independent on intragranular slip. Each of them can make own contribution to total strain during superplastic deformation in the optimum region. Theoretical assessment of a ratio between intragranular slip and slip induced sliding shows that contribution of the latter to total strain is small. Therefore, intragranular slip and grain boundary sliding operating in the optimum region of superplasticity can be considered to some extent as independent and competing processes.

REFERENCES

1. O.A. Kaibyshev, Superplasticty of Alloys, Intermetallides, and Ceramics, Springer-Verlag, Berlin, 1992, 317 p.
2. T.G. Nieh, J. Wadsworth and O.D. Sherby, Superplasticity in Metals and Ceramics, Cambridge University Press, 1996, 273 p.
3. H. Gleiter and B. Chalmers, Progr. Mater. Sci., **16,** p. 189 (1972).
4. O.A. Kaibyshev, V.V. Astanin, R.Z. Valiev and V. G. Khairullin, Phys. Met. Metallogr., **51,** p. 166 (1981).
5. H. Fukutomi, H. Takatori and R. Horiuchi, Trans. Japan Inst. Met., **23,** p. 579 (1982).
6. A.D. Sheikh-Ali, F.F. Lavrentyev and Yu.G. Kazarov, Acta Mater., **45,** p. 4505 (1997.
7. A.D. Sheikh-Ali, Scripta Metall. Mater., **33,** p. 795 (1995).
8. R.C. Pond, D.A. Smith and P.W.J. Southerden, Philos. Mag. A, **37,** p. 27 (1978).
9. A.P. Horton, N.B.W. Thompson and C.J. Beevers, Metal Sci. J., **2,** p. 19 (1968).
10. A.D. Sheikh-Ali and J.A. Szpunar, Philos. Mag. Lett., **79,** p. 545 (1999).
11. A.D. Sheikh-Ali and J.A. Szpunar, Mater. Sci. Forum, **294-296,** p. 645 (1999).
12. H. Naziri and R. Pierce, J. Inst. Metals, **98,** p. 71 (1970).
13. Shu-En Hsu, G.R. Edwards and O.D. Sherby, Acta Metall., **31,** p. 763 (1983).
14. O.A. Kaibyshev, I.V. Kazachkov and N.G. Zaripov, J. Mater. Sci., **23,** p. 4369 (1988).
15. D. Lee, Acta Metall., **17,** p. 1057 (1969).
16. R.B. Vastava and T.G. Langdon, Acta Metall., **27,** p. 251 (1979).
17. K. Matsuki, N. Hariyama and M. Tokizawa, J. Japan Inst. Metals, **45,** p. 935 (1981).
18. K. Matsuki, N. Hariyama, M. Tokizawa and Y. Murakami, Metal Sci., **17,** p. 503 (1983).

EFFECTS OF CRYSTALLOGRAPHIC TEXTURE ON INTERNAL STRESS SUPERPLASTICITY INDUCED BY ANISOTROPIC THERMAL EXPANSION

K. KITAZONO*, R. HIRASAKA**, E. SATO*, K. KURIBAYASHI*, T. MOTEGI**

*The Institute of Space and Astronautical Science, 3-1-1 Yoshinodai, Sagamihara, Kanagawa 229-8510, Japan
**Chiba Institute of Technology, 2-17-1 Tsudanuma, Narashino, Chiba 275-8588, Japan

ABSTRACT

Polycrystalline materials having crystallographic anisotropy show internal stress superplasticity under thermal cycling conditions. The deformation mechanism is analyzed using a theoretical model based on continuum micromechanics including the effects of crystallographic texture. The model is experimentally verified through the thermal cycling creep tests using polycrystalline zinc having two kinds of fiber-texture, and agrees well with the experimental results.

INTRODUCTION

When a small external stress is applied on some kinds of materials under a thermal cycling condition, they experience high internal stresses and deform with an average strain rate which is proportional to the applied stress and much faster than the isothermal creep rate. This linear creep behavior is known as "internal stress superplasticity" [1].

Internal stress superplasticity is classified depending on the type of materials as follows: (i) Transformation superplasticity [2,3] in materials having phase transformations in a thermal cycling regime, (ii) Composite coefficient of thermal expansion (CTE)-mismatch superplasticity [4,5] in metal matrix composites, (iii) Anisotropic CTE-mismatch superplasticity [6,7] in polycrystalline materials with anisotropic CTE.

Several deformation models for internal stress superplasticity have been proposed in order to explain the mechanism of the linear relationship between the average strain rate and the applied stress. Assuming an ideal plastic material, Greenwood and Johnson [3] derived a linear equation for transformation superplasticity. Zwigl and Dunand [8] recently extended the model by Greenwood and Johnson and derived a non-linear solution valid over the whole range of stress. Sato and Kuribayashi [5] proposed a quantitative and self-consistent model based on multiaxial continuum micromechanics for composite CTE-mismatch superplasticity.

In anisotropic CTE-mismatch superplasticity, though the uniaxial deformation model by Wu et al. [6] can describe the linear creep behavior at low stress and power-law creep at high stress qualitatively, it cannot predict the important value of the generated internal stress and no attention is paid to the polycrystalline texture.

The authors have recently discussed the effects of crystallographic texture on anisotropic CTE-mismatch superplasticity [9] through developing a new theoretical model based on continuum micromechanics. The purpose of the present study is to reexamine the theoretical model and to verify experimentally using polycrystalline zinc having two kinds of fiber-texture.

THEORETICAL MODEL

In this section, the newly proposed model [9] is discussed briefly. First, the polycrystalline material D consisting of crystal grains Ω_m is considered [Fig. 1(a)]. The material D is applied an

Mat. Res. Soc. Symp. Proc. Vol. 601 © 2000 Materials Research Society

external stress σ_{ij}^0 during the temperature change with a rate of $\dot{T}$. Since the material D consists of large number of Ω_m, the CTE tensor in D-Ω_m, $\alpha_{ij}^{D-\Omega_m}$, is approximated as α_{ij}^D.

The material D deforms according to multiaxial power-law creep under the isothermal condition which is written as

$$\dot{\varepsilon}_{ij} = \frac{3}{2} K \exp(-\frac{Q_C}{RT})(\sigma_e)^{n-1} \sigma_{ij}',\tag{1}$$

where σ_e and σ_{ij}' are the equivalent and the deviatoric stresses, respectively, n is the stress exponent ($n>1$) and Q_C is the activation energy of the power-law creep. It is assumed that the effects of grain boundary sliding and interface diffusion are negligible since the grain size is sufficiently large. Second, we modify the polycrystalline material D into the composite material having a spherical crystal grain Ω_m surrounded by continuum matrix D-Ω_m [Fig. 1(b)]. This modification enables to apply the continuum micromechanics on the polycrystalline material. Throughout the present study, the elastic strain remains unchanging because we only analyze quasi-steady state creep deformation during the temperature change with a rate of $\dot{T}$.

In the present study, the material has anisotropic CTE of $\alpha_a = \alpha_b \neq \alpha_c$ such as h.c.p. or tetragonal crystals where the indexes of a-b-c show crystallographic axes, and x_1-x_2-x_3 shows specimen axes [Fig. 2]. Then, the CTE tensor in Ω_m is expressed as

$$\alpha_{ij}^{\Omega_m} = \begin{pmatrix} \alpha_a + \Delta\alpha^0 \cos^2\psi \sin^2\phi & -\Delta\alpha^0 \sin\psi \cos\psi \sin^2\phi & \Delta\alpha^0 \cos\psi \sin\phi \cos\phi \\ & \alpha_a + \Delta\alpha^0 \sin^2\psi \sin^2\phi & -\Delta\alpha^0 \sin\psi \sin\phi \cos\phi \\ \text{Sym.} & & \alpha_a + \Delta\alpha^0 \cos^2\phi \end{pmatrix},\tag{2}$$

where $\Delta\alpha^0 = \alpha_c - \alpha_a$, and angles ϕ and ψ are the zenith and azimuth angles, respectively [Fig. 2]. Using the polar density $P(\phi, \psi)$ of the c-axis which is described as a function of the angles ϕ and ψ, α_{ij}^D can be written as

$$\alpha_{ij}^D = \frac{1}{4\pi} \int_0^\pi \int_0^{2\pi} \alpha_{ij}^{\Omega_m} P(\phi, \psi) \sin\phi \, d\phi d\psi .\tag{3}$$

Under the temperature change with a rate of $\dot{T}$ in a period of δt, the mismatch strain increment $\delta\varepsilon_{ij}^{mis}$ in Ω_m is described as

$$\delta\varepsilon_{ij}^{mis} = (\alpha_{ij}^{\Omega_m} - \alpha_{ij}^D)\dot{T}\delta t .\tag{4}$$

This is the non-elastic eigenstrain increment and must be accommodated simultaneously by the

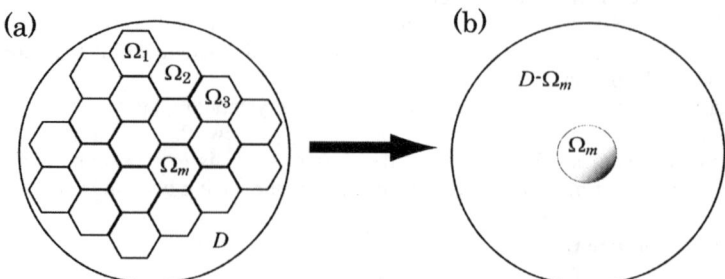

Fig. 1 Illustrations of the equiaxial polycrystalline material D (a) and the matrix containing a spherical crystal grain Ω_m (b).

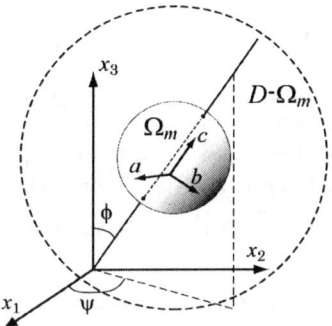

Fig. 2 Crystal grain Ω_m in the specimen coordinate system x_1-x_2-x_3.

plastic flow in Ω_m ($\delta\varepsilon_{ij}^{I}$) and in D–Ω_m ($\delta\varepsilon_{ij}^{M}$) in order to achieve the steady state condition. The plastic strain increment for the accommodation, $\delta\varepsilon_{ij}^{I}$, are given by

$$\delta\varepsilon_{ij}^{I} = (S_{ijkl} - I_{ijkl})\delta\varepsilon_{kl}^{mis}, \tag{5}$$

where S_{ijkl} is Eshelby's tensor [10] with $\nu = 0.5$ [11] and I_{ijkl} is identity tensor. Because the shape of Ω_m is spherical, the plastic strain increment in Ω_m is given by

$$\delta\varepsilon_{ij}^{I} = -\tfrac{3}{5}\delta\varepsilon_{kl}^{mis}. \tag{6}$$

When the external stress σ_{ij}^{0} and internal stress σ_{ij}^{I} are applied on the material at the same time, the quasi-steady state creep rate $\dot{\varepsilon}_{ij}^{\Omega_m}$ in Ω_m is described as

$$\dot{\varepsilon}_{ij}^{\Omega_m} = \frac{3}{2}K\exp(-\frac{Q_C}{RT_{eq}})(\sigma_e^{0} + \sigma_e^{I})^{n-1}(\sigma_{ij}^{0'} + \sigma_{ij}^{I'}), \tag{7}$$

where σ_{ij}^{I} and $\dot{\varepsilon}_{ij}^{I}$ can be expressed as a function of the power-law creep [12]:

$$\sigma_{ij}^{I'} = \frac{2}{3}K^{-1/n}\exp(\frac{Q_C}{nRT_{eq}})(\dot{\varepsilon}_e^{I})^{-(1-1/n)}\dot{\varepsilon}_{ij}^{I'}, \tag{8}$$

where $\dot{\varepsilon}_e^{I}$ is the equivalent strain rate of $\dot{\varepsilon}_{ij}^{I'}$. Here, $\dot{\varepsilon}_{ij}^{I'} = \dot{\varepsilon}_{ij}^{I}$ and $\sigma_{ij}^{I'} = \sigma_{ij}^{I}$ because $\dot{\varepsilon}_{ij}^{I}$ and σ_{ij}^{I} have no isotropic components.

We consider the condition that uniaxial external stress $\sigma_{ij}^{0} = \sigma^{0}\delta_{i3}\delta_{j3}$ (δ_{ij} is the Kronecker delta) is applied parallel to the x_3–axis. The strain rate in D in the direction of the applied stress is given by averaging $\dot{\varepsilon}_{33}^{\Omega_m}$ in the entire volume and is also given by averaging it in the stereographical sphere:

$$\dot{\varepsilon}_{33}^{D} = \frac{1}{V}\int_{V}\dot{\varepsilon}_{33}^{\Omega_m}\,dV = \frac{1}{4\pi}\int_{0}^{\pi}\int_{0}^{2\pi}\dot{\varepsilon}_{33}^{\Omega_m}P(\phi,\psi)\sin\phi\,d\phi d\psi. \tag{9}$$

If σ_{ij}^{I} is much larger than σ_{ij}^{0}, $\dot{\varepsilon}_{33}^{\Omega_m}$ in Eq. (7) reduces to the following:

$$\dot{\varepsilon}_{33}^{\Omega_m} = \frac{3}{2}K\exp(-\frac{Q_C}{RT_{eq}})\{(\sigma_e^{I})^{n-1}\sigma_{33}^{0'} + \{(\sigma_e^{I})^{n-1} + (n-1)(\sigma_e^{I})^{n-2}\sigma_e^{0}\}\sigma_{33}^{I}\}, \tag{10}$$

where all terms except the first order of σ_{ij}^{0} were neglected. Substituting Eq. (10) into Eq. (9), the average strain rate per a cycle is expressed as

$$\bar{\dot{\epsilon}}_{33}^D = K^{1/n} \exp(-\frac{Q_C}{nRT_{eq}})\sigma^0 \frac{1}{4\pi} \int_0^\pi \int_0^{2\pi} (\dot{\epsilon}_e^1)^{1-1/n} P(\phi, \psi) \sin\phi \, d\phi d\psi \, , \quad \dot{\epsilon}_{ij}^1 = -\frac{3}{5}(\alpha_{ij}^{\Omega_m} - \alpha_{ij}^D)\dot{T} \, . \quad (11)$$

Above equations indicate the general constitutive equation of anisotropic CTE-mismatch superplasticity which includes the effect of the polycrystalline texture.

We explain the simple examples. In the case of the axial symmetrical material with fiber-texture around x_3-axis which is parallel to the loading axis, $\bar{\dot{\epsilon}}^D$ becomes

$$\bar{\dot{\epsilon}}^D = K^{1/n} \exp(-\frac{Q_C}{nRT_{eq}})\sigma^0 \int_0^{\pi/2} (\dot{\epsilon}_e^1)^{1-1/n} P(\phi) \sin\phi \, d\phi \, ,$$

$$\dot{\epsilon}_e^1 = \frac{3}{5}|\Delta\alpha^0\dot{T}|\sqrt{\frac{2}{3}\left\{2\sin^2\phi + (3\cos^2\phi - 1)\int_0^{\pi/2}\sin^3\phi P(\phi)\, d\phi\right\}} \, . \quad (12)$$

If $P(\phi)=1$, that is no texture, $\bar{\dot{\epsilon}}^D$ becomes a simple formula expressed as

$$\bar{\dot{\epsilon}}^D = K^{1/n} \exp(-\frac{Q_C}{nRT_{eq}})\sigma^0 (\frac{2}{5}\sqrt{2}|\Delta\alpha^0\dot{T}|)^{1-1/n} \, . \quad (13)$$

This shows of the maximum strain rate in anisotropic CTE-mismatch superplasticity.

EXPERIMENTAL PROCEDURE

Specimens

Polycrystalline pure zinc was selected as a model material to verify the proposed model because it possesses large anisotropy of CTE and shows power-law creep at high temperature [13]. 99.99% zinc was cast in the ambient atmosphere. The ingot was hot-extruded with the reduction ratio of 12. Specimens for compression creep tests, which were named the extruded zinc, and the rod-axis is parallel to the extruded direction. The other specimens, which were named the compressed zinc, were machined after 50% hot of the hot-extruded ingot in the direction of the rod-axis at 673 K. The fabricated specimens were annealed in the salt bath at 673 K for 24 h. Optical microscopy revealed that both specimens have equiaxial crystal grains in shape and the size of about 1 mm.

Creep tests

The strain rates of isothermal and thermal cycling creep were measured by special dilatometer. The uniaxial load was applied to the specimen through a quartz rod. The specimen was heated using a high-frequency induction coil and cooled by cooling gas. The temperature was controlled and was measured through R-type thermocouple spot-welded on a specimen. The temperature profile for thermal cycling was triangular waveform between 523 and 623 K with the heating and cooling rates of ±5 K/s. The maximum controlling error, which occurred at points where heating was reversed to cooling, was less than ±3 K. The thermal cycling creep rates were measured as an average strain rate.

The polar densities were measured using a reflection method by Field and Merchant [14]. The X-ray beam was irradiated to the transverse section of the specimens. Since the specimens have fiber-texture, the intensities were averaged around the rod-axis and were obtained as a function of ϕ.

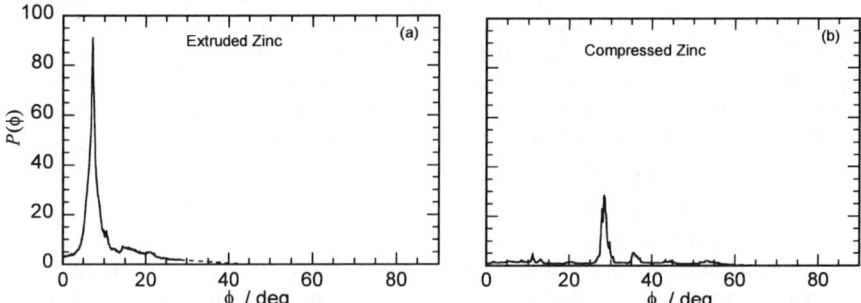

Fig. 3 Polar densities of the c-axis in (a) extruded and (b) compressed zinc specimens are plotted as a function of the angle ϕ. The broken lines at high angles show the graphical extrapolation.

RESULTS AND DISCUSSION

The polar densities of the c-axis (the basal plane) in the extruded and compressed specimens are shown in Fig. 3. Because the reflection method cannot measure the intensities at high angle of ϕ, the graphical interpolation was performed [broken lines]. The extruded specimen shows sharp peak of intensity in the range of low ϕ, while the compressed specimen shows small peak at high ϕ. These results indicate that the extruded specimen has a typical fiber-texture and the compressed specimen has comparatively random texture.

Isothermal and thermal cycling compression creep tests were performed and the strain rates were plotted as a function of the applied stress in Fig. 4. The temperature of the isothermal creep is 577 K which corresponds to the equivalent temperature of the thermal cycling creep.

Under the thermal cycling condition, each specimen shows a linear relation between the strain rates and the applied stresses at low stresses, and their creep rates were much higher than

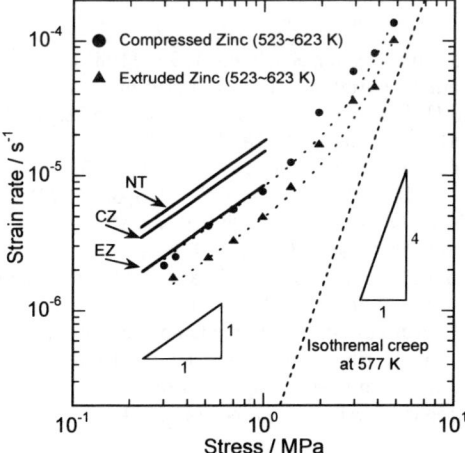

Fig. 4 Results of isothermal and thermal cycling creep tests in polycrystalline zinc. Three solid lines (NT, CZ and EZ) at low stresses show the theoretically calculated average strain rates of the specimen with no texture, extruded and compressed specimens, respectively.

the corresponding isothermal creep rates. The strain rates of compressed zinc are higher than the extruded one, which is caused by the difference in the crystalline texture.

Using Eq. (12) and the polar densities obtained in Fig. 3, we can estimate the thermal cycling creep rate of each specimen at low stresses. Three solid lines (NT, CZ and EZ) in Fig. 4 indicate theoretically calculated average strain rates of the specimen with no texture, extruded and compressed specimens, respectively. These equations are expressed as

$$\bar{\varepsilon}^{NT} \ /s^{-1} = 1.8 \times 10^{-5} (\sigma^0 / MPa),$$

$$\bar{\varepsilon}^{CZ} \ /s^{-1} = 1.5 \times 10^{-5} (\sigma^0 / MPa),$$

$$\bar{\varepsilon}^{EZ} \ /s^{-1} = 8.5 \times 10^{-6} (\sigma^0 / MPa).$$

Theoretical predictions of the extruded and compressed specimens agree well with the experimental results within the factor of two. These experimental results show the quantitative validity of the proposed model.

SUMMARY

Theoretical model for anisotropic CTE-mismatch superplasticity was developed using continuum micromechanics and the constitutive equation without any unknown parameters was obtained including the effects of the crystallographic texture and the equivalent temperature. The model was experimentally verified using polycrystalline pure zinc having two types of fiber-texture and quantitatively agreed well with the experimental results.

It is predicted that the present model is applicable to many other polycrystalline materials having crystallographic anisotropy such as ceramics and intermetallics.

ACKNOWLEDGEMENTS

This work was supported in part by Research Fellowships of the Japan Society for the Promotion of Science for Young Scientists and the Light Metal Educational Foundation.

REFERENCES

1. O. D. Sherby and J. Wadsworth, Mater. Sci. Tech. **1**, 925 (1985).
2. M. de Jong and G. W. Rathenau, Acta metall. **9**, 714 (1961).
3. G. W. Greenwood and R. H. Johnson, Proc. Roy. Soc. London **283A**, 403 (1965).
4. M. Y. Wu and O. D. Sherby, Scripta metall. **18**, 773 (1984).
5. E. Sato and K. Kuribayashi, Acta metall. mater. **41**, 1759 (1993).
6. R. C. Lobb, E. C. Sykes and R. H. Johnson, Met. Sci. J. **6**, 33 (1972).
7. M. Y. Wu, J. Wadsworth and O. D. Sherby, Metall. Trans. **18A**, 451 (1987).
8. P. Zwigl and D. C. Dunand, Acta mater. **45**, 5285 (1997).
9. K. Kitazono, R. Hirasaka, E. Sato, K. Kuribayashi and T. Motegi, Acta mater. (submitted).
10. J. D. Eshelby, Proc. Roy. Soc. London **A241**, 376 (1957).
11. E. Sato, T. Ookawara, K. Kuribayashi and S. Kodama, Acta mater. **46**, 4153 (1998).
12. K. Kitazono, E. Sato and K. Kuribayashi, Acta mater. **47**, 1653 (1999).
13. J. E. Flinn and D. E. Munson, Phil. Mag. **10**, 861 (1964).
14. M. Field and E. M. Merchant, J. Appl. Phys. **20**, 741 (1949).

Superplasticity in Industry

THE INDUSTRIAL APPLICATION OF ALUMINUM
SUPERPLASTIC FORMING

A. J. BARNES
Superform USA, Riverside, CA 92517-5375

ABSTRACT

Over the past thirty years superplastic aluminum forming has advanced from a laboratory curiosity to a well established manufacturing route. Applications are now found in many and diverse markets, including: medical systems, specialist automobiles, rail cars, architectural panels and all types of aircraft.

This paper examines the reasons for success in these markets and explores future prospects including the potential for fast forming alloys currently under development.

INTRODUCTION

Although the phenomena we now know as superplasticity in metals was first reported almost ninety years ago by Bengough [1] and studied most conscientiously by Pearson in the 1930's [2] and again in the 'hallmark' investigations undertaken in Russia during the 1940's, and 50's, [3] [4] it was not until the early 1960's that interest began to be shown in the potential of forming superplastic materials. [5] By the late 1960's prototype sheetmetal components were being formed by Fields in the USA, Hundy in the UK and others [6] [7] using the highly superplastic Zn-Al eutectoid. Figure 1 illustrates the remarkable formability of this alloy. The prospect of producing complex shaped components using low cost tooling and forming times of just a few minutes was sufficiently attractive for the first commercial superplastic forming company to be started in 1971 (ISC Alloys, Avonmouth, UK) using 'superplastic zinc' – SPZ, based on the ZnAl eutectoid. This pioneering venture had limited success as industrial applications were restricted by the modest mechanical properties and poor creep resistance of the duplex eutectic and eutectoid type alloys then available.

Figure 1 - Deep formings made using ZnAl eutectoid alloy-SPZ illustrates the remarkable formability of this alloy.

The development that most influenced the direction and industrial application of superplastic aluminum forming in the 1970's and 80's began in 1969 with the break through work of Stowell, Watts and Grimes. This was focused on rendering dilute aluminum alloys superplastic. This work, not reported in detail until 1975 [8], lead to the development of a family of alloys known by their trade name SUPRAL. These dilute alloys containing zirconium, were heavily cold worked to sheet and dynamically recrystallized to a fine stable grain size, typically 4-5μm, during the initial stages of hot deformation. Through special processing the zirconium, in the form of extremely fine ZrAl 3 precipitates effectively inhibited grain growth by grain boundary pinning – the Zener drag effect. The remarkable superplastic ductility of the most commonly used of these alloys, SUPRAL 100(international alloy designation –2004)is illustrated in Figure 2

Figure 2.　　The neck free superplastic ductility of SUPRAL 100 (Al-6% Cu – 0.4% Zr) contrasted with non-superplastic hot tensile specimens.

The commercial significance of this major innovation did not go unnoticed by it's sponsors, Tube Investments Ltd. (a UK based group of companies) and culminated in the formation of the company Superform Metals in late 1973, based in Worcester England. The project management of this innovation – from research through to industrial implementation – was reported in detail by Buchanan [9]

Figure 3 illustrates one of the first production runs of electronic enclosures made for IBM at Superform Metals in early 1975 utilizing SUPRAL 100. Much has happened since these pioneering days and the remainder of this paper focuses on the materials, processing, engineering and techno – economics that have lead to today's successful industrial applications in different market sectors and takes a look forward to the challenges and opportunities ahead.

Figure 3　　Early production run of male formed electronic enclosures formed from SUPRAL 100.

COMMERCIALLY AVAILABLE SP ALLOYS

In contrast to the earlier duplex eutectic and eutectoid alloys investigated in the late 1960's and early 70's (e.g. ZnAl, AlCu, AlSi, AlCa & AlPd) all contemporary commercially produced superplastic aluminum alloys are dilute alloys which utilize an ultra fine dispersoid of intermetallic particles to pin and stabilize grain size. The TEM micrograph shown in Figure 4 illustrates this. Table 1 lists the composition and typical mechanical properties of six of these alloys.

Figure 4: TEM micrograph showing grain boundary pinning by the fine disposoid distribution in SP5083.

Table 1: Composition and typical properties of superplastic aluminum alloys currently in use.

Alloy Composition	0.2%PS MPa	UTS MPa	Modulus Gpa	Density Mgm-3
2004 Al-6%Cu-0.4%Zr	300	420	74	2.83
2090 Al-2.5%Cu-2.3%Li-0.1Zr	340	450	79	2.57
2095 Al-4.7%Cu-0.37%Mg-1.3%Li0.4%Ag-.14%Zr	585	620	78	2.7
5083 Al-4.5%Mg-.1%Cr	150	300	72	2.67
7475 A1-5.7%Zr-2.3%Mg-1.5%Cu-.2%Cr	500	550	70	2.8
8090 A1-2.4%Li-1.2%Cu-0.6%Mg-0.1Zr	350	450	78	2.55

Alloy Selection

Three of the above alloys represent more than 90% of the current usage; they are 2004 [10] 7475 [11] and 5083 [12]. Each alloy has specific characteristics that influence selection, these are shown in Table 2:

Table 2: Factors effecting material selection

	2004	7475	5083
• Cost	-	M	L
• Superplasticity	M	-	L
• Heat Treatable	Y	Y	N
• Strength	-	M	L
• Weldable	Y	N	Y
• Corrosion resistance	L	-	M

(L = Least,   M = Most,   Y = Yes,   N = No)

The designer, in consort with other concerned team members (eg; procurement, marketing, stress, producability, quality, program management, etc.), must weigh the pros and cons of each to establish which is the most cost effective and capable of fulfilling specific functional/market needs, as well as, comparing with alternative materials and processes. Functional needs may range from just 'looking good' to complex interrelated issues such as; specific strength, fire resistance, electronic screening, surface finishing etc.

Availability

Availability is another important issue in terms of gage and size limitations (mainly width) and minimum purchase quantity to avoid heavy surcharges. Although new and better alloys can be expected to become available in the future they will have to demonstrate volume potential before becoming commercially available. It is important to realize that aluminum sheet producing companies are volume sensitive businesses. For each there is a 'critical mass' viable tonnage below which sustained production is not worthwhile. Exotic alloys for limited markets are no longer attractive to the aluminum industry. In this context not only is the cost and time of developing new superplastic alloys important so is the time to generate a sufficient market demand to meet 'viable tonnage' if aluminum sheet producers are to remain enthusiastic about superplasticity.

SUPERPLASTIC PROCESSING TECHNIQUES

The principal deformation characteristics of superplastic materials at appropriate temperature are; **high strain rate sensitivity, low flow stress and remarkable neck free tensile elongation**. Although sheet forming is the most common usage of superplasticity both tube 'blow forming' and bulk forging have [13] and are [14] under investigation. Here consideration is limited to sheet forming.

Much of the success of superplastic aluminum sheet forming has resulted from the pioneering innovations undertaken by Superform (both in their UK and USA facilities) over the past 25 years [15]. The fact that Superform was set-up as the world's first 'stand alone' superplastic aluminum forming company, and given the resources to be able to specialize in this specific market without the distractions of other business, created and has sustained the highly motivated and technology focused team necessary for success.

The range of sheet-forming techniques developed and used by Superform and other's include:

1. **The Simple female forming** in which the heated superplastic sheet is clamped around its edge and stretched into the heated cavity tool using gas pressure.
2. **Drape forming** where the heated and clamped superplastic sheet is stretched into a cavity containing one or more male form blocks.
3. **Male forming** in which gas pressure and tool movement are combined enabling deeper more uniform thickness parts to be made.
4. **Back pressure forming** (BPF) utilizes gas pressure on both sides (front and back) of the deforming superplastic sheet which produces a hydrostatic confining pressure capable of suppressing cavitation. Gas control creates a positive pressure differential enabling forming to be achieved. BPF is usually applied to female and drape forming methods when structural applications demand cavitation levels be contained below 0.5% volume fraction.
5. **Diaphragm forming** uses one or two 'slave' sheets of superplastic alloy to urge a smaller unclamped blank to be augmentedly drawn and draped into tool contact. The process can shape non-superplastic metals having limited formability at room temperature and in certain cases superplastic alloys are shaped this way when limited thickness variation is demanded. This process has also been successfully applied to shaping advanced thermoplastic graphite composites. [16]

The choice of which forming method should be used for a particular component application is a complex one. Laycocks detailed review of forming methods [17] helps to delineate the problem and a previous paper by the author [18] outlines the selection criteria for the female, drape and male forming methods. The following figures 5-9 illustrate classic examples of each forming technique.

Figure 5   8 ft. dingy hull female formed from a single sheet of 5083 illustrates the depth of forming possible by this method. Generous radii and the absence of angular features limit the local material thinning often associated with female forming associated with female forming.

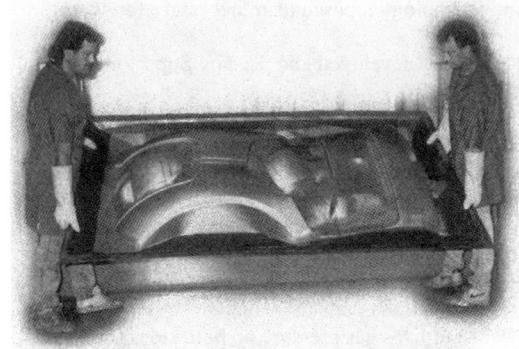

Figure 6   Six automotive body panels drape formed in a single pressing from SP5083 illustrate the cost effectiveness of multipart drape forming and it's ability to realize a class 'A' airside finish.

Figure 7   A very deep formed landing gear hydraulics cover made from 2004. Quite uniform thickness distribution is achieved using Superform's unique male forming technique.

Figure 8   Structural aircraft door inner panel produced in 7475 using back pressure forming to suppress cavitation and facilitate remarkable complex geometry.

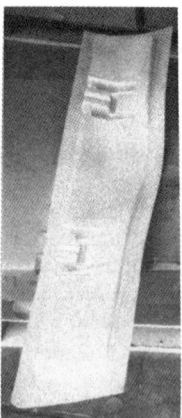

Figure 9   This PEEK/graphite thermoplastic composite aircraft landing gear door is diaphragm formed between thin superplastic aluminum sheets. The graphite fibers are augmentedly drawn and draped into shape between the confining superplastic diaphragms.

PROCESS MODELING

For many years numerical simulation of the superplastic forming process has been evolving. Significant advances have been made; notably by Sadeghi [19] and Rebelo[20] in USA, Bellet, Massoni and Chenot [21] in France and Bonet and Wood [22] in the UK. Non-linear finite element methods with contact and strain rate control pressure algorithms and friction modeling are now capable of predicting material thickness distributions and processing pressure profiles for 3D superplastic forming from flat sheet. CPU times have been significantly reduced by the use of adaptive meshing and the ever increasing computing speeds available. However, constitutive equations that adequately describe the behavior of the particular superplastic material to be used are needed. These equations must capture the influence that composition, temperature, strain history, strain rate, grain size, etc. have on the associated flow stress and strain rate sensitivity parameters.

It can be anticipated that as computer based process modeling advances and becomes more flexible 'virtual forming' will become common place, enabling part and tool designs to be tested and iteratively adapted to achieve minimum thinning, eliminate unwanted wrinkles and define optimum forming conditions before committing to final designs and costly tool making. There are, however, some challenges to be overcome before this becomes a reality, notably;

- simplifying inputting procedures for CAD data entry.
- accommodating 'batch to batch' material variability into constitutive equations via some simple test procedures.
- capture electronic descriptions of the pragmatic 'tricks of the trade' employed in tool design and process control by the leading industrial practitioners of superplastic aluminum forming.

FORMING EQUIPMENT & TOOLING

Specialized forming equipment (SPF Presses) is needed if superplastic forming is to be successful. The presses used must be capable of uniformly heating tools and the superplastic sheet, accurately controlling gas pressure and in the case of male forming

213

tool velocity and position, to achieve appropriate superplastic strain rates and safety features to prevent press opening under pressure.

The press capacity (tonnage) needed is directly related to component size, the superplastic sheet thickness and the flow stress associated with the superplastic strain rate(s) for the alloy being processed. Commercially available superplastic aluminum alloys (2004, 5083, and 7475) have relatively low flow stresses <15 Mpa and the sheet thickness' used do not usually exceed 4.0 mm. Accordingly gas pressures used do not normally exceed 300 psi. Component geometry, size, thickness, and strain rate dictate the actual gas pressure profile required for 'optimized forming conditions.' Figures 10 & 11 show computer controlled male and female press installations

Figure 10   One of Superform's unique male forming machines. It is of interest to note that this machine has been operational since 1974 – with control system upgrades.

Figure 11   Large modern female drape forming press installation producing non-planar clampline aerospace panels from 2004 alloy.

Fundamental to the success of most component manufacturing processes is tooling, **tooling** and **TOOLING!** Superplastic forming is no exception, in fact, it is at the heart of the economic viability for commercial superplastic aluminum forming.

Tooling for the lifetime of the product being made, needs to be:

- **Durable**: able to continuously operate at elevated temperature without degrading and safely contain the gas pressures and mechanical forces applied.
- **Accurate**: to consistently produce formings to the required dimensions and surface quality.
- **Productive**: yielding maximum production output commensurate with the superplastic strain rate limits of the alloy being formed and the pressure containment capacity of the forming equipment (forming presses) being used.
- **Cost effective:** manufactured at minimum cost for 'optimized forming conditions'.

With the advent of computer aided design (CAD) almost all SPF tooling is now NC machined from electronic data. However, much proprietary know-how and 'tricks of the trade' are incorporated at all stages of tooling from tool design, material selection, manufacturing methods, through to their mode of operation. Integrating these into a coherent computer based expert system is a key task still needing to be tackled.

CURRENT MARKETS FOR SPF ALUMINUM

Figure 12   shows the estimated percentage breakdown, by market segments for superplastic aluminum formed components. Successful markets characteristically have a common profile, which includes:

High added value end products ($K to $M)
- Medium to low volume production (10's to low 1000's)
- Need for the properties and attributes of aluminum alloys
- Innovative design philosophy
    - part designs incorporating thin gages (.020" to .160")
    - complex shape (not simple)
    - medium size parts (1 to 40 sq ft)

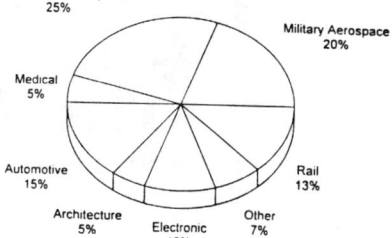

Figure 12. Market segment breakdown for superplastic aluminum formed components.

215

Successful innovators of superplastic aluminum forming answer the question **"Why superplastic forming?"** by saying it:

- reduces costs associated with tooling and fabrication
- increases design flexibility
- yields weight savings and reduced parts count
- increases structural integrity
- improves consistency/interchangeability
- offers shorter lead times from design to first off parts production.

Unsuccessful innovators say:

- unable to handle thickness variations
- to expensive compared with plastic parts
- not competitive for their volume production needs
- challenged by the need to integrate aluminum into 'other material' dominated products
- unable to change design or accept new material

Clearly superplastic aluminum is not the panacea for every component application, however, as the following examples show (Figures 13-16) SPF is often a cost-effective solution to a specific niche market need.

Figure 13   Limited series Panoz Esperante sports car, body and some interior panels formed from SP 5083 – benefits from low cost/short lead time tooling.

Figure 14   The unique "Nauticus" National Maritime Center utilized SP 5083 close out panels.

Figure 15   This deep formed Xray housing utilizing 2004 is just one of many Medical system parts successfully superplastically formed.

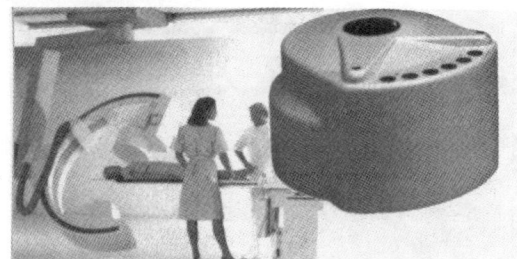

Figure 16 Structural aircraft door inner skin back pressure formed from SP7475. One of more than twenty SPF parts used in this corporate jet.

THE TECHNO-ECONOMICS OF SPF ALUMINUM

An oversimplified view of the superplastic aluminum parts manufacturing process is:

- it takes a relatively expensive sheet of material
- forms it relatively slowly into a complex shape
- using relatively inexpensive tooling.

These conditions define the viable cost-effective niche shown for SPF aluminum in Figure 17. Material cost and forming time offset the benefit of low cost tooling as annual production quantity increases. To overcome this techno-economic barrier to higher production quantities either material costs and /or forming times must be reduced. This is shown schematically in Figure 18. The impact of metal cost and forming times on breakeven quantity for autobody panels has been previously reviewed by the author [23] but it is worthwhile to review this again using the graph shown in Figure 19. Current forming technology and material costs are compared with targets for fast forming (high strain rate) alloys; if realized annual breakeven forming quantities, compared to steel stampings could increase ten fold! Such a quantum change would transform the SPF aluminum business from its present specialized niche position into the main stream of competitive manufacturing.

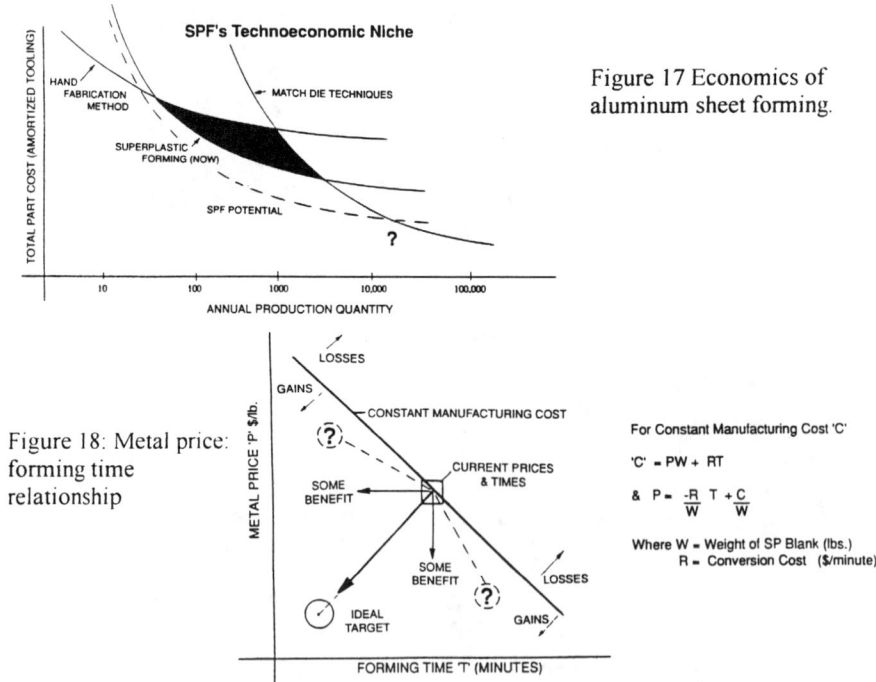

Figure 17 Economics of aluminum sheet forming.

Figure 18: Metal price: forming time relationship

For Constant Manufacturing Cost 'C'

'C' = PW + RT

& $P = \frac{-R}{W} T + \frac{C}{W}$

Where W = Weight of SP Blank (lbs.)
R = Conversion Cost ($/minute)

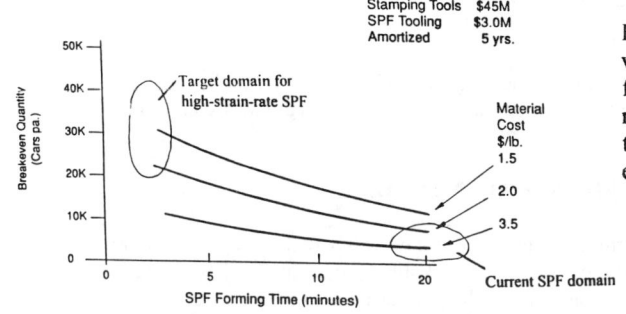

Figure 19: Steel stamping verus superplastic aluminum forming-influences of material cost and forming time on breakeven economics.

FAST FORMING 'HIGH-STRAIN-RATE' SUPERPLASTICITY

The recent developments associated with high-strain-rate superplasticity are most exciting and have opened up a whole new era in the study of superplasticity. A comprehensive review by Mabuchi and Higashi [24] of this subject reiterated the importance of grain size to strain rate- finer grain size higher strain rates. This insight did not escape Pearson back in 1934 [2] and in 1982 Sherby [25] saw the economic potential of seeking to reduce grain size in SUPRAL 100 (Al-6%Cu-0.4 % Zr) to achieve greater productivity.

To date most aluminum based materials reported to exhibit high-strain-rate superplasticity have been produced by non-conventional processing such as mechanical alloying, physical vapor deposition and others are metal matrix composites produced by thermo mechical treatments. In the author's opinion it is very unlikely that these 'exotic' materials will prove to be suitable or economically viable for industrial superplastic sheet forming. More likely candidates are; ingot cast alloys that either dynamically recrystallize to a ultra fine grain size during the early stages of hot deformation after conventional processing, or are processed by intense plastic straining such as ECA pressing [26]. A more direct PM powder roll compaction route has also been proposed [27].

A 'viable' fast forming 'high-strain-rate' alloy suitable for commercial application in high volume markets, such as the automotive industry, would need to have the following characteristics:

- Suitable services properties; preferably an alloy of similar composition to that currently in use
- Low cost when produced in reasonable quantity.
  - Sheet < $2.0/lb
  - Tonnage > 5000 tons pa.

- Superplastic elongation to failure of 500% or more at a strain rate of at least 2×10^{-2} /sec.
- Flow stress under 'high-strain-rate' forming conditions not greater than 20 Mpa., preferably lower.

- Forming temperature <500°C.
- Readily recycled.

Whether such challenging criteria can be met remains to be seen. The author believes that the potential rewards and commercial significance of success are sufficiently great to justify pursuing this goal.

THE FUTURE PROSPECTS FOR SPF ALUMINUM

The outlook for SPF aluminum has never looked better with new applications coming on stream at an increasing pace. The remarkable formability of these materials combined with the low cost/short lead-time tooling continuous to create unique opportunities in markets where the attributes of aluminum are needed and where low to medium quantities of complex shaped components are required.

Recently, however, some in the superplasticity community have expressed disappointment in the lack of Industrial application of superplasticity [28]. In one sense this is true in that SPF has not, so far, become a main stream mass production process. It is also the case that from the outset more than 25 years ago the techno economics of SPF were understood so that over the years focus has been on those 'high added value end product' industries (Aerospace, specialist Automotive alike) that could benefit from the technology.

Although there are many challenges ahead, with the expectation of greater technology transfer, more engineering improvements and process optimization, the prospect of 'smart' computer modeling allowing designers to integrate accurate thickness predictions alongside product-design and finally the possibility of achieving cost effective high volume production via fast forming 'high-strain-rate' materials technology, the future is as exciting as it was 25 years ago.

REFERENCES

[1]   G.D. Bengough, J. Inst. Metals 7, 123 (1912).
[2]   C.E. Pearson, J. Inst. Metals 54, 111 (1934).
[3]   A.A. Bochvar and Z.A. Sirderskaja, Izvest. Akad. Nauk SSSR, OTN, N9, 881 (1945).
[4]   A.A. Presnjakov and V.V. Chevyakova, Izvest. Akad. Nauk SSSR OTN, N3, 120. (1958).
[5]   W.A. Bachofen, I.R. Turner and D.H. Avery, Trans. ASM 57, 980 (1964).
[6]   D.S. Fields Jr., IBMJ. Res. Dev. 9. 134 (1965).
[7]   B.B. Hundy: *Plasticity and Superplasticity* Inst. Met. London, Review Course Series 2 (3) 73 (1969).
[8]   R. Grimes, M.J. Stowell and B.M. Watts: Aluminium, 51, 720 (1975).
[9]   J.A.F. Buchanan: *Innovation Management in Metallurgy*. Inst. Met. Review Course Series 3 (4) 811-75-T (1975).
[10]   B.M. Watts, M.J. Stowell, B.L. Baikie and D.G. E. Owen Met. Sci. 1189-206, June. (1976).

[11]   G.J. Mahon, D. Warrington, R. Grimes and R.G. Butler. Materials Science
         Forum Vols. 170-172 (1994).
[12]   G.J. Mahon: Alcan Int. Report No. GTA04-90 (1994).
[13]   T.Y.M. Al-Naib and J.L. Duncan. Inst. J. mech. Sci. vol 12 pp 463-477
         (1970).
[14]   H. Yamagata, Mat.Sci. Forum, vol 304-306 pp 797-804 (1970).
[15]   Superform Aluminium, Inform Brochure-25 year anniversary edition.
[16]   A.J. Barnes, J.B. Caltanch: *Advances in Thermoplastic Composite
         Fabrication, Technology*, Proceedings 2nd Conf. in Materials Engineering.
         IN Eng., London. (1985).
[17]   D.B. Laycock: *Superplastic Forming of Structural Alloys*, N.E. Paton and
         C.H. Hamilton ed. AIME pp 257-271. 1982).
[18]   A.J. Barnes *Design Optimization for Superplasticity* TMS/AIME Conf.
         San Diego (1982).
[19]   R.S. Sadeghi and Z.S. Pursell. Mat. Sai. Forum vols. 170-172 pp 571-576
         (1994).
[20]   N. Rebelo and T.B. Wertheimer. *Finite Element Simulation of
         Superplastic Forming* Proc. 16th North American Manufacturing Research
         Donf. Urbana, IL (1998).
[21]   E. Massoni, M. Bellet and J.L. Arenot. *Thin Sheet Forming Numerical
         Analysis* Modeling of Metal Forming Processes, Sophia Antipolis, France.
         pp 187-196 (1988).
[22]   J. Bonet, R.D. Wood and A.H.S. Wargadipura Int. J. Numerical Methods
         in Aug. Vol 30, 1719-1737 (1990).
[23]   A.J. Barnes, Mater. Sci. Forum 170-172 pp 701-714. (1994).
[24]   M. Mabuchi and K. Higashi. JOM. June 1998 pp 34-39.
[25]   O.D. Sherby and O.A. Ruano: *Superplastic Forming of Structural Alloys*,
         N.E. Paton and C.H. Hamilton ed. AIME pp 241-254 (1982).
[26]   T.G. Langdon, M. Furukawa, Z. Honita and M. Nemoto. JOM June 1998
         pp 41-45.
[27]   A.J. Barnes, Mat. Sci Forum vols. 304-306 1999 pp 785-796.
[28]   L.A. Times, Business section P 3 May 17th 1999.

MANUFACTURING ALUMINIUM SANDWICH STRUCTURES BY MEANS OF SUPERPLASTIC FORMING

P. IMPIÖ *, J. PIMENOFF **, H. HÄNNINEN **, M. HEINÄKARI ***
* Leinovalu Oy, Finland
** Helsinki University of Technology, Laboratory of Engineering Materials, P.O. Box 4200, FIN-02015 HUT, Finland
*** Turku Polytechnic, Finland

ABSTRACT

This paper examines the possibilities of manufacturing large-scale aluminium sandwich structures using superplastic forming. The materials tested were Mg-alloyed production quality aluminium (Al 5083-O, Al 5083-H321) and Aluminium 1561. Tensile tests at elevated temperatures were performed in order to establish the suitability of the test materials for superplastic forming. The microstructural changes in the test material specimens were examined. Thereafter numerical simulation of the Al 5083-O forming process was conducted based on the tensile test results. The numerical simulation results were subsequently used to estimate forming parameters and the feasibility of manufacturing large-scale structures by superplastic forming.

The results indicate that higher strain can be reached at higher temperatures for the test materials. Aluminium alloy 1561 exhibited the largest elongation to fracture and Al 5083-H321 the smallest. Strain appeared to be temperature dependent but not much affected by the strain rate. Metallographic examination clarified that Al-5083-O and Al-5083-H321 showed susceptibility for cavity forming whereas Aluminium 1561 formed relatively few voids. The numerical simulation indicated that Al 5083-O can be superplastically formed using relatively low forming pressure (0.9 - 1.4 bar).

INTRODUCTION

The continued need for lightweight, yet inexpensive, manufacturing techniques in the transportation industry has lead to an increased interest in the use of sandwich structures. The main advantage of sandwich structures is their superior combination of lightness and stiffness, making them an attractive option for all weight dependent transport applications [1].

Superplastic forming (SPF) is one way of manufacturing sandwich structures. It is, however, a competitive alternative when the desired structure is time consuming and difficult to manufacture by conventional means. One obvious drawback of SPF though is its severe material orientation and the high cost of the superplastic grade materials. This has effectively caused the industry to consider the use of commercial grade alloys in superplastic manufacturing processes. Several studies have been conducted to determine the superplastic behaviour of commercial aluminium alloys (especially AA5083) [2, 3, 4]. The results of these studies have mostly been promising; considerable elongations have been achieved at moderate forming temperatures.

The present paper is devoted to the study of the superplastic forming potential of three as-received commercial grade aluminium alloys and their use in manufacturing multi-sheet sandwich structures. The effect of various forming temperatures and strain rates is focused on determining their effects on the microstructure and mechanical behaviour. Suitable joining methods for the sandwich sheets are also dealt with.

Mat. Res. Soc. Symp. Proc. Vol. 601 © 2000 Materials Research Society

EXPERIMENTAL

Materials

The materials used in this study were annealed Al 5083-O (supplied by ALCAN Aluminium-Walzproducte), cold-rolled Al 5083-H321 (supplied by Hoogovens Aluminium) and extruded Al 1561 (supplied by a Russian manufacturer). The chemical compositions of the alloys can be seen in Table 1. Al 5083-O is a typical marine aluminium with reasonable mechanical properties (see Table II) and good welding and corrosion properties. Al 5083-H321 (supplied in work-hardened state) is also intended for marine conditions but has better strength properties than the annealed Al 5083-O. The Russian alloy Al 1561 is supplied in a non-recrystallised state. It has better strength values and a smaller grain size than the other two tested alloys. Dog-bone tensile test bars with a gauge length of 100 mm and width of 4 mm were machined from the as-received material billets. No static recrystallisation heat treatment was performed prior to tensile loading.

Table I. The chemical composition of the tested alloys (all values in wt-%).

Alloy	Si	Fe	Cu	Mn	Mg	Cr	Zn	Ti	Zr	Ni
Al 5083-O	0.14	0.33	0.03	0.7	4,69	0.07	0.01	0.01	-	-
Al 5083-H321	0.29	0.26	0.09	0.56	4,62	0.11	0.12	0.03	-	-
Al 1561	0.16	0.22	0.04	0.76	5,96	0.01	0.03	0.04	0.04	0.011

Table II. Room temperature tensile properties of the tested alloys.

Alloy	0.2 % yield strength (MPa)	Ultimate tensile strength (MPa)	Elongation (%)
Al 5083-O	153 - 155	318 - 320	23 - 26
Al 5083-H321	251 - 288	348 - 361	> 5 *
Al 1561	210 - 227	353 - 359	> 11 *

* All test specimens broke at the ends, leaving the point of fracture beyond the measured area and making it impossible to estimate the elongation.

Tensile testing

Hot tensile testing was carried out at temperatures between 430 °C and 510 °C using an induction heating apparatus connected to an MTS servo-hydraulic tensile testing machine. A water-cooled induction-heating coil was manufactured for the testing. The coil heated the test specimens over a length of 50 mm. The heating effect of the coil was calibrated prior to the tensile testing of the bars. The strain rate, that was varied from 2.5×10^{-4} s^{-1} to 6.0×10^{-3} s^{-1}, and other necessary data were fed to the testing machine by a computer. Heating of the bars was carried out using fixed heat effect settings while the real-time temperature of the bars was measured and monitored by a pyrometer. Testing was started 5 min after the pyrometer readings had stabilised, giving a total heat-up time of approximately 10 min. The strain during testing was measured from the movement of the lower grip of the tensile testing machine. The elongation to fracture values were calculated from line markings on the gauge sections of the bars that were within the heated area during testing.

Metallography

The sides and fracture surfaces of the tensile bars were characterised by scanning electron microscopy (SEM). The side surfaces were polished electrolytically in a solution of ethanol, distilled H$_2$O, 2-butoxyethanol and HClO$_4$ (60%) prior to SEM and additionally evaluated by a back-scattered electron detector coupled to a SEM. The back-scattered electron technique was intended to reveal the grain boundary phases and grain sizes.

<u>Joining</u>

In SPF, the predominant method of joining sheet material before forming is diffusion bonding (DB). Aluminium, however, does not suit well to DB techniques from an industrial point of view and as such needs alternative methods. The research work on joining in this paper was restricted to plasma arc (PAW), robotised tandem MIG/MAG and mechanised MIG/MAG welding. Pairs of 2 mm thick Al 5083-O plates were joined by seam welds without joint preparation.

<u>Numerical simulation</u>

The numerical simulation was performed by non-linear finite element modelling. The objective of the simulations was to find optimal pressure parameters for the superplastic forming of the sandwich structure in Figure 1.

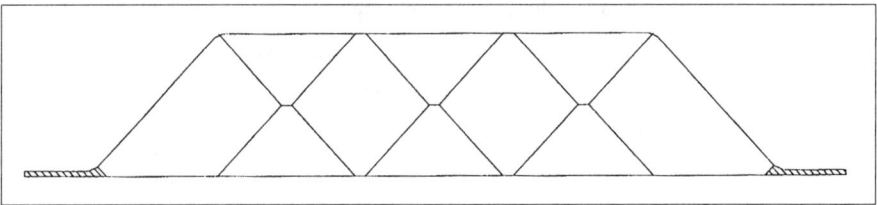

Fig. 1. The simulated sandwich structure.

The simulation was carried out on a four-sheet and two-dimensional profile made of Al 5083-O. The elongation to fracture results of Al 1561 were greater than those for Al 5083-O but due to the availability of Al 1561 it was decided to use Al 5083-O in the simulation. The bottom face sheet in Fig. 1 was 3.0 mm thick and the three other sheets were 2.0 mm thick prior to forming. The cross-section was assumed to be of infinite length, thus, enabling a two-dimensional examination.

RESULTS

<u>Tensile testing</u>

Tensile elongations of the materials tested measured at four constant strain rates and five different temperatures are presented in Figures 2-4. The results show a clear distinction between the superplastic behaviour of the materials. The fine-grained alloy Al 1561 exhibited the highest elongations to fracture and Al 5083-H321 the lowest.

Al 5083-O exhibited the highest values of elongation to fracture at 470 °C, Al 5083-H321 at 450 °C and Al 1561 at 450 °C. The elongation to fracture of the materials did not depend much on the strain rate. The temperature, however, influenced the elongation to fracture greatly. Al 1561 exhibited a nearly complete loss of strength and ductility at temperatures above 490 °C. Specimens tested at 510 °C broke immediately intergranularly and therefore showed little or no strain. This phenomenon may have been a result of segregation or dislocation behaviour caused by dynamic strain ageing. The same kind of weakening was observed in Al 5083-O and Al 5083-H321.

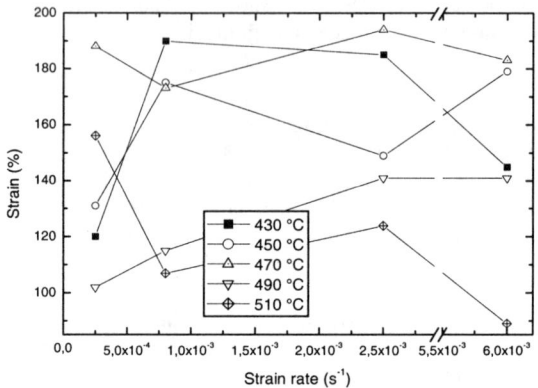

Figure 2. Comparison of the elongation to fracture properties of alloy Al 5083-O.

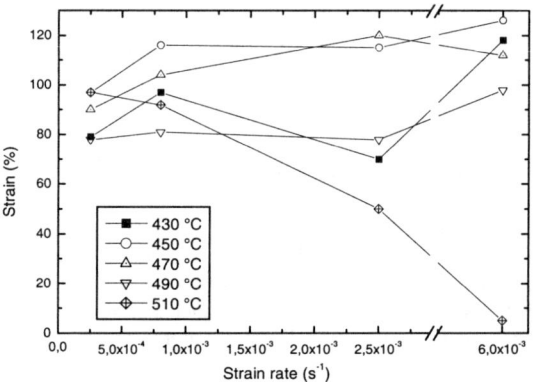

Figure 3. Comparison of the elongation to fracture properties of alloy Al 5083-H321.

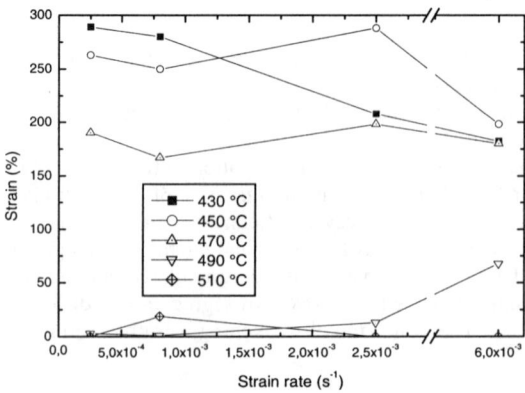

Figure 4. Comparison of the elongation to fracture properties of alloy Al 1561.

Metallography

The highly elongated areas adjacent to the fracture surfaces of all Al 5083-O specimens showed a large number of voids and cracks (see Fig. 5). It appeared that failure of the specimens was due to coalescence of cavities followed by ductile failure of the neighbouring material. Particles containing varying amounts of aluminium, magnesium, chromium, iron, manganese, copper and silicon were found in the cavities. Several phases, e.g. Mg_2Si, $FeAl_3$ and $(FeMn)Al_6$, can form in magnesium-alloyed aluminium alloys [5].

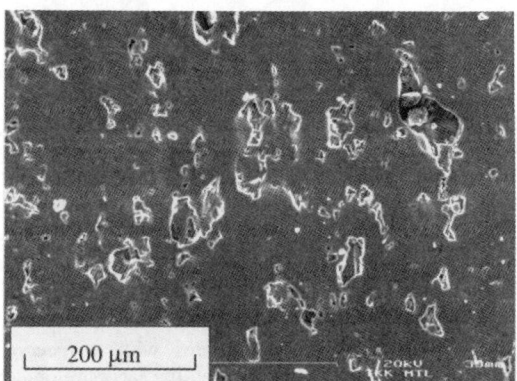

Fig. 5. Structure of Al 5083-O after hot tensile testing. Strain rate 6.0×10^{-3} s^{-1}, T = 470 °C and elongation to fracture 183 %.

The large particles found in the voids indicate that they act as nucleation sites for the voids. This is further supported by the observation that voids with particles were larger than voids without. The grain structure of the materials was vaguely revealed, but with not enough contrast to ensure a reliable grain size estimate. Therefore the grain sizes of the materials were not defined.

Joining

Plasma arc welding of 2 mm thick aluminium plates turned out to be difficult in cases were only the two top-most plates were to be joined. Control of the penetration was difficult and the welding mode turned frequently from keyhole to melt-in mode. The heat input caused distortion of the plates making subsequent welding difficult. Plasma arc welding was also proven to be a slow welding technique.

Welding the plates with robotised tandem MIG/MAG apparatus enabled a high welding speed and deposition rate. A high welding speed is an obvious advantage in manufacturing sheet packages for SPF formed large-scale sandwich structures. The heat input was smaller than in plasma arc welding, thus, causing less distortion of the plates.

Mechanised MIG/MAG welding had lower welding speed and deposition rate. Use of the apparatus was, however, simple when mounted on a linear conveyor. Heat input is smaller than in plasma arc welding. The root penetration was smaller than in tandem MIG/MAG.

Numerical simulation

The sandwich structure formed well in the beginning of the simulation sequence. At a critical stage, seen in Fig. 6, the upper core sheet started to neck extensively near the outer edge of the structure. At that stage the thickness of the sheet was 0.64 mm. As forming continued, the sheet became thinner until it finally ruptured. The upper face sheet also became thinner, but showed no signs of rupturing. The forming pressure was 0.9 bar in Fig. 6 and 1.4 bar at the point of rupture. The joints between the sheets in the simulated structure were not defined.

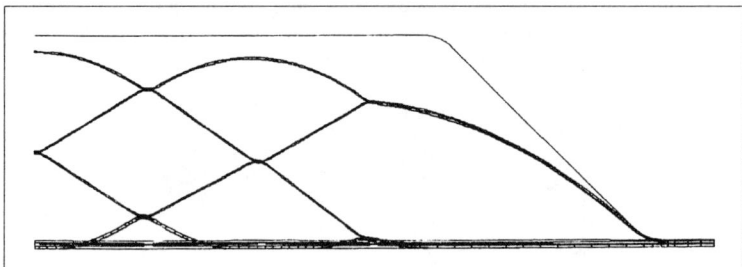

Fig. 6. The simulated sandwich structure at the onset of necking (P = 0.9 bar).

CONCLUSIONS

High strains can be reached at high temperatures with as-received commercial grade magnesium alloyed aluminium alloys. Al 1561 showed the largest elongation to fracture of the materials tested (~290 %) and Al 5083-O exhibited better elongation to fracture (~190 %) than Al 5083-H321 (~130 %). The elongation to fracture achieved was dependent on the test temperature but not much on the strain rate.

The highest temperature of forming Al 1561 was 470 °C. At higher temperatures the formability and mechanical properties of the alloy dropped significantly. Al 5083-O and Al 5082-H321 exhibited reasonable plastic behaviour in temperatures above 470 °C. Void nucleation was extensive in both Al 5083-O and Al 5082-H321, whereas Al 1561 was less prone to void nucleation at the test temperatures.

The results of the numerical simulation demonstrated that Al 5083-O could be formed with relatively low forming pressure (0.9 - 1.4 bar). Forming of the intended sandwich structure, however, failed since the tested material began to neck locally and ruptured finally before the desired shape was achieved.

ACKNOWLEDGEMENTS

This work has been performed in connection with the Finnish national technology program KENNO - Lightweight Panels. This work was supported by Kvaerner Masa-Yards Technology, Finland.

REFERENCES

1. P. Impiö, MSc thesis, Helsinki University of Technology, 1998.

2. J.S. Vetrano, C.A. Levander, C.H. Hamilton, M.T. Smith and S.M Bruemmer, Scripta Met. **30**, 565 (1994).

3. I.C. Hsiao and J.C. Huang, Scripta Mat. **40**, 697 (1999).

4. M.H. Hojas and W. Kühlein, in *Proc. 3rd Int. Conf. Aluminium Alloys*, L. Arnberg, O. Lohne, E. Nes, N. Ryum, Ed., Vol II, 1992, pp. 151-156.

5. J.E. Hatch, *Aluminium: Properties and Physical Metallurgy*, (American Society for Metals, OH, 1984), p. 424.

THE MICROSTRUCTURE AND SUPERPLASTIC BEHAVIOUR OF CLEAN MECHANICALLY ALLOYED TITANIUM – TITANIUM BORIDE ALLOYS

A.P. Brown[*], R Brydson[*], C. Hammond[*], A. Wisbey[**], and T.M.T. Godfrey[**].

[*]Department of Materials, School of Process, Environmental and Materials Engineering, University of Leeds, Leeds, LS2 9JT, UK.
[**]Structural Materials Centre, Defence Evaluation Research Agency, Farnborough, Hants, GU14 0LX, UK.

ABSTRACT

The superplastic forming (SPF) of titanium alloys is an established technology. A reduction in grain size from that of the typical sheet materials would lead to enhanced SPF properties and hence a reduction in production cycle times. This study describes the microstructural development and superplastic behaviour of fine-grained Ti-6%Al-4%V alloys. Ball-milling Ti-6%Al-4%V powder produces a nanocrystalline material; however on consolidation by hot isostatic pressing rapid grain growth occurs. Addition of boron powder during milling leads to boride precipitates in the matrix of the consolidated alloy. The precipitates are dispersed inhomogeneously, resulting in localized grain refinement. Superplastic testing revealed cavitation formation but in comparison to conventional sheet material, large elongations were achieved at relatively high strain rates.

INTRODUCTION

Many titanium alloys can be superplastically formed (SPF) and Ti–6%Al-4%V (wt.%) is the most widely used commercially, with applications in fields as diverse as the aerospace industry and dentistry.

At typical superplastic forming temperatures (850 – 930 °C) Ti–6%Al-4%V consists of two phases; a hexagonal close packed (α) phase and a body centred cubic (β) phase. The large-scale (micro)structural reorganization required for superplastic flow principally takes place by grain boundary sliding and migration, which is accommodated by diffusion and/or dislocation activity [1]. For Ti–6%Al-4%V the normally observed maximum in superplasticity occurs when the α:β phase ratio is 50:50 [2].

Models of superplastic flow predict that there is an inverse relationship between strain rate and the square or cube of grain size [3]. Therefore a reduction in grain size from the typical 5 µm grains observed in sheet materials could dramatically enhance the SPF properties of Ti–6%Al-4%V. This effect of grain size reduction has been demonstrated in submicron titanium alloys, which were prepared by mechanical milling [4]. Furthermore, particulate reinforcement of the alloy matrix may control the grain growth normally observed during SPF, and limit the loss in tensile strength of the post SPF material. Mechanical alloying is a possible route to produce nanocrystalline solid solutions that would be ideal precursors to fine-grained particle reinforced alloys.

The clean mechanical alloying of Ti-6%Al-4%V with 0, 0.1 and 0.5 wt. % boron powder and the consolidation of this milled material by hot isostatic pressing (HIPing) at temperatures between 600 and 900 °C has been carried out. It has been shown that this process leads to the precipitation of a ceramic phase (TiB) within the matrix of the HIPed alloys and some grain-size refinement [5].

EXPERIMENTAL

Material preparation has been described previously [6]. X-ray diffraction (XRD) data was obtained using a Philips diffractometer with Cu $K\alpha_1$ radiation. The consolidated material was prepared for SEM by mechanical polishing and then dip etching in a solution of 10% HNO_3 and 2% HF (in distilled water). The samples were examined in a CamScan series 3 SEM operating at 20 kV. Specimens for TEM were prepared from 3 mm diameter discs ground to 100 μm thickness. The final polishing and perforation of the discs was carried out by electropolishing at 20 V and 120 mA in a polishing solution of 5 % perchloric acid, 35 % 2-butoxyethanol and 60 % ethanol held at –20 °C. The foils were examined using a Philips CM20 TEM operated at 200 kV and fitted with a LaB_6 filament. The milled powders were prepared for TEM examination by crushing in a pestle and mortar under acetone and then pipetting onto holey carbon support films.

For high temperature tensile testing square necked test-pieces with a cross-sectional area of 2 mm by 3 mm and a gauge length of 5 mm were prepared. High temperature tensile testing was performed in a three zone furnace, at 700, 800, and 900 °C with an environmental retort flushed with argon. To enable nominal constant strain rate testing to failure the cross head displacement was recalculated for every 20 % elongation up to 400 % there after it was recalculated for every 50 % elongation.

RESULTS

Powders

The *gas atomized powder* particles are < 250 μm in size with some (argon) gas porosity (0.5 ppm). X-ray diffraction reveals that only the α phase is present. It is expected that the powder would consist of martensitic α [7].

The *mechanically alloyed powders* agglomerated into particles ~ 1 mm in size. XRD reveals only the α phase and size-broadening analysis suggests crystallite sizes of 10 – 50 nm (dependent on strain). These crystallite sizes are confirmed by TEM (Fig. 1). Contamination analysis showed the oxygen (1100 ppm), nitrogen (110 ppm) and iron (0.13 %) content to be minimal. The argon content (1.5 ppm) is about three times that of the gas atomized powder.

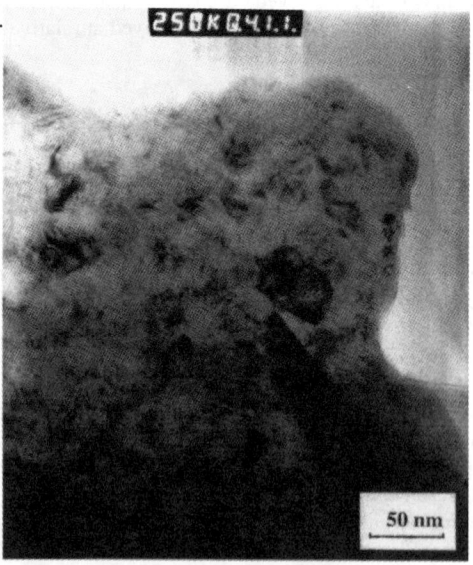

Figure 1, a) Bright field TEM image of the mechanically milled Ti-6%Al-4%V powder. The polycrystalline nature of the powders is evident with crystallite sizes of 10 – 50 nm.

600 °C HIP, 0 % boron

Prior particle boundaries within this alloy can easily be resolved optically. The α grains are equiaxed, and ~ 500 nm in size (Fig. 2). There is little sign of dislocations in the α grains. The β grains have been preferentially etched in the TEM sample preparation but it is evident that β grain nucleation has occurred at the α grain boundaries during consolidation. The volume fraction of β grains is ~ 15 % and this is broadly consistent with quantitative TEM EDX analysis, which indicates a compositional difference between the α and β grains (Table I).

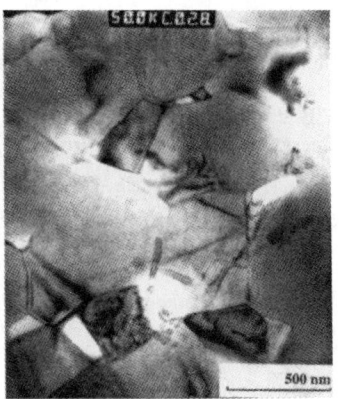

Figure 2, Bright field TEM image of the 0 % B 600 °C HIPed material. Equiaxed α grains ~ 500 nm in size are visible whilst the β phase has been preferentially etched.

Table I, The composition of the α and β phases in the 600 and 900 °C 0 % boron HIPed samples (the balance is titanium).

HIP Temperature	at. % V in α	at. % V in β	at. % Al in α	at. % Al in β	at. % Fe in β
600 °C	2	22	12	4	1 - 2
900 °C	2	17	12	4	1

600 °C HIP, 0.5 % boron

This alloy consists of regions similar to the 0 % boron material (above) plus regions of grain refinement with α grain sizes of < 100 nm. Preliminary electron diffraction analysis from the fine-grain regions shows evidence of TiB and TiB₂ precipitates.

900 °C HIP, 0 % boron

Prior particle boundaries within this alloy cannot be resolved optically (in contrast to the above). The α grains are equiaxed and ~ 5 μm in size (right hand side, RHS, of Fig. 3). TEM reveals that there is a significant dislocation density in these α grains. Further β grain growth at the boundaries between the α grains is evident, the volume fraction of β grains is ~ 30%, and the compositional difference between the α and β grains decreases (Table I).

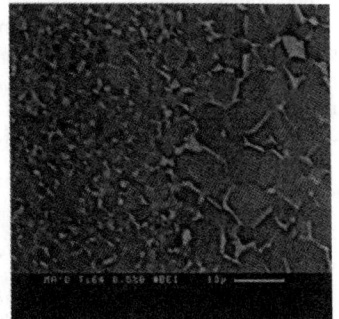

Figure 3, Backscattered SEM image of the 0.5 % B 900 °C HIPed material highlighting refined (LHS) and unrefined (RHS) regions and the α (dark grains) / β (light grains) compositional difference.

900 °C HIP, 0.5 % boron

Inhomogeneous grain refinement is evident in this alloy (Fig. 3). Highly faulted/twinned needle and plate-shaped precipitates are present in the refined regions (Figs. 4 and 5). Electron diffraction analysis of these precipitates indicates that they are twinned TiB crystals forming in the B27, B_f, or a mixture of the two orthorhombic structures. Energy-filtered TEM confirms the presence of boron in both types of precipitates.

Figure 4, Bright field TEM image of the 0.5 % B 900 °C HIPed material showing needle-shaped and plate-like boride precipitates.

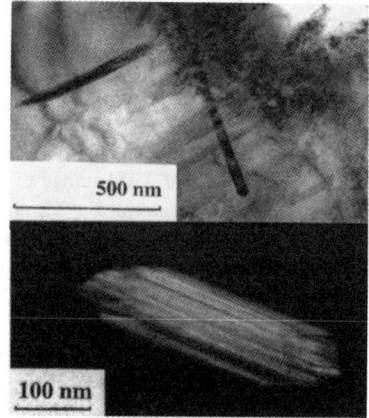

Figure 5, Dark field TEM image of the 0.5 % B 900 °C HIPed material showing the growth faulting of the needle precipitates.

High Temperature Tensile Tests

All of the ball-milled Ti-6%Al-4%V based materials, tested at 900 °C and a strain rate of 6×10^{-4} s^{-1}, exhibit relatively low elongations to failure (maximum 940 %) compared with conventional sheet material (1200 %) tested under identical conditions (Fig. 7) [6]. Indeed the HIPed (800 and 900 °C) gas atomized Ti-6%Al-4%V exhibit higher elongations to failure than the milled materials HIPed at the same temperatures.

There is extensive cavitation in all of the material produced by powder metallurgy (Fig. 6) and qualitatively this is highest in the milled material. Finally, consolidated 0 % B samples annealed at 1300 °C do exhibit residual porosity.

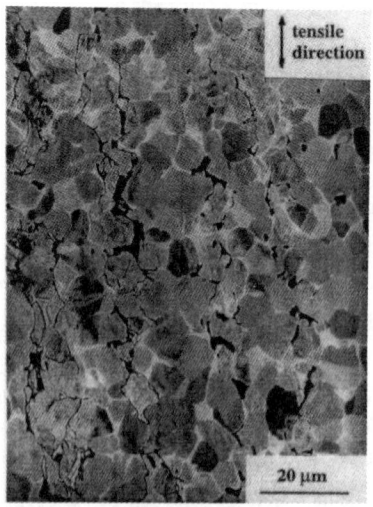

Figure 6, Backscattered SEM image showing cavitation in a milled Ti-6%Al-4%V sample HIPed at 700 °C and tensile tested at 900 °C and 6×10^{-4} s^{-1}.

Testing at a higher strain rate, of 6 x10^{-2} s^{-1} (Fig. 7), results in some decrease in elongation to failure, however all samples tested achieved elongations of ~ 460 %. This elongation is larger than that for sheet material (~ 200 %) tested at similar strain rates [8].

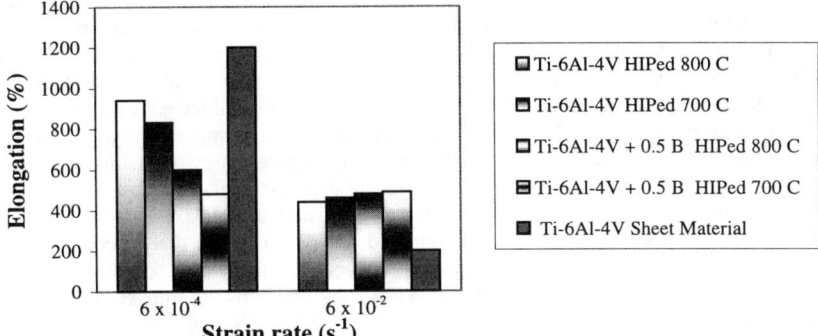

Figure 7,   The elongation to failure of milled and consolidated Ti-6Al-4V and Ti-6Al-4V + 0.5 % B when tested at two different strain rates at 900 °C. Also shown are the elongations for standard sheet materials tested under similar conditions [6, 8]. The milled materials outperform the sheet material at the high strain rate.

DISCUSSION

Mechanical milling produces 10 - 50 nm grain size Ti-6%Al-4%V powders. Only the α phase is evident in the milled powders (Fig. 1) suggesting that the vanadium is in solid solution with the aluminium and titanium. This super saturated solid solution is retained from the original martensitic composition of the gas atomized powder, even though the grain size and shape is not [7]. During consolidation, β grains nucleate at the α grain boundaries (Fig. 2) and as expected the vanadium strongly segregates to the β phase [9]. This produces a low volume fraction of β grains that are highly enriched in vanadium and a high volume fraction of α grains retaining aluminium, and consequently both these enrichments tend to stabilize their respective phases [10]. Clearly this, in turn, could be a hindrance to further β production during HIPing and may explain why the volume fraction of β in the 900 °C HIPed material (~ 30 %) is lower than that in conventional Ti-6%Al-4%V (~ 50 %) heated at 900 °C [8]. Furthermore the β volume fractions of the milled materials are significantly lower than the optimum superplastic behaviour volume of 50 % [2]. However, minor iron contamination from the ball-mill segregates to the β phase and this may be beneficial to the SPF properties as Fe enhances the diffusivity of the β phase [11].

Generally on consolidation rapid grain growth occurs which, whilst being undesirable, does confirm the purity of the milling process i.e., there is little contamination to cause grain refinement. The 600 °C HIPed alloys exhibit a submicron grain structure ideal for enhanced SPF properties, however the observed prior particle boundaries may weaken this material. The 900 °C HIPed alloys have larger grain sizes, comparable to that of sheet material. The boride precipitates do successfully result in grain refinement, however their distribution is inhomogeneous. The faulting of the precipitates may give an indication of their growth mechanism.

Despite the promising equiaxed microstructures of the milled alloys, tensile test results at standard Ti-6%Al-4%V strain rates yield poor elongations to failure, which may be attributed to the observed cavitation. The source of this cavitation is thought [6] to be a combination of the low volume fraction of β (discussed above), argon entrapment and consolidation defects. The observed argon entrapment in the gas-atomized and milled powders (~ 1 ppm) is enough to cause 1 vol.% porosity [12] (evident in the samples annealed at 1300 °C). Certainly, the argon levels are highest in the milled materials and this is consistent with the cavitation (qualitatively) being highest in the milled and HIPed alloys. Finally, consolidation defects/delamination may come from incomplete bonding of the powder particles especially in the samples HIPed at low temperatures. All three of these problems are, to some extent, inherent to the milling and then consolidation route.

However, in all the Ti-6%Al-4%V based materials tested at high strain rates, relatively large elongations to failure, of 480 %, were achieved in comparison to that of conventional sheet material (~ 200 %) [8]. The average grain size of each of the tested samples is smaller than that of sheet material. This result is encouraging, and confirms the impact of grain size reduction in Ti-6%Al-4%V.

CONCLUSION

The mechanical milling route has successfully demonstrated the production of nanocrystalline Ti-6%Al-4%V based powders. Consolidation of these powders results in rapid grain growth, however a near-ideal microstructure for SPF is produced. Elongations of the milled material during hot tensile testing are limited by cavitation formation, but despite this large elongations in comparison to conventional sheet material have been achieved at relatively high strain rates.

ACKNOWLEDGEMENT

This work was supported by an EPSRC research grant (Grant No. GR/M28927).

REFERENCES

[1]   J. Pilling, N. Ridley, Superplasticity in Crystalline Solids, Int. of Metals, p. 65, (1989).
[2]   C.H. Hamilton, A.K. Ghosh, M.M. Mahoney, Advanced Processing Methods for Titanium, TMS-AMIE, p. 129, (1982).
[3]   M.F. Ashby, R.A. Verrall, Acta Metall., **21**, 149, (1973).
[4]   H. Inagaki, Zeitschrift fur Metallkunde, **86**, 643, (1995).
[5]   P.S. Goodwin, T.M.T. Hinder, A. Wisbey, C.M. Ward-Close, Mat. Sc. Forum, **269**, 53, (1998).
[6]   T.M.T. Godfrey, A. Wisbey, P.S. Goodwin, C.M. Ward-Close, A. Brown, R. Brydson, C. Hammond, Proceedings of 9th World Conference on Titanium, (1999).
[7]   G. Itoh, T.T. Cheng, M.H. Loretto, Mat. Trans., JIM, **35** (8), 501, (1994).
[8]   D. Lee, W.A. Backofen, Trans. TMS-AIME, **239**, 1034, (1967).
[9]   J.R. Leader, D.F. Neal, C. Hammond, Metall. Trans. A, 14A, 93, (1986).
[10]   A.D. McQuillan, M.K. McQuillan, Titanium, Butterworths Sci. Pub., p. 172 273, (1955).
[11]   J.A. Wert, N.E. Paton, Metall. Trans. A, **14A**, 2535, (1983).
[12]   R. Gerling, R. Leitgreb, F.P. Schimansky, Mat. Sci. and Eng., **A252**, 239, (1998).

CAVITY NUCLEATION IN AL 5083 ALLOYS

N. CHANDRA[*] and Z. CHEN[**]

[*]Professor, and [**]Graduate Student
Department of Mechanical Engineering, FAMU-FSU College of Engineering, Florida State University,
Tallahassee, Florida 32310, USA
E-mail: chandra@eng.fsu.edu

ABSTRACT

In this paper we address the controversial issue of nucleation of cavities in Al 5083 alloys. We focus on the origin of cavities during the manufacture of these alloys into SPF (superplastic forming) sheet form. Experimental observations on the pre-existing cavities in this alloy are made using optical and electron microscopy. The effects of rolling direction and state of stress during superplastic deformations on the formation of cavities are also discussed. Numerical simulations of the sheet manufacturing process are carried out to understand the effect of hard phase/matrix, mechanical properties and interfacial strength on the origin of cavities. Based on the numerical results, a simplified model relating the process, material parameters and the cavity nucleation is presented.

INTRODUCTION

The presence or absence of pre-existing cavities during thermo-mechanical processing of superplastic sheet manufacture is subjected to controversies. Some researchers have assumed that there should be pre-existing cavities because the manufacturing processes involve extensive plastic deformation during thermo-mechanical processing [1], and a few have actually observed pre-existing cavities [2-4]. Others argued that pre-existing cavities are totally absent from superplastic metal sheets since they could not observe cavities in the as-received materials [5-7], and very small cavities (< 0.5 μm), even if present will be sintered during the initial thermal exposure.

Cavitation behavior of superplastic sheet metal depends on alloy chemistry especially the second phase particles [8]. The addition of Mn, Ti, and Zr to Al matrix introduces dispersoids (e.g. Al_2Mg_2Cr in Al 7475 and $AlMn_6$ in Al 5083) that prevent grain growth during superplastic forming process, retaining a fine equi-axed grain structure. However, these very same dispersoids are responsible for the nucleation of cavities. Because dispersoids have both beneficial and deleterious effects, they must be chosen in such a way that they are effective in preventing grain growth while resisting nucleation of cavities. Cavitation has been experimentally observed to develop at grain boundary particles in quasi-single phase alloys. Stowell [9] proposed that the volume fraction of pre-existing cavities could be determined when cavity volume fraction is measured as a function of strain. However, Chokshi et al. [10] and Jiang et al. [11] reported that there were no observations to support the concept of pre-existing cavities. Ghosh et al. [12] assume that voids can occasionally pre-exist in alloy similar to Al 5083 but the concurrent nucleation during superplastic deformation is the determinant step.

There are ample experimental evidences to suggest that the relative strengths of the matrix and second phase particles constitute an important parameter in the nucleation and growth of cavities. During the deformation in the sheet manufacturing process, very sharp heterogeneity in the state of strain exist at the matrix/hard phase particle interfaces, and act as potential cavitation sites. For example, the deliberate addition of hard particle of Ag_3Sn to the superplastic Pb-Sn alloy causes the cavities to nucleate at the matrix-particle interface. Typically Pb-Sn alloy do not exhibit cavitations [13]. Similar observations have also been made for Ti(C, N) particles in micro duplex stainless steels [14]. Also when Fe particles were added to α / β brass cavitation began to occur [15]. Thus the strength differential between two hetero phases (alpha and beta) or hard phase particles/inclusions causes the separation to occur preferably in the interfacial region. Separation signals the nucleation of cavities and is strongly influenced by the external thermo-mechanical loading conditions (temperature, time and state of stress), geometrical features of the phases (size, shape, distribution of second phase), the relative strength of the two phases, and thermo-mechanical properties of the inter-phases.

In this paper we examine the origin of cavities during the thermo-mechanical processing of Al 5083 alloy, the alloy being a serious contender for automotive applications. Origin of pre-existing cavities in the other popular aluminum based superplastic alloy Al 7475 is also briefly addressed in this paper. We examined the as-received materials for the presence of pre-existing cavities and the size of the cavities. We have numerically simulated the final cold rolling process to understand the origin of cavities and the effect of matrix, particle and particle-matrix interfacial mechanical properties on the size of the cavities. We also developed an expression to relate the cavity size to the final thickness reduction based on a simplified analysis. The influence of the rolling direction on the formation of the cavity stringers is also explored.

EXPERIMENTAL RESULTS ON Al 5083 AND Al 7475

Al 5083-SPF is a relatively new solid solution strengthened alloy for superplastic forming application. The Al 5083 sheet metal used in this work was obtained from Sky Aluminum, Japan. The nominal chemical compositions are shown in Table 1. The as-received alloy shows large elongated grains with the appearance of the dendritic structure (Figure 1). Complete static recrystallization can be reached by heating up to 555°C in 40 minutes to produce an equiaxed grain structure with an average grain size of 17 μm (Figure 2).

Table 1. Chemical composition of Al 5083 alloy (wt. %)

Mg	Mn	Cr	Fe	Si	Ti	Al
4.70	0.65	0.13	0.04	0.04	0.03	bal

In the base Al-Mg-Mn alloy, Cr, Zn, Ti and Zr are added for strengthening the alloy and producing second phase particles. During thermo-mechanical processing of Al 5083-SPF sheets, the ingot undergoes extensive plastic deformation at elevated temperature with a final finish rolling at room temperature. The heat treatment cycle is designed to obtain fine equi-axed grain structure, a pre-requisite for superplastic deformation. The details of the thermo-mechanical processing can be found elsewhere [16]. In the present context, we focus our attention on the final cold rolling from 10 mm to 2 mm with a thickness reduction of 10-15% per pass and its effect on the formation of cavities. Like Al 5083, Al 7475 is also a statically recrystallized alloy with a similar thermo-mechanical processing (including the final cold roll) aimed to achieve fine equi-axed grain structure for superplastic applications [17]. In both the alloys, grain growth during subsequent high temperature superplastic forming is retarded by the introduction of sub-micron dispersoids.

Figure 1. DIC Optical Micrograph of the as-recieved materials showing elongated dentritic structures.

Figure 2. Equi-axed grain structure after 40 minutes exposure at 555 °C.

In this work, as-received samples for optical examination were prepared in three different orientations as shown in Figure 3. Observations were made on facial, transverse and longitudinal sections on both the material systems. The samples were polished up to 0.05μm solution but with no subsequent chemical etching. The etching sometimes introduces black chemical marks that can interfere with the observation of cavities.

It should be noted that the lower limit of the optical examination (with a magnification of 500) is about 0.1 to 0. 5μm, i.e., any particle (or cavity) smaller than this value even if present, cannot be observed. A few large particles of size as much 5 to 10 μm were observed in some locations. In Al 5083 superplastic alloy, second phase particles of 0.1 to 2 μm are uniformly distributed throughout specimen.

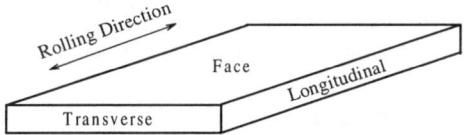

Figure 3. Sectioning terminology.

Figure 4 (a) shows large pre-existing cavities in Al 5083, where rolling direction is horizontal and thickness direction vertical. It can be seen that cavities are present both ahead and behind the particle with respect to the rolling direction. This observation of cavities on either ends of particle (along rolling direction) is very typical. Also, the size of the cavities seemed to depend on the size of the particle, larger cavities accompanying larger particles. Some large particles were broken into pieces and cavities usually surrounded (along rolling direction only) the broken pieces. The tapered section of the cavity is possibly due to the healing of cavities during subsequent thermo-mechanical rolling where the matrix (aluminum) freely flows into the cavities. After heating the specimen to 555°C for 40 minutes, the size and number of pre-existing cavities in the specimen reduced due to sintering. Figure 4 (b) shows the presence of pre-existing cavities in Al 7475, with very similar features as that of Al 5083.

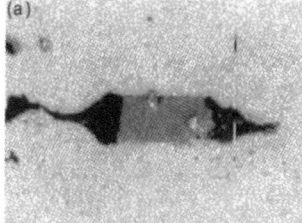 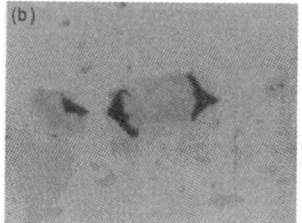

Figure 4. Pre-existing cavity in (a) Al 5083 and (b) Al 7475. In (b) cavities is seen on both sides of the particles along rolling direction.

Effect of thermal exposure

Figure 5 is the optical microphotograph of the as-received Al 5083. Particles with radius up to 10 µm are aligned in a direction parallel to the rolling direction. The particles were identified as Al$_6$Mn by Philips Xpert X-ray diffractometer. Observations under higher magnification showed that large particles were broken up. Cavities formed around the large particles (B), as well as between the cracked sections (A). These pre-existing cavities aligned predominantly along the rolling direction.

By heating to 555°C for 40 minutes, the size and the number of pre-existing cavities were reduced, when complete recrystallization occurs. Measurements showed that density of the material in the deformed area decreased by about 0.06% compared to the grip area. In the grip area the static hydraulic pressure during deformation eliminate most of the pre-existing cavities. Figure 6 shows the same specimen observed by scanning electronic microscope before any applied strain. It is seen that cavities are around large particles but absent in the vicinity of small precipitates.

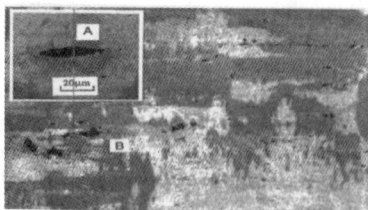 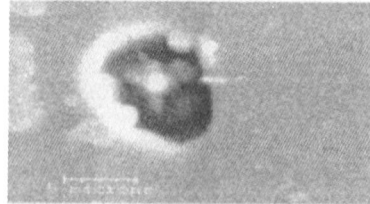

Figure 5. Particle alignment in a direction parallel to the rolling direction, pre-existing cavities can be found both around the large particles (a) and between the broken sections of large particles (b). The rolling direction is horizontal.

Figure 6. SEM micro-photograph shows that the pre-existing cavity was not completely sintered out after recrystallization.

INFLUENCE OF ROLLING-CAVITY STRINGERS

Several specimens were examined to determine the influence of the rolling direction, the tensile axis, and the stress states on the cavity stringers. Figure 7 shows the montage of the polished specimen stretched to $\varepsilon = 0.4$ under the equibiaxial stress state, in which the rolling direction is horizontal. The formation of cavity stringer is very evident, and the cavities are essentially rounded with the maximum radii of 8 - 10 μm. Inspection at higher magnification gave limited evidence for cavity interlinkage along a direction both parallel to and perpendicular to the rolling direction.

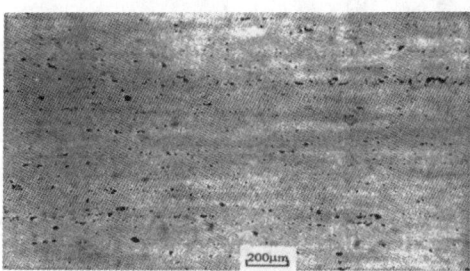

Figure 7. Cavitation after equi-biaxially formed to e = 0.4, showing cavity stringers aligned in a direction parallel to the rolling direction. The rolling direction is horizontal.

Figure 8 depicts the development of cavitation of tensile specimens pulled to, (a) $\varepsilon = 0.8$ with the gauge length parallel to the rolling and (b) $\varepsilon = 0.6$ with the gauge length perpendicular to the rolling direction. The tensile axis is horizontal in both of these two microphotographs. Cavity stringers parallel to the rolling direction still could be identified in both of these two microphotographs.

In all these three figures (Figure 7, 8(a) and 8(b)), the alignment of cavities in stringers parallel to the rolling direction can be observed. It is noted that at lower strain levels ($\varepsilon < 0.6$), under both the stress states, the morphology of the cavities appear to be spherical and limited interlinkage along the rolling direction can be observed (Figure 7 and Figure 8 (b)). At higher strain level, $\varepsilon = 0.8$, the cavity interlinkage occurs along both direction (Figure 8(a)). Comparing the three figures it can be found that with the increasing of strain levels the cavity stringers become more and more unclear.

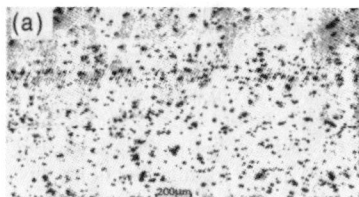 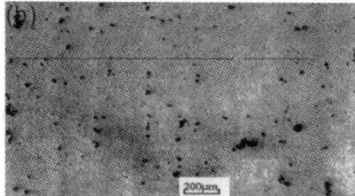

Figure 8. Micro-photograph of tensile specimen pulled to (a) e=0.8 with gauge length parallel to the rolling direction, (b) e=0.6 with the gauge length perpendicular to the rolling direction, revealing that the cavities always aligned in the rolling direction.

239

The appearances of cavity at higher strain, $\varepsilon = 0.85$, under (a) uniaxial stress state, and (b) biaxial stress state are presented in Figure 9 with the rolling direction being horizontal. The cavity stringers are masked due to the interlinkage of cavities, which can be observed in both microphotographs. It can be also noted that the interlinkage appears not only in the direction parallel to the rolling direction (A-A), but also in the transverse direction (B-B). A comparison of (a) and (b) reveals that the interlinkage is more serious in uniaxial state than that in biaxial state, but the number density of cavities appears more in the biaxial stress state.

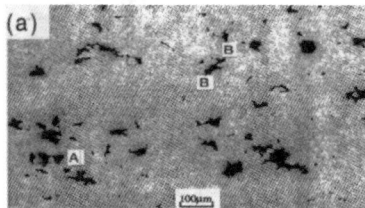 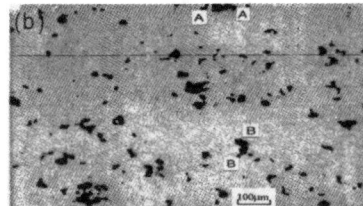

Figure 9. After a strain of e = 0.85 under (a) uniaxial stress state and (b) biaxial stress state, cavity interlinkage occurred in directions both parallel and perpendicular to the rolling direction. The rolling direction is horizontal.

An important question is the role of pre-existing cavities in the cavity growth and eventual fracture during superplastic deformation. Caceres and Silvetti [18] while studying the cavitation damage of Zn-22% Al superplastic alloy concluded that cavities grew from the pre-existing microvoids. However, Chokshi and Langdon [6] argued against pre-existing cavities leading to damage, since they did not observe them in the as-received material in the first place and they argued that even if present, the micro-voids may be sintered because of the thermo-mechanical conditions during superplastic deformation. We tend to agree with Caceres and Silvetti [18] for the following reasons.

When the Al 5083 was superplastically deformed, measurements indicated that the density in the grip region (subjected to thermal excursion and compressive stress), was almost the same as that of as-received material (ratio being 0.9994). This is in contrast to the superplastically deformed regions where the density varied by about 5%. Cavitation damage is the primary mechanism of final fracture in these materials; in most of the cases, the cavitation is in the vicinity of particles. Though we cannot unequivocally state that pre-existing cavities cause the final fracture, we know that both pre-existing cavities and the fracture inducing cavities are associated with particles. This can be attributed to the vastly differing flow properties of the particles and matrix material (eg. Aluminum).

The fact that cavities pre-exist at least in some materials system is indisputable. The question then is whether these cavities heal (or disappear) completely before superplastic deformation starts to occur; cannot be clearly answered. Since we observe cavities in the pre-existing stage as well as final fracture stage around the same region (mostly in the rolling direction), this observation leads us to surmise that the pre-existing cavities do play a role in the growth of cavities. Duality in the literature about the absence or presence of cavities (or micro voids) should be a consequence of the inability or ability to observe them under a given set of conditions. Some may not be able to observe cavities under some experimental conditions when in fact the micro-voids that can be observed under a set of different conditions may in fact exist. This statement will be further clarified when we look at the results of the numerical simulations.

FEM SIMULATION OF THE ROLLING PROCESS

Since it was suspected that cavities might originate during the final cold-rolling process of the SPF sheet manufacture, a numerical simulation of this stage was conducted using a nonlinear finite element method. According to sheet metal manufacturing process, the last step of thermo-mechanical process is cold-rolling with a reduction in thickness of 10-15% [15-16]. Since the final cold rolling process in the manufacture of superplastic sheet is the most vulnerable step, it was analyzed using FEM. It was assumed that the material was completely cavitation free prior to this step. A 15% reduction in thickness was applied during the FEM analysis. Mechanical properties of Aluminum at room temperature were used for the matrix material, whereas the particle was assumed to be rigid elastic with a value of Young's modulus equal to ten times that of aluminum. Both the matrix and the particle were assumed to be isotropic. As shown later, the mechanical/fracture properties of the particle-matrix interface play a critical role in the debonding and nucleation of cavities. Though interfaces can be modeled using shear based (debond strength), or energy based (fracture toughness) criteria, interfacial stiffness E_i was used in this work to describe its behavior. This concept can be easily implemented in the code as a spring element separating the particle and the matrix. A high value of E_i (equal to that of matrix, E_m signifies a strong bond, whereas a low value $E_i = 0.05E_m$ denotes a weak bond. The effects of variation of E_i on the results are also analyzed.

The rolling process was simulated for a single pass with a reduction in thickness as a given parameter. Rolling friction between the work (sheet metal) and the die (roller) was assumed. Feeding was forced initially at one end at a give rate after which the sheet is automatically fed due to frictional condition between the work and the die. The model was implemented using MARC/MENTAT FEM package [20]. Young's modulus of Al matrix was assumed to be 10 GPa with a initial yield value of 140 MPa and work-hardening afterwards. Four-node plane strain element was used. There are 2434 nodes and 2324 elements in the model, as shown in Figure 10 (a) where 46 elements are used as a particle. Further, MARC contact algorithm was implemented to prevent two materials (the Al matrix and the particle) from penetrating each other and denser meshing was used along particle/matrix interface, as shown in Figure 10 (b). A total of 65 increments were necessary to achieve the thickness reduction process.

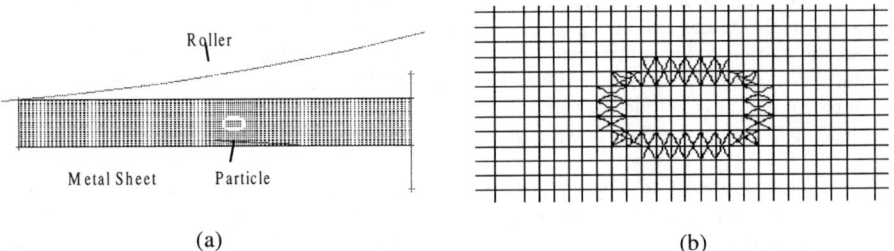

(a)   (b)

Figure 10. FEM model for simulating for a single pass rolling (a) during cold rolling and (b) local FEM meshing surrounding particle.

In the initial simulation very low bond strength was used and the results are shown in Figure 11. During the early stages of deformation (increment 20), cavities are seen to form as shown in Figure 11 (a). With further deformation of the sheet, cavities change both in size and in shape. However, the cavities are always associated with the particle at the both ends along the rolling direction. This is not surprising considering the fact that tensile stresses at the particle-matrix interface occurs only along that direction. At increment 25, pre-existing cavities grow and can now be clearly seen in Figure 11 (b). Further deformation continues to grow the cavities. When the rolling is finally completed as in Figure 11(e) (increment 50), the cavities are fully developed, Figure 11 (c) indicating the origin of processing induced preexisting cavities.

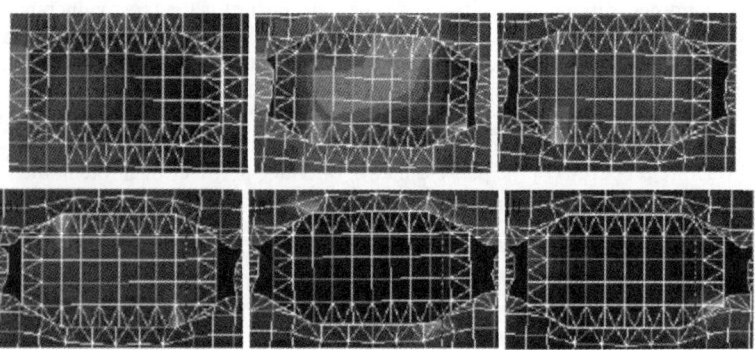

Figure 11. Post-FEM processing showing pre-existing cavities formed during a single pass rolling process, 46-element particle. (a) increment 20, (b) increment 25, (c) increment 30, (d) increment 35, (e) increment 50 and (f) increment 65.

Effect of particle-matrix interface and particle size on cavities

The size of the particles plays a critical role in determining the cavity initiation and further growth and coalescence. Though shapes may also play some role, they have not been analyzed in this paper. In modeling a sheet metal of 2-10 mm thick, with size of the particles ranging from sub-microns to a few microns, it is important to understand the effect of scales. Three orders of magnitude difference exists between the thickness of the sheet and that of the particle. However, for most of the rolling process a homogeneous state of deformation was observed in the thickness direction for a single material system. Also the strain and stress fields were affected only a few length scales from the particle location as shown later. Thus when the size effect was studied, the particle to sheet thickness was so chosen that edge effects do not play a role. Same original sheet thickness was selected with particles of different sizes by varying the number of elements from 1, 2 to 6. Under identical rolling conditions the results of the simulations were examined. The interface bonding was assumed to be constant of $E_i = 0.05 \ E_{Al}$.

For one element particle, Figure 12 shows the effect of thickness reduction and particle size on cavity fraction in terms of the area fraction of particle. At low spring constant (0.05% E_{Al}), area fraction of pre-existing cavity linearly increases with the thickness reduction. When particle size decreases, area fraction of pre-existing cavity also decreases slightly.

It is observed that when the strength of interface bonding is increased to 50% of E_{Al}, then cavitations does not occur. Figure 13 shows that there is no cavity when the stiffness of the interface equals the Young's modulus of matrix. The higher the spring constant, the better interface bonding between particles and matrix. Therefore, it is understandable that a better interface bonding will prevent cavities to occur during thermo-mechanical process. Figure 14 shows detailed numerical results of the effect of spring constant on area fraction of pre-existing cavity for a 6-element particle. With a spring constant of 50% Young's modulus of matrix, area fraction is only 0.3%, which is virtually zero. Decreasing spring constant will increase cavity area fraction of cavities.

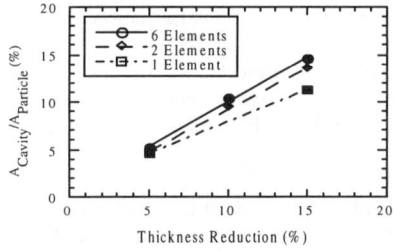

Figure 12. The effect of thickness reduction and particle size on pre-existing cavity, with E_i of 0.05% E_{Al}

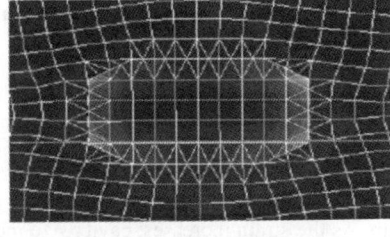

Figure 13. Higher E_i (50% E_{Al}) prevents pre-existing cavity

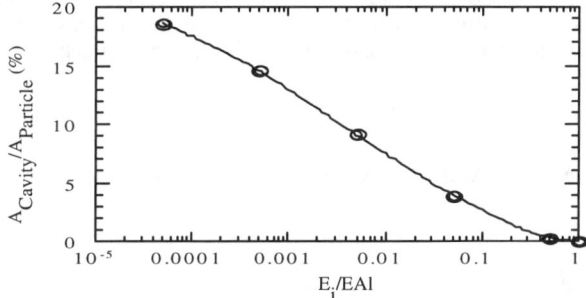

Figure 14. Variation of area fraction of pre-existing cavity with E_i for 6-element particle.

DISCUSSION

Although presence or absence of pre-existing cavities has been a controversial topic for decades, several models were established on the assumption that pre-existing cavities nucleate during the thermo-mechanical process. In our experimental observation on Al 5083 and Al 7475, we have directly obtained evidence that cavities pre-exist and nucleate during the thermo-mechanical processing cycles in the manufacture of Al 5083 and Al 7475 sheet metals. The pre-existing cavities are found to be always associated with second phase particles. Intuitively we can conclude that the origin and distribution of pre-existing cavities depend on the size, shape, distribution of particles, chemical and mechanical properties of matrix and second phase particles, and more importantly interfacial bonding between matrix and particles. For larger size of second phase particles, pre-existing cavities are unavoidable. To eliminate pre-existing cavitation, particle size of much less than a micron are essential.

During the cold rolling process, these particles break up and align into particle stringers parallel to the rolling direction. These particle stringers are most obvious on longitudinal section. It is possible that the previous reported researches might have paid attention only to the face section, and not to the longitudinal or transverse sections. The plasticity of the matrix is not so high that the discrepancy or void produced at the interface was rehealed. The broken brittle particles will also induce pre-existing cavities between the parts. These cavities with the radii up to 5 μm were observed around very large particles, and these large cavities were significantly sintered during the process of static recrystallization prior to superplastic deformation.

FEM simulation of cold rolling, the last step during the thermo-mechanical process, shows the possibility of occurrence of pre-existing cavities. Pre-existing cavities were always initiated on the matrix/particle interface. Cavities are always associated with both ends of particle along rolling direction where tensile stress predominates.

On the other hand, pre-existing cavities initiate immediately whenever there is a plastic strain. As the strain increases, cavity size increases. The area fraction of pre-existing cavities increases with the thickness reduction or with increase of plastic deformation. At the final increment the relationship between cavity growth and thickness reduction is linear (Figure 12). A simple model for one element particle can explain this phenomenon as shown in Figure 15. In the FEM model, rectangular elements with height (h_0) and width (w_0) are used. Particle occupies one element. After rolling deformation, the meshes are deformed and distorted. Here we neglect small elastic deformation of particle. It is further assumed that the elements directly above and below the particle remain rectangular in shape. From the definition of thickness reduction (δ),

$$\delta = \frac{h_0 - h}{h_0} \times 100\% \tag{1}$$

Since the area of each element keeps constant during plastic deformation, we have $wh = w_0 h_0$.

Therefore,

$$w = \frac{w_0 h_0}{h} = \frac{w_0}{\frac{h}{h_0}} = \frac{w_0}{1-\delta} = (1 + \delta + \delta^2 + \delta^3 + ...)w_0 \tag{2}$$

or the area fraction of pre-existing cavity as defined above,

$$\text{Cavity Area Fraction} = (\frac{wh_0 - w_0 h_0}{w_0 h_0}) = (\delta + \delta^2 + \delta^3 + ...) \tag{3}$$

If we neglect higher order components, we get the area fraction of pre-existing cavity is approximately equal to the thickness reduction. The numerical results are consistent with this simple model for 6-element particle.

Particle size plays in an important role in pre-existing cavitation. With particles of smaller, it is difficult to find pre-existing cavities. From our simple model, the area fraction of pre-existing cavities is of the same order as that of thickness reduction. If particle size is submicron, the size of pre-existing cavities has only tenth of particle size. Only under SEM we can observe cavities.

If there are particles with size larger than 5 μm, it is very much likely that pre-existing cavities to be easily observed with optical microscopy. If there is strong bonding between particles and matrix, we expect no pre-existing cavities.

SUMMARY AND CONCLUSIONS

Although the subject of pre-existing cavitation is controversial, cavities do pre-exist in the two important Aluminum based superplastic materials that we examined. The evidence is provided by the direct observation on as-received Al 5083 and Al 7475 sheet metal samples. Pre-existing cavities are typically associated with second phase particles, especially when the size of the particle is large. Numerical simulation supports the theory that pre-existing cavities originate during cold rolling process of the superplastic sheet manufacture. The volume of pre-existing cavities depends on the level of thickness reduction, size of the particle in relation to thickness, and the bonding strength between particles and the matrix.

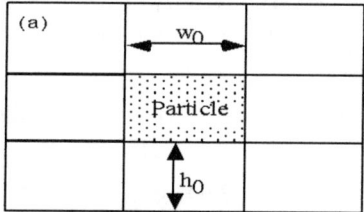 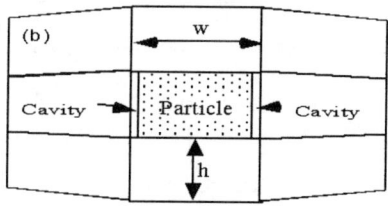

Figure 15. Schematic relationship between cavity area fraction and thickness reduction, (a) before and (b) after rolling deformation.

References

1. M.J. Stowell, *Superplastic Forming of Structural Alloy*, The Metallurgical Society of AIME, ed. by N.E. Paton and C.H. Hamilton, p. 321-336, 1982.

2. C.H. Caceres and D.S. Wilkiinson, *Acta Metallurgia*, 35, p. 897-906, 1987.

3. S.H. Goods and L.M. Brown, *Acta Metallurgia*, 37, p. 35, 1984.

4. K. Kannan, C.H. Johnson and C.H. Hamilton, *Materials Science Forum*, 243-245, p. 125-130, 1997.

5. A.H. Chokshi and A.K. Mukherjee, *Materials Science and Engineering*, A110, p.49-60, 1989.

6. A.H. Chokshi and T.G. Langdon, *Acta Metallurgica*, 37, p. 715-723, 1989.

7. X. Jiang, J. Cui and L. Ma, *Superplasticity in Metals*, Ceramics, and Intermetallics, Vol. 196, MRS, ed. by M. J. Mayo, M. Kobayashi and J. Wadsworth, p. 51-56, 1990.

8. C.C. Bampton and R. Raj, *Acta Metallurgica*, 30, p. 2043 - 2053, 1982.

9. M.J Stowell, Cavitation in superplasticity, in N. E. Paton and C. H. Hamilton ed. *Superplastic Forming Of Structural Alloys*, p. 321-336.

10. A. H. Chokshi and A. K. Mukherjee, Acta Metallurgica, 37, p. 3007-3017, 1989.

11. Jiang, X.G, Cui, J. Z, and Ma, L. X, *Acta metall.*, 41, p. 2721-2727, 1993.

12. A.K. Ghosh and D.H. Bae, Materials Science Forum, 243-254, p 89-98, 1997.

13. Humphries and Ridley, *J. Materials Science*, 12, p. 851-855, 1977.

14. C. I. Smith, B. Norgate and N. Ridley, Materials Science, 10, p. 182-188, 1976.

15. C.W. Humphries and N. Ridley, *J. Material Science*, 13, p. 2477-2482, 1978.

16. H. Iwasaki, K. Higashi, S. Tanimura, T. Komatubara and S. Hayami, *Superplasticity in Advanced Materials*, ed. by S. Hori, M. Tokizane and N. Furushiro, The Japan Society for Research on Superplasticity, p. 447-452, 1991.

17. J. Pilling and R. N. Ridley, *Superplasticity in Crystalline Solids*, The Institute of Metals, p. 20, 1989.

18. C.H. Caceres and S.P. Silvetti, Acta Metallurgica, 35, p. 897-906, 1987.

19. C.C. Bampton and J.W. Edington, *Metallurgical Transactions A*, 13A, p. 1721 -1727, 1982.

20. MARC/MENTAT user guide, MARC Analysis Research Corporation,Palo Alto, CA 94306, USA, 1997

HYPERPLASTIC FORMING:
PROCESS POTENTIAL AND FACTORS AFFECTING FORMABILITY

Glenn S. Daehn, Vincent J. Vohnout and Subrangshu Datta
Department of Materials Science and Engineering, The Ohio State University, Columbus, OH 43210, Daehn.1@osu.edu

ABSTRACT
This paper has two distinct goals. First, we argue in an extended introduction that high velocity forming, as can be implemented through electromagnetic forming, is a technology that should be developed. As a process used in conjunction with traditional stamping, it may offer dramatically improved formability, reduced wrinkling and active control of springback among other advantages. In the body of the paper we describe the important factors that lead to improved formability at high velocity. In particular, high sample velocity can inhibit neck growth. There is a sample size dependence where larger samples have better ductility than those of smaller dimensions. These aspects are at least partially described by the recent model of Freund and Shenoy. In addition to this, boundary conditions imposed by sample launch and die impact can have important effects on formability.

INTRODUCTION AND MOTIVATION
Aluminum has an outstanding potential for reducing the mass of automobiles. One of the key problems is that it is very difficult to form without tearing as illustrated in the photos in Figure 1. Presently this is overcome by assembling aluminum sheet structures from a number of components or using a sub-optimal design. Over the past several years it has become clear that metals can be stretched to much higher strains at high velocity versus conventional quasi-static stretching. We refer to this extended ductility in high velocity conditions as hyperplasticity. Much of the past work of the Ohio State group, as well as recent research activities, is posted at a website [1].

One of the most attractive methods of high velocity metal forming is electromagnetic forming. Electromagnetic forming actuators can be fabricated into a wide variety of configurations and used in conjunction with stamping operations. We believe that the integration of electromagnetic forming and traditional stamping will enable many components to be fabricated with a smaller number of operations and that it will enable the fabrication of many components from difficult materials such as aluminum. This ultimately might ultimately lead to lighter weight and more cost effective land and aerospace vehicles.

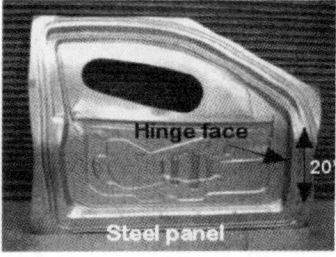 

Figure 1. Results of forming steel and forming grade aluminum into a typical door shape using tooling designed for steel. The aluminum panel weighs 11.5 pounds whereas the steel panel weighs 26.

Electromagnetic forming is based on the Lorentz force developed between an actuator that has a current pulse run through it and a nearby conductive workpiece. Basically whenever an electrical current is rapidly imposed within an electrical conductor, it will develop a magnetic field. The initial change in magnetic field will induce eddy currents in any nearby conductor that generally run in a direction opposite to the primary current (like in a transformer). These eddy currents develop their own magnetic field and cause a

247

mutual repulsion between the workpiece and actuator. This technique is quite general and is suitable for any workpiece made from a good conductor, provided the current pulse is of a sufficiently high frequency (usually pulse rise times under 30μs). The spatial distribution of pressure can be controlled by the configuration of the actuator and the overall force magnitude is largely controlled by the discharge energy. The basics of this technology are described well in Moon's book [2].

There is a long history in the use of both high velocity forming in general (much on explosive forming) and electromagnetic forming in specific. Hundreds of capacitor banks have been manufactured and used in long-run production since the 1960's to fabricate millions of parts without operator injury. The overwhelming majority of operations that have been carried out in the past involve simple axisymmetric compression or expansion as part of an assembly operation. Improvements in formability and suppression of wrinkling have received scant attention in the older literature. Several reviews of research performed in the heyday of high velocity forming have been written [3,4].

There are some fundamental aspects to high velocity forming that make it particularly attractive for use in practical metal forming. These include:
1) *Formability is improved.* We have observed over 100% plane strain elongation in aluminum subjected to a single room temperature high velocity forming event. This is discussed in the latter part of this paper.
2) *Impact has benefits.* When sheet metal strikes a tool at high velocity, large compressive impact stresses are developed. These coin the sheet into the die surface. This can reduce springback, improve surface finish and enhance formability.
3) *Wrinkling is suppressed.* When a sheet is launched with a particular velocity profile, each part of the sheet would like to travel along its launch path. Wrinkling usually necessitates a change in direction. Thus, at high velocity, wrinkling is inhibited by material momentum. As an example, one can reduce the diameter of slender rings with an electromagnetic impulse by a 2:1 ratio or more.

The integration of traditional stamping with electromagnetic forming holds some potentially exciting advantages. In principle, both operations can be carried out in one nearly-traditional press stroke. Electromagnetic actuators can either be pulsed while the tool punch is advancing, or a single impulse can be delivered at the bottom of the press stroke. Several actuators might be independently controlled in one press stand as suggested in Figure 2.

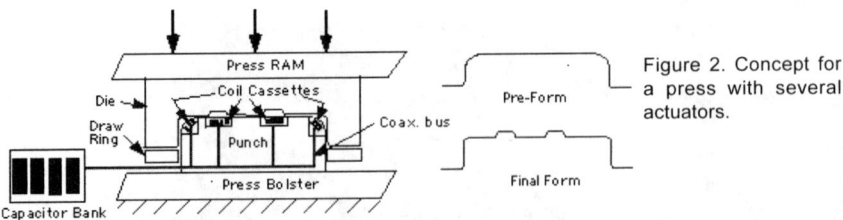

Figure 2. Concept for a press with several actuators.

Such an integrated electromagnetically assisted stamping operation can have many potential benefits. These include:
Improved Formability – Forming limits can be significantly increased at high velocity.
Improved strain distribution – In addition to improved formability, the freedom afforded by electromagnetically assisted stamping allows modification of strain distributions. For example small pulses can produce strain in an area that might otherwise be locked out by friction.

248

Reduction in wrinkling – This has several advantages, for example, it greatly widens the window of operation for shrink flanging.

Active control of springback – The use of small pulses at the bottom of a press stroke can be used to 'tune' springback behavior to accommodate differences in material.

Distortions at local features are minimized – The high contact pressures and high velocities associated with electromagnetic forming can minimize the distortions that are typical with restrike operations.

Local coining – When a driven sheet hits a die, impact pressures are high enough that coining, embossing and hitting corners to minimize springback are possible.

This approach has been demonstrated in a limited way in cooperation with the USAMP materials forming team. The project examined how an electromagnetic forming operation combined with a softened initial shape could enable the fabrication of a one-piece aluminum door inner. A pre-form with softened corners was formed by Milford Fabricating, then this was electromagnetically re-formed into a GM Cavalier door inner die. The electromagnetic operation took the corner back to its design radius. The results of this demonstration are shown in Figure 3. This part could not be formed by traditional means with aluminum, as was demonstrated in Figure 1.

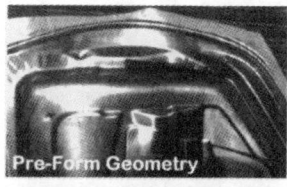

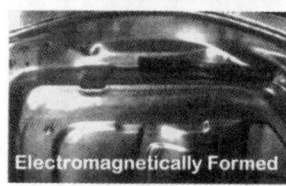

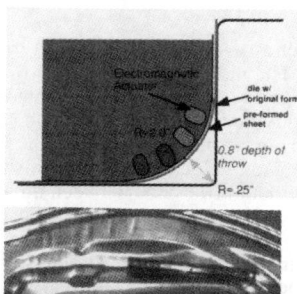

Figure 3. The Electromagnetic re-forming process carried out on the hinge face of an alum-inum door inner. The photographs show a softened initial shape that is electromagnetically re-formed to very near the original product shape.

FORMABILITY AT HIGH VELOCITY

In order to follow the vision outlined above, it would be quite useful to have a fundamental understanding of the mechanisms of high velocity formability and also to have some simplified criteria that can be used in an engineering sense to design high velocity forming processes. The progress in the areas of traditional sheet forming and superplastic forming can be used to guide the development of the technical base and engineering rules that will be needed with high velocity forming

In traditional sheet metal forming the challenge usually is centered on designing forming tools that don't stretch the metal outside its forming limits while also not allowing so much draw of metal into the die cavity so as to cause wrinkling. The theoretical background on sheet instability with respect to stretching goes back to classic work by researchers such as Considere [4], Swift [5] and Marciniak and Kuczynski [6]. This was extended and made more practically useful by the empirical insights of others such as Keeler and Goodwin in the development of the forming limit diagram [7-9]. These diagrams are, of course, fundamentally based on the strain-hardening behavior of the sheet, its anisotropy, microstructure and many other issues (such as the important complication of non-proportional strain paths). Despite the complications, this approach largely forms the basis of how sheet metal forming operations are analyzed from a practical point of view.

Similarly in superplastic forming there are several noteworthy analytical studies that relate the material strain rate sensitivity to elongation to failure [10, 11]. And an empirical view of data has also shown that strain rate sensitivity is a leading-order factor in determining uniaxial strain to failure [12-14]. From this background the apparent accepted engineering method of developing superplastic forming processing conditions is to select the strain rate and temperature that maximize the material strain rate sensitivity. Usually there is more elongation than needed. To study the process in a more detailed way one must also consider the evolution of the material microstructure with time and strain, understand the issues related with cavitation and possibly account for non-uniaxial strain states that are encountered in actual parts. Again, the accepted engineering approach successfully accepts many simplifications in order to provide an easily-used approach that correctly identifies the leading order effect.

If high velocity electromagnetic forming is to become commonly used a simplified design approach will be necessary. This paper does not attempt to propose such rules, but hopes to point towards the possible fundamentals of an approach. In the remaining paragraphs promising theoretical approaches are identified and it is shown that they correspond well to empirical studies of data, and other important practical complications are noted.

Three analytical approaches to material formability at high strain rates have recently become available. First, Hu and Daehn used a 1-D finite element approach to study the fracture of slender rings and tensile samples subjected to high velocity extension. This study was restricted to studying instability by a single neck at an imperfection. This work demonstrated that in ring expansion, strain to failure is expected to increase monotonically with expansion velocity, as has been shown experimentally by Altynova et. al. [16]. Also an important difference between uniaxial extension and ring expansion was noted. In the former case, the loading wave can cause the sample to fail at the driven end at very low average strains. This phenomena has been well studied by vonKarman and Duez[17]. This points out that varied sample launch conditions can produce very different failure with respect to formability. Similarly, impact on sample deceleration can positively or negatively influence formability, as discussed later.

Recently, Shenoy and Freund [19] developed an analysis of necking bifurcations in high strain rate, plane strain extension of a rate-independent material. This is based on a formalism developed by Hill [19]. Basically their analysis shows that for a material that is not subjected to stress waves during its initial acceleration, instability is delayed by inertial forces. In particular, as velocity increases the wavelength at which instabilities grow decreases such that multiple necks are expected at increasing sample velocity. Also the stress that a sample can support increases with increasing extension velocity. There are two components to this. First, as a neck grows, it will tend to unload the adjacent regions, thus continued deformation will require an increased stress. Second, as a neck forms, a radial inward acceleration is required, this again increases the stress needed to extend the sample. Their analysis considers a sample that has a length to width aspect ratio and is infinite in the third dimension. They found that as samples became wider at a fixed length the periodicity of the dominant necking mode decreased and the dynamic stress a sample can support increased. For these reasons ductility is expected to increase with both deformation velocity and cross sectional dimensions.

Experimental observations are largely in very good accord with the model. In particular the work of Altynova et. al. [16], has shown that strain to failure in ring expansion can be increased by about two fold relative to quasi-static ductility in 6061-T4, 6061-T6 and annealed OFHC copper. Also, the necking periodicity decreases drastically as expansion velocity is increased. It should be noted that while high velocity expansion leads to fragmentation of the rings into a number of pieces, this is still useful ductility.

Experiments have been performed where a steel die is used to arrest the sample before it fails. In such cases, the sample can be stopped immediately from high velocity and high strains can be obtained in unfailed samples. In unarrested ring expansion if the ring is to remain unfailed, it must come back to zero velocity (by conversion of kinetic energy to plastic work). This points out: 1) that it is difficult (at best) to perform experiments at constant extension velocity and 2) use of a die to arrest the sample may be required to access the improved material formability at high velocity.

Tamhane et al [20] have studied the effect of ring and tube height for 1mm thick, 31 mm diameter rings of 1, 2, 4, 8 and 16 mm in height. The series of experiments showed that the sample ductility of unarrested samples increases strongly with increasing ring height. Circumfrential strains over 60% have been observed in the longer tube samples, whereas only 30% is available in the 1mm tall samples*. Although the samples tested here are clearly geometrically different than the idealized plane strain samples of Shenoy and Freund, similar size effects in necking are seen. In particular the number of fragments generally decreases with increasing sample height. At a launch velocity of 135 m/s a 4 mm tall sample will fragment into about 10 pieces, whereas a the 16 mm tall sample will remain unfailed.

Although the correlation between experiment and observation for the Shenoy and Freund model is impressive, the model is a bit difficult to directly compare to experiment. First, experiments suggest the actual values of all the sample dimensions are important and cannot be considered in the present model. Second, it is important that complicated non-uniform launch profiles can be handled (i.e., vonKarman effects), Thirdly, velocity is not constant with time. Such issues can be approached with a finite element approach. Recently Pandolfi, Krystl and Ortiz [22] have demonstrated the application of rate-independent finite elements with embedded cohesive elements to the expanding ring problem. This approach was shown able to make predictions of both the number of ring fragments as well as the fracture strain as a function of ring velocity.

One last complication that requires attention is the effect of workpiece-die impact. The die can simply arrest sample motion before failure, as mentioned earlier. Also, when a workpiece impacts a die at a velocity of 10 m/s or more, the impact pressures may be quite high (on the order of the workpiece flow stress). In the presence of a discontinuity this can produce an inertial-blanking effect, where the workpiece pierces and the slug carries on past a corner. However, on a smooth gently curved surface the impact can produce a large through-thickness compressive stress that may act to extend the sample in-plane in a manner similar to ironing. It seems that such an effect is necessary to explain the extraordinary ductility (over 100%, near plane strain) seen in aluminum, copper and ferrous alloys in the experiments of Balanethiram and Daehn [23, 24]. In these experiments flat sheets were essentially thrown into conical dies at high velocity. The sample impacted the conical die with velocities on the order of 100 m/s and fine details from the die surface were imprinted upon the sample, demonstrating a coining-like effect. Whereas the arguments of stability and ductility extended by inertial necking resistance can only account for approximately a two-fold increase in forming limits, here an approximate 5-fold increase in formability was observed. It seems benefits of the compressive stress-state on impact are required to account for the formability. The examination of such more complex forming operations may be ideally suited to the cohesive element finite element approach.

---

* In this case there was also a systematic decrease in the ratio of the ring thickness strain to height strain with increasing launch velocity. A modified launch system now being used by S. Datta, seems to remove this experimental complication [21].

CONCLUDING REMARKS

There is considerable experimental and analytical evidence that high velocity metal forming can considerably extend material forming limits. Also, strategies for using this effect in manufacturing are taking shape. In order to effectively design actual manufacturing processes a quantitative understanding of formability in complex inertially driven situations is needed. It appears that such an understanding is now beginning to develop.

ACKNOWLEDGEMENTS

This work is largely sponsored by the National Science Foundation, Division of Design Manufacture and Industrial Innovation and the Center for Advanced Materials and Manufacturing of Automotive Components (CAMMAC) at Ohio State. The J-Car Door demonstration was financed by the USAMP Materials Forming Team, and Larry DuBois of GM was active in all phases of that work.

REFERENCES

1. http://www.osu.edu/hyperplasticity.
2. F.C. Moon, *Magneto-Solid Mechanics*, John Wiley and Sons Inc., 1984.
4. *High Velocity Forming of Metals*, Revised Edition, E.L. Bruno, ed., ASTME, 1968
3. R. Davies and E. R. Austin, *Developments in High Speed Metal Forming*, Industrial Press, New York, 1970.
4. A. Considere, *Ann Ponts Chaussees,* **9**, 574-75 (1885).
5. H. W. Swift, *J. Mech. Phys. of Sol.,* **1**, pp. 1-18 (1952).
6. Z. Marciniak and K. Kuczynski, *Int. J. Mech. Sci,,* **9**, 609 (1967).
7. S. P. Keeler and W. A. Backofen, *Trans ASM,* **56**, 25 (1963).
8. G. M. Goodwin, *SAE Automotive Congress,* Paper 680093 (1968).
9. S. S. Hecker, *Metals Eng. Quart,* **14**, 30 (1974).
10. E. W. Hart, *Acta Met,* **15**, 351 (1967)
11. A. K. Ghosh and R. A. Ayres, *Met Trans,* **7A**, 1589 (1976).
12. C. E. Pearson, *J. Inst. Metals.* **54**, 111 (1934).
13. D. H. Avrey and W. A. Backofen, *Trans Q. ASM,* **58**, 551-62 (1965).
14. W. B. Morrison, *Trans Met Soc AIME,* **242**, pp. 2221-2227 (1968).
15. X. Hu and G.S. Daehn, *Acta Mater.,* **44**, pp 1021-33, 1996.
16. M. Altynova, X.Hu and G. S. Daehn, *Met and Mat Trans,* **27A**, 1837-43 (1996).
17. T. vonKarman and P. Duwez, *J. Appl. Phys.,* **21**, 987 (1950).
18. V. B. Shenoy and L. B. Freund, *J. Mech. Phys. Solids,* **47**, 2209-2233.
19. R. Hill, *In: Problems in Continuum Mechanics,* SIAM, Philadelphia, 155-164 (1961).
20. A. Tamhane , M. Altynova and G. S. Daehn, *Scrpta Mater,* **34**, 1345-50 (1996).
21. S. Datta, work in progress, Department of Materials Science and Engineering, The Ohio State University.
22. A. Pandolfi, P. Krystl and M. Ortiz, *Int. J. of Fracture,* **95**, 279-97 (1999).
23. V. S. Balanethiram and G. S. Daehn, *Scripta Met et Mat,* **30**, 515-20 (1994).
24. V. S. Balanethiram and G. S. Daehn, *Scripta Met et Mat,* **27**, 1783-88 (1992).

SUPERPLASTIC DOME FORMING OF MACHINED SHEETS

Nihat Akkus, Toshihiro Usugi, Masanori Kawahara, Ken-ichi Manabe, Hisashi Nishimura
Tokyo Metropolitan University, Faculty of Engineering,
1-1 Minamiosawa, Hachioji-shi, Tokyo, 192-0397, JAPAN

ABSTRACT

An experimental work on the superplastic bulge forming of machined sheets is presented in this study. Unlike the previously employed incremental-iterative method, a reverse deformation model was used to estimate the initial thickness distribution of the machined sheets from which a constant final thickness can be obtained when the shape of the bulged sheet is hemisphere. The reverse deformation model was obtained by modifying previously-known models, which were based on the axisymmetric membrane and the incremental strain theory.

Bulge forming experiments were conducted on machined sheets of Al alloy, A5083, at about 530°C under constant pressure control mode. The result of this simulation to estimate the final constant thickness distribution agreed well with the experiment, and confirmed that the reverse deformation model can be successfully applied in optimizing the thickness distribution of the starting sheets in order to obtain the desired final thickness distribution of the free bulged hemispherical product.

1. INTRODUCTION

The non uniform thickness distribution of the final products produced by superplastic deformations remain a problem, although several researchers have published results of investigations on reducing the non uniformity of deformed parts. In 1970, Al-Naib and Duncan [1] mentioned that pre-deformed or pre-machined sheets could be a solution to increasing the thickness uniformity of the final product. In 1972, Johnson et al [2] showed experimental methods for improving the thickness distribution of cylindrical caps. It was demonstrated that forming caps with a retarding die gives a better thickness distribution than forming them by plug-assisted forming or by female forming.

Takahashi et al [3] presented a useful model for thickness control in the superplastic bulge forming of a spherical tank shell fabricated from pre-machined sheets. An incremental-iterative approach, based on the modification of the initial distribution according to the non uniform final thickness distribution, was employed to optimize the initial thickness distribution of the free bulged hemisphere. However, the incremental-iterative model requires more repetitions in estimating the initial sheet's distribution by calculation, thus making the simulations more complicated. The number of repetition of the calculation for obtaining the initial thickness distribution also related to the final thickness tolerance between the apex and the rest of the bulged shape. If a smaller tolerance is chosen, a larger number of repetition is needed.

Highly accurate simulation results can be obtained by Finite Element Method (FEM), however, some problems such as unsatisfied convergence and mesh refining were encountered during the trials of reverse modeling in FEM by using commercially available codes. At present, only special FEM codes which can overcome the above difficulties may be used for reverse modeling.

In the present study, an analytical model whose details are explained in Ref. [4] is modified to simulate reverse forming and to estimate the initial thickness distribution of the sheet, which will have constant thickness distribution when deformed to a dome shape. Machined test pieces then were dome formed experimentally and the final thickness distribution of the domes were compared to the simulation results to examine the validity of proposed reverse method. Some simulation results of the machined test pieces were also compared to those of the uniform test pieces to see the difference in bulging time and deformation rates.

2. METHOD OF OBTAINING INITIAL THICKNESS DISTRIBUTION

2.1 Incremental-iterative method

An incremental-iterative approach was reported in a previous study by Takahashi et al [3] to estimate the initial thickness distribution of the sheet. The method will be briefly explained here for the sake of easy comparison with the reverse modeling. In this method, the initially uniform sheet is simulated into a dome shape and the thickness distribution is obtained. The thickness of each element then compared with the desired constant final thickness and a difference, Δt, is obtained. If the thickness of an element is not within a thickness tolerance, Δt_t, then the initial thickness of the elements is modified by $\Delta t/2$, and subsequently a new dome forming simulation is performed. The thickness modification followed by simulation from sheet to hemispherical shape is repeated until the thickness of each annuli is within the thickness tolerance of the desired bulging height.

The disadvantage of this method is that it requires repeated calculations that makes the computer programs more complicated and takes a longer time to obtain initial thickness distribution. The smaller the tolerance, Δt_t, the longer the calculation time needed. A tolerance of $\Delta t_t = 0.001$ was used in the incremental-iterative simulation of a sheet with 100 elements in the present study. The number of modifications of the initial sheet thickness was found about 24 times if an element thickness was modified by $\Delta t/2$ each time, and 10 times if modified by Δt. Figure 1a shows the illustration of the incremental-iterative simulation.

2.2 Reverse method

A set of equations, explained in detail in Ref. [4] were used to reverse simulate the superplastic bulge forming process. Figure 1b shows the illustration of the reverse method.

In this method, the superplastic bulge forming process of sheet was reversely simulated from hemispherical shape to the initial sheet form . The aim of the reverse simulation is to obtain the initial thickness distribution without any repetitions. First the input parameters such as time step ΔT, material parameters m, K and shape parameters such as half diameter R_0 and initial thickness t_0 are defined. A perfect hemispherical shape with a constant thickness distribution from the apex to the periphery is then divided into 100 elements on which the stresses, strains and thickness are assumed constant. Then the fully bulged spherical initial shell shape is reversed to a initial sheet by increasing time step by a proper value.

One of the most important tasks in the formulation of superplastic forming is the definition of the strain rate. Equation (1), which was obtained from the material model, $\sigma_{eq} = K \dot{\varepsilon}_{eq}{}^m$, and from von Mises equations, was used to obtain the meridional strain rate of each element in the present study.

$$\dot{\varepsilon}_m = \{(2 - \lambda)/2\} (p R/2 t K)^{1/m} (1 - \lambda + \lambda^2)^{(1-m)/2m} \tag{1}$$

where λ is stress ratio between meridional and circumferential direction, p is pressure, R is the radius of curvature in the meridional direction, t is the thickness and K is the strength coefficient of the material used. Although Eq. (1) is similar to the one used by Guo and Ridley [5], the stress ratio of the above equation was obtained from experiments in their study. Considering that the strains are the results of σ_c and σ_m in a membrane, the relationship between stresses and strains given by the Levy-Mises equation can be utilized to obtain the stress ratio theoretically as follows:

$$d\varepsilon_c / \sigma'_c = d\varepsilon_m / \sigma'_m = 3 d\varepsilon_{eq} / 2 \sigma_{eq} \tag{2}$$

The stress ratio λ can be obtained by solving $d\varepsilon_c / \sigma'_c = d\varepsilon_m / \sigma'_m$ of the above equation as:

$$\sigma_c / \sigma_m = \lambda = (1 + 2 d\varepsilon_c / d\varepsilon_m) / (2 + d\varepsilon_c / d\varepsilon_m) \tag{3}$$

$d\varepsilon_m$ and $d\varepsilon_c$ are the incremental strains in the meridional and circumferential directions, respectively. In the present study, the bulge surface assumed always in a part of circular arc in the meridional and

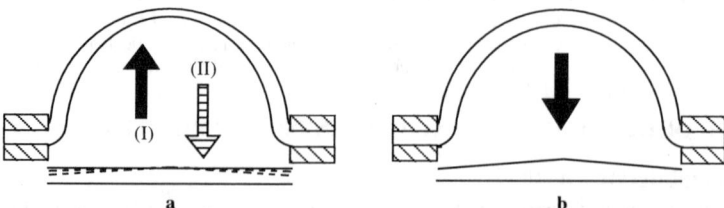

a
(I) Simulation of bulge forming
(II) Modification of initial sheet thickness

b
Reversing the dome shape
to initial sheet

Fig. 1 Illustration of a) incremental-iterative b) reverse modeling

circumferential directions. This assumption can provide the strain values in the above equation by comparing the shape changes between two steps. The location of each element can be obtained from the geometrical relations shown in Fig. 2.

An element length defined by $s_{(n,j)}$ at the step n will be shortened in the next step by:

$$\Delta s_{(n+1,j)} = - (\dot{\varepsilon}_{m(n,j)} \times \Delta T) \times s_{(n,j)} \quad (4)$$

Then the new half length of bulge profile, $L_{(n)}$, is obtained by summing each element's length in the meridional direction. Finally the strain in the thickness direction can be obtained by applying the volume constancy rule to calculate the thickness of each element. These calculations are repeated until the work piece becomes flat sheet.

Figure 3 shows the comparison of the estimated initial thickness distribution of reverse and iterative modeling. This figure reveals that there is no difference between the incremental-iterative and reverse methods in the estimation of initial thickness distribution within the limit of present modeling. Therefore, the final thickness distribution of the hemispheres whose initial thickness distribution was obtained from incremental-iterative and reverse modeling should be the same.

3. EXPERIMENTS

Aluminium alloy sheets of A5083 were used as specimens. Iwasaki et al [6] studied the superplastic deformation characteristics of this material. Table 1 shows the chemical composition of the A5083 alloy and properties related to superplasticity. The simulation results obtained for constant pressure mode revealed that the machined surface of the sheet should be of a parabolic arc shape. The parabolic surface of the test pieces were approximated by four straight lines at point 0-10, 10-15, 15-25 and 20-25 (Fig. 3), which best fit the parabolic shape because of the availability of the machining process. The desired final uniform thickness was 0.5 mm when the bulge shape was hemispherical. The initial geometry of Al sheets were $80^{\phi} \times 1.5^t$ mm ($R_0 = 50$ mm) prior to machining. Experiments were conducted at the superplastic temperature of 530℃ with a strain rate sensitivity index and strength coefficient of $m = 0.5$ and $K = 162$ MPa s^m, respectively. The experimental setup used for bulge forming tests is shown in Fig. 4.

Forming pressure of 0.4 MPa for constant pressure mode was chosen to keep the strain rate in a range where the optimum m and K values are valid. Ar gas was used as a means of forming

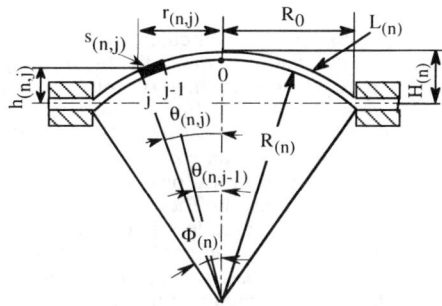

Fig. 2 Illustration of the bulge shape

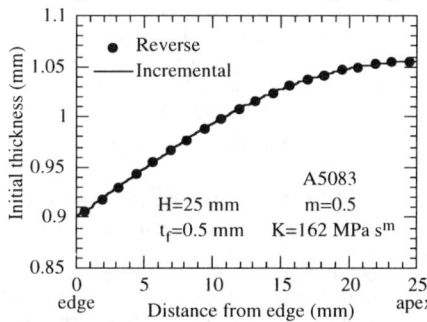

Fig. 3 Comparison of the estimated initial thickness distribution of reverse and incremental-iterative modeling

Table 1 Chemical composition of the A5083 (wt.%) and properties related to superplasticity

Mg	Mn	Cr	Fe	Si	Ti	Al
4.7	0.65	0.13	0.04	0.04	0.03	Remainder

SP temp./℃	Strain rate range s^{-1}	m	K, MPa s^m
530	1.0×10^{-4} 1.0×10^{-3}	0.5	162

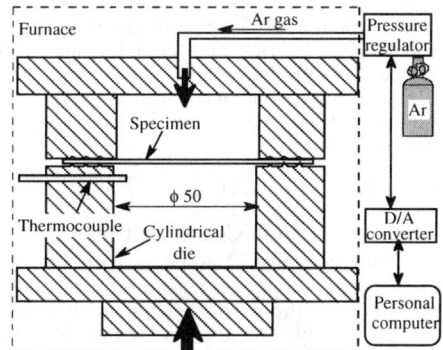

Fig. 4 Illustration of the experimental setup for superplastic bulge forming of hemisphere

255

pressure and controlled by a pressure regulator connected to personal computer. An important task of the simulation is to predict the forming time or pressure-time cycling. A sample output of the pressure-time cycling of the *constant pressure* mode and the pressure-bulging height of the *constant maximum strain rate* mode for machined and uniform test pieces are given in Figs. 5 and 6, respectively. These figures indicate that machined sheets need less deformation time under the *constant pressure* or less deformation pressure under the *constant maximum strain rate* mode than do non machined initially uniform sheets.

Bulged hemispheres were measured to obtain thickness distribution and bulging heights after the forming tests. A point micrometer with an accuracy of 0.01 mm was employed for the measurement of the thickness.

4. RESULTS AND DISCUSSION

4.1 Results of reverse modeling experiment

Severe thinning occurs at the apex of the hemispheres formed under superplastic deformation. About a 40 % thickness difference between the apex and periphery is commonly seen in the hemispherical bulge forming of ordinary sheets. The excessive thickness towards the periphery of the hemisphere can be avoided by modifying the initial thickness in the same area in a proper way, such as machining. The amount of thickness reduction of the initial sheet can be calculated by simulating the deformation process instead of the trial and error method, which takes a longer time and has a higher cost.

Unlike incremental-iterative modeling, the initial thickness distribution can be obtained in a single simulation using reverse modeling. This reduced the time of the simulation needed to estimate the initial thickness distribution of the sheet. Figure 7 shows an acceptable agrement between the simulation and experiment of the dome forming of machined sheet at the relative bulge height of $H_r=0.85$, especially around the apex of the bulge.

4.2 Weight reduction and effect of *m* on

estimating initial thickness distribution

The other useful point of the machined sheets is the weight reduction which is very important in the aerospace industry where superplastically formed parts are mostly used. Domes formed from machined and uniform sheets are compared to see how much weight reduction can be obtained from the test pieces

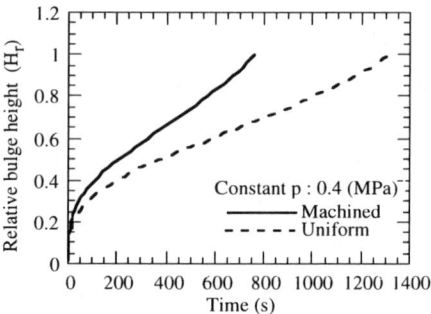

Fig. 5 Relative bulging height vs time curves of machined and uniform sheets deformed under *constant pressure*

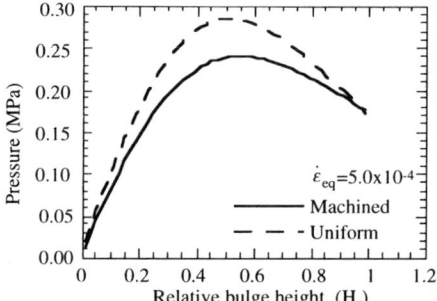

Fig. 6 Comparison of pressure path for machined and uniform sheets deformed under *constant strain rate* deformation

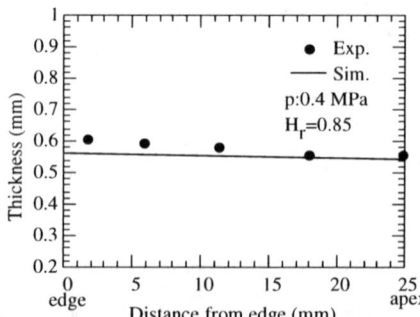

Fig. 7 Comparison of thickness distribution between experiment and simulation of machined sheet

used in the present study. A uniform test piece and machined test pieces with a initial thickness given in Fig. 8 was simulated to dome shape for the comparison of weight reduction. The volumes of the test pieces were calculated as 2349 mm³ for uniform test pieces and 1945 mm³ for machined test pieces after dome forming simulations which revealed a 17.39 % weight reduction. This figure also indicates that higher initial thickness is needed for initially uniform sheets to reach 0.5 mm thickness at the dome apex. Thickness strain is reduced at the apex of the bulge during deformation if the periphery of the initial sheet is machined.

In an early study, Cornfield and Johnson [7], reported that the thinning over the bulge profile is more uniform as m increases. Figure 9 shows the effect of the strain rate sensitivity index, m, on the estimation of initial thickness distribution by present modeling. A higher m value provides a smaller thickness difference between the apex and periphery of the sheet in case of machined sheets. A higher weight reduction will be achieved in the superplastically formed parts produced from the material with low strain rate sensitivity index value.

4.3 Variation of strain rate during bulge forming

Changes in deformation rates along the bulge profile in the machined sheets are of interest, since the initial shape is non uniform. The maximum strain rate can take place at any point of the bulge profile, unlike initially uniform sheets, in which maximum strain rate always occurs at the apex of the bulge. The location of the maximum strain rate is directly influenced by the level of machining in the case of machined sheets. This was confirmed by plotting normalized strain rate vs element number from apex to edge at various relative bulging heights in Fig. 10. This figure shows that in the early stage of deformation, the machined sheet is deformed faster at the vicinity close to the periphery until bulging height reaches a point such as relative bulge height $H_r = 0.4$, and then the deformation rates around the apex becomes higher while the bulging height continues to increase.

Material parameters such as m and K were assumed constant in the strain rate range of $\dot{\varepsilon}_{eq}=1.0 \times 10^{-4}$ and 1.0×10^{-3} s^{-1}. Variation of the maximum and minimum strain rates under *constant pressure* and *constant maximum strain rate* modes should be checked whether the above assumption is satisfied in case of reverse simulated machined sheets. The equivalent strain rate variation curves with the relative bulging

a - Initial thickness of uniform sheet
b - Initial thickness distribution of machined sheet
c - Final thickness distribution of dome
d - Final constant thickness of dome

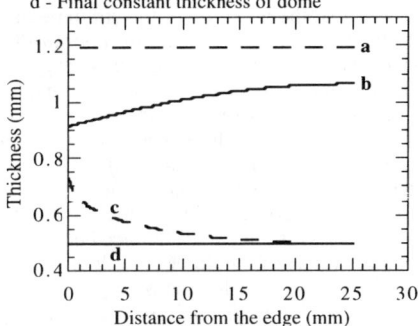

Fig. 8 Comparison of the thickness distribution of the uniform and machined test pieces before and after dome forming (simulation results)

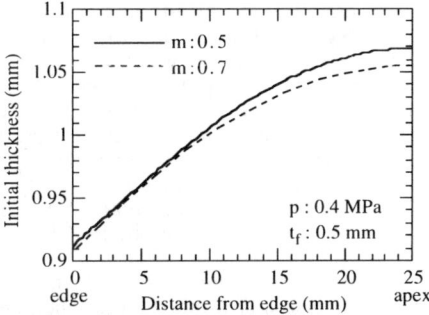

Fig. 9 Effect of m value on the estimation of initial thickness distribution in machined sheet

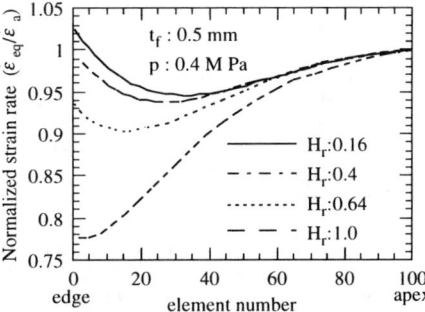

Fig. 10 Distribution of maximum and minimum strain rates along bulge profile at different relative bulge height

height of *constant strain rate* mode and *constant pressure* mode were plotted in Fig. 11 and 12, respectively. A comparison of the two figures reveals that the variation of the strain rate along the bulge profile is less under the *constant strain rate process* than under the *constant pressure* deformation. The difference between maximum and minimum strain rate under the *constant strain rate* mode could be kept within the strain rate range of 3.5×10^{-4} and 5×10^{-4}. Because of the machining, the deformation rates are changed around the periphery of the machined sheets resulting less strain rate difference between apex and other part of the dome. The less strain rate difference is better for the optimization of the material parameters such as m and K values. Thus the deformabilty of the bulged sheet is improved around the periphery.

5. CONCLUSIONS

The deformation process of the superplastic bulge forming of machined sheet was reversely simulated to estimate the initial thickness distribution of sheet from which a final constant thickness distribution can be obtained when the shape is hemispherical shell. The simulated parabolic curve like shape of the sheets were then approximated to four straight lines for the sake of easy machining and then superplastically deformed to a hemispherical shape. The following conclusions are obtained from the present study:

1- The proposed reverse modeling can be successfully used to simulate the superplastic bulge forming process and to estimate the initial thickness distribution of machined sheets which will provide constant final thickness distribution.

2- Simulation results revealed that 17% weight reduction is achieved by machining Al 5083 test pieces which would be very useful for the superplastically produced parts used in the aerospace or transportation industry.

3- Higher weight reduction can be achieved in the superplastically formed parts produced from the material with low strain rate sensitivity index value, because of higher thickness difference between apex and periphery of dome.

4- Machined sheets provides smaller strain rate difference between apex and other part of the dome if compared with the uniform sheets. Thus the difference in m value becomes smaller and the deformabilty around the periphery is improved.

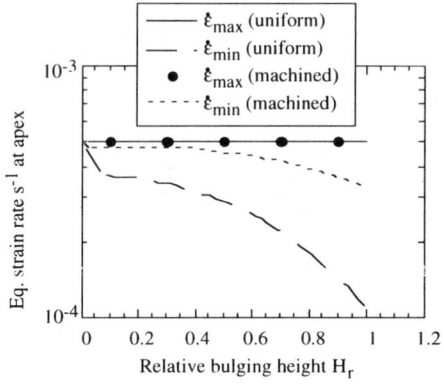

Fig. 11 Strain rate difference between apex and edge under *constant strain rate* process

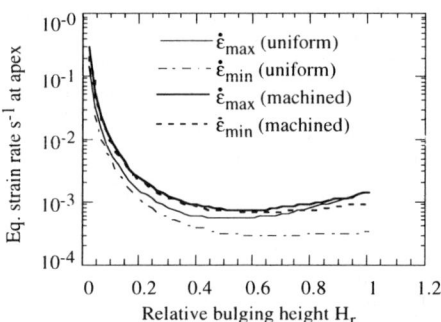

Fig. 12 Strain rate difference between apex and edge under *constant pressure*

References

1. T. Y. M. Al-Naib and J. L. Duncan, Int. J. for Mec. Sci., Vol. 12-6, pp. 463-477 (1970).
2. Johnson, W., Al-Naib, T. Y. M., Duncan, J. L., J. of the Ins. of Met., Vol. 100, pp. 45-50 (1972).
3. A. Takahashi, S. Shimizu, and T. Tsuzuku, Japan Soci. for Tech. of Plasticity, Vol 31, No. 336, pp.1128-1134 (1990),(In Japanese).
4. T. Usugi, N. Akkus, M. Kawahara, H. Nishimura, Materials Science Forum, Vols. 304-306, pp 735-740 (1999)
5. Z. X. Guo, and N. Ridley, Material Science and Engineering, A114, pp. 97-104 (1989).
6. H. Iwasaki, K. Higashi, S. Tanimura, T. Komatubara, and S. Mayami, Superplasticity in Advanced Materials, edited by S. Hori, M. Tokizane, N. Furushiro, pp 447-452 (1991).
7. G. J. Cornfield and R. H. Johnson, Int. J. for Mech. Sci., Vol. 12, pp. 479-490 (1970).

High Strain Rate
Superplasticity

MATERIALS DESIGN AND PROCESSINGS
FOR INDUSTRIAL HIGH-STRAIN-RATE SUPERPLASTIC FORMING

H. HOSOKAWA and K. HIGASHI
Department of Metallurgy and Materials Science, Osaka Prefecture University,
Sakai, Osaka 599-8531, Japan.

ABSTRACT

The optimum materials design in microstructural control could be developed for the high-strain-rate superplastic materials in the industrial scale. In the present work, it is reported that the high-performance-engine pistons with near-net-shape can be fabricated by the superplastic forging technology in the high-strain-rate superplastic PM Al-Si based alloy, which is produced by using this optimum materials design.

INTRODUCTION

The possibility of high-strain-rate superplasticity was first recognized over 10 years ago in a series of experiments conducted on a metal matrix composite consisting of an Al-2124 alloy reinforced with SiC whiskers where high elongations were achieved at strain rates above 10^{-2} s^{-1} [1]. High-strain-rate superplasticity was defined as superplasticity occurring at strain rates at or above 10^{-2} s^{-1} and there have been numerous reports of the occurrence of high-strain-rate superplasticity in a limited range of metal matrix composites, mechanically alloyed materials and alloys fabricated using powder metallurgy (PM) procedures [2-13]. These reports have prompted much interest because of the potential for using the high-strain-rate superplasticity phenomenon to achieve a substantial reduction in the processing time associated with commercial superplastic forming operations. It was noted that ultrafine grain sizes and high-strain-rate superplasticity could be achieved in materials produced by a combination of powder metallurgy and/or advanced processing methods as shown in **Figure 1** [14, 15]. Actually, it was reported [16] that the high-performance-engine pistons with near-net-shape of the PM Al-Si based alloys could be fabricated by using the high-strain-rate superplastic forging technology, as shown in **Figure 2**. That was because the grain size in this material was sufficiently refined.

In the materials having large particles, the particle size refinement as well as the grain size refinement is required to obtain large elongations during superplastic flow, because cavities

Mat. Res. Soc. Symp. Proc. Vol. 601 © 2000 Materials Research Society

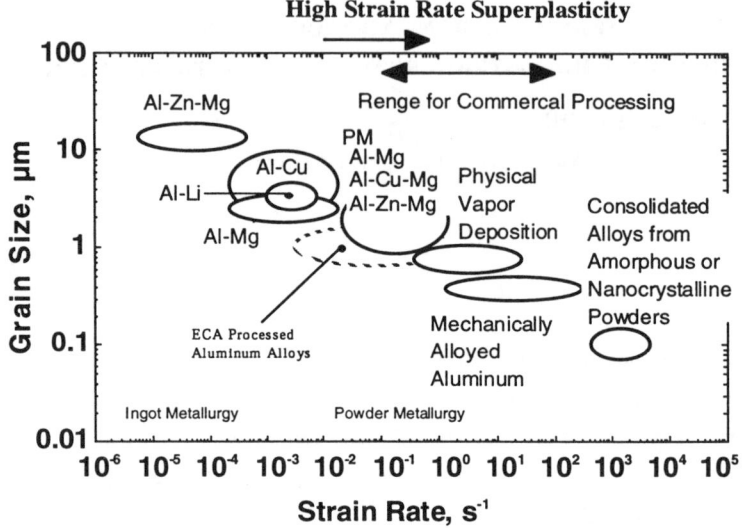

Figure 1   Variation of Optimum superplastic strain rate for several aluminum-based alloys with different grain sizes produced by different processing routes[14,15].

may nucleate from large particles in the materials.   Cavitation could cause the premature fracture. This perennial problem, and the role of cavitation in superplastic materials with particles, are undoubtedly topics that should be considered to achieve enhanced superplasticity, especially at high strain rates.

In the present work, the optimum materials design in microstructural control for high-strain-rate superplasticity is developed in the PM Al-Si system alloy.

Figure 2 The high-performance-engine pistons with near-net-shape of the PM Al-Si based alloy[16].

DETERMINATION OF CHEMICAL COMPOSITION

It is known that Al-Si system alloys have good mechanical properties, owing to addition of Si, Fe, Mg and so on. The variation in the various mechanical properties as a function of the addition of elements, such as Si, Fe and Mg, is shown in **Figure 3, Figure 4** and **Figure 5**, respectively. The increasing Si content improves the mechanical properties in aluminum alloys: the decreasing thermal expansion coefficient, the increasing Young's modulus and the decreasing specific gravity, as shown in Figure 3. The strength of the aluminum alloys increases with Fe addition, but ductility decreases with large amount of more than a few Fe content. Also, the large tensile strength can be obtained in Al-Si-Fe alloys with Mg addition above 1 %. Hence, it is concluded that Al-Si system alloys used in the present work should contain an addition of Si between 15 and 25 %, with about 1% Fe and 2 % Mg.

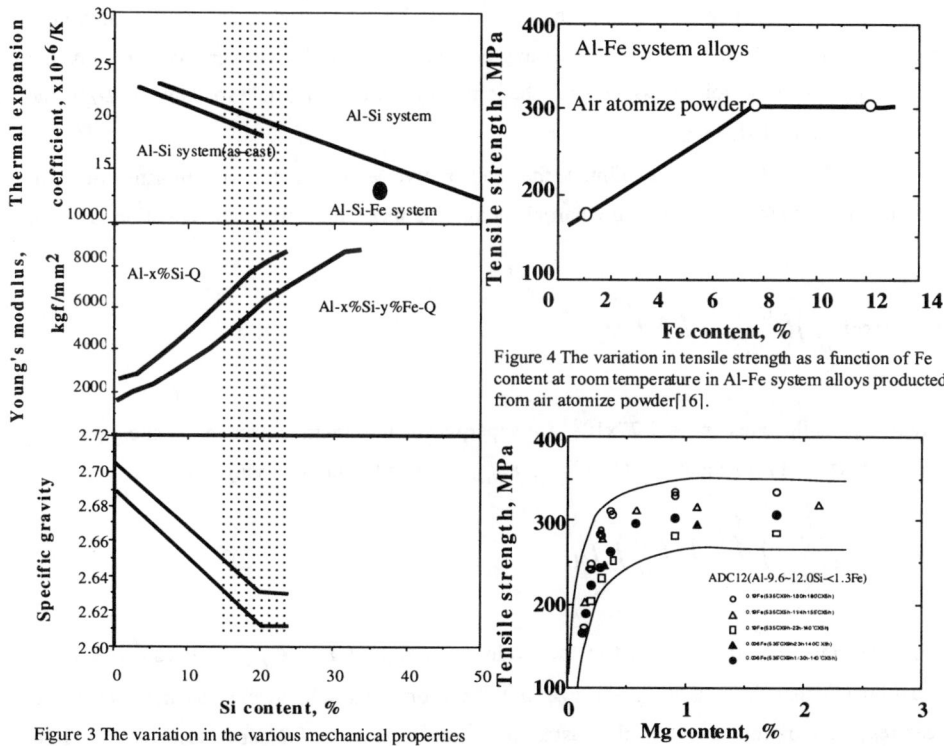

Figure 3 The variation in the various mechanical properties as a function of Si contents[16,17]:
(a) Thermal expansion coefficient, (b) Young's modulus and (c) Specific gravity.

Figure 4 The variation in tensile strength as a function of Fe content at room temperature in Al-Fe system alloys produced from air atomize powder[16].

Figure 5 The variation in tensile strength as a function of Mg content at room temperature in ADS12[17].

DETERMINATION OF MICROSTRUCTURAL CONDITION

The most important requirement for high-strain-rate superplasticity is the refinement of grain size in matrix of the materials, because it is clearly concluded that high-strain-rate superplasticity requires a very small grain size. It is well known that the strain rate for superplastic flow increases with the decreasing grain size. The constitutive equation for creep flow and superplastic flow has been developed in order to estimate the maximum strain rate for superplastic flow in the materials including large particles, such as composites, $\dot{\varepsilon}_{max}$.

The strain rate for aluminum matrix composites exhibiting creep behavior may be given by [18]

$$\dot{\varepsilon} = A_{creep} \left(\frac{\sigma - \sigma_{th}}{aG} \right)^8 \left(\frac{\lambda}{b} \right)^3 D_L$$

(1)

where $\dot{\varepsilon}$ is the strain rate, A_{creep} is the constant (= 2.32×10^{23} for creep flow in aluminum matrix composites), a is the strengthening coefficient, b is the Burgers vector, σ is the stress, σ_{th} is the threshold stress, G is the shear modulus, λ is the particle spacing and D_L is the lattice diffusion coefficient.

Mabuchi and Higashi exhibited the constitutive equation for superplastic flow in aluminum matrix composites below partial melting temperature [19]. The equation is given by

$$\dot{\varepsilon} = A_{SP} \left(\frac{b}{d} \right)^2 \left(\frac{\sigma - \sigma_{th}}{aG} \right)^2 \left(\frac{\lambda}{b} \right)^{1/2} D_L$$

(2)

where A_{SP} is the constant (= 1.77×10^{25} for superplastic aluminum matrix composites) and d is the grain size. From equation (1) and (2), $\dot{\varepsilon}_{max}$ is given by the following equation:

$$\dot{\varepsilon}_{max} = \left(\frac{A_{SP}^4}{A_{Creep}} \right)^{1/3} \left(\frac{b}{\lambda} \right)^{1/3} \left(\frac{b}{d} \right)^{8/3} D_L$$

(3)

Superplasticity region is obtained in the strain rate range of $\dot{\varepsilon} < \dot{\varepsilon}_{max}$. It can be seen in equation (1) that the smaller grain size and the shorter particle spacing as microstructural features are required to obtain the faster maximum strain rate for superplastic flow. It is recognized, however, that grain boundary sliding, which is the dominant deformation mechanisms for superplastic flow, causes stress concentrations at triple points, ledges of grain boundaries and large particles. Therefore, it is required to relax the stress concentrations in order to continue grain boundary sliding without excessive development of cavitation and thereby attain large elongations.

The local tensile stress at the interfaces, σ_i, caused by the stress concentration from sliding, is given by [20,21]

$$\sigma_i = \frac{0.92kTd_p\dot{\varepsilon}dV_f}{\Omega D_L\left(1+5\frac{\delta D_{GB}}{d_pD_L}\right)}$$

(4)

where k is the Boltzmann constant, T is the absolute temperature, Ω is the atomic volume, dp is the particle diameter, δ is the grain boundary width, D_{GB} is the grain boundary diffusion coefficient and V_f is the volume fraction of particles. Therefore when σ_i in equation (4) is equal to σ in equation (2) the critical strain rate, $\dot{\varepsilon}_c$, below which the stress concentrations around reinforcements are sufficiently relaxed, can be defined in the form

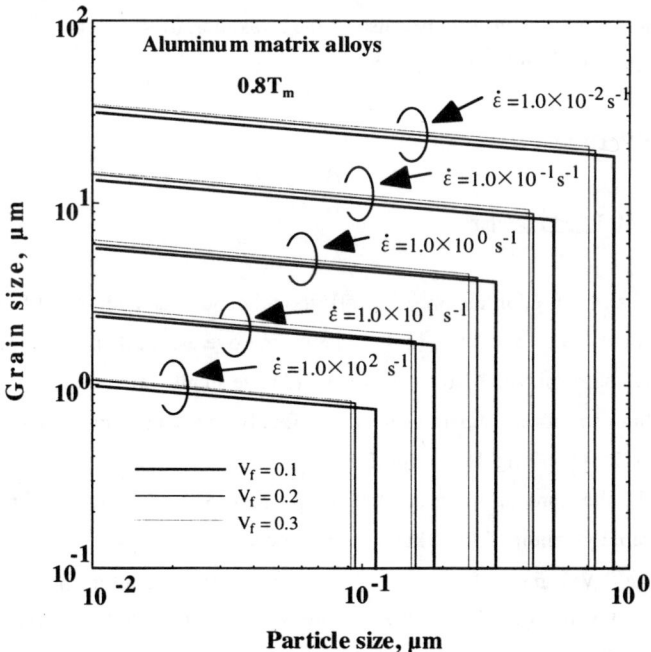

Figure 6   The variation in grain size as a function of particle size at a normalized temperature of 0.8Tm for Al matrix composites[18].

$$\dot{\varepsilon}_c = 8.99x10^{-12}\left(\frac{\Omega G}{kT}\right)^2\frac{(1+5\delta u/d_p)^2}{d_p^{3/2}}\frac{a^2}{V_f^{3/2}}D_L$$

$$(5)$$

where u is the ratio of the diffusion coefficient ($= D_{GB}/D_L$). It is noted that σ_{th} is not taken into consideration in this case because the origin of σ_{th} is not understood. Accommodation mechanisms for superplastic flow in the composites can be approximately estimated using the critical strain rate. The accommodation mechanism is diffusional flow or diffusion-controlled dislocation movement in a strain rate range below the critical strain rate. It is revealed from Eq. (1) and Eq. (2) that the maximum strain rate for superplastic flow and the critical strain rate for relaxation of the stress concentration in a solid state strongly depend on grain size and particle size. The variation in grain size as a function of particle size at a normalized temperature of $0.8T_m$ in the composites is shown in **Figure 6**. Where a value of T_m is taken to be the absolute melting point of pure aluminum. When microstructural features, grain size and particle size, are inside squares formed x, y axis and solid lines, it is possible for the composites to obtain large elongations in a solid state. In this case only superplastic flow can be obtained without cavity nucleation because of the relaxation in stress concentration around large particles even if in a solid state.

MATERIAL PROCESS

Super-Rapidly-Solidified Powder

Al-Si system alloys used in the present work should contain an addition of Si between 15 and 25 %, with about 1% Fe and 2 % Mg. in order to improve the mechanical properties of the high-performance-engine piston. Also the alloy should have small grain size and particle size in order to attain high-strain-rate superplasticity. So finally the composition of the alloy was determined as Al-17Si-2Fe-2Mg-1Cu-1Ni-0.5Mo.

Very recently, the spinning water atomization process (SWAP) in the industrial scale has produced super-rapidly-solidified (SRS) aluminum powders of Al-Si system alloys [33]. The cooling rate in the SWAP method is above 10^4 K/s. Therefore, the ultra-fine microstructures can be obtained in this method. SEM microstructures of the Al-17Si-2Fe-2Mg-1Cu-1Ni-0.5Mo alloy are shown in **Figure 7**. The microstructure produced by the SWAP contains Si particles less than 1 μm and homogeneous distribution of Si particles, as observed in Figure 7(a). On the contrary, the microstructure produced by the air atomization technique contains Si particles more than 5 μm and inhomogeneous distribution of Si particles, as shown in Figure 7(b).

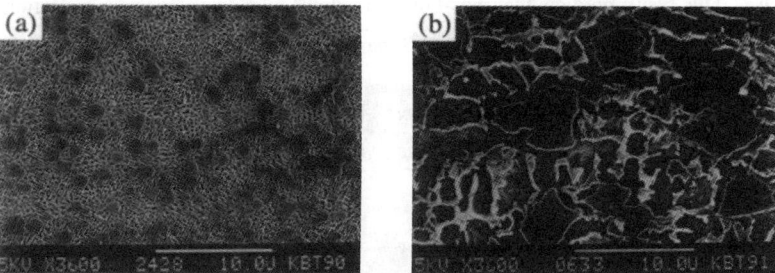

Figure 7 SEM microstructures of the Al-Si-Fe-Mg-Cu-Ni-Mo alloy: (a)by the spinning water atomization process (SWAP) and (b) by air atomization techniques[23]

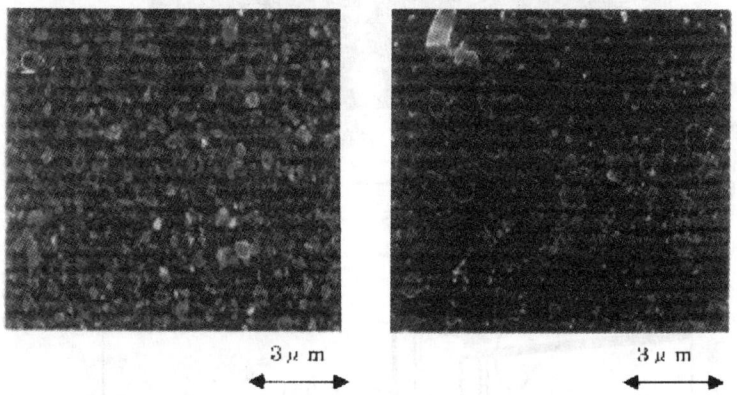

Figure 8   SEM photomicrographs after preforming:(a) discharge plasma forming at 703 K and (b) extrusion forming at 753 K[23].

Powder Sintering Process

It is important to determine the sintering method because the different sintering process may cause the differences in microstructure.   Microstructures after sintering by discharge plasma forming and by extruded forming in Al-17Si-2Fe-2Mg-1Cu-1Ni-O alloy are shown in **Figure 8**.   Si particles is less than 1 μm after sintering by discharge plasma forming and homogeneous distribution of Si particles are observed in Figure 8(a).   On the contrary, Si particles are more than 5 μm after sintering by extruded forming and the distribution of Si particles is inhomogeneous, as shown in Figure 8(b).   Hence, it would be better for refinement in microstructural features to produce by the discharge plasma forming. A TEM microstructure after sintering by discharge plasma forming consists of very fine recrystallized grains (gray or white color) with a mean size of 1.4 μm, as shown in **Figure 9**.   Comparison between the microstructural features of the alloy and theoretical values at a temperature of 793 K is shown in **Figure 10**.   This result predicts that the alloy can exhibit high-strain-rate superplasticity at 793 K without large stress concentration around Si particles.

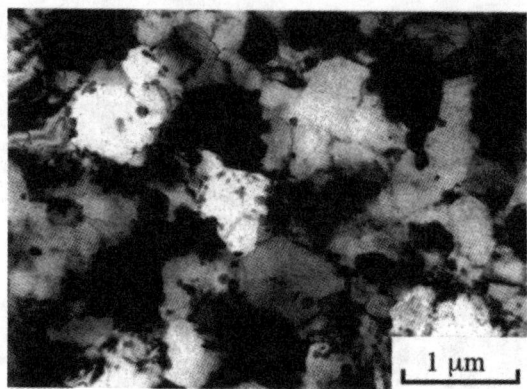

Figure 9   A typical TEM microstructure of the Al-Si-Fe-Mg-Cu-Ni-Mo alloy[24].

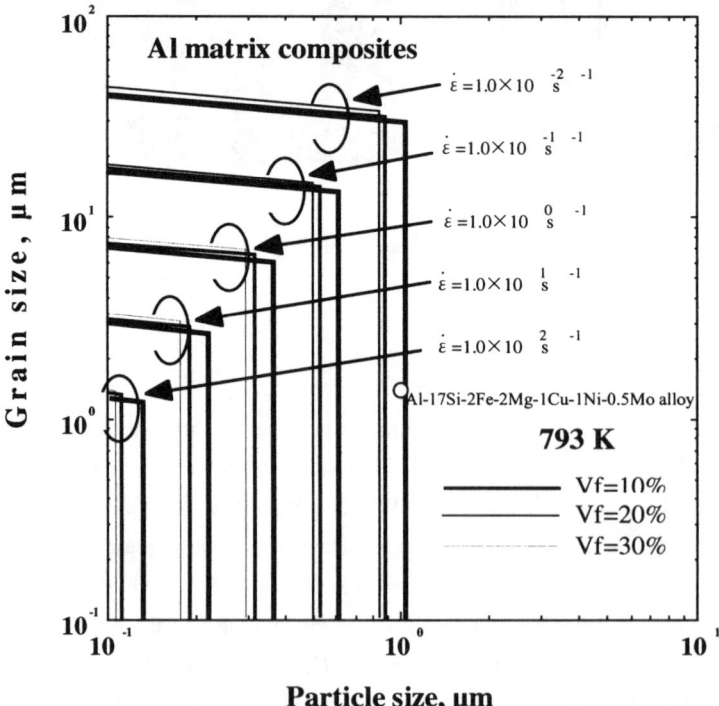

Figure 10   The variation in grain size as a function of particle size at a temperature of 793 K for Al matrix composites, where the open symbol is the microstructure features in Al-17Si-2Fe-2Mg-1Cu-1Ni-0.5Mo alloy.

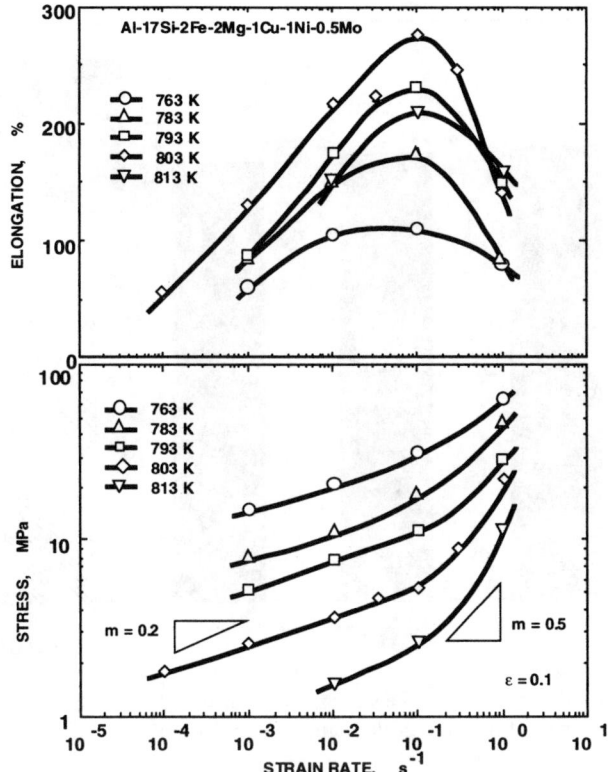

Figure 11 The variation in elongation (top) and flow stress (bottom)
for the Al-Si-Fe-Mg-Cu-Ni-Mo alloy as a function of strain rate[24].

Final Forming

The variation in elongation (top) and flow stress (bottom) for the Al-17Si-2Fe-2Mg-1Cu-1Ni-0.5Mo alloy as a function of strain rate is shown in **Figure 11** for the testing temperature ranging from 763 to 813 K. Large elongations above 200 % are obtained at 1×10^{-1} s^{-1} from 793 to 813 K. Therefore, final forming condition of 1×10^{-1} s^{-1} and 793 K was determined to produce the high-performance-engine pistons.

POST DEFORMATION TENSILE PROPERTIES

Post deformation tensile properties were investigated in each parts of the high-performance-engine piston produced at 793 K and 1×10^{-1} s^{-1}. As shown in **Figure 12**, tensile strength of each part is almost constant and better than that of IM Al-Si alloy.

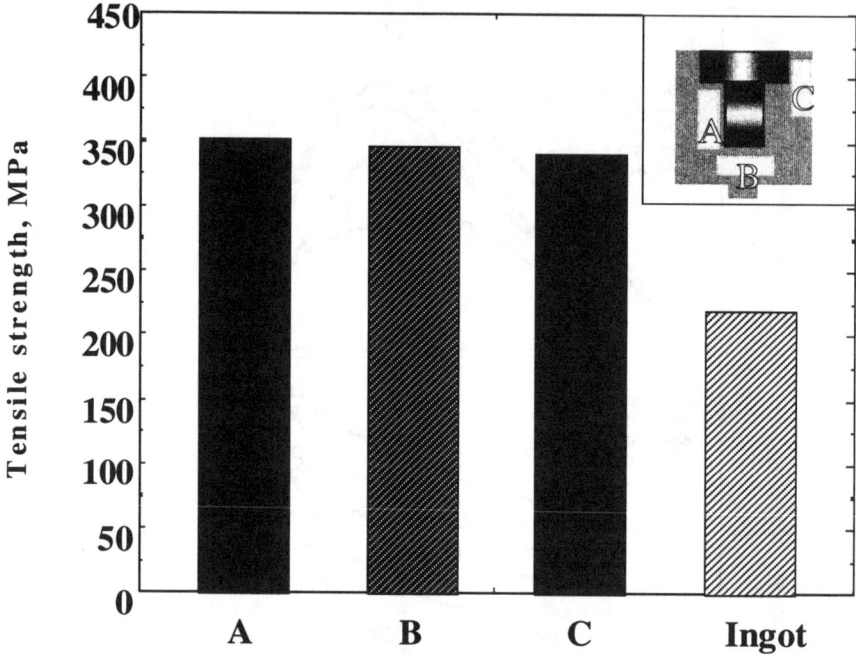

Figure 12   Post deformation tensile strength fablicated at 793 K and 1×10^{-1} s^{-1} in each part of the high-performance-engine piston.

CONCLUSION

In the present work, the optimum materials design in microstructural was discussed to obtain the high-strain-rate superplasticity for powder metallurgical processed Al-17Si-2Fe-2Mg-1Cu-1Ni-0.5Mo alloy. The observed microstructural features, grain size of the matrix and particle size in the processed Al-17Si-2Fe-2Mg-1Cu-1Ni-0.5Mo alloy were in good agreement with the predicted microstructural features in the current investigation. Post deformation tensile strength in each part of the high-performance-engine piston is almost constant and stronger than that of IM Al-Si alloy.

ACKNOWLEDGMENT

The authors express their thanks to the financial support of the Priority Area "Innovation in Superplasticity" from the Ministry of Education, Science, Culture and Sports, Japan.

REFERENCES

1. T.G.Nieh, P.S.Gilman and J.Wadsworth, *Scripta metall.*, **19**, p. 1375 (1985).

2. T.R.Bieler, T.G.Nieh, J.Wadsworth and A.K.Mukherjee, *Scripta metall.*, **22**, p. 81 (1988).

3. M.Mabuchi and K.Higashi, in *Key Eng. Mater.*, edited by G.M.Newaz, H.Neber-Aeschbacher and F.H.Wöhlbier (Trans Tech Publications, Zürich, Switzerland, 1995), vol. 104-107, p. 225.

4. M.Mabuchi, K.Higashi and T.G.Langdon, *Acta metall. mater.*, **42**, p. 1739 (1994).

5. M.Mabuchi and K.Higashi, *Mater. Sci. Eng.*, **A179/A180**, p. 625 (1994).

6. M.Mabuchi and K.Higashi, *Mater. Trans. JIM*, **35**, p. 399 (1994).

7. M.Mabuchi and K.Higashi, *Mater. Trans. JIM,* **36**, p. 420 (1995).

8. K.Higashi, T.Okada, T.Mukai, S.Tanimura, T.G.Nieh and J.Wadsworth, *Scripta metall.*, **26**, p. 185 (1992).

9. K.Higashi, T.Okada, T.Mukai and S.Tanimura, *Scripta metall.*, **25**, p. 2053 (1991).

10. K.Higashi, T.Okada, T.Mukai and S.Tanimura, *Scripta metall.*, **26**, p. 761 (1992).

11. K.Higashi, T.Okada, T.Mukai and S.Tanimura, *Mater. Sci. Eng.*, **A159**, p. L1 (1992).

12. K.Matsuki, G.Staniek, H.Nakagawa and M.Tokizawa, *Z. Metallkde.*, **79**, p. 231 (1988).

13. N.Furushiro, S.Hori and Y.Miyake, in *Superplasticity in Advanced Materials*, edited by S.Hori, M.Tokizane and N.Furushiro (The Japan Society for Research on Superplasticity, Tokyo, 1991), p. 557.

14. K. Higashi, *Superplasticity, 60 Years after Pearson*, Ed. by N. Ridley, The Institute of Materials, London, England, (1995), p. 93-102.

15. T.G. Langdon, *Mater. Sci. Forum*, **304-306**, p. 13(1999).

16. *Aluminum powder alloys*, Kubota Co.(1998).

17. *Microstructures and properties in aluminum*, J. Inst. Metals. (1998) p. 239 and p. 244.

18. H. Hosokawa and K. Higashi, private communication.

19. M. Mabuchi and K. Higashi, *J. Mater. Res.* **13**, p. 640 (1998).

20. R. Raj and M.F. Ashby, *Metall. Trans.* **2**, , p. 1113 (1971).

21. R. Raj and M.F. Ashby, *Acta Metall.* **23**, p. 653 (1975).

22. M. Yoshino, T. Aoki, F. Kasai and K. Kobiki, Kubota Tech. Rept., No.29, p. 7 (1995).

23. *The development of the partial manufacture technic in high strain rate superplasticity on an explanatory meeting* , Kubota Co.(1998)p. 8 and p.12.

24. S. Fujino, N. Kuroishi, M.Yoshino, T. Mukai, Y. Okanda and K. Higashi, *Scripta Mater.*, **37**, p. 673 (1997).

HIGH STRAIN RATE SUPERPLASTICITY IN THREE CONTRASTING FINE GRAINED ALUMINIUM ALLOYS

R.I. TODD[*], J.S. KIM[**], G.H. ZAHID[**], P.B. PRANGNELL[**]
[*]University of Oxford, Department of Materials, Parks Road, Oxford, OX1 3PH, UK.
[**] Manchester Materials Science Centre, Grosvenor Street, Manchester, M1 7HS, UK.

ABSTRACT

The superplastic properties and microstructures of three contrasting fine grained aluminium alloys were investigated. These included (i) a powder metallurgy MMC, (ii) a severely deformed spray cast alloy, and (iii) the Zn-22%Al eutectoid alloy. The results showed some differences in the details of behaviour between the alloys. One of these was the presence of true work hardening, associated with dislocation activity, in the MMC, and its absence in the microduplex Zn-Al eutectoid alloy. In addition, the powder route MMC had the high threshold stress (up to 10MPa) commonly encountered in such materials, whereas this was not the case with the cast and severely deformed alloy, indicating that the threshold stress was associated with the presence of ceramic particles in the MMC (oxide + reinforcement). In all the alloys, however, the m value tended to increase with temperature, and this led to a corresponding increase in elongation with temperature until the microstructure became destabilised. In the two predominantly single phase alloys studied this destabilisation corresponded to the solvus temperature. The constantly changing m value indicates that there is no unique value of this parameter. The same was found to be true of the grain size exponent, p. The direct interpretation of apparent activation energies in terms of simple physical processes should be made with care in the light of the present results, as (i) the microstructures of two of the alloys were found to change continuously and significantly with small changes in temperature, and (ii) the fact that m is a function of temperature necessarily implies that the apparent activation energy is a function of stress.

INTRODUCTION

One of the disadvantages of superplastic forming as an industrial process is that it is slow compared to conventional sheet forming, with each forming operation typically taking from a few minutes to tens of minutes, depending on the complexity of the part and the material used. Inspection of the usual constitutive equation,

$$\dot{\varepsilon} \propto \frac{\sigma^n}{L^p} e^{-\frac{Q}{RT}} \qquad (1)$$

describing the relationship between the strain rate, $\dot{\varepsilon}$, and the applied stress, σ, grain size, L and temperature, T, shows that the strain rate of the material can be increased by (i) increasing the stress, (ii) increasing the temperature, and (iii) reducing the grain size. Methods (i) and (ii) are limited in scope, however, because they generally lead to either a different deformation mechanism with an increased value of stress exponent, n, and therefore a reduction in the stability of tensile elongation, or in the case of (ii) to microstructural instability, or ultimately melting. Grain size refinement does not suffer from such drawbacks, and recent interest in increasing the strain rate of superplastic deformation has concentrated on alloys with fine grain sizes.

273

The main difficulties in achieving superplastic flow at fine grain sizes are: (a) the production of a small grain size in the first place, and (b) the stabilisation of the microstructure against subsequent grain growth. This paper compares the superplastic characteristics of three aluminium alloys with fine grain sizes which overcome these problems in different ways, with the aims of examining whether there are any 'new' phenomena in such materials (i.e. which are not in accord with the classical eq. 1), and, if so, of identifying their origin. The materials include: a powder metallurgy MMC, in which the fine grain size was produced by rolling and recrystallising and was stabilised by fine oxide particles; a cast alloy in which the fine grain size was produced by severe plastic deformation and stabilised by intermetallic particles; and the Zn-22%Al eutectoid alloy, in the which the fine grain size was produced by heat treatment and stabilised by virtue of its microduplex structure.

EXPERIMENTAL

Metal Matrix Composite. A 2124Al-18% 3μm SiC$_p$ MMC was supplied by AMC Ltd. as a forged billet made by a powder metallurgy route. The material was hot rolled (530°C) to a reduction of 40%, and then warm rolled at 350°C to a further reduction of 45%. On heating to 450°C a fine equiaxed grain structure was developed, and this was stable against static annealing at all the deformation temperatures used (470°C - 510°C). Microstructures and grain sizes were measured using SEM of etched specimens and TEM. Tensile specimens were stamped from the rolled sheet and tested at constant true strain rate in air using an Instron machine.

Spray Cast and Severely Deformed 7010 Alloy. A modified 7010 alloy was supplied as a spray cast billet. The alloy composition was Al-5.7Zn, 2.2Mg, 1.6Cu, 0.5Ti, 0.4Zr. The Ti and Zr additions were intended to provide intermetallic particles to stabilise the grain size. Cylinders of the billet were solution treated and aged in order to develop additional precipitates to hinder grain growth. At this stage the grain size was ~25μm. The cylinders were equi-channel angular pressed (ECAP) at 300°C in a die with a 120°C angle. Specimens with 5, 10 and 15 passes were produced, with equivalent strains of 3.4, 6.7 and 10.1. The specimens were then annealed at 450°C. The microstructural development of the material during ECA pressing was investigated by high resolution electron back scattered diffraction (EBSD). Tensile specimens were machined from the ECAP cylinders and tested in a similar manner to the MMC specimens.

Zn-22%Al Eutectoid Alloy. High purity Zn and Al were melted in Argon at 660°C and chill cast into a copper mould. The ingot was homogenised for 48h at 400°C and rolled at 250°C. Tensile specimens were punched from the strip and annealed at 405°C for 24h, after which they were quenched into an ice/water mixture and left for 3 hours. The specimens were then annealed for 90mins at 20°C before transfer to a freezer for storage. Grain sizes of 0.3-3.5μm were obtained by annealing. The tensile specimens were tested using both the constant true strain rate technique used for the other specimens, and a constant true stress rig for lower stresses. The initial elongation before taking measurements was 25% in all cases, and a strain of 5% was allowed for each subsequent strain rate used. The initial strain rate was repeated at the end of the test series in all cases to check that significant grain growth had not occurred. Conditions in which there was grain growth are not reported here.

RESULTS

Metal Matrix Composite. The results from this alloy have been described in detail previously [1], and will therefore only be summarised here. After processing most of the matrix had a grain size of 2-4µm. Areas which were deficient in SiC reinforcement tended to have larger grains and this was attributed to the consequent reduction in both the inhomogeneous plastic deformation around the particles during rolling and the particle stimulated nucleation of new grains during recrystallisation. The grains were stable with respect to growth during static annealing at all the superplastic testing temperatures used.

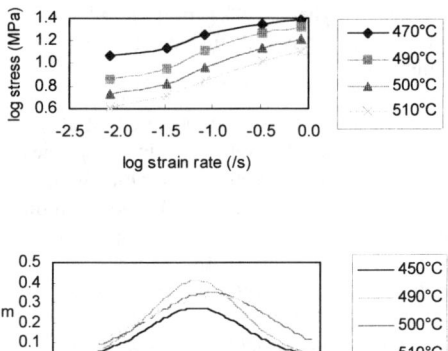

Figure 1. Flow properties of the MMC at various temperatures

Strong work hardening was observed under all conditions, with the flow stress increasing by about a factor of 2 during straining. Measurement of the grain size after deformation showed that there was insufficient growth to explain the work hardening in terms of dynamic grain growth, and a careful analysis of the testing method showed that the hardening was not an artefact of the nominally true strain rate test method. One possible source of the work hardening is the involvement of lattice dislocations in the deformation mechanism. The dislocation density was observed to increase during deformation at all strain rates used, and particularly so in larger grains.

The relationship between stress and strain rate for the MMC is shown in fig. 1, along with the corresponding m value as a function of strain rate. It is evident that whilst the curves have the classic sigmoidal shape, region II is limited in extent, largely because of the occurrence of Region I at relatively high stresses (up to 10MPa). This behaviour has been reported in a number of similar materials [2,3]. Figure 1 shows that there is a distinct optimum temperature as far as the m value is concerned of 490°C. The elongation showed the same trends as the m value. The maximum elongation (430%) was achieved at 490°C with a strain rate of $8.3 \times 10^{-2} s^{-1}$.

The stress-strain rate curves in fig. 1 show that a true activation energy is difficult to define in this material, because changes in temperature appear to lead more to a change in flow stress than to a change in strain rate. One reason for this may be that Region I is the result of a temperature-dependent threshold stress, and it has been suggested [2,3] that meaningful values of activation energy can be obtained by replacing the stress, σ, in eq. 1 by the effective stress, $\sigma - \sigma_0$, and performing the usual activation energy plot at constant effective stress. This procedure has been carried out previously [1], and showed that there was a discontinuity in the apparent activation energy at the optimum temperature, 490°C. At temperatures below this value, the apparent activation energy was 72kJ mol^{-1}, which is rationalisable in terms of the grain boundary diffusion coefficient of aluminium. At higher temperatures, however, the activation energy was much higher (298kJ mol^{-1}), which is higher than any diffusion process in aluminium.

The previous study [1] looked in detail at the reason for the discontinuity in properties which occurred at 490°C. Careful thermal analysis and microstructural observation showed that this temperature corresponded to the solvus temperature of the material, and it was observed that approximately 70% of the intermetallic particles in the material dissolved as the temperature was increased from 490°C to 510°C. Melting of the matrix was ruled out. The reduction in elongation above the solvus was attributed to the acceleration of Region III, the lattice dislocation dominated part of the flow curve, which is inimical to superplasticity. The change in activation energy is clearly an apparent one and must be attributed to the fact that the microstructure of the material was changing rather than to a change in rate controlling mechanism.

Spray Cast and Severely Deformed 7010 Alloy. Figure 2 shows the microstructure after 5, 10 and 15 passes through the equichannel die + annealing, in the form of EBSD orientation maps in which the tone of a grain represents its orientation. After 5 passes, the grain size estimated from the orientation map is approximately 3μm, but most of the grain boundaries are low angle boundaries, as shown by the fact that neighbouring grains are of similar shade. After 10 passes, the grain size has

5 passes ECAP
L *(MLI)* = 3.0μm
16% high angle gbs

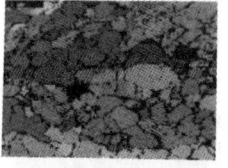

10 passes ECAP
L *(MLI)* = 2.1μm
24% high angle gbs

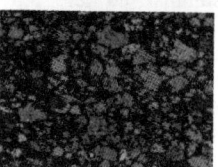

15 passes ECAP
L *(MLI)* = 2.1μm
31% high angle gbs

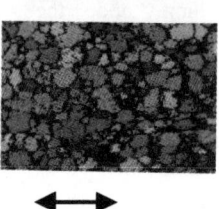

⟷
15μ

Figure 2. EBSD orientation maps showing grain structure after ECAP + annealing

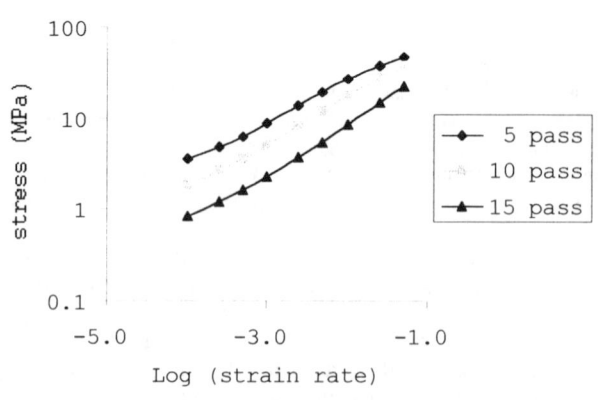

Figure 3. Stress-Strain Rate Plot for Material ECA Pressed to Different Deformations. T = 450°C

reduced to 2.1μm, and there are more high angle boundaries. After 15 passes the grain size was still 2.1μm, but there was a further increase in the fraction of high angle boundaries.

The influence of fractio n of high angle boundaries on superplastic properties is shown in fig. 3. As the number of ECAP passes and the proportion of high angle grain boundaries increases the flow stress decreases, and the m value increases. It is also evident from these curves that the high threshold stress observed in the MMC is not present in this material, with deformation continuing at stresses as low as 1MPa. The stress strain curves in figure 4 show the effect on flow

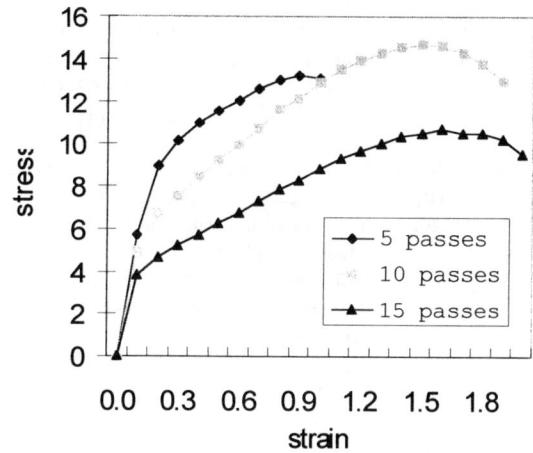

Figure 4. Stress Strain Curves as a Function of Number of ECAP Passes. $T = 450°C$, **Strain Rate = $5 \times 10^{-3} \text{s}^{-1}$**

stress in more detail. Although deforming the material more severely reduces the flow stress at a given strain, it does not prevent work hardening from occurring, with the flow stress rising by more than a factor of 2 during deformation for the 15 pass specimen.

The influence of testing temperature on superplastic deformation is shown in fig. 5. It is evident that the strain rate at a given stress increases as the temperature rises, as is predicted by eq. 1. It is also evident, however, that m ($=1/n$ in eq. 1) is not constant, as is often assumed, but gradually increases with temperature. At 500°C, high m values are observed, and m remains above 0.5 at strain rates well in excess of 10^{-2}s^{-1}. The elongation, however, did not follow the same pattern as the m value, with the maximum elongation of 830% occurring at 450°C with a strain rate of

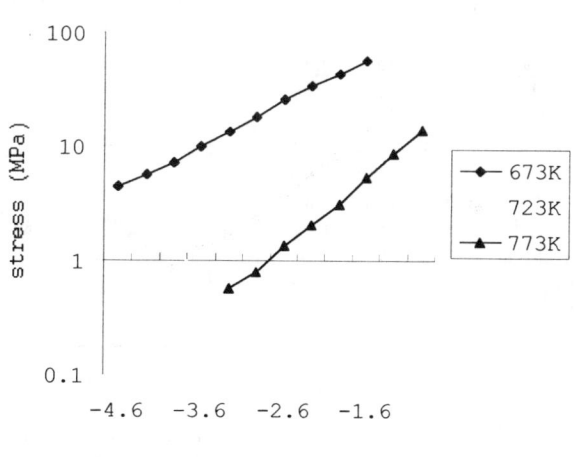

Figure 5. Influence of Temperature on Stress-Strain Rate Curve. 15 ECAP Passes.

$8.3 \times 10^{-2}\text{s}^{-1}$. Although this strain rate is not as high as many that have been reported, and is no higher than can be achieved with SUPRAL dynamically recrystallising alloys, it is certainly significantly higher than can be achieved with 'conventionally' produced SP grade 7000 series alloys. This is presumably due to the fact that conventional alloys have grain sizes between 10 and 15µm, which is more than 5 times larger than the present ECAP alloy.

As with the MMC, there was a discontinuity in the apparent activation energy at the optimum temperature (fig. 6), although this time the activation energy showed the higher value of 293kJ mol^{-1} at *low* temperatures, and the value at high temperature (144kJ mol^{-1}) can be rationalised in terms of lattice diffusion in aluminium. In common with the MMC, however, this optimum temperature corresponded to the *solvus* temperature for many of the precipitates, with copper rich precipitates being present at 450°C, and absent at 500°C. This is consistent with the published solvus for the alloy of 475°C.

The reason that exceeding the solvus temperature reduces the elongation is not clear in this case, as we have not yet measured the grain size after deformation. It seems plausible, however, that in this cast alloy, which in contrast to the MMC does not contain an oxide dispersion to stabilise the grain size, grain growth is likely to limit the elongation attainable if a significant proportion of the precipitates dissolve.

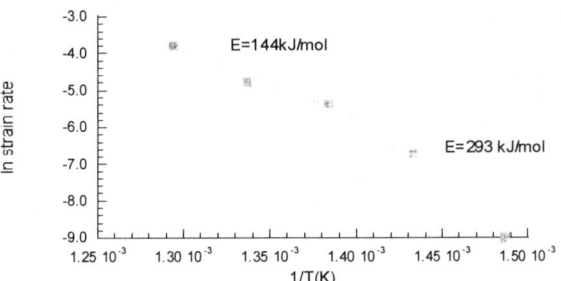

Figure 6. Activation Energy Plot for ECAP Material. Stress = 10MPa

Zn-22%Al Eutectoid Alloy.
The only difference between the conventional processing route for this alloy and that used here is that after quenching from above the eutectoid temperature into iced water to develop a microduplex structure, some relatively mild annealing treatments were used in order to retain a finer grain size (0.3-2µm) than normal (≥2µm). The fine grain sizes enabled the use of an unusually wide range of testing temperatures (60-200°C) at practicable strain rates.

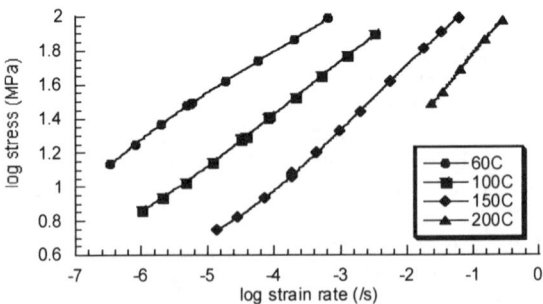

Figure 7. Stress – Strain Rate Plots for Various Temperatures. Zn-22%Al, Grain Size = 1µm

Figure 7 shows the stress as a function of strain rate for material with a grain size of 1µm tested at four different temperatures. The behaviour resembles that of the ECAP alloy, in that whilst the main effect of increasing the temperature was to increase the strain rate, the m value is also a monotonically increasing function of temperature. The stress-strain curves for this alloy did not exhibit the work hardening observed in the alloys consisting predominantly of a single phase matrix, and so exhibited a true steady state. Careful examination of the results enabled grain growth and the effect of the temperature dependent threshold stress, which is known to exist in this alloy [4], to be ruled out as explanations for the reduction in m at low temperatures. The range of conditions covered by the data in fig. 7 makes it clear that the reduction in m is not a consequence of a fundamental change of mechanism; m changes gradually with temperature, and little variation is seen with stress or strain rate.

Whilst m depends on temperature, it does not depend on grain size. Figure 8 shows that m is constant for grain sizes varying by more than a factor of 3. The grain size exponent, p in eq. 1, was, like m, a function of temperature, however, as shown in fig. 9. The gradual change in both the stress and grain size exponents with temperature are evident in this figure, and the extent of the changes in $n=1/m$ and p are similar, so that $n+p$ is approximately constant, with a value of ~4.

DISCUSSION

Comparison of the similarities and differences of the three alloys investigated gives clues at a phenomenological level as to the nature of the main features of superplastic deformation. Starting with the differences in behaviour, the fact that only the microduplex Zn-Al alloy was free from work hardening, for example, suggests that there is a fundamental difference between superplasticity in

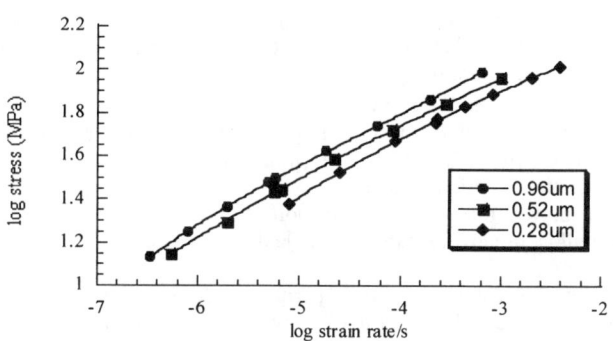

Figure 8. Influence of Grain Size on m. $T = 60°C$.

microduplex structures and in mainly single phase aluminium alloys (although grain growth cannot yet be unequivocally ruled out as an explanation for the work hardening in the ECAP alloy). The striking difference in behaviour at low stresses between the powder route MMC and the cast ECAP alloy strongly suggests that the strong threshold stress in the former is a consequence of the ceramic particles present in the MMC, which include both fine oxide particles from the original surface of the powder used to make the matrix, and the SiC reinforcement.

The main similarity between the three alloys concerns the m value, the single most important parameter in superplastic deformation. In all three alloys the m value increased as the temperature was raised from the lowest level at which superplasticity could be reasonably be observed. This seems to be a general feature of the mechanism of superplastic deformation, and the fact that it has not been emphasised previously is probably because grain sizes have previously been too coarse to observe a sufficiently wide range of temperatures. In the Zn-Al eutectoid alloy, for example, most previous studies have used grain sizes in excess of 2μm, and the mechanical properties have been measured at temperatures of 140°C-200°C, over which range the variation in m value is small (figs. 7 and 9). The use of submicron grain sizes, however, enables the deformation to be studies at temperatures as low as 60°C, however, and this makes clear the variation in m (and in p).

The dependence of m on temperature is important from both a fundamental point of view, and at the level of application. In the former regard, the fact that there is no special value of m (e.g. 0.5) is at odds with most simple models for superplastic flow, and demonstrates that the phenomenon is more complex than has been suggested. As far applications of superplasticity are concerned, the results show that whilst refining the grain size certainly does lead to the possibility of higher strain rates, there may be a limit to how much grain size

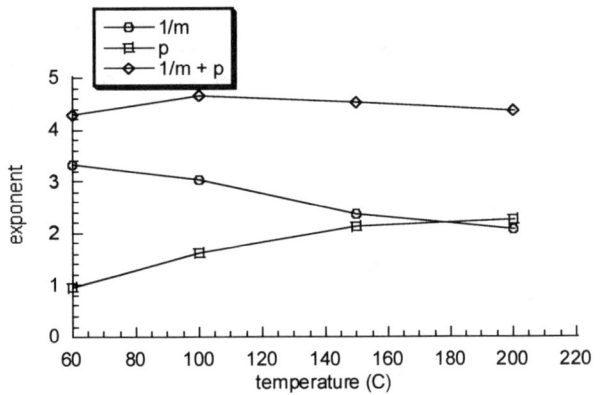

Figure 9. The Variation of the Stress and Grain Size Exponents with Temperature

refinement will allow the deformation temperature to be reduced, because eventually the m value will be too low to stabilise tensile deformation.

Although the details of the superplastic deformation mechanism may be complex, the phenomenological description of how large elongations may achieved is relatively simple according to these results. To get maximum elongation at maximum strain rate, it is clearly important to have a fine grain size. Then, as m increases with temperature, the elongation (and strain rate) will tend to increase as the temperature is increased. The optimum condition comes at the temperature above which the microstructure is destabilised in some way. In the Zn-Al alloy studied here, for example, destabilisation would occur above the eutectoid temperature, when the microstructure would become single phase and rapid grain growth would ensue. Ultimately, the destabilisation may correspond simply to the gross melting of the alloy, and this certainly seems to be true of some high strain rate compositions which have been studied by Higashi and co-workers [5]. In the two predominantly single phase alloys studied here, however, the destabilisation corresponded to the dissolution of precipitates above the solvus temperature. The details of how exceeding the solvus affects superplastic flow are complex, and may differ between different alloys. In the MMC studied here, for example, the main effect seemed to be the acceleration of Region III because of the reduction in the number of precipitates hindering dislocation motion. This was not the case for the ECAP material, however, as deformation was possible well away from the influence of Region III, and indeed the effect on the apparent activation energy was different to that observed in the MMC. It is reasonable to speculate that the dominant effect in the case of the ECAP alloy was grain growth caused by the unpinning of the grain boundaries. In general, other effects may be anticipated as well, as the consequence of dissolving the particles is not simply to remove their direct effect on dislocation motion, grain boundary sliding and grain growth, but also to change the composition of the matrix as the constituent elements of the particles dissolve.

This discussion also makes clear the danger in the direct interpretation of the apparent activation energy in terms of simple physical processes. This can only be attempted when the microstructure does not vary with temperature, and it is clear that this is far from the truth in many of the complex aluminium alloys being investigated at present. It should be noted that even below the solvus temperature, the volume fraction and composition of precipitates can be expected to vary. Furthermore, even when the variation is slight, as with the microduplex Zn-Al eutectoid alloy studied here, the fact that m is a function of temperature necessarily implies that the apparent activation energy is a function of stress (even when the temperature dependent threshold stress is allowed for).

CONCLUSIONS

1. Superplastic deformation at higher than usual strain rates has been compared in three contrasting aluminium alloys with fine grain sizes. These were a powder route MMC, a severely deformed 7010 alloy, and the Zn-Al eutectoid alloy.

2. Only the MMC showed a high threshold stress, indicating that this was associated with the presence of ceramic particles in the material.

3. The m value increased with temperature, and at the same time the grain size exponent decreased where measured (Zn-Al eutectoid). There is no unique value of m.

4. The use of fine grain sizes is effective in increasing the strain rates which can be achieved, but the reduction in m value may limit the extent to which it can lead to lower forming temperatures.

5. Because m increases with temperature, the optimum temperature often represents the point above which the microstructure is severely destabilised. In the two predominantly single phase alloys studied here this point corresponded to the solvus temperature.

6. Apparent activation energies need to be interpreted with caution because (i) in many alloys of current interest the microstructure (e.g. the presence of precipitates) changes significantly with temperature, and (ii) because the fact that m is a function of temperature necessarily implies that the apparent activation energy is a function of stress.

REFERENCES

1. G. H. Zahid, R. I. Todd and P. B. Prangnell, *Materials Science Forum* 304-306 (1999) 233
2. M. Mabuchi and K. Higashi, *Scripta Mater.* 34 (1996) 1893
3. R. S. Mishra, T. R. Bieler and A. K. Mukherjee, *Acta Metall. Mater.* 43 (1995) 877
4. P. K. Chaudhury and F. A. Mohamed, *Acta Metall.* 36 (1988) 1099
5. K. Koike, M. Mabuchi and K. Higashi, *Acta Metall. Mater.* 43 (1995) 199

THE EFFECT OF Fe-ADDITION TO Al-10Ti ALLOY ON SUPERPLASTICITY AT HIGH STRAIN RATES

D. KUM*, W. J. KIM**
* Korea Institute of Science and Technology, P.O. Box 131, Seoul 130-650, KOREA,
dkum@kistep.re.kr
** Hong-ik University, 72-1 Mapo-ku, Sangsu-dong, Seoul 121-791, KOREA

ABSTRACT

Ultra-fine microstructure consisting of equiaxed Al-grains and aluminide particulate was produced by powder metallurgy process using gas-atomized powders of Al-10wt%Ti-2wt%Fe alloy. High strain rate superplasticity (HSRS) has been investigated at 873 – 923K and strain rates higher than $10^{-3}s^{-1}$ in tension, and total elongation up to 500% was observed at the strain-rate of $10^{-1}s^{-1}$. The strain rate vs. flow stress behavior exhibits the typical aspect of HSRS such as the increase of strain-rate sensitivity exponent with increase in strain-rate and an apparent activation energy higher than that for lattice diffusion in aluminum. The concept of threshold stress has been incorporated to illustrate the HSRS behavior, where the stress exponent of 3 describes the experimental data. The determined threshold stress showed strong temperature dependency as in the case of a similarly processed Al-10wt%Ti alloy, which exhibited the stress exponent of 2 in the same testing conditions. Solute drag mechanism has been postulated for the Al-Ti-Fe alloy.

INTRODUCTION

Hyper-peritectic Al-Ti alloys could be made into ultra-fine grain microstructure by powder metallurgy (P/M) consolidation of gas atomized powders, and exhibit high strain-rate superplasticity(HSRS) at very high temperatures(e.g., above 550°C)[1,2]. That is because the P/M processed microstructure is similar to that of the powder metallurgy aluminum-matrix composites exhibiting HSRS [3-5]: the high melting point Al$_3$Ti flakes are uniformly distributed in the ultra-fine and equiaxed aluminum grains. Murty et al. [1] firstly reported high strain-rate-superplasticity in a Al-4wt%Ti alloy, and recently Kum et al.[2] observed the similar behavior and analyzed the HSRS behavior by incorporating the concept of threshold stress a generalized form,

$$\dot{\varepsilon} = AD(\frac{b}{d})^p (\frac{\sigma - \sigma_o}{E})^n \qquad (1)$$

where b is the Burgers vector, D the relevant diffusivity, E the Young's modulus, d the grain size, p the grain size exponent, σ the flow stress, σ_o the threshold stress, n the stress exponent and A is a geometrical constant.

The threshold stress was highly temperature dependent in such a way that an Arrhenius-type interpretation could be applied [6]. Kum has extended this concept to reanalyze creep behavior of other fine-grained Al-Ti alloys [7]. The Al-Ti alloys provide a new opportunity to understand the nature and role of threshold stress in the HSRS materials because the issue of incipient melting. It needs to be noted that the dispersion of Al_3Ti reinforcement is accomplished without external mixing step and cohesion between matrix and reinforcement is rather perfect, and thus this material provides an advantage over conventional aluminum matrix composites to understand the HSRS behavior.

When Fe is added into a hyper peritectic Al-Ti alloy, Fe has a very limited solubility in aluminum and alloying content higher than the solubility limit results in high melting point aluminides. $Al_xFe(X = 3$ or $6)$ aluminide forms by an eutectic reaction at 655^oC [8], and thus the Al_xFe phase could be an additional reinforcement to the P/M processed Al-Ti alloys. The presence of additional eutectic Al_xFe particles to the peritectic Al_3Ti flakes is expected to enhance superplasticity by retarding grain growth during deformation at elevated temperatures. The purpose of this study is to investigate the effect of the additional Al_xFe particles on the HSRS behavior, and it was hoped to see any evidence and role of solute drag effect which was previously suggested by the current authors. With this in mind, the HSRS behavior of Al-10wt%Ti-2wt%Fe alloy has been discussed with those of similarly processed Al-10wt%Ti alloy and Al-10wt%Ti + 5v/o SiC_p composite reported elsewhere [9].

EXPERIMENT AND ANALYSIS

Gas atomized powders of Al-10wt%Ti-2wt%Fe (less than 240 μm in diameter) were cold compacted in an aluminum can, vacuum degassed and then hot extruded to a rod of 15mm in diameter at 450^oC. The chemical composition of the alloy analyzed by wet chemical method is listed in Table 1 with other materials compared. The microstructure of the Al-10wt%Ti-2wt%Fe alloy after the extrusion consists of ultra-fine aluminum grains of about 1.2 μm in average and contains two kinds of particles(**Figure 1**). The major ones are the Al_3Ti phase and, being formed by pro-peritectic reaction rapid cooling in liquid, they are mostly in flake-shaped particles of less than 5 μm in length and about 0.2 μm in thickness. The minor ones are finer globular particles of Al_xFe phase formed by eutectic reaction (diameter of 0.2-0.5 μm). The Al_3Ti flakes are mostly dispersed between and across grains and all globular Al_xFe particles are uniformly located inside grains.

The strain rate vs. stress characteristics and the strain-rate-sensitivity exponent, m, were evaluated by the strain-rate-change tests in tension, wherein step changes of cross head speed in increasing order were employed from $10^{-4}s^{-1}$ to $10^{-1}s^{-1}$. Elongation-to-failure tests were carried out under conditions of constant cross-head speeds with values of nominal strain rate of $10^{-3}s^{-1}$ to 10^0s^{-1}.

Table 1. Chemical Compositions

(unit: wt%)

Specimen	Ti	Fe	Si	Cu	C	Al
Al-10Ti-2Fe	9.95	2.01	0.06	<0.01	<0.03	balance
Al-10Ti	9.53	0.12	0.02	0.002	<0.01	balance
Al-10Ti+5v/oSiC$_p$	9.53	0.12	0.02	0.003	<0.01	balance

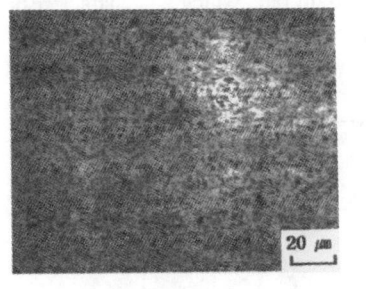 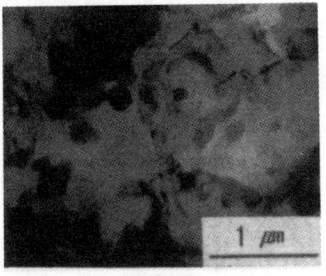

(a)   (b)

Figure 1. Microstructures of the Al-10Ti-2Fe alloy: (a) optical micrograph and (b) transmission electron micrograph.

Threshold stress was determined by the linear extrapolation method used by Lin and Sherby [10] for steady state creep and Mohamed [11] for Region I analysis of superplastic deformation. Bieler and Mukherjee[12] adapted this method to interpret the HSRS behavior of IN90211 alloy. When all data fall upon the same line with a fixed stress exponent(i.e., n), the extrapolation data to zero strain rate on (strain-rate)$^{1/n}$ vs. (stress) graph on linear scale was considered to be the threshold stress for that temperature. If the stress exponent does not represent the exact behavior, data near zero strain rate show a tail of asymptotic approach.

RESULTS

Superplasticity of the Al-10wt%Ti-2wt%Fe is shown in **Figure 2**, where total elongation-to-failure at different testing temperatures is represented as a function of the initial strain rate. Below 500°C, tensile elongation is about 100% or less. At 620°C and 635°C, the material exhibits the typical behavior of high strain rate superplasticity as expected from earlier observation from a similarly processed Al-10wt%Ti alloy: tensile ductility increased with strain rate and the maximum elongation of 600% could be achieved at an extremely high strain rate of $8\times10^{-1}\text{sec}^{-1}$. However, tensile elongation at 650°C drops significantly. This reduction is attributable to the start of incipient melting at 650°C, which was confirmed by differential scanning calorimetry. Comparison of this elongation-to-failure result with the data for the Al-10wt%Ti alloy indicates that the optimum strain rate for superplasticity shifts from 10^{-1}s^{-1} to $8\times10^{-1}\text{s}^{-1}$. This trend of increase in the optimum strain rate with decrease in grain size agrees with that often noted in many superplastic metallic alloys.

The strain rate-stress curves constructed based upon the result of strain rate change tests are shown in **Figure 3**. The stress exponent that can be measured from the slopes of the curves shows the trend to increase with decreasing strain rate, which is a characteristic often noted when a threshold stress exists for plastic flow.

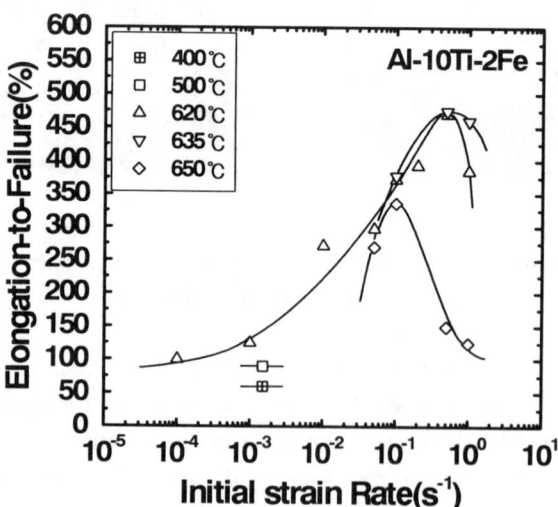

Figure 2. Elongation-to-failure test result given as a function of strain rate at different temperatures, which displays typical behavior of high strain rate superplasticity.

Data at 600°C, 620 °C and 635°C were used for the threshold stress analysis and compared with those of reference material (i.e., Al-10wt%Ti). The best linear fits on the $\dot{\varepsilon}^{1/n}$ vs. σ graph was obtained at n=3, and the threshold stresses were determined by the linear extrapolation to zero strain rate at each temperature. They are listed in Table 2 with those of Al-10wt%Ti alloy [7] and Al-10wt%Ti + 5v/o SiC$_p$ composite [9] for comparison. Note that the stress exponent for Al-10wt%Ti alloy and Al-10wt%Ti + 5v/o SiC$_p$ composite was 2, while it was 3 for the present material. This aspect is further discussed in **Figure 4**.

Table 2. Threshold Stress

(unit: MPa)

Temperature(°C)	600	620	635	640	650
Al-10Ti-2Fe	2.40	2.02	0.34	-	-
Al-10Ti	3.6	2.5	-	1.5	1.0
Al-10Ti+5v/oSiC$_p$	4.85	2.65	-	1.76	0.8

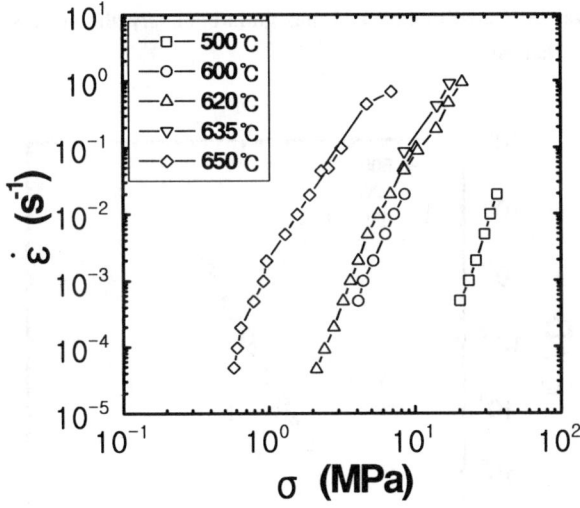

Figure 3. Strain rate-stress relationship at different temperatures.

When 2wt%Fe is added, threshold stress has decreased more than 20%. When microstructural difference between current specimen and the Al-10wt%Ti alloy is considered, the reduction of grain size could be postulated. Grain sizes for the present alloy and the Al-10wt%Ti alloy were about 1.2 μm and about 2.6 μm, respectively. If this hypothesis is correct, threshold stress for HSRS needs to be depending upon the grain size and thus it could be possible to bring new insight for the nature of threshold stress in HSRS deformation. Recently, Kim and Kum [14] found that when the grain size effect on threshold stress is considered with the solidus temperature, the threshold-stress data for many superplastic aluminum alloys including the Al-10Ti-xFe alloys with different grain size, chemical composition and nature of dispersoids are well correlated by a single relation. According to them, the grain size effect on threshold stress is represented by the following equation

$$\frac{\sigma_o}{E} \propto d^{1.1} \qquad (2)$$

The relation between temperature-compensated strain rate and modulus-compensated stress is shown in **Figure 4**. The lattice self-diffusivity and Young's modulus of aluminum are used in this correlation. All data at 600°C ~ 635°C fall on a line demonstrating that the stress exponent of 3 describes deformation mode well. This result attests that the 2wt%Fe-addition to the Al-10wt%Ti alloy changes the deformation mode in such a way that the solute drag creep is the governing mechanism. This observation is different from the Al-10wt%Ti alloy, where true stress exponent was estimated to be 2 and thus grain boundary sliding was believed to be the rate-controlling mechanism.

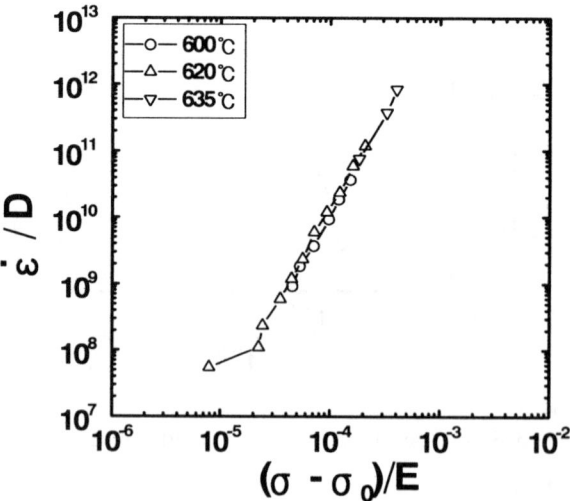

Figure 4. The relation between temperature-compensated strain rate and modulus-compensated stress.

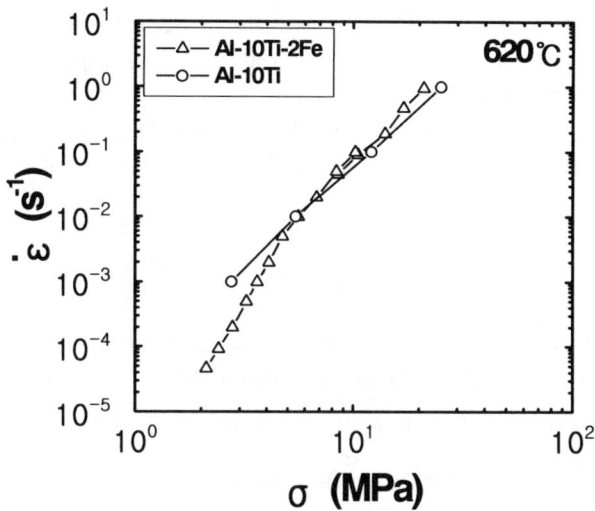

Figure 5. The strain rate-stress relationship for the Al-10wt%Ti-2wt%Fe and Al-10wt%Ti alloys compared at 620 °C.

Referring to the fact that solute drag creep is independent of grain size [13], the solute drag mechanism itself is not sufficient to describe the HSRS behavior of the Al-10wt%Ti-2wt%Fe specimen. The strain rate-stress relationship for the Al-10wt%Ti-2wt%Fe and Al-10wt%Ti alloys is compared at 620°C (**Figure 5**). The former alloy is seen to be weaker than the latter alloy at high strain rates over $10^{-2}s^{-1}$. This trend is anticipated at a grain-size reduction when plastic flow is governed by grain boundary sliding. Therefore, the deformation mode of the present material could be described as follows. Grain boundary sliding takes place extensively in the ultra-fine grain solid solution material and stress concentration is released by solute drag type deformation inside grains to sustain further plastic flow. The effect of Fe to the change in deformation mode needs further investigation.

SUMMARY AND CONCLUSION

1. When gas atomized powders were P/M consolidated, 2wt%Fe addition to the Al-10wt%Ti alloy significantly decreases the grain size and ultra-fine grain material of about 1.2 μm could be manufactured. Most of the added Fe exist as spherical $Al_xFe(X = 3$ or $6)$ aluminide of $0.2 \sim 0.5$ μm in diameter inside aluminum grains.

2. The ultra-fine grain size Al-10wt%Ti-2wt%Fe alloy exhibits the typical high strain rate

superplasticity at 600°C ~ 635°C, and its behavior can be described by usual phenomenological equation for superplasticity with the stress exponent of 3 and by incorporating the concept of threshold stress.

3. The high strain rate superplasticity of the Al-10wt%Ti-2wt%Fe alloy has been interpreted by grain boundary sliding mechanism accommodated by the solute drag deformation inside grains.

REFERENCES

1. G.S. Murty, M.J. Koczak and W.E. Frazier, Scripta Metall. **21**, 141 (1987).
2. Dongwha Kum and Hyeseung Kim, Mater. Sci. Forum **170-172**, 543 (1994).
3. T. G. Nieh, G. A. Henshall and J. Wadsworth, Scripta metall. **18**, 1040 (1984).
4. T. R. Bieler, T. G. Nieh, J. Wadsworth and A. K. Mukherjee, Scripta metall. **22**, 81 (1988).
5. T. G. Nieh and J. Wadsworth, Mater. Sci. Engr. **A147**, 129 (1991).
6. Dongwha Kum and G. Frommeyer, Met. Mater. **3**, 239 (1997).
7. Dongwha Kum, Mater. Sci. Forum, **243-245**, 287 (1997).
8. T. B. Massalski, *in Binary Alloy Phase Diagrams (2nd edition)*, (ASM International 1990), p. 147.
9. H. Kim, W. Kim and D. Kum, Mater. Sci. Forum, **304-306**, 321 (1999).
10. R.J. Lin and O.D. Sherby, Res. Mechanica **2**, 251 (1981).
11. F.A. Mohamed, J. Mater. Sci., **18**, 582 (1983).
12. T.R. Bieler and A.K. Mukherjee, Mater. Sci. Engr. **A128**, 171 (1990).
13. W.R. Cannon and O.D. Sherby, Metall. Trans. **1**, 1030 (1970).
14. W. J. Kim and D. W. Kum, Metal Trans. JIM, in press.

ROUTES TO DEVELOP FINE-GRAINED MAGNESIUM ALLOYS AND COMPOSITES FOR HIGH STRAIN RATE SUPERPLASTICITY

Toshiji Mukai *, Hiroyuki Watanabe *, T. G. Nieh **, Kenji Higashi ***
*Osaka Municipal Technical Research Institute, Osaka 536-8553, Japan,
mukai@inet-osaka.or.jp
**Lawrence Livermore National Laboratory, Livermore, California, USA
***Dept. of Metallurgy and Materials Science, Osaka Prefecture University, Osaka, Japan

ABSTRACT

Superplastic properties of magnesium alloys and their composites were reviewed with a special emphasis on the achievement of high strain rate superplastic forming. The role of grain size on superplastic deformation mechanisms was particularly addressed. Commercial Mg-Al-Zn alloys and a ZK60-based composite are used as model materials to illustrate the underlining principles leading to the observation of high strain rate superplasticity. In this paper, experimental results from several processing routes, including thermomechanical processing, severe plastic deformation, and extrusion of machined chips and rapidly solidified powders, are presented. High strain rate superplasticity (HSRS) is demonstrated in ZK60-based composites.

INTRODUCTION

Recent activities in the research of magnesium become higher in order to reduce the weight of components such as motor vehicles from the economical and ecological point of view. A number of magnesium alloys have been demonstrated to exhibit excellent mechanical properties, such as high specific strength at room temperature and superplasticity at moderate temperatures [1,2]. There are many potential opportunities for the use of magnesium alloys in motor vehicle components [3]. This is not only a result of magnesium's relatively low density, which can directly and substantially reduce vehicle weight, but is also a result of its good damping characteristics, dimensional stability, machinability, and low casting cost. These attributes enable magnesium to economically replace many zinc and aluminum die castings, as well as cast iron and steel components, and assemblies in motor vehicles [3].

Despite these advantages, magnesium alloys normally exhibit limited ductility (~ 7%). Therefore, recent fabrications of magnesium-based material were mainly performed by die casing with hot- or cold- chamber, or semi-solid processing of thixomolding technique. These technologies enable mass-production of small components of handheld device, such as mobile phone or note PC case. However, the microstructure of the fabricated products contains cast dendritic structures [4], thus the strength and ductility are poor. In order to use for structural components, which require high toughness, their microstructures must be modified. Figure 1 illustrates the relationship between strength and ductility (elongation-to-failure) in cast- and wrought-magnesium alloys. Wrought magnesium is noted to have a higher strength and ductility than cast magnesium. This primarily results from the fact that the grain size of wrought magnesium is normally finer than that of the cast material. Therefore, there is a strong need to develop fine-grained magnesium for structural applications.

In order to exploit the benefits of fine-grained magnesium, it is important to develop secondary processing which can effectively produce complex engineering components directly from wrought products. Superplastic forming is a viable technique that can be used to fabricate magnesium into complex shapes. Superplasticity, i.e. high tensile elongation, usually occurs

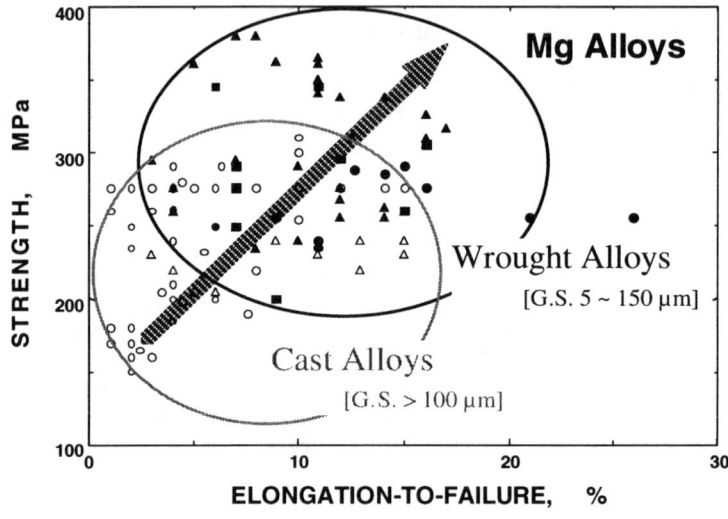

Fig. 1 The relations between strength and elongation-to-failure in cast- and wrought-magnesium alloys at room temperature.

when grain size is small, typically less than ~10 μm [5]. At this small grain size, grain boundary sliding is readily taking place at elevated temperatures. It is pointed out that, for magnesium alloys, there is another mechanism which can also lead to large tensile elongation, i.e. Class I solid solutions. Class I solid solutions are a group of dilute alloys in which the glide segment of the glide/climb dislocation process is rate controlling because solute atoms impede dislocation motion. Since glide-control mechanism is grain size independent, Class I solid solution alloys are of interest especially in coarse-grained conditions. Also, this group of alloys have an intrinsically high strain rate sensitivity of about $m = 0.33$, thereby should exhibit high elongation of over about 200% [6]. The elongation-to-failure for magnesium-based materials at elevated temperatures as a function of grain size is shown in Fig. 2. It can be seen that the elongation-to-failure typically increases with decreasing grain size.

In the following section, the deformation mechanisms at elevated temperatures in magnesium alloys with various grain sizes are first summarized and discussed. From the discussion, the grain size requirement for high strain rate superplasticity (~ 10^0 s^{-1}) is derived. The strain rate of 10^0 s^{-1} is chosen because it corresponds to the forming rate in commercial mass-production. Four different processing routes for grain size refinement will be presented and the developed grain structures are compared. Finally, a fine-grained magnesium-based composite fabricated by powder metallurgy route is selected to demonstrate the proposed HSRS at a strain rate of ~ 10^0 s^{-1}.

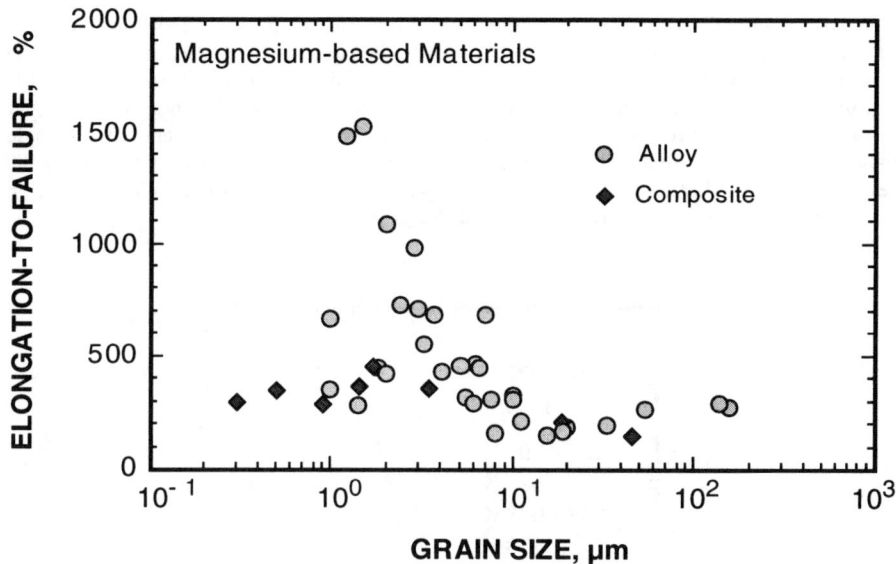

Fig. 2   Elongation-to-failure at elevated temperatures as a function of grain size
for Mg-based materials.

MATERIALS AND EXPERIMENTS

To investigate the grain size effect, three materials were prepared by a hot extrusion or rolling. A coarse-grained (>50 μm) material of AZ31(Mg-3Al-1Zn-0.2Mn, by wt.%) was received in the as-rolled sheet with an average grain-size of ~ 130 μm [7]. In contrast, an extruded AZ31 bar has a fine-grained structure with an average grain size of ~ 5 μm [8]. For the intermediate-grained material, AZ61(Mg-6Al-1Zn-0.2Mn, by wt.%) extruded sheet has an average grain size of 17 μm [9]. All these materials have been tested at different temperature and strain rate ranges where their respective superplastic behavior was shown.

In general, the constitutive equation to describe the superplasticity of magnesium alloys can be expressed by [10, 11]:

$$\dot{\varepsilon} = A \frac{D_0 G b}{kT} \left(\frac{b}{d} \right)^p \left(\frac{\sigma - \sigma_0}{G} \right)^n \exp\left(-\frac{Q}{RT} \right) \tag{1}$$

where $\dot{\varepsilon}$ is the strain rate, A a constant, D_0 the pre-exponential factor for diffusion, σ the stress, σ_0 the threshold stress, G the shear modulus, k the Boltzmann's constant, n the stress exponent ($=1/m$), d the grain size, b the Burgers vector, p the grain size exponent, R the gas constant, T the absolute temperature and Q the activation energy which is dependent on the rate controlling process. The three variables of n, p and Q in equation (1) are often used to identify the deformation mechanism.

By annealing AZ31 alloy to different grain sizes [7], we found that, for the coarse-grained material (>50μm), the deformation rate was independent of grain size, i.e. the grain size exponent p was virtually zero. The activation energy was estimated to be 127 kJ/mol, which is close to the value of activation energy for Al diffusion in Mg ($Q_S=143\pm10$ KJ/mol [12]). The relationship between the normalized strain rate, ($\dot{\varepsilon}/D_S$) and normalized stress, $(\sigma-\sigma_0)/G$ is shown in Fig. 3. Also included are data from other magnesium alloys with grain sizes over 50 μm [13,14]. It is readily seen that the normalized relations can be represented by a single straight line, suggesting the deformation mechanism for coarse-grained alloys is essentially the same. Also, from the facts that $n=3$, $p=0$, and $Q\sim Q_S$, the dominant deformation mechanism in the coarse-grained alloy is probably the glide of dislocations in grains.

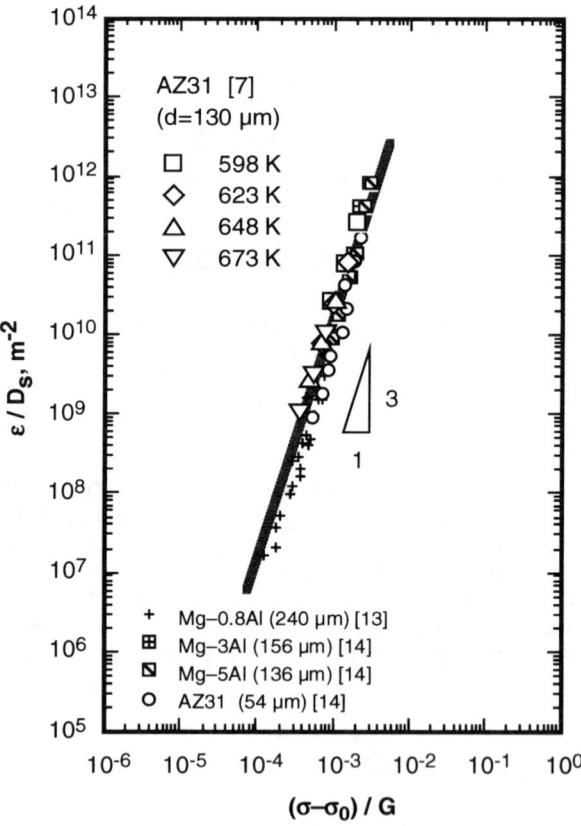

Fig. 3 Relation between normalized stress and normalized strain rate for coarse-grained AZ31 alloy [7]. Also included are data from some coarse-grained magnesium [13,14].

For fine-grained magnesium alloys (grain sizes < 10 μm), the grain size exponent was found to be approximately 3 [15,16]. The activation energy for fine-grained AZ31 was estimated to be 96 kJ/mol; this is close to the value of activation energy for grain boundary diffusion in Mg (Q_{gb} = 92 kJ/mol [17]). The relationship between the normalized strain rate, ($\dot{\varepsilon}/D_{gb})(kT/Gb)(d/b)^3$ and normalized stress, $(\sigma-\sigma_0)/G$ is shown in Fig. 4. Also included are data from other magnesium alloys with grain sizes less than 10 μm [16, 18–20]. Again, the normalized relations can be represented in a single straight line. The nice fit in Fig. 4 indicates that the deformation mechanism for fine-grained materials is the same. From the results of $n=2$, $p=3$, and $Q \sim Q_{gb}$, the dominant deformation mechanism in the fine-grained magnesium alloy is probably grain boundary sliding accommodated by slip controlled by grain boundary diffusion.

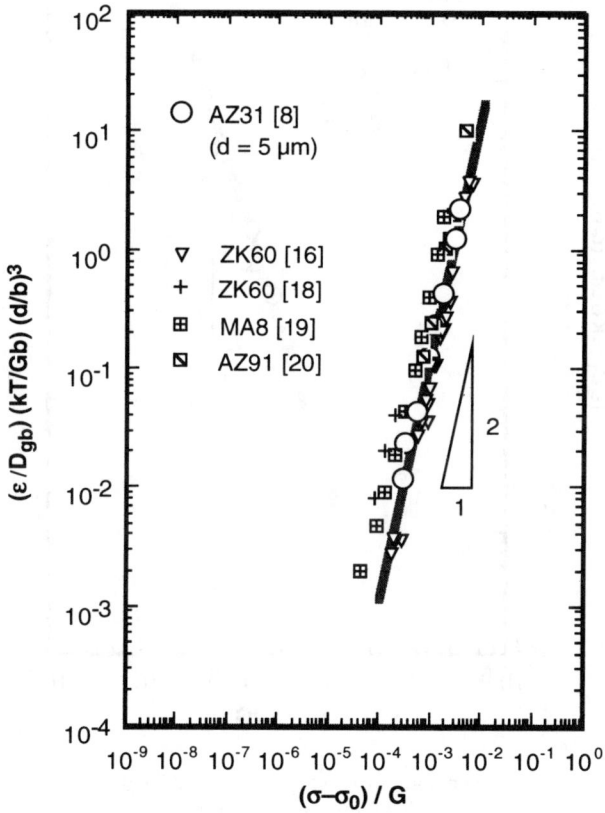

Fig. 4 Relation between normalized stress and normalized strain rate for fine-grained AZ31 alloy [8]. Also included are data from other fine-grained magnesium [16,18–20].

The experimental results obtained from intermediate-grained materials were found to be a mixture of the two extremes: coarse- and fine-grained. For example, for the intermediate-grained AZ61 alloy, the grain size exponent was found to be 2 [9]. The activation energy was calculated to be 143 kJ/mol, which is close to the value of activation energy for lattice diffusion in Mg (Q_L = 135 kJ/mol [17]). The relationship between the normalized strain rate, $(\dot{\varepsilon}/D_L)(kT/Gb)(d/b)^2$ and normalized stress, $(\sigma - \sigma_0)/G$ is shown in Fig. 6. From the results of n=2, p=2, and $Q \sim Q_L$, the dominant deformation mechanism in the intermediate-grained alloy is probably grain boundary sliding accommodated by slip controlled by lattice diffusion.

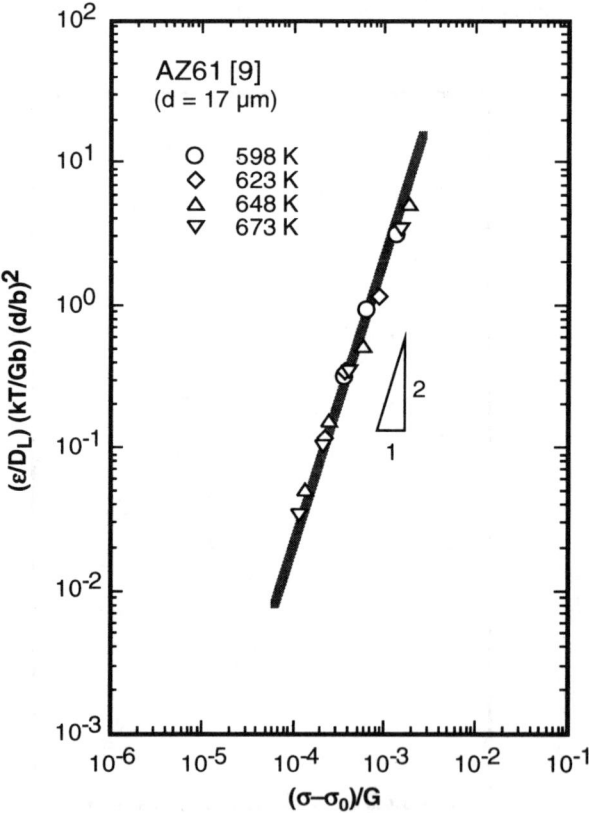

Fig. 5 Relation between normalized stress and normalized strain rate for the
intermediate-grained AZ61 alloy [9].

The deformation characteristics of magnesium alloys with various grain sizes are summarized in Table 1. It is evident in Table 1 that grain size refinement is essentially to effectively promote grain boundary sliding in magnesium.

Table 1 Deformation characteristics of magnesium alloys with different grain sizes.

Alloys	Grain Size (μm)	Stress Exponent [n]	Grain Size Exponent [p]	Activation Energy [Q]
Coarse-grained	$d > 50$	3	0	$\sim Q_s$
Intermediate-grained	$10 < d < 50$	2	2	$\sim Q_L$
Fine-grained	$d < 10$	2	3	$\sim Q_{gb}$

As can be seen in Table 1, superplastic behavior exhibits a strong dependence on grain size in fine-grained magnesium alloys. Figure 6 shows the optimal superplastic strain rate against the inverse of grain size for fine-grained magnesium alloys. The relations can be well described by a single straight line. The slope of the curve represents the grain size exponent p which is about 3. For a target strain rate of 10^0 s^{-1} for mass production, the grain size must be about 0.5 μm, as indicated in Fig. 6. Therefore, magnesium alloys with a grain size of ~0.5 μm must be developed.

FINE GRAIN PROCESSING

Mabuchi *et al.* have demonstrated that grain size developed by hot extrusion in AZ91 machined-chip increased from 7.6 to 66.1 μm when extrusion temperature was raised from 573 to 753 K [20]. They also found that grain size decreased with an increase in Zener-Hollomon parameter, i.e. $\dot{\varepsilon} \exp(Q/RT)$. This provides useful guides for the grain-refinement in magnesium: (1) increase the strain rate (or stress) in the process, or (2) decrease the temperature below the recrystallization temperature.

Following these guides, four thermomechanical processes (TMP) were subsequently selected for refining the grain structures; (a) extruding cast alloys with a high ratio of 100:1, (b) severe plastic deformation, e.g., Equal-Channel-Angular-Extrusion (ECAE), (c) compaction of machined chip by extrusion, and (d) powder-metallurgy (P/M) processing. Experiments of refining grain size were conducted for commercial magnesium alloys of AZ31 (process a,b,c), ZK60 (process a,d) and ZK61(process d). The developed grain sizes for the above four processes are summarized in Table 2. It can be seen that the target grained-structures ($d \sim 0.5$ μm) can be achieved by (b) ECAE of cast alloy [26] or (d) P/M process [27].

Table 2 Grain size produced in four different thermomechanical processes (TMP).

TMP	Developed grain size
(a) Extrusion of cast alloy	$2 \sim 5$ μm
(b) ECAE of cast alloy	$0.5 \sim 1$ μm
(c) Extrusion of machined chips	$1 \sim 3$ μm
(d) Extrusion of rapidly-solidified powders (RSP)	$0.5 \sim 1$ μm

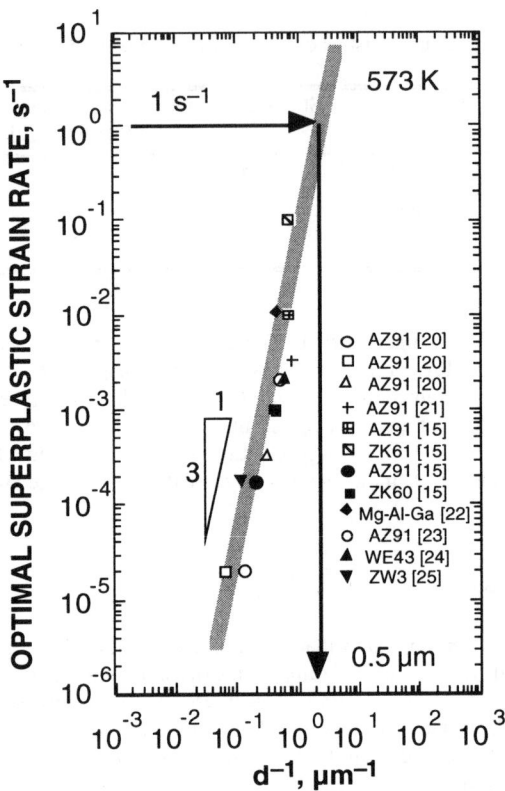

Fig. 6 Optimal superplastic strain rate against the inverse of grain size for fine-grained
magnesium alloys [15,20–25].

HIGH-STRAIN-RATE-SUPERPLASTICITY IN MAGNESIUM-BASED COMPOSITE

In addition to TMP, there is yet another method to produce fine-grained magnesium, namely, by composite techniques. A typical microstructure developed for ZK60-based composite by P/M route [28] is shown in Fig. 7. Equiaxed fine grain structure ($d \sim 0.5$ μm) is readily observed. Fine-grained structures have been reported in several magnesium-based composites [29–35]. Resulting from fine-grained structures, these composites exhibited high strain rates superplasticity (HSRS). The HSRS characteristics of these magnesium-based composites are summarized in Table 3. The extensive ductility, high m-value are reported at relatively lower temperatures.

Superplastic elongation of the fine-grained ZK60/SiC/17p composite as a function of temperature is shown in Fig. 8 [28]. Also included are data from HSRS Al-based composites [36,37] and fine-grained ZK60 magnesium alloy [16,38]. It is particularly noted that for Al-based composites the maximum tensile elongation occurs near the partial melting temperature

[39]. (Arrows indicate the partial-melting temperatures of Al-based composites and melting temperature of ZK60/SiC/17p.) In contrast, the optimum superplasticity in magnesium-based composites is observed at a considerably lower temperature than the melting point. This is also true for the unreinforced ZK60 alloy. In the case of HSRS Al-based composites, it was argued that the presence of liquid phases at the matrix/reinforcement interfaces and grain triple junctions is necessary to relax the local stress concentration [40]. Apparently, for magnesium alloys and composites there is no need for the presence of a liquid phase. This is because diffusion processes are much faster in magnesium as compared to that in aluminum. The result that superplasticity in magnesium alloys and composites occurs at relatively low temperatures are quite encouraging. This suggests that magnesium-based materials have a great potential to superplastically form at high strain rates and at a reasonably low temperature (< 673 K).

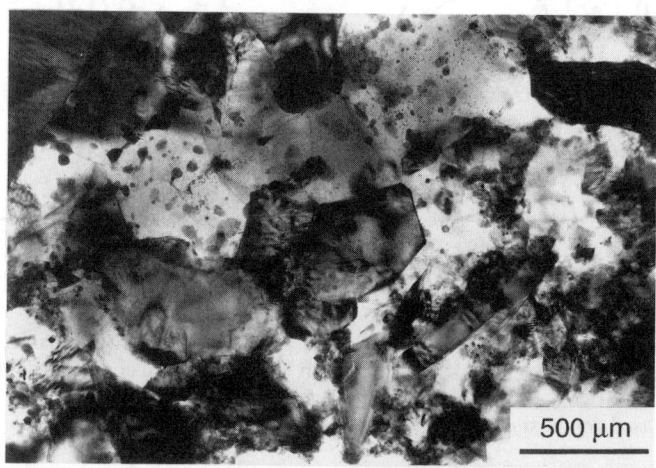

Fig. 7 Typical fine-grained microstructure of ZK60/SiC/17p composite.

Table 3 High-strain-rate superplasticity in magnesium-based composites.

Material	Grain Size d, µm	Temperature T, K	Strain Rate ε, s^{-1}	Elongation εf, %	Flow Stress σ, MPa	m	Ref.
Mg-5Zn/TiC/20p	2	743	6.67×10^{-2}	295	8.0	0.43	29
ZK51/TiC/20p	-	743	3×10^{-2}	136	4.1	0.3	30
Mg-Al/Mg2Si/28p	1.4	788	1×10^{-1}	368	8.0	0.5	31
Mg-Zn/Mg2Si/28p	0.9	713	1×10^{-1}	290	10.0	0.5	31
ZK60/SiC/17p	0.5	673	1	350	35.9	0.5	32
Mg-5Al/AlN/15p	< 2	673	5×10^{-1}	236	33.8	0.39	33
Mg alloy/AlN/20p	-	673	7.05×10^{-2}	231	19.0	0.43	34
AZ31/SiC/10p	-	798	2.08×10^{-1}	228	19.8	0.36	35
ZK60/SiC/17p	2.1	623	1×10^{-1}	450	15.4	0.5	28

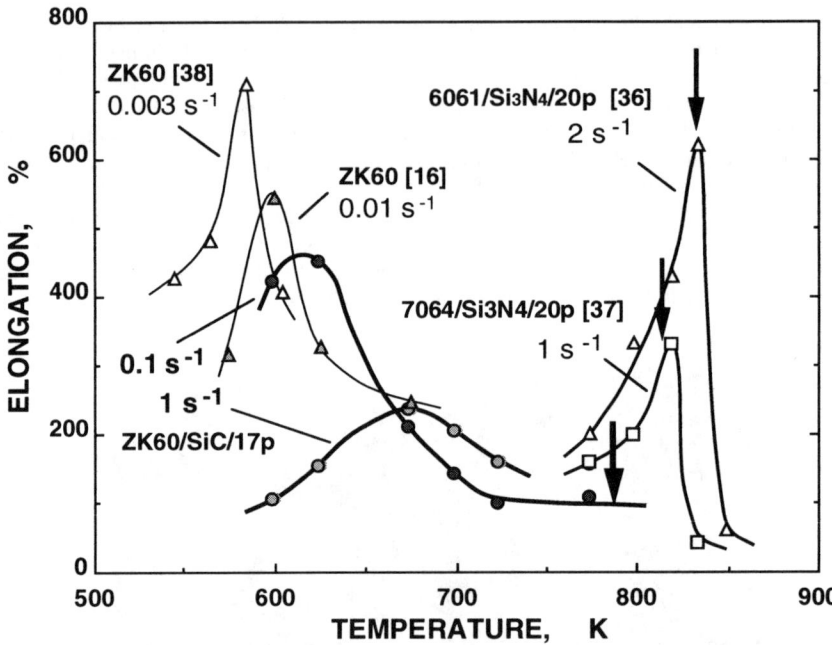

Fig. 8 Superplastic elongation of ZK60/SiC/17p composite as a function of testing temperature. Also included are data from HSRS Al-based composites [36,37] and ZK60 alloy [16,38]. Arrows indicate the partial-melting temperature of Al-based composites and melting temperature of ZK60/SiC/17p.

CONCLUSIONS

Superplastic behavior of magnesium alloys and composites are reviewed, and the deformation mechanisms in materials with different grain sizes are identified. High strain rate superplasticity was observed in ultrafine-grained (grain size ~0.5 μm) magnesium alloys and composites. The submicronmeter fine grains can be produced via severe plastic deformation of ECAE or P/M route. This creates opportunities for the potential application of wrought magnesium for structural components. It is demonstrated that the P/M processed magnesium-based composite can exhibit HSRS at a considerably lower temperature as compared to Al-based composites. This result suggests the possibility of mass-production of near-net-shape magnesium parts.

ACKNOWLEDGMENT

This work was partially supported by the Army Research Office and performed under the auspices of the U.S. Department of Energy by LLNL (contract No. W-7405-Eng-48). One of the authors (KH) gratefully acknowledges the financial support of the Ministry of Education

Science and Culture of Japan as a Grant-in-Aid for Scientific Research on Priority Area "Innovation in Superplasticity" (08242107).

REFERENCES

1. K. Higashi and J. Wolfenstine, Mater. Lett. **10**, 329 (1991).
2. K. Higashi, O.D. Sherby, G. Gonzalez-Doncel and J. Wolfenstine in *Superplasticity in advanced Materials*, edited by Hori, M. Tokizane and N. Furushiro (ICSAM–91, Osaka, 1991) pp. 235–242.
3. J. Davis, Technical Paper No. 910551, (SAE International, Warrendale, PA, 1985)
4. G. Schindelbacher and R. Rösch in *Magneium alloys and their applications*, edited by B.L. Mordike and K.U. Kainer (Werkstoff-Informationsgesellschaft, Frankfurt, 1998), p. 247.
5. T.G. Langdon, Metall. Trans. **13A**, 689 (1982).
6. O.D. Sherby and J.Wadsworth, Prog. Mater. Sci. **33**, 169 (1989).
7. H. Watanabe, H. Tsutsui, T. Mukai, H. Kohzu, S. Tanabe and K. Higashi, submitted.
8. H. Watanabe, T. Mukai, K. Ishikawa, Y. Okanda, M. Kohzu and K. Higashi, J. Japan Inst. Light Metal **49**, 401 (1999).
9. H. Watanabe, T. Mukai, M. Kohzu, S. Tanabe and K. Higashi, Acta Mater. **47**, 3753 (1999).
10. R.S. Mishra, T.R. Bieler and A.K. Mukherjee, Acta Mater. **43**, 877 (1995).
11. R.S. Mishra, T.R. Bieler and A.K. Mukherjee, Acta Mater. **45**, 561 (1997).
12. G. Moreau, J.A. Cornet and D. Calais, J. Nucl. Mater. **38**, 197 (1971).
13. S.S. Vagarali and T.G. Langdon, Acta Metall. **30**,1157 (1982).
14. J. Saeki and M. Otsuka, in *Proceedings of IMSP '97*, edited by T. Aizawa, K. Higashi, M. Tokuda (Mie University Press, Tsu, 1997), pp. 69–74.
15. M. Mabuchi, T. Asahina, H. Iwasaki and K. Higashi, Mater. Sci. Tech. **13**, 825 (1997).
16. H. Watanabe, T. Mukai and K. Higashi, Scripta Mater. **40**, 477 (1999).
17. H.J. Frost and M.F. Ashby, *Deformation-Mechanism Maps*, (Pergamon Press, Oxford, 1982), p. 44.
18. A.-U. Karim and W.A. Backofen, Metall. Trans. **3**, 709 (1972).
19. R.Z. Valiev and O.A. Kaibyshev, Phys. Stat. Sol. (a) **44**, 65 (1983).
20. M. Mabuchi, K. Kubota and K. Higashi, Mater. Trans., JIM **36**, 1249 (1995).
21. J.K. Solberg, J. Tørklep, Ø. Bauger and H. Gjestland, Mater. Sci. Eng. **A134**, 1201 (1991).
22. A. Uoya, T. Shibata, K. Higashi, A. Inoue and T. Masumoto, J. Mater. Res. **11**, 2731 (1996).
23. T. Sato, J. Kaneko and M. Sugamata, J. Japan Inst. Light Metal **42**, 345 (1992).
24. T. Mohri, M. Mabuchi, N. Saito and M. Nakamura, Maer. Sci. Eng. **A257**, 287 (1998).
25. M.M. Tilman and L.A. Neumeier, RI8662, Bureau of Mines, U.S. Department of the Interior (1982).
26. T. Mukai, K. Moriwaki, H. Watanabe and K. Higashi, submitted.
27. H. Watanabe, T. Mukai, M. Mabuchi and K. Higashi, Scripta Mater. **41**, 209 (1999).
28. T. Mukai, T.G. Nieh, H. Iwasaki and K. Higashi, Mater. Sci. Tech. **14**, 32 (1998).
29. S.-W. Lim, T. Imai, Y. Nishida and T. Choh, Scripta Metall. Mater. **32**, 1713 (1995).
30. T. Imai, S.-W. Lim, D. Jiang, Y. Nishida and T. Imura, Mater. Sci. Forum **304–306**, 315 (1999).
31. M. Mabuchi and K. Higashi, Phil. Mag. A **74**, 887 (1996).
32. T.G. Nieh, A.J. Schwartz and J. Wadsworth, Mater. Sci. Eng. **A208**, 30 (1996).
33. T. Imai, S.-W. Lim D. Jiang and Y. Nishida, Scripta Mater. **36**, 611 (1997).

34. A.B. Ma, T. Imura, M. Takagi, Y. Nishida, J.-H. Jiang, T. Imai, S.-W. Lim and J.-Q. Jiang, Mater. Sci. Forum **304–306**, 279 (1999).
35. A.B. Ma, J.-Q. Jiang, J.-H. Jiang, Y.-S. Sun, Y. Nishida, T. Imai, P.S. Chen, T. Imura and M. Takagi, Mater. Sci. Forum **304–306**, 285 (1999).
36. M. Mabuchi, K. Higashi, Y. Okada, S. Imai and K. Kubo, Scripta Metall. Mater. **25**, 2517 (1991).
37. M. Mabuchi, K. Higashi, K. Inoue, S. Tanimura, T. Imai and K. Kubo, Mater. Sci. Eng. **A156**, L9 (1992).
38. M.M. Tilman, R.L. Crosby and L.A. Neumeier, RI8382, Bureau of Mines, U.S. Department of the Interior (1979).
39. K. Higashi, T.G. Nieh, M. Mabuchi and J. Wadsworth, Scripta Metall. Mater. **32**, 1079 (1992).
40. M. Mabuchi and K. Higashi, Phil. Mag. Lett. **70**, 1 (1994).

FLOW STRESS AND ELONGATION OF SUPERPLASTIC DEFORMATION IN $La_{55}Al_{25}Ni_{20}$ METALLIC GLASS

Y. KAWAMURA, A. INOUE
Institute for Materials Research, Tohoku University, 2-1-1 Katahira, Aoba-ku, Sendai 980-8577, Japan, rivervil@imr.tohoku.ac.jp

ABSTRACT

We have investigated the flow stress and elongation of superplastic deformation in a $La_{55}Al_{25}Ni_{20}$ (at%) metallic glass that has a wide supercooled liquid region of 72 K before crystallization. The superplasticity that appeared in the supercooled liquid region was generated by the Newtonian viscous flow that exhibits the m value of unity. The elongation to failure was restricted by the transition of the Newtonian flow to non-Newtonian one and the crystallization during deformation. We succeeded in establishing the constitutive formulation of the flow stress in the supercooled liquid region. Its formulation was expressed very well by a stretched exponential function $\sigma_{flow}=D\dot{\varepsilon}\exp(H^*/RT)$ $[1-\exp(E/\{\dot{\varepsilon}\exp(H^{**}/RT)\}^{0.82})]$. Formulations describing the elongation to failure in constant-strain-rate and constant-crosshead velocity tests were, moreover, established. It was found from the simulation that the maximum elongation in the constant-strain-rate test reached more than 10^6 % which was two orders of magnitude larger than that in the constant-crosshead-velocity test.

INTRODUCTION

We have previously reported that the Zr-, La-, Pd- and Fe-based metallic glasses having a temperature interval of supercooled liquid above 60 K before crystallization exhibited a superplastic-like behavior at high strain-rates above 1×10^{-2} s^{-1} and at temperatures above the glass transition temperature [1-6]. Although the word of "superplasticity" is generally used for polycrystalline solids [7, 8], the metallic glasses with large supercooled liquid region are accepted as one of high-strain-rate superplastic materials [1, 9-12]. Most of metallic glasses have a limitation of product size and a lack of workability and machinability. The metallic glasses are, however, settling the disadvantages by powder consolidation and working through the superplasticity inherent in the supercooled liquid [13-16]. The consolidated Zr-based bulk metallic glasses are put into practical use as a face material of golf clubs [17]. However, little is known about the mechanism of the superplasticity in metallic glasses.

In this study, we will discuss the mechanism of the superplastic deformation in a $La_{55}Al_{25}Ni_{20}$ (at%) metallic glass that exhibits an excellent superplasticity [3, 4]. The glass transition temperature (T_g) and the crystallization temperature (T_x) of the metallic glass are 479 K and 551 K, respectively, resulting in the supercooled liquid region ($\Delta T_x = T_x - T_g$) of 72 K [4].

EXPERIMENT

$La_{55}Al_{25}Ni_{20}$ (at %) metallic glassy ribbons with a cross section of about 1.4×0.04 mm^2 were produced by a single-roller melt-spinning method in an argon atmosphere. X-ray-diffractometry (XRD) and transmission electron microscopy (TEM) confirmed the formation of a single glassy phase. The glass transition temperature (T_g) and the crystallization temperature (T_x) were measured by differential scanning calorimetry (DSC) at scanning rate of 0.67 K/s. The time-teperature-transormation (T-T-T) diagram for the onset of crystallization was constructed by isothermal DSC measurements. The tensile tests were conducted using an Instron-type tensile test machine at various crosshead velocities that remained unchanged during the tests. The strain rates represented in this study, therefore, corresponds to the initial strain rates. The gauge length was 10 mm. A silicone oil bath was used for heating the samples. The tensile tests were started after allowing the sample to equilibrate for 200 s. The deformed samples were examined by TEM observations. Ductility of the samples stretched to fracture was examined by a 180 degrees bend method at room temperature. The samples that were unable to bend through 180 degrees were judged to be brittle.

RESULTS AND DISCUSSION

Flow Stress in Superplasticity

Figure 1 shows the strain-rate dependence of the flow stress (σ_{flow}) for the $La_{55}Al_{25}Ni_{20}$ metallic glass. The flow stress is defined as a manner that has previously been described [18-21]. The flow stress increased with increasing strain rate, and the curves shifted towards higher strain-rates with increasing temperature. Figure 2 shows the strain-rate dependence of the viscosity (η). The viscosity was calculated from the flow stress by $\eta = \sigma_{flow}/3\,\dot{\varepsilon}$. The viscosity of the glassy solid decreased with increasing strain rate, showing a non-Newtonian flow where the m value was less than 0.3. On the other hand, a Newtonian flow, in which the viscosity is essentially independent of strain rate and the m value is unity, was observed in the supercooled liquid state at lower strain-rates. The higher strain-rates, moreover, led to a decrease in the viscosity, showing the transition from the Newtonian to the non-Newtonian flow in which the m value declined rapidly to about 0.4. This critical strain-rate of the Newtonian flow increased with increasing temperature in accordance with the Arrhenius relation.

The viscosity of the supercooled liquid can be expressed by one master curve as shown in Fig. 3. The master curve was produced by normalizing the viscosity (η) of each temperature with the Newtonian viscosity (η_0), and then shifting the each normalized viscosity curves towards the curve of the reference temperature (483 K). The entire master curve could be fitted very well with the stretched exponential function $\eta/\eta_0=1-\exp(A/\dot{\varepsilon}_m^{\beta})$, where $\dot{\varepsilon}_m$ is the strain rate in the master curve and the parameter A was 32.4 ± 4.8. The parameter β was 0.82 ± 0.01. The inset of Fig. 3 shows the Arrhenius plots of the Newtonian viscosity (η_0) and shift factor (a_T). The Newtonian viscosity of the neighborhood of the glass transition temperature could be fitted by $\ln\eta_0=B+H^*/RT$, where the activation energy H^* was 296 ± 9 kJ/mol and the parameter B was -52.0 ± 2.3.

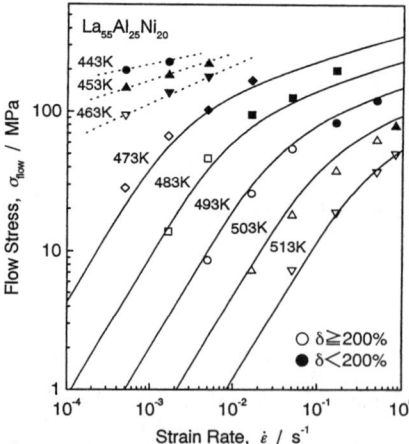

Figure 1. Strain-rate dependence of the flow stress (σ_{flow}) in $La_{55}Al_{25}Ni_{20}$ metallic glass. The solid lines are simulated by the eq. (1). The open and solid symbols represent the samples with and without elongations of more than 200 %, respectively.

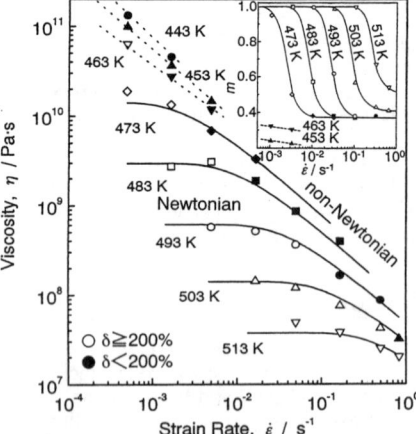

Figure 2. Strain-rate dependence of the viscosity (η) in $La_{55}Al_{25}Ni_{20}$ metallic glass. The open and solid symbols represent the samples with and without elongations of more than 200 %, respectively.

The shift factor $a_T = \dot{\varepsilon}_m / \dot{\varepsilon}$ of the strain rate could be expressed by $\ln a_T = C + H^{**}/RT$, where the activation energy of the shift factor H^{**} was 265 ± 13 kJ/mol and the parameter C was -65.7 ± 3.2. Accordingly, we can constitute the formulation of the flow stress for the superplasticity in the $La_{55}Al_{25}Ni_{20}$ metallic glass. The constitutive formulation was expressed by the following relation

$$\sigma_{flow} = 3\eta\dot{\varepsilon} = D\dot{\varepsilon}\exp(H^*/RT)\,[1-\exp(E/\{\dot{\varepsilon}\exp(H^{**}/RT)\}^{0.82})]\,, \qquad (1)$$

where the parameters D and E were 7.09×10^{-23} and 7.71×10^{-21}, respectively. It is clear that the formulation gives a good approximation to the experimental data of the supercooled liquid, as shown in Fig. 1.

Elongation to Failure in Superplasticity

Figure 4 shows the strain-rate dependence of the elongation to failure (δ) at various temperatures. As the strain rate increased, the elongation increased, reached a peak and then decreased. The peak in the elongation of the supercooled liquid shifted towards higher strain-rates with temperature. The large elongations of more than 200 % were obtained in the supercooled liquid state. The samples retained a ductile nature at the higher strain-rate side of the peak. On the other hand, the lower strain-rate side of the peak resulted in embrittlement. It was confirmed by transmission electron micrograph observations that the ductile samples consisted of a glassy single phase and crystallization occurred in the brittle samples.

Now, we shall consider the origin of the peak in the elongation. It must be kept in mind that the strain rates in this study correspond to the initial strain-rates and decrease with stretching. Figure 5 exemplifies the change in the m value with stretching for the deformation at 503 K. It was found that the m value of the ductile samples exhibiting large elongation reached unity in the course of the stretching. The smaller m value resulted in smaller elongation. The flow exhibiting the m value of unity is the Newtonian flow. The inset of Fig. 5 shows the strain-rate intervals

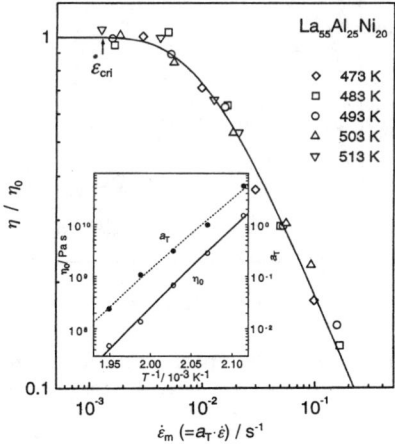

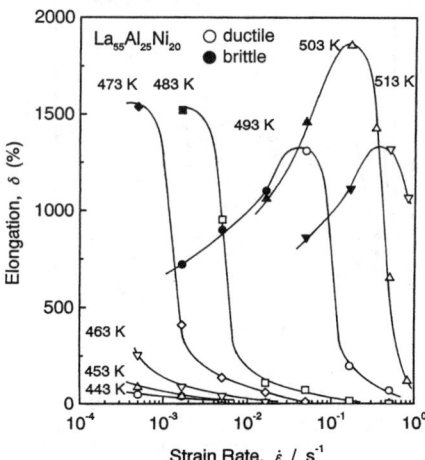

Figure 3.　Master curve of the normalized viscosity η/η_0 for the supercooled liquid in $La_{55}Al_{25}Ni_{20}$ metallic glass. The Arrhenius plots of the Newtonian viscosity (η_0) and the shift factor (a_T) are shown in the inset.

Figure 4.　Strain-rate dependence of the elongation to failure (δ) in $La_{55}Al_{25}Ni_{20}$ metallic glass. The samples indicated with the open and solid symbols exhibit the ductile and brittle nature, respectively, at ambient temperature after the tensile tests.

between the commencement of stretching and the failure for samples exhibiting maximum elongations at each temperature. It is clear that the maximum elongations were obtained under the conditions of the Newtonian flow. In the Newtonian flow, the growth of a neck is suppressed because of the ultimate m value of unity. The decrease in the m value, that is the prominent characteristics of the non-Newtonian flow, may contribute to the decrease in the elongation at higher strain-rates. On the other hand, the elongation of the embrittled samples in which the crystallization occurred was independent of the m value that was unity. The m value is, therefore, unable to describe the decrease in the elongation at the lower side of strain-rate. Figure 6 shows the relationship between the temperature and duration to failure. The time to failure was calculated by $t=\delta/100\dot{\varepsilon}+200$ taking account the retention time of 200 s. The failure time of the samples retaining the ductility increased with increasing temperature and the tendency shifted towards higher temperature side with increasing strain rate. On the contrary, the embrittled samples were found to lie in the T-T-T line for the onset of the crystallization, which was measured by isothermal DSD measurements. This describes that the crystallization during the deformation brought about the decrease in the elongation at the lower side of the strain rate. As described above, it may be concluded that the Newtonian viscosity generates the superplasticity and that both the non-Newtonian flow and crystallization restrict the elongation.

Consider the ideal maximum elongation for the tensile tests at constant strain rate and constant crosshead velocity. The elongation to failure (δ) is expressed by $\delta=100\dot{\varepsilon}_0t$, in which $\dot{\varepsilon}_0$ and t are the initial strain rate and time, for the constant-crosshead-velocity test, and by $\delta=100(\exp\dot{\varepsilon}t-1)$, where $\dot{\varepsilon}$ is the strain rate, for the constant-strain-rate test. As described above, the maximum elongation at each temperature can be obtained at the critical strain-rate of the Newtonian flow. As shown in the inset of Fig. 3, the critical strain-rate $\dot{\varepsilon}_{cri}$ can be expressed

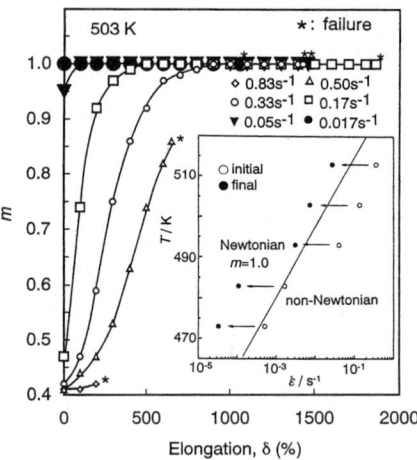

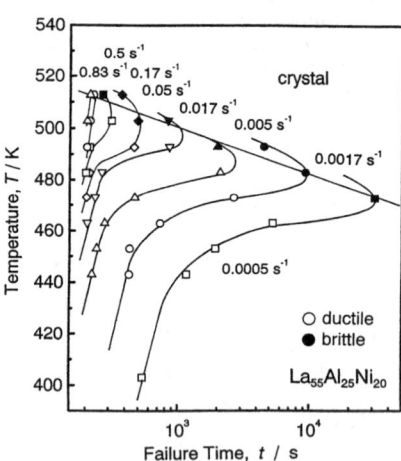

Figure 5. Change in the m value with stretching at 503 K in La$_{55}$Al$_{25}$Ni$_{20}$ metallic glass. The samples indicated with the open and solid symbols exhibit the ductile and brittle nature, respectively, at ambient temperature after the tensile tests. The inset shows initial and final strain-rates for the maximum elongation of La$_{55}$Al$_{25}$Ni$_{20}$ metallic glass at each temperature. The critical strain-rate of the Newtonian flow is also represented for reference.

Figure 6. Relationship between the failure time and temperature in La$_{55}$Al$_{25}$Ni$_{20}$ metallic glass. The T-T-T diagram of the crystallization obtained by DSC measurements is represented for reference. The samples indicated with the open and solid symbols exhibited the ductile and brittle nature, respectively, at ambient temperature after the tensile test.

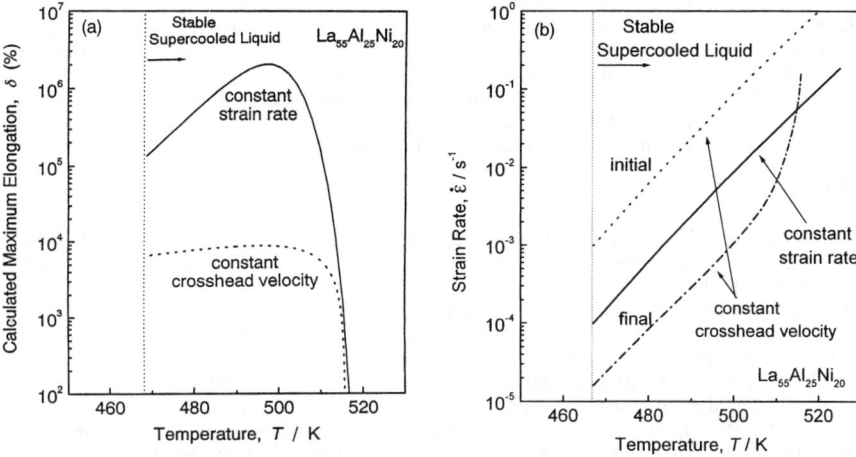

Figure 7. Simulated results of the maximum elongation (a) and strain rate (b) for the maximum elongation in constant-strain-rate and constant-crosshead-velocity tensile tests considering the holding time of 200 s.

by ln $(\dot{\varepsilon}_{m\text{-}cri}/\dot{\varepsilon}_{cri})=C+H^{**}/RT$, where $\dot{\varepsilon}_{m\text{-}cri}$ was the critical strain-rate of 1.3×10^{-3} s^{-1} in the master curve. The incubation time for the onset of the crystallization can be expressed by ln $t=F+H^{***}/RT$ where the parameter F was -49.5 ± 1.5 and the activation energy H^{***} was 235 ± 6 kJ/mol, as shown in Fig. 5. Figure 7 (a) shows the temperature dependence of the calculated ideal maximum elongation for the constant-strain-rate and constant-crosshead-velocity tests, where the holding time of 200 s for temperature equalization was considered. For the constant-crosshead-velocity test, we postulated from the inset of Fig. 5 that the initial strain-rate is ten times as large as the $\dot{\varepsilon}_{cri}$ because of the decreasing in strain rate with stretching. The strain rates for the maximum elongation were shown in Fig. 7 (b). The ideal maximum elongation in the constant strain-rate test was estimated to be 2×10^{6} % at 497 K and at 6.5×10^{-3} s^{-1}. It is clear that the elongation in the constant-strain-rate test is two orders of magnitude larger than that in the

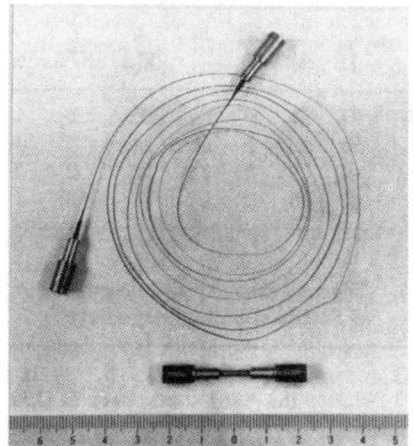

Figure 8. Photograph of La$_{55}$Al$_{25}$Ni$_{20}$ bulk metallic glasses before and after the deformation up to about 15000 % in elongation.

constant-crosshead-velocity test. Figure 8 shows the photograph of the $La_{55}Al_{25}Ni_{20}$ bulk metallic glass deformed up to an elongation of more than 15000 % at about 500 K, where the crosshead velocity increased from about 0.5 to 20 mm/s during stretching [22]. The tensile samples of 2 mm in diameter and 10 mm in gauge length were produced by machining the bulk metallic glass of 5 mm in diameter and 50 mm in length that was synthesized by copper-mold casting. The elongation is, therefore, expected to approach the ideal maximum elongation by controlling the crosshead speed so as to keep the strain rate constant.

CONCLUSIONS

We have examined the superplastic deformation of a $La_{55}Al_{25}Ni_{20}$ metallic glass having wide supercooled liquid region of 72 K before crystallization. The results obtained are summarized as follows.

1. The superplasticity observed in the supercooled liquid region was generated by the Newtonian viscous flow that exhibits the m value of unity. Non-Newtonian flow and crystallization restricted the elongation to failure.

2. The constitutive formulation of the flow stress was established, which could be expressed well by the stretched exponential function $\sigma_{flow}=D\dot{\varepsilon}\exp(H^*/RT)\,[1-\exp(E/\{\dot{\varepsilon}\exp(H^{**}/RT)\}^{0.82})]$.

3. The constitutive equation describing the maximum elongation in constant-strain-rate and constant-crosshead-velocity tests were established. It was found from the simulation that the maximum elongation in the constant-strain-rate test reached more than 10^6 % which was two orders of magnitude larger than that in the constant-crosshead-velocity test.

ACKNOWLEDGEMENT

The authors wish to thank Mr. T. Nakamura for his contributions to this paper.

REFERENCES

1. Y. Kawamura, T. Shibata, A. Inoue and T. Masumoto, Scripta Metall. Mater. **37**, 431 (1997).
2. Y. Kawamura, T. Nakamura and A. Inoue,   Scripta Metall. Mater. **39**, 301 (1998).
3. Y. Kawamura, T. Shibata and A. Inoue in *The Third Pacific Rim International Conference on Advanced Materials & Processing (PRICM 3)*, edited by M.A. Imam, R. DeNale, S. Hanada, Z. Zhong and D.N. Lee (TMS, Warrendale, PA, 1998), p. 490-493.
4. T. Nakamura, Y. Kawamura and A. Inoue, Mater. Sci. Forum **304-306**, 379 (1999).
5. Y. Kawamura, T. Nakamura and A. Inoue, Mater. Sci. Forum **304-306**, 349   (1999).
6. Y. Kawamura, T. Nakamura and A. Inoue, Mater. Trans. JIM **40**, 794 (1999).
7. J. Wadsworth, T. G. Nieh and O. D. Sherby in *Superplasticity in Advanced Materials*, edited by S. Hori, M. Tokizane and N. Furushiro (The Jpn. Soc. of Research on Superplasticity, Japan 1991), p. 13.
8. H. Ohsawa, J. Mater. Process. Technol. **68**, 193 (1997).
9. T. Sakuma, Mater. Sci. Forum **304-306**, 3 (1999).
10. O. M. Smirnov, Mater. Sci. Forum **304-306**, 341 (1999).
11. T. G. Nieh, T. Mukai, C. T. Liu and J. Wadsworth, Scripta Mater. **40**, 1021 (1999).
12. T. Sakuma and K. Higashi, Mater. Trans. JIM **40**, 702 (1999).
13. Y. Kawamura, H. Kato, A. Inoue and T. Masumoto, Appl. Phys. Lett. **67**, 2008   (1995).
14. Y. Kawamura, H. Kato, A. Inoue and T. Masumoto, Int. J. Powder Metall. **33**, 50 (1997).
15. Y. Kawamura, H. Kato, A. Inoue and T. Masumoto, Mater. Sci. Eng. **A219**, 39   (1996).
16. Y. Kawamura, T. Shibata, A. Inoue and T. Masumoto, Acta Mater. **37**, 253   (1998).
17. V. I. P. Vintage Model Amorphous Face Golf Club Catalog, Dunlop, Tokyo (1998).
18. Y. Kawamura, T. Shibata, A. Inoue and T. Masumoto, Appl. Phys. Lett. **69**, 1208 (1996).
19. Y. Kawamura, T. Shibata, A. Inoue and T. Masumoto, Jpn. J. Appl. Phys. **37**, L666 (1998).
20. Y. Kawamura, T. Shibata, A. Inoue and T. Masumoto, Appl. Phys. Lett. **41**, 779   (1997).
21. Y. Kawamura, T. Shibata, A. Inoue and T. Masumoto, Mater. Trans. JIM **40**, 336 (1999).
22. A. Inoue, Y. Kawamura and Y. Saotome, Mater. Sci. Forum **233-234**, 147 (1997).

Developments Using Severe Plastic Deformation

GRAIN REFINEMENT OF ALUMINUM USING
EQUAL-CHANNEL ANGULAR PRESSING

Z. HORITA*, M. FURUKAWA**, M. NEMOTO*, T.G. LANGDON***
*Department of Materials Science and Engineering, Faculty of Engineering, Kyushu University, Fukuoka 812-8581, Japan, horita@zaiko.kyushu-u.ac.jp
**Department of Technology, Fukuoka University of Education, Munakata 811-4192, Japan
***Departments of Materials Science and Mechanical Engineering, University of Southern California, Los Angeles, CA 90089-1453

ABSTRACT

Using the technique of equal-channel angular (ECA) pressing, it is possible to reduce the grain size of polycrystalline materials to the submicrometer level. Thus, this processing technique has the potential for producing materials which may exhibit superplasticity. This paper describes various factors affecting the development and evolution of the microstructure produced by ECA pressing. Optimization of such factors is then presented for the advent of superplasticity.

INTRODUCTION

There are important requirements for the attainment of superplasticity. They are (a) a small grain size less than ~10 μm, (b) an equiaxed grain structure with high-angle boundaries and (c) a small grain structure stable at high temperatures. The requirements of (a) and (b) are due to the reason that superplasticity occurs through grain boundary sliding and its accommodation. The small grain size increases the fraction of grain boundaries in a given volume and thus gives a larger contribution from grain boundary sliding. The geometrical accommodation may occur more easily with an equiaxed grain structure. The requirement (c) arises because grain boundary sliding and its accommodation are a thermally activated process via atomic diffusion.

Equal-channel angular (ECA) pressing is a useful technique for grain refinement in polycrystalline metals: the grain size is often reduced into the submicrometer range. The principle of the ECA pressing technique was initially developed by Segal et al. [1] and was applied to grain refinement by Valiev et al. [2]. With this technique, a sample is pressed through a channel having an equal cross-section but bent at an angle of Φ in a solid die as illustrated in Fig.1 [1, 3]. Shear strain is introduced when the sample passes through the bending point of the channel. Repetitive pressing is feasible as the sample cross-section remains the same throughout the channel. Grain refinement is achieved during a process of repetitive pressings. The significance of the ECA pressing technique is that suprplasticity may appear at high strain rates and/or lower testing temperatures as this processing technique is capable of producing grain sizes in the submicrometer range. Furthermore, the process does not rely on complex thermomechanical treatments but the grain refinement is achieved by simple shear.

There are several factors influencing the development of fine and equiaxed grain structures with high-angle grain boundaries in ECA pressing. These factors may be the amount of imposed strain, the shearing direction, the channel angle, the ECA-pressing speed, the ECA-pressing temperature and the alloying elements. This paper presents an overview of

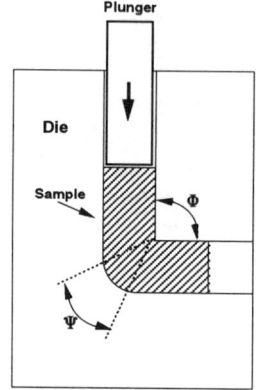

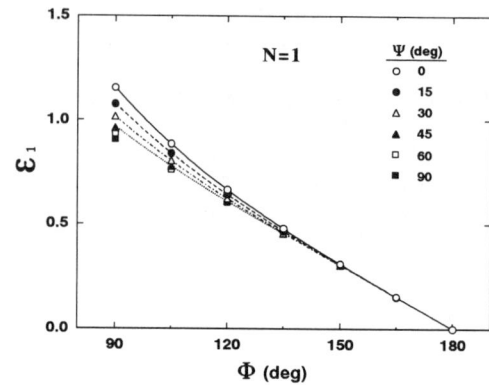

Fig.1 Schematic illustration for the ECA pressing facility.

Fig.2 Variation of equivalent strain with channel angle for N=1.

the factors including those affecting the thermal stability of the fine grain structure. Optimized conditions are then suggested for the achievement of superplasticity.

OPTIMIZATION FOR FINE GRAINED STRUCTURE

Imposed Strain

The strain introduced through the ECA pressing die is given by the following equation [3]

$$\varepsilon_N=(N/\sqrt{3})[2\cot\{(\Phi/2)+(\Psi/2)\}+\Psi\csc\{(\Phi/2)+(\Psi/2)\}] \qquad (1)$$

where N is the number of cycles in repetitive pressings, ε_N is the equivalent strain after N cycles and, as illustrated in Fig.1, Φ and Ψ are the channel angle and the angle defining the outer arc of curvature at the bending points of the channel, respectively. Figure 2 shows the variation of the strain created on a single passage through the die, ε_1, with the angle of Φ for a range of values of Ψ. This plot demonstrates that a single passage leads to a strain of ~1 for Φ=90° and this is insensitive to the value of Ψ.

Figure 3 shows microstructures and selected area electron diffraction (SAED) patterns of pure Al (99.99%) after (a) a single pass, (b) two passes, (c) three passes and (d) four passes, respectively [4]. The microstructures were taken on the y plane defined in Fig.4 and the SAED patterns were recorded from an area of 12.3 μm. The SAED pattern exhibiting a net after a single pass reveals that the microstructure consists of an array of subgrains separated by low angle boundaries. These subgrains are elongated in a band structure lying approximately parallel to the shearing direction at 45° to the top and bottom edges. The net pattern remains visible after two passes, indicating that the subgrain structure is retained in the sample. After further pressing, the SAED patterns tend to form rings and it is demonstrated that the subgrain structure becomes less evident after three passes and grains with high-angle grain boundaries are developed after four passes. Furthermore, many grains become equiaxed

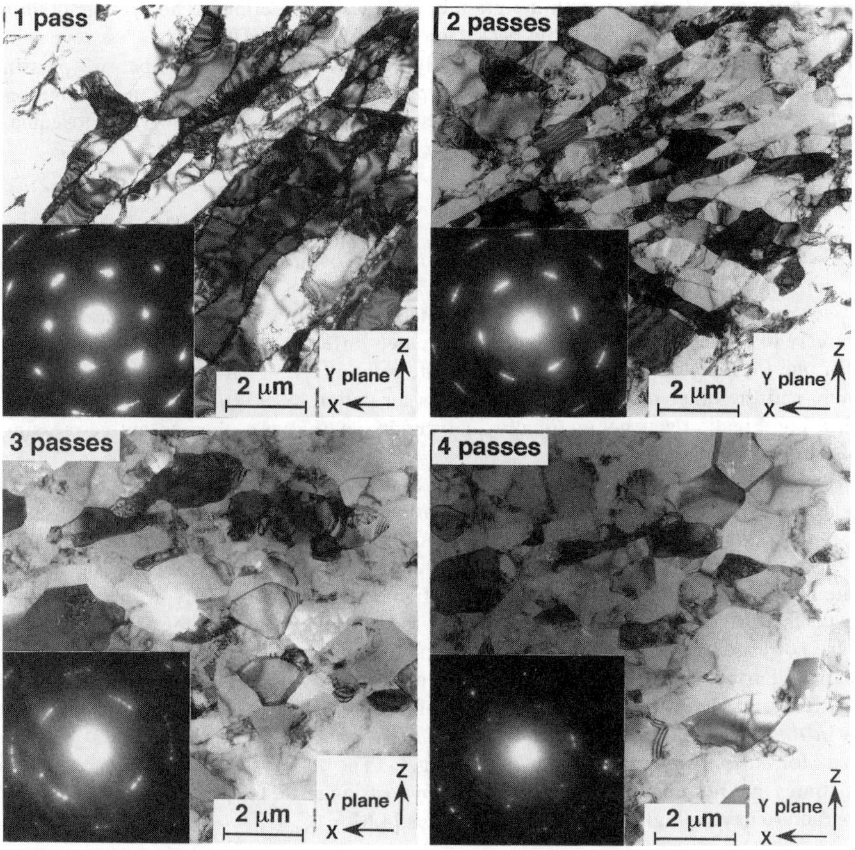

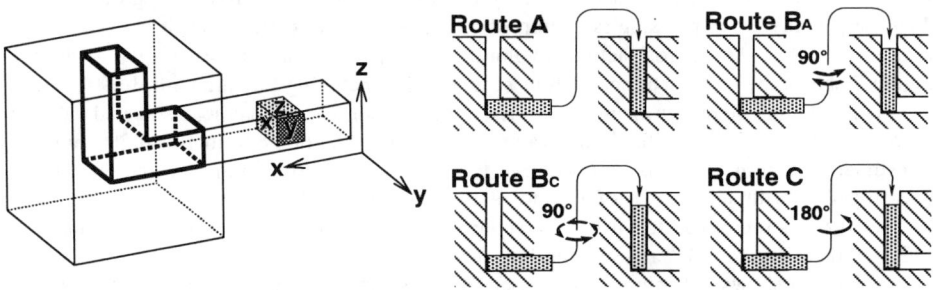

1 pass   **2 passes**

2 μm   Y plane   Z X

3 passes   **4 passes**

2 μm   Y plane   Z X

Fig.3 Microstructures of pure Al after 1 pass, 2 passes, 3 passes and 4 passes
through a die having a channel angle of 90°.

Fig.4 Definition of x, y and z planes.   Fig.5 Schematic illustration of the four
routes used for ECA pressing.

and contain fewer dislocations with the grain boundaries well-defined. The average grain size after four passes is ~1.3 μm. Electron backsctter diffraction (EBSD) analysis has revealed [5] that the boundary misorientation progressively increases with the number of pressings, although four passes may not be sufficient to produce a completely random distribution of misorientations. It is concluded that the microstructure evolves from subgrains to equiaxed grains with high-angle boundaries as the number of pressings increases.

Shearing Direction

The shearing direction in ECA pressing may be controlled by rotation of the sample about the longitudinal axis as illustrated in Fig.5: route A, route B and route C denote repetitive pressing without rotation, with rotation by 90° and with rotation by 180°, respectively [4]. It should be noted that there are two different ways of rotating the sample in route B: one (designated route B_A) is an alternate rotation by 90° in a forward and backward direction, and the other (designated route B_C) is a rotation by 90° in the same direction [6]. The shearing direction was geometrically analyzed in detail after each consecutive pressing in the four different routes [7-9].

Figure 6 shows microstructures including SAED patterns after 4 passes through the die following the four routes shown in Fig.5. The microstructures and their SAED patterns were taken on the x, y and z planes defined in Fig.4. All SAED patterns were recorded from an area of 12.3 μm. The microstructures obtained using routes A, B_A and C consist of subgrains separated by low-angle boundaries as the diffracted beams essentially form net patterns. However, the diffracted beams lie around rings for route B_C in any of the x, y and z planes. Thus, the corresponding microstructures are found to consist of grains having high-angle grain boundaries. Furthermore, most of the grains produced by route B_C are equiaxed with a homogeneous distribution having a grain size of ~1.3 μm whereas there are many elongated grains in the structures of routes A, B_A and C. It should be noted that route B_C has been adopted for the examination shown in Fig.3. The microstructural observation thus demonstrates that route B_C is the most efficient processing route for production of an equiaxed grain structure having high-angle grain boundaries [4,6].

Channel Angle

The microstructures shown in Figs 3 and 6 were obtained using a die having a channel angle of $\Phi=90°$. In a practical sense, it may be easier to press a sample through a die with $\Phi>90°$ especially when pressing is attempted on harder materials. However, it is necessary to examine whether the microstructural evolution takes place in a fashion similar to $\Phi=90°$. Figure 7 shows the cross-sections of four separate dies including the one with $\Phi=90°$, which were prepared for use in this examination. Each die was constructed having a selected value of Φ from 90° to 157.5° and with similar values for Ψ within the range from 10° to 30°. In order to estimate the numbers of passes required to give the same total strains, equation (1) was used together with the values of Φ and Ψ recorded in Fig.7. From these calculations, it follows that a strain close to ~4 can be achieved by 4 pressings for $\Phi=90°$, 6 pressings for 112.5°, 9 pressings for 135° and 19 pressings for 157.5°.

Figure 8 shows the microstructures and the associated SAED patterns for the samples pressed with the four dies shown in Fig.7. The microstructures were taken in the x plane which is perpendicular to the longitudinal axis of the sample. The SAED patterns were

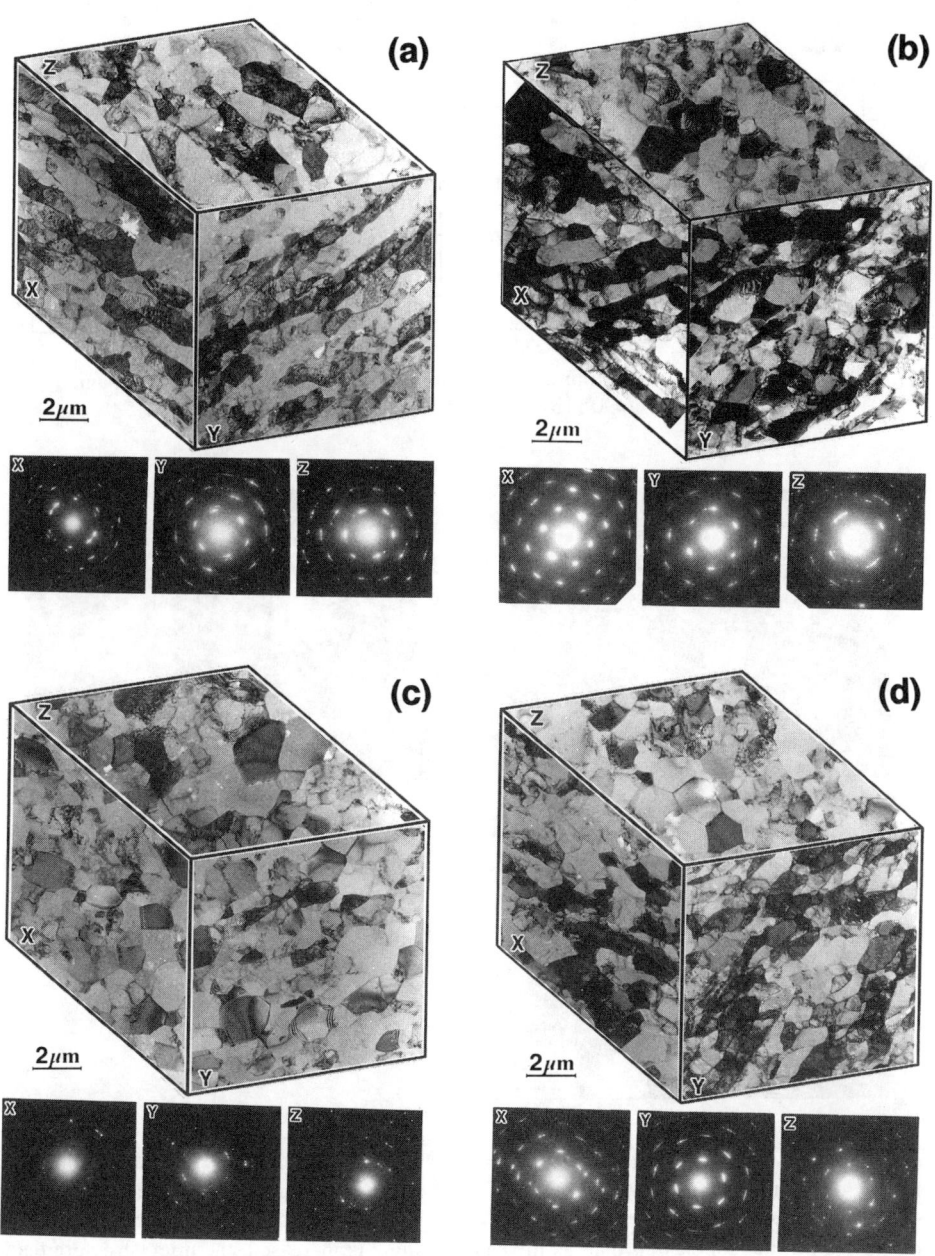

Fig.6 Microstructures and SAED patterns in pure Al after 4 pressings
using (a) route A, (b) route B_A, (c) route B_C and (d) route C.

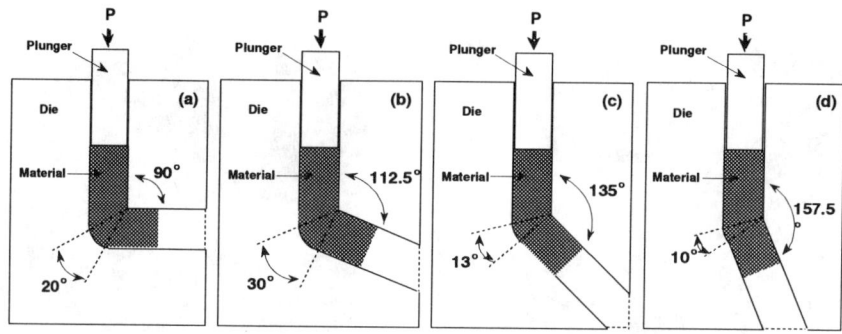

Fig.7 Cross-sections of four different dies used to examine the effect of channel angles of (a) Φ=90°, (b) Φ =112.5°, (c) Φ =135° and (d) Φ =157.5°.

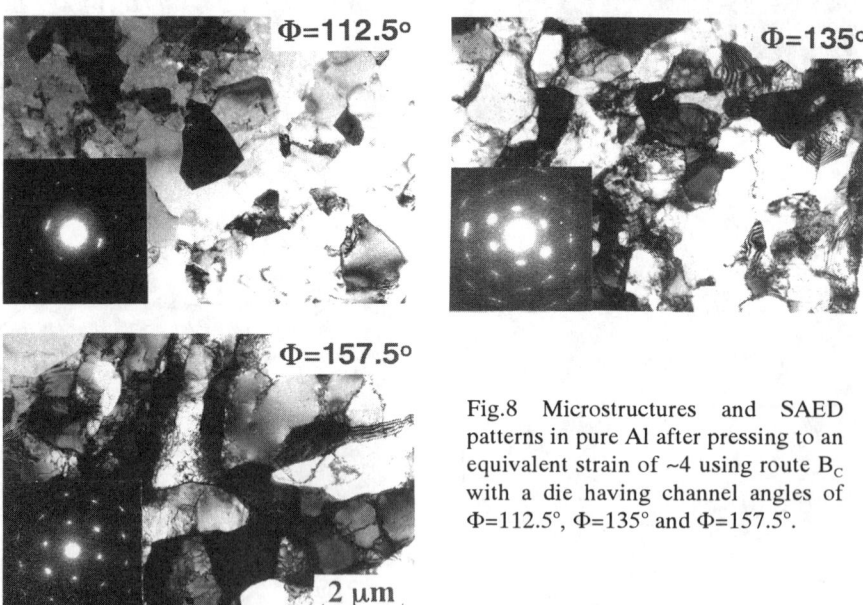

Fig.8 Microstructures and SAED patterns in pure Al after pressing to an equivalent strain of ~4 using route B_C with a die having channel angles of Φ=112.5°, Φ=135° and Φ=157.5°.

recorded from an area of 12.3 μm. The sample rotation was made by following route B_C. Inspection of the SAED patterns reveals that a ring pattern tends to form for Φ=90° and Φ=112.5° but well-defined net patterns are developed for Φ=135° and Φ=157.5°. Thus, these results lead to the conclusion that, as the channel angle is increased, the miocrstructure tends to consist of subgrains with low-angle boundaries [10]. In order to produce fine-grained structure with high-angle grain boundaries, it is therefore necessary to use a die having a channel angle equal to or close to 90°.

Pressing Speed

ECA pressing was conducted with pressing speeds ranging from ~8.5x10⁻³ mm s⁻¹ to ~7.6 mm s⁻¹ [11]. For these experiments, samples of pure Al were pressed with a die having a channel angle of $\Phi=90°$ to a total equivalent strain of ~4 (4 passes) through route B_c. Figure 9 shows representative microstructures and associated SAED patterns in the x plane for the samples pressed at the slowest and fastest speeds. The SAED patterns were taken from an area of 12.3 μm. Comparison shows that there is little difference in both the fine-grained structure and the appearance of the SAED patterns. The average grain size for both structures were measured to be ~1.3 μm. However, close observation of the microstructures reveals that there are more extrinsic dislocations both within the grain interiors and in the grain boundaries after pressing at the fastest speed. This suggests that recovery occurs more easily at the slower speed and the corresponding microstructures are in a more equilibrated condition.

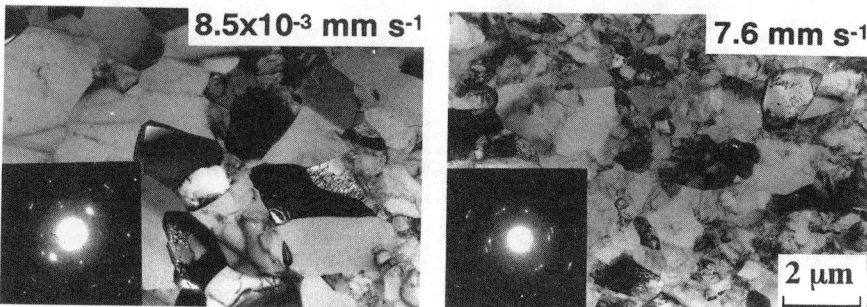

Fig.9 Microstructures and SAED patterns in pure Al after pressing to an equivalent strain of ~4 using route BC and pressing speeds of 8.5x10-3 mm⁻¹ s⁻¹ and 7.6 mm⁻¹ s⁻¹.

Pressing Temperature

All ECA pressings presented above were conducted at room temperature. Then, ECA pressing was operated at elevated temperatures to examine the effect of pressing temperature on the microstructural evolution. Samples of pure Al were pressed at temperatures of 373, 473 and 573 K using a die having a channel angle of 90°. All samples were subjected to 6 pressings through route B_c. Figure 10 shows the microstructures and the associated SAED patterns taken from the samples pressed at the corresponding temperatures. The ECA pressing at 373 K yielded an equiaxed grain structure containing high-angle grain boundaries. This structure is essentially similar to that for room temperature pressing, although there is a slight increase of the average grain size: ~1.5 μm for the 373K pressing in comparison with ~1.3 μm for the room temperature pressing. The grain structures produced by ECA-pressing at 473 and 573 K are also exquiaxed but the difference is that the grain boundaries are now low-angle boundaries because both SAED patterns form nets. These microstructural observations reveal that subgrain structures are developed during ECA pressing at temperatures of 473 K or higher. The grain size or the subgrain size is plotted in Fig.11 against the ECAP temperature including for the room temperature pressing. Figure 11 also plots the 0.2% proof stress which was obtained by compression testing at room temperature with an initial strain rate of 3.3x10⁻⁴ s⁻¹. The grain size increases with an increase of ECA-pressing temperature and this trend is opposite to that for the 0.2% proof stress.

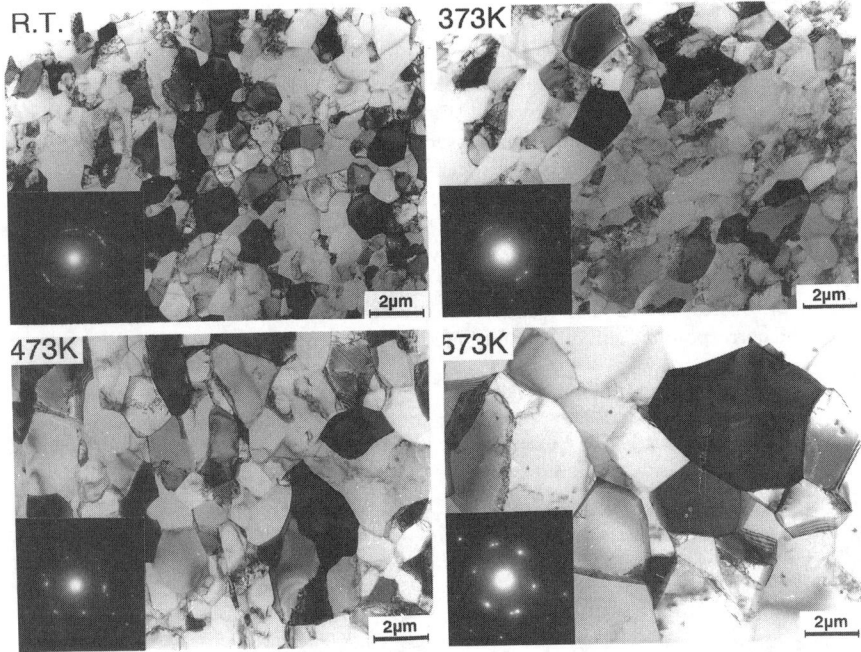

Fig.10 Microstructures and SAED patterns in pure Al after pressing at room
temperature, 373 K, 473 K and 573 K to an equivalent strain of ~6
through route B_C using a die having a channel angle of Φ=90°.

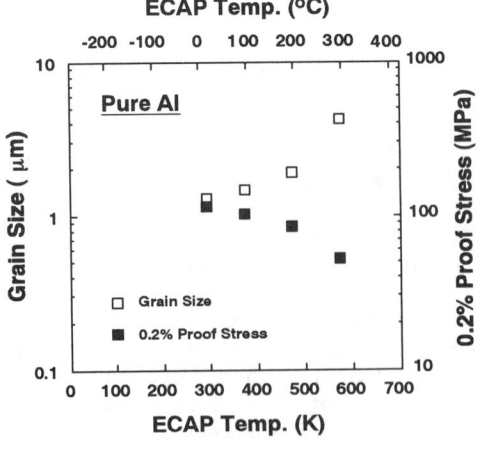

Fig.11 Grain size and 0.2% proof
stress with respect ECA
-pressing temperature.

Alloying Elements

Two kinds of elements, Mg and Zr, were added individually to Al to make alloys with compositions of Al-1%Mg, Al-3%Mg and Al-0.2%Zr. It is well known that Mg is an important element for Al because it contributes to a significant enhancement of Al strength due to the effect of solid solution hardening. It is also known that Zr plays an important role in retaining a fine-grained structure at higher temperatures through the formation of stable Al_3Zr particles.

Each alloy was subjected to ECA pressings at room temperature with a die having a channel angle of $\Phi=90°$. Route B_C was adopted for these pressings. Figure 12 shows the microstructures together with the SAED patterns taken from an area of 12.3 μm for Al-1%Mg and Al-3%Mg alloys [12]. The SAED patterns for both alloys exhibit well-defined rings and thus indicate that the selected areas contain many grains separated by high-angle grain boundaries. When compared with Figs 3 and 6 for pure Al, it is concluded that the grain size is significantly reduced with the addition of Mg. The measurements of grain sizes result in ~0.45 μm for Al-1%Mg and ~0.27 μm for Al-3%Mg in comparison with ~1.3 μm for pure Al. It should be noted, however, that a more number of pressings is required for the attainment of homogeneous equiaxed grain structures with high-angle grain boundaries as the amount of Mg is increased: at least 4 pressings are needed in pure Al but 6 pressings or more in the Al-1%Mg alloy and 8 pressings or more in the Al-3%Mg alloy [12].

Figures 13 and 14 show the microstructures after statically annealing the ECA-pressed pure Al and an Al-0.2%Zr alloy at 573 K for 1 hour, respectively [13]. Whereas significant grain growth has occurred in the pure Al sample, the fine-grained structure is essentially retained in the Al-0.2%Zr sample. Figure 15 is a more complete summary of the grain size variation with respect to annealing temperature, including the results after annealing of ECA-pressed Al-1%Mg, Al-3%Mg and Al-0.12%Zr alloys [13]. All samples except the alloy containing Zr exhibit significant grain growth at ~500 K. Therefore, although Mg has an important effect for reducing the grain size, it is not effective to retain the small grain size at the temperature higher than ~500 K. By contrast, it is evident that the Zr addition shifts the critical temperature for significant grain growth by ~100 degrees to a higher temperature. The stability of the fine-grained structure is maintained at higher temperatures by the addition of Zr and thus it may be possible to anticipate the advent of superplasticity with such Zr containing Al alloys. In fact, there have been reports that ECA-pressed fine-grained Al-6%Cu-0.5%Zr and Al-5.5%Mg-2.2%Li-0.12%Zr alloys exhibit superplasticity of ~1000% at a testing temperature of 573 K with an initial strain rate of 10^{-2} s^{-1} [14,15]. It has also been demonstrated [16] that when the grain size of an Al-3%Mg alloy containing 0.2%Sc is refined by ECA pressing, superplasticity of ~1000% has also been attained at 573 K with an initial strain rate of 3.3×10^{-2} s^{-1}. This work showed that Sc, as Zr, can suppress significant grain growth and retain the ECA-pressed fine-grained structure as high as ~700 K. More details are described elsewhere in this volume.

SUMMARY

(1) Equal-channel angular (ECA) pressing technique is capable of producing grain sizes of metallic materials in the submicrometer range.
(2) There are several factors influencing the development of fine and equiaxed grain structures with high-angle grain boundaries in ECA pressing.
(3) The microstructure evolves from subgrains to equiaxed grains with high-angle grain

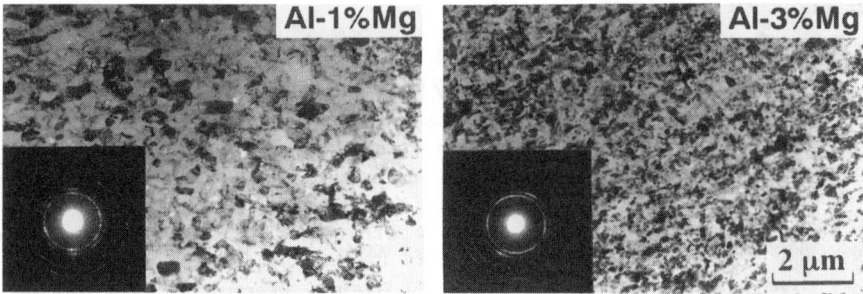

Fig.12 Microstructures and SAED patterns in Al-1%Mg and Al-3%Mg after 6 and 8
pressings through route BC.

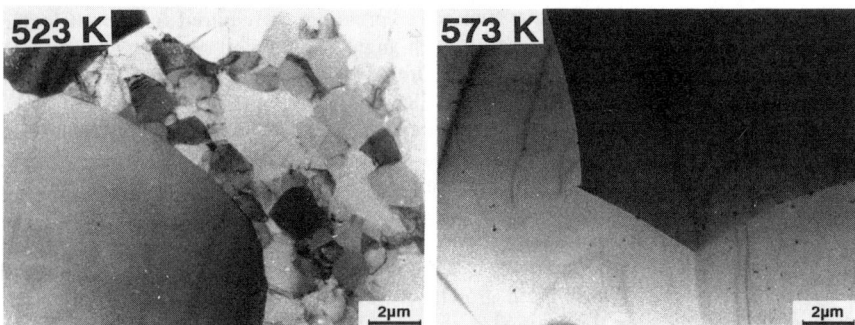

Fig.13 Microstructures of ECA-pressed Al after annealing at 523 K and 573 K
for 1 hour.

Fig.14 Microstructures of ECA-pressed Al-0.2%Zr after annealing at 523 K and 573 K
for 1 hour.

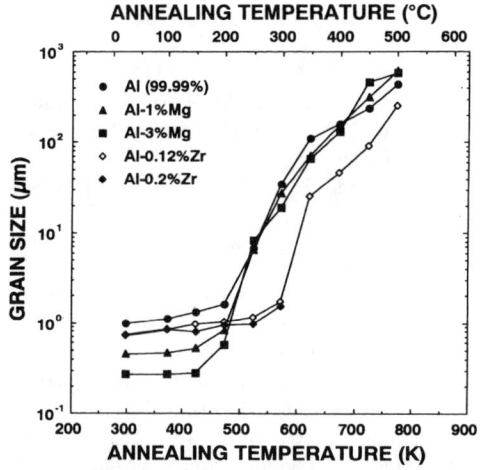

ANNEALING TEMPERATURE (°C)

- Al (99.99%)
- ▲ Al-1%Mg
- ■ Al-3%Mg
- ◇ Al-0.12%Zr
- ◆ Al-0.2%Zr

Fig.15 Variation of average grain size with annealing temperature

boundaries as the number of pressings increases.

(4) Route BC is the most efficient processing route for the production of an equiaxed grain structure having high-angle grain boundaries.

(5) It is necessary to use a die having a channel angle equal to or close to 90° in order to produce a fine-grained structure with high-angle grain boundaries.

(6) There is little difference in the fine-grained structures produced by high and low pressing speeds.

(7) The grain size increases and the subgrain structure is developed as the ECA-pressing temperature increases.

(8) The grain size is reduced with the addition of Mg but a more number of pressings is required for the attainment of equiaxed grain structures with high-angle grain boundaries.

(9) The addition of Zr or Sc is effective to retain the ECA-pressed fine grain structure at higher temperatures.

(10) The ECA-pressed Al-Mg alloys containing Zr or Sc have great potential for superplasticity.

ACKNOWLEDGMENTS

This work was supported in part by the Light Metals Educational Foundation of Japan, in part by a Grant-in-Aid for Scientific Research from the Ministry of Education, Science, Sports and Culture of Japan, in part by the Japan Society for the Promotion of Science and in part by the National Science Foundation of the United States under Grants No. DMR-9625969 and INT-9602919.

REFERENCES

1. V. M. Segal, V. I. Reznikov, A. E. Drobyshevskiy and V. I. Kopylov: Metally 1, p.115 (1981).
2. R. Z. Valiev, N. A. Krasilnikov and N. K. Tsenev,Mater. Sci. Eng. A137, p.35 (1991).

3. Y. Iwahashi, J.Wang, Z.Horita, M.Nemoto and T.G.Langdon, Scripta Mater. 35, p.143 (1996).
4. Y. Iwahashi, Z. Horita, M. Nemoto and T. G. Langdon, Acta Mater. 46, p.3317 (1998).
5. S.D. Terhune, Z. Horita, M. Nemoto, Y. Li, T.G. Langdon and T.R. McNelley in The fourth International Conference on rcrystallization and Related Phenomena, edited by T Sakai and H.G. Suzuki (The Japan Institute of Metals, Sendai, Japan, 1999), p.515.
6. K. Oh-ishi, M. Furukawa, Z. Horita, M. Nemoto and T. G. Langdon, Metall. Mater. Trans. 29A, p.2011 (1998).
7. M. Furukawa, Y. Iwahashi, Z. Horita, M. Nemoto and T. G. Langdon, Metall. Mater. Trans. 29A, p.2245 (1998).
8. M. Furukawa, Y. Iwahashi, Z. Horita, M. Nemoto and T. G. Langdon, Mater. Sci. Eng. A257, p.328 (1998).
9. M. Furukawa, Z. Horita, M. Nemoto and T. G. Langdon in Proceedings of the International Symposium on Light Metals, edited by M Bouchard and A. Faucher (The Canadian Institute of Mining, Metallurgy and Petroleum, Montreal, Quebec, Canada, 1999), p.583.
10. K. Nakashima, Z. Horita, M. Nemoto and T. G. Langdon, Acta Mater. 46, p.1589 (1998).
11. P.B.Berbon, M.Furukawa, Z.Horita, M.Nemoto and T.G.Langdon, Metall. Mater. Trans. 30A, p.1989 (1999).
12. Y. Iwahashi, Z.Horita, M.Nemoto and T.G.Langdon, Metall. Mater. Trans. 29A, p.2503 (1998).
13. H. Hasegawa, S.Komura, A.Utsunomiya, Z. Horita, M.Furukawa, M. Nemoto and T.G. Langdon, Mater. Sci. Eng. A265, p.188 (1999).
14. P.B.Berbon, N.K.Tsenev, R.Z,Valiev, M.Furukawa, Z.Horita, M.Nemoto and T.G.Langdon, Metall. Mater. Trans. 29A, p.2237 (1998).
15. P.B.Berbon, N.K.Tsenev, R.Z,Valiev , M.Furukawa, Z.Horita, M.Nemoto and T.G.Langdon, in Advanced Light Alloys and Composites, edited by R.Ciach, (Kluwer Academic Publishers, Netherlands, 1998), p.477.
16. P.B.Berbon, S.Komura, A.Utsunomiya, Z.Horita, M.Furukawa, M.Nemoto and T.G.Langdon, Mater. Trans. JIM 40, p.772 (1999).

HIGH RESOLUTION EBSD ANALYSIS OF SEVERELY DEFORMED SUBMICRON GRAINED ALUMINIUM ALLOYS

P. B. PRANGNELL[*], J. R. BOWEN[*] A. GHOLINIA[*], AND M. V. MARKUSHEV[**]

* Manchester Materials Science Centre, UMIST, Grosvenor St., Manchester, M17HS, UK. philip.prangnell@umist.ac.uk.
** Institute for Metals Superplasticity Problems, Russian Academy of Science, 450001, 39, St. Khalturin St. Ufa, Russia.

ABSTRACT

There is currently much interest in the use of severe plastic deformation techniques for the production of submicron grained Al-alloys. Because of their small grain sizes, previous investigations have relied on the TEM to study the grain structures after processing. However, the TEM is not that well suited to obtaining statistically significant data on the misorientations of the boundaries present in the deformed state. Recent improvements in the resolution of SEM-EBSD systems means that this powerful technique can now be used to quantitatively analyse severely deformed structures more reliably than was previously possible. Results using this method have shown that the deformation structure is very sensitive to the strain path, material, and processing conditions. With a constant strain path fibrous grain structures tend to be formed and homogeneous submicron grain structures can only be produced at very high plastic strains, of greater than ten. Materials deformed with cyclic strain paths are more isotropic, but generally have bimodal grain structures and contain a lower density of high angle boundaries and only bands of submicron grains.

INTRODUCTION

Most commercially produced aluminium alloys have grain sizes in the range of 10 to 100 μm's. With superplastic Al-alloys, specialised thermomechanical treatments can reduce the grain size to 5-10 μm's and this appears to be close to the limit that can be achieved by conventional processing methods. There are, however, many potential property advantages from reducing the grain size of aluminium alloys to submicron levels, including increased strength [1] and enhanced superplasticity [2]. At the extreme end of the scale amorphous and nanocrystalline materials can be produced, but this generally requires exotic processing methods that are not suitable for the commercial production of bulk materials.

In conventional processes, like rolling and extrusion, one or more of the work piece's dimensions are reduced with strain, and large plastic strains (>5) can only be achieved in foils or filaments which have few structural applications. Recently, a great deal of interest has arisen in methods for deforming *bulk* materials to ultra-high plastic strains under conditions where the net-shape of the billet remains constant during processing, so that there is no restriction to the maximum strain that can be achieved. Such techniques include; reciprocating extrusion [3], repeated roll bonding [4] and equal channel angular extrusion (ECAE) [5]. These methods have become known collectively as "Severe Deformation Processes", because they can potentially be used to deform bulk materials to unlimited plastic strains. By using severe deformation techniques it has been widely claimed that submicron, and in some cases nanocrystalline, grain structures can be obtained in a range of alloys [4,6-9]. Severe deformation techniques therefore have a great deal of potential for the commercial production of submicron grained alloys; in that they can process conventional alloys, can be scaled up to produce relatively large billets, and are cheaper than most other processing methods.

During conventional deformation processes, like plane strain compression in rolling, the development of the deformed state to moderate strains has been extensively investigated and the main features are now reasonably well understood (e.g. [10-12]). However, there have been few systematic studies of how the deformation structure further evolves to form submicron grains at the far higher strains relevant to severe deformation processing. Furthermore, because of the requirement of maintaining a constant billet geometry, some severe deformation processes involve non-conventional strain paths. For example, in reciprocating extrusion the strain is reversed each alternate extrusion cycle and in ECAE the strain path is often varied by rotating the billet between each repeated extrusion [13]. Although in general there is little published information on the affect of varying the strain path on the development of severely deformed microstructures, at lower strains it has been found that in some cases the deformation structures can be partially reversed by redundant straining [14]. In contrast, with ECAE, it has been claimed that processing routes that vary the strain path can be advantageous, in terms of producing a more homogeneous and finer grain structure [13]. This suggests that in ECAE an additional mechanism of grain refinement may be occurring that is related to the intense nature of the shear within the die.

Because of the fine scale of their deformation structures, the TEM has been extensively used to examine the microstructure of severely deformed alloys (e.g. [7,13,15]). Most published observations have relied on the spread in selected area diffraction rings to *qualitatively* interpret the density of high angle boundaries in the deformed state (e.g. [7]). While this has provided much important information, the TEM is not well suited to obtaining statistically significant quantitative data on the misorientations of the boundaries present. It is therefore often not clear in the literature whether the materials observed are genuinely submicron grained, or are comprised of a mixture of low and high angle boundaries (e.g. [16]). Furthermore on annealing severely deformed alloys it has been reported they can recrystallise either discontinuously, or continuously [9,17] and this range of observations is probably related to the poor current understanding of the homogeneity of the deformed state, prior to annealing.

The aim of this paper is to obtain a better understanding of the structure of severely deformed alloys and the mechanisms of the formation of submicron grains, by obtaining quantitative measurements of the boundaries present in the deformed state. To this end, an electron-back-scattered-diffraction (EBSD) system attached to a field emission gun (FEG) SEM has been used to analyse a range of severely deformed materials, as a function of their processing histories. As most severely deformed alloys have well developed cellular structures [15], a quantitative description of the boundary misorientations present, and their distribution within the deformation structure, can provide a very valuable insight into how ultra-fine grain structures evolve, and the effects of the processing conditions and material variables. For example, it is the rate of formation of high angle boundaries (HAGBs) that determines how rapidly a large grained material becomes refined during plastic deformation.

EQUAL CHANNEL ANGULAR EXTRUSION ECAE

The results presented here are based on data obtained from alloys severely deformed using equal channel angular extrusion (ECAE). This processing technique has been adopted because the metal flow behaviour in the tooling used has been well characterised (e.g. [18]) and the strain path can be systematically varied by rotating the billet around its axis before each extrusion cycle [7]. The ECAE process has been adequately described elsewhere and will not be considered in detail here [5,18]. The technique involves extruding a sample through a die with two channels of equal cross-section that meet at an angle (typically 90°). Theoretically in ECAE the sample is sheared homogeneously across its width at the juncture between the two die channels and the magnitude of the strain is determined by the angle between the entrance and exit channel. More normally the sample is slightly bent as well as being sheared and the deformation zone is spread through an arc resulting in a lower overall shear strain and an inhomogeneously deformed layer on the bottom ~ 20% of the billet [18]. The ECAE press used can deform 15 mm diameter, 100 mm long, rods with a constant extrusion rate and die temperature (controlled to ± 2°). 120° and 90° dies were used with sharp internal corners (fillet radii < 1 mm), giving a von Mises effective strain of 0.67 and 1.15 respectively, per extrusion cycle. Samples for EBSD analysis were only taken from regions within the billet subjected to a homogenous shear, and were sectioned in the orientation of the symmetry plane of the die.

HIGH RESOLUTION EBSD ANALYSIS

Electron back-scattered diffraction in the SEM is now a well established technique for obtaining crystal orientation maps from materials (e.g.[19]). However, it is only recently that the introduction of FEG-SEMs has led to an improvement in resolution sufficient for it to be possible to apply this powerful technique to the analysis of sub-micron deformation structures. The instrument used, to obtain the results discussed in this paper, was a commercial EBSD system comprised of a low light CCD camera and an image processor attached to a Philips XL30 FEG-SEM. HKL Technology acquisition software was used to obtain the EBSD orientation maps, which were post-processed with software developed in house (Humphreys ©). The optimum conditions for obtaining high resolution orientation maps with this instrument have been investigated by Humphreys and Brough [20] and have been found to be dependent on a compromise between the minimum diffracting volume and the signal to noise ratio in the back scattered patterns. With small probe currents and low accelerating voltages the pattern quality deteriorates and the software can not as readily deconvolute overlapping patterns, whereas with high beam currents and accelerating voltages the resolution is reduced by the larger beam size and interaction volume. Because of the 70° sample tilt, necessary for EBSD analysis, the spot on the sample is elliptical and with aluminium under optimum operating conditions the resolution of the current system has been shown to be 20 by 60 nm [20]. To obtain reliable data with EBSD analysis at least ten measurements are required across each grain. When analysing severely deformed alloys very fine step sizes are therefore required, which can lead to problems with contamination from the previous beam position reducing the pattern quality. To maximise the speed of each measurement, the EBSD system was configured to use the minimum level of frame integration at the expense of a small reduction in pattern quality. With the existing set-up the system can make 5 orientation measurements per second. This means that 18000 data points can be obtained per hour, allowing grain structure measurements to be made over statistically significant areas.

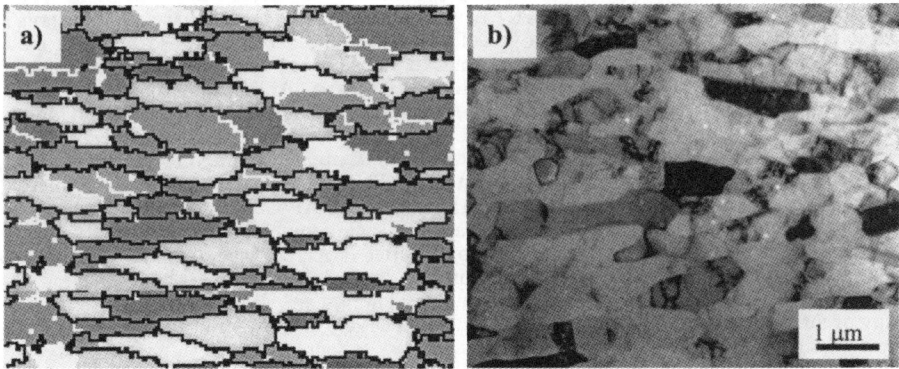

Figure 1 *A comparison of a reconstructed EBSD map (a), obtained with a 0.06 μm step size and showing high (black) low (white) angle boundaries, with a TEM image (b) from the same Al-0.13%Mg sample deformed to a strain of 10 by ECAE.*

In figure 1 an example is given where a cleaned EBSD map, showing the high and low angle boundaries, is compared to a TEM image from the same Al-0.13%Mg sample, deformed to a strain of 10 by ECAE. It can be seen from figure 1 that in terms of the boundary spacings the two images are extremely similar. However, the EBSD map is constructed from data points where the crystal orientation is defined at each location and can be interrogated to obtain quantitative information, such as the boundary misorientations, and can distinguish between grains and subgrains.

In the context of severely deformed alloys, the main draw back of the EBSD technique is that the pattern quality is affected by local lattice distortions. Most dilute Al-alloys exhibit significant recovery at room temperature and can be readily mapped in the deformed condition. However, higher Mg content alloys (e.g. 3%Mg) deformed at room temperature could not be successfully mapped, due to their greater levels of internal stresses and substantially finer grain sizes. The other disadvantage is that the accuracy with which misorientations can be measured deteriorates with poor pattern quality and for deformation structures is typically of the order of 0.5 to 1° [20]. Boundary misorientations less than 1.5° must therefore be excluded from the data. From comparing more accurate TEM measurements of boundary misorientations with the EBSD data, for the same severely deformed sample, it has been found that this can lead to a small underestimation of the fraction of low angle boundaries present (< 4%) [21]. In line with convention, in the EBSD maps presented below, high angle boundaries (shown as dark lines) have been defined as having misorientations greater than 15° and the low angle boundaries therefore have misorientations from 1.5 – 15° (shown as white lines).

THE MICROSTRUCTURE OF SEVERELY DEFORMED ALLOYS

The deformation structure of severely deformed materials has been reported to be strongly influenced by the alloy composition, strain path and deformation characteristics of the process used [6,7,22]. Unfortunately there is currently no definition of when a high plastic strain can be described as severe deformation. Below, the deformation structures of a range of alloys, analysed from EBSD orientation maps, have been compared after deformation to a minimum effective strain of ten at temperatures below their recrystallisation temperature. Although this choice of minimum strain level is somewhat arbitrary, it is substantially higher than can be achieved by conventional processes and is at a level where submicron grains can be formed in most alloys. In the following section the mechanisms of the formation of submicron grain structures will be discussed.

Severely Deformed Microstructures Produced with a Constant Strain Path

In figure 2 an example is shown of an Al-0.3%Mn alloy deformed at room temperature to effective strains of ~ 10 and 16 by ECAE with a 120° die. A summary of the grain size measurements made on a range of alloys deformed to a strain of 10 is also given in table 1. These measurements are based on the average dimensions of grains with greater than 15° misorientation boundaries reconstructed from the EBSD maps. At the same strain level of ten all the alloys investigated appeared to have fairly similar deformation structures, although the scales of the microstructures were dependent on the alloy composition. The deformation structures typically consisted of grains with an aspect ratio of 2.5-3, aligned in the direction of shear. The higher solute alloys deformed at room temperature could genuinely be described as submicron, as their average grain sizes were less than 1µm. However, pure Al had an average grain size greater than 1 µm and all of the materials still contained a significant proportion of low angle grain boundaries. The fraction of high angle grain boundaries (HAGBs) was in the range of 0.6 - 0.8, which is significantly less than that expected for a random grain assembly (HAGB fraction = 0.96 [23]). (Some typical boundary misorientation distributions are shown below in figure 4). In many of the materials, even after a strain of ten some larger, fibrous grains were still present (see arrow in figure 2a). In general the average grain size decreased with higher Mg levels and reduction in processing temperature [24]. The data for the Al-3%Mg alloy in table 1 had to be estimated from SEM back-scattered images and the TEM, because after room temperature deformation this alloy had not recovered as much as the other aluminium alloys and could not be mapped successfully. Some of the boundary spacings in this alloy were of the order of 100 nm and are therefore approaching the nanocrystalline size range.

In figure 2 and table 1 data is included on the effect of further increasing the effective strain from ten to sixteen with an Al-0.3% Mn alloy. From these results it can be seen that, although the grain width did not change significantly with higher strain, the fraction of HAGBs has increased from 0.75 to 0.79 and the average grain length has reduced from 1.9 to 1.5. These measurements show that, even after strains as high as ten, with a further increase in the strain the average grain size can still be refined (all-be-it slowly). This continued refinement can thus be mainly attributed to the progressive break up of the remaining larger fibrous grains, which results in an increasingly homogeneous grain structure.

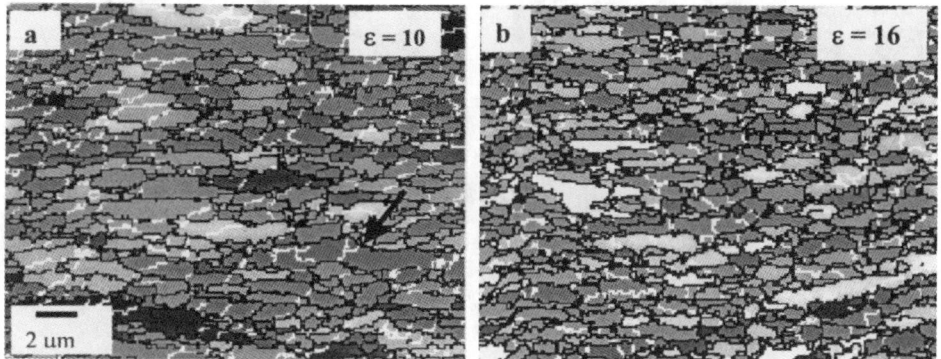

Figure 2 *EBSD maps for an Al-0.3%Mn alloy processed at room temperature by ECAE, with a 120 °die to a strain of (a) ten and (b) sixteen, with a constant strain path.*

Material	Effective Strain	Temp.	Grain Size (μm) Width	length	Fraction HAGBs
pure Al -99.8	10	20 °C	0.9	2.3	0.65
Al-0.13% Mg	10	20 °C	0.4	1.1	0.72
Al-0.13% Mg	10	100 °C	0.8	2.0	0.75
Al-3%Mg0.6Fe0.2Zr	10	200 °C	0.5	0.8	0.6
Al-3%Mg	10	20 °C	~ 0.1-0.2	~ 0.3-0.6	0.7-0.8
Al-0.3% Mn	10	20 °C	0.8	1.9	0.75
Al-0.3% Mn	**16**	20 °C	0.7	1.5	0.79

Table 1 *Grain size measurements from EBSD maps, using reconstructed grains > 15 ° in misorientation, in a range of alloys deformed at different temperatures by ECAE, with a constant strain path using a 120 °die.*

Severely Deformed Microstructures with a Cyclic Strain Path

It is well established that deformation with a cyclic strain path can lead to the formation of different deformation structures compared to deformation with a constant strain path [7,22,25]. In ECAE processing it is possible to change the strain path by rotating the billet around it's axis between each extrusion cycle. Furthermore techniques like reciprocating extrusion, are based on reversing the strain every alternate extrusion cycle. In ECAE three main processing routes have been investigated [7,22,25].

Route A (No Rotation) Where a constant strain path is maintained.

Route B (+ 90° Rotation) Where the billet is rotated through 90° between each extrusion cycle. This is normally continuously in one direction (hence +90°) but can also be in alternate directions (±90°).

Route C (180° Rotation) Where the billet is rotated through 180° between each extrusion cycle.

With 180° rotation shear occurs in one plane and is reversed every alternate extrusion cycle, giving a redundant strain every second cycle. With + 90° rotation the billet is sheared in two orthogonal planes and in the second extrusion the billet is sheared at 90° to the original shear direction. The third extrusion then reverses the first shear and the forth the second, so that strain becomes redundant every four extrusion cycles.

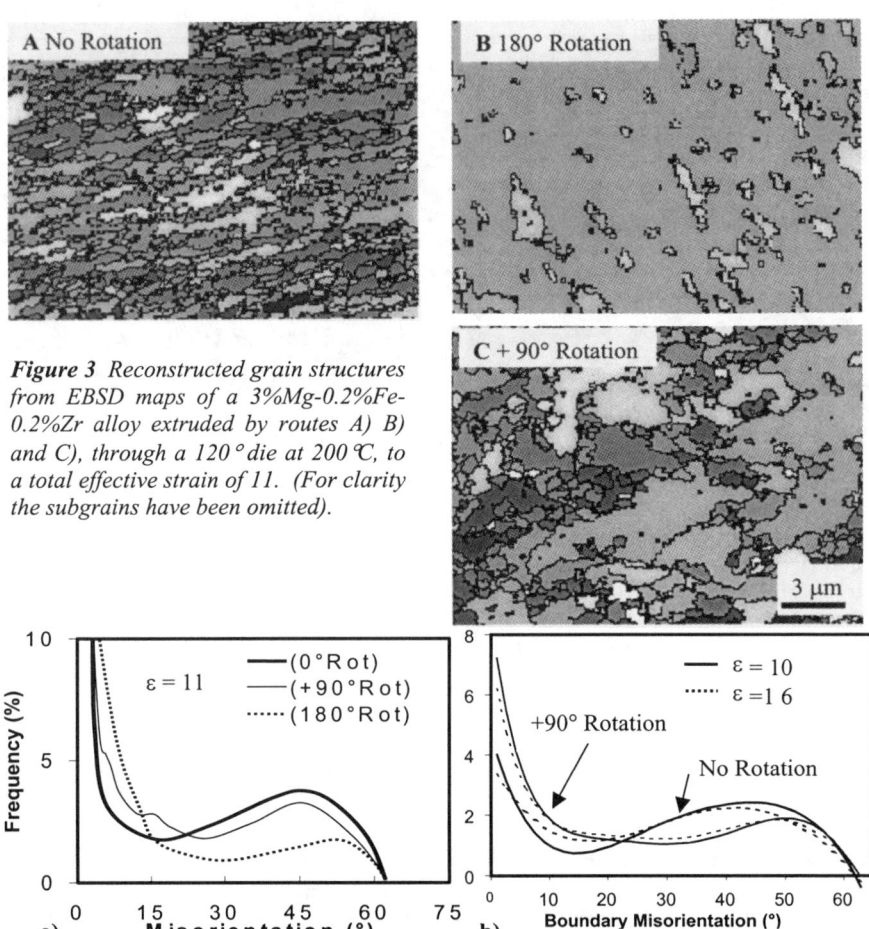

Figure 3 *Reconstructed grain structures from EBSD maps of a 3%Mg-0.2%Fe-0.2%Zr alloy extruded by routes A) B) and C), through a 120° die at 200°C, to a total effective strain of 11. (For clarity the subgrains have been omitted).*

Figure 4. *Boundary misorientation distributions measured for (a) a 3%Mg-0.2%Fe-0.2%Zr alloy extruded to a total effective strain of 11 through a 120° die, at 200°C, by routes A, B, and C and (b) a 0.3%Mn alloy extruded at room temperature by routes A and B to total strains of 10 and 16. Each distribution includes measurements from over 1000 boundaries.*

In figure 3 reconstructed grain structures are shown that have been obtained for a 3%Mg-0.2%Fe-0.2%Zr alloy extruded, by routes A, B and C, 16 times through a 120° die, at 200°C, giving a total effective strain of 11. With a constant strain path the grain structure is similar to that discussed above. With 180° rotation the structure was still found to be largely comprised of subgrains. The original grains were still distinguishable and only a few new isolated fine grains could be seen, appearing in bands within the original grains (shown in fig. 3b). In comparison the +90° rotated sample (fig. 3c) was much more refined and contained a much higher density of fine grains, again appearing in bands, resulting in the original grains being split into a bimodal distribution comprised of submicron grains and grains of 4-6μm in size. With the cyclic strain paths the average grain shape was more equiaxed than for constant strain path deformation. However, from the boundary misorientation distributions, shown in figure 4a, it can be seen that the sample deformed *without* rotation has the highest density of high angle grain boundaries and the sample processed with a 180° rotation by far the lowest, while the +90° rotated sample falls between these two extremes. Although these results were obtained at an elevated temperature, similar observations have been found from comparing EBSD maps for an 0.3%Mn alloy deformed at room temperature by routes A and B (figure 2b and 5a) and the corresponding boundary misorientation distributions shown in figure 4b. In figure 5b a further example is given of a Al-3%Mg alloy deformed at room temperature by route B using a 90° die to a total effective strain of eight. The maximum strain was limited because the rotated sample fractured after seven extrusion cycles. As has been explained previously, EBSD mapping is not possible on alloys containing high Mg levels deformed at room temperature. The map in figure 5b was therefore obtained after lightly annealing the samples at 175° C for one hour. Although the annealing treatment does cause some recovery, it should not significantly alter the distribution of boundaries within the deformed state. From figure 5b it can be seen that again the 3%Mg materials contains only bands of submicron grains.

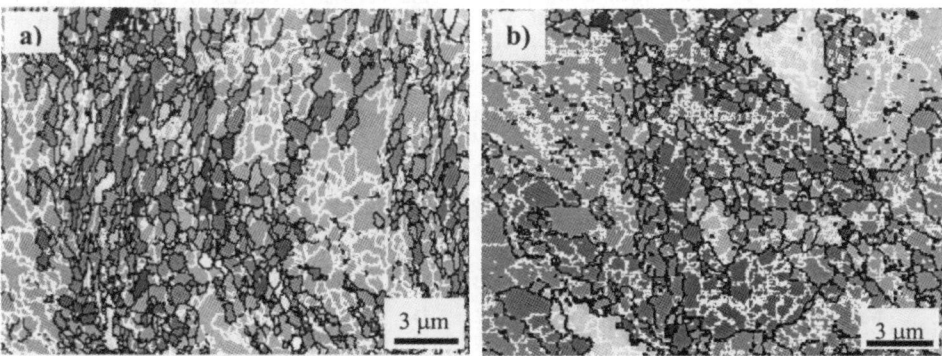

Figure 5 *EBSD orientation maps of two materials deformed by route B at room temperature, showing the high and low angle boundaries. (a) a 0.13%Mn alloy deformed at room temperature through a 120° die to a total effective strain of 16 and (b) a 3%Mg alloy deformed to a total effective strain of 8 through a 90° die. (figure 5a should be compared to 2b).*

The grain structures produced with a cyclic strain path are therefore quite different to those with a constant strain path. Although the cyclic strain paths do not tend to form the directional structures found with a constant strain path, the deformation structures produced are still often inhomogeneous and generally contain a lower density of high angle boundaries at similar high strains. For the conditions investigated, it appears that the 180° rotation route (C) produces the least refinement. Similar observations have been reported from qualitative TEM analysis [7], and more recently boundary misorientations measured in a AA5083 aluminium alloy deformed by 180° rotation at 150°C in a 90° die have also shown a high proportion of low angle boundaries [16]. With +90° rotation, it appears that much more refined grain structures can be produced, but the submicron grains appear to be concentrated in bands, resulting in a bimodal grain size distribution.

MECHANISMS OF GRAIN REFINEMENT DURING SEVERE DEFORMATION

Constant Strain Path

In this section an example of a simple Al-0.13%Mg alloy deformed by ECAE to high plastic strains will be used to illustrate the evolution of the deformation structure with a constant strain path. At low strain levels, the main features of the deformation structure were similar to those found with more conventional deformation processes and theories well established for moderate strains can thus be extended to explain the evolution of structures at very high plastic strains (e.g. [10-13]). It should be noted that the following analysis should apply to most dilute Al-alloys and the Al-3%Mg alloys deformed above ~ 200°C. However, at moderate strains high Mg content alloys deformed at room temperature are known to develop different substructures to lower solute alloys and may behave differently during severe deformation [12]. In particular such alloys are susceptible to the localisation of plastic flow [26], which can be promoted by the intense shear strain developed in ECAE processing.

The formation of a submicron grain structure requires an enormous increase in the density of high angle grain boundaries within a material. During severe deformation, possible mechanisms that can form new high angle grain boundaries are 1) the distortion and 2) the subdivision of the original grains [27]. At ambient temperatures grain subdivision is normally thought to be the dominant mechanism. Grain subdivision is complex and occurs at several microstructural scales. The deformation pattern that develops within an individual grain depends on its crystallographic orientation and the constraint of the surrounding grains. Coarse grain scale primary deformation bands are produced from quite low strains. Much finer scale, deformation bands, or "cell blocks" [11], also develop within each coarse band. This behaviour occurs because it is usually energetically easier for a grain to deform by splitting into bands, which individually deform on a lower number of slip systems, but collectively retain the constrained grain shape, than by the five independent slip systems required by Taylor theory [23]. As the strain increases the deformation bands progressively rotate to diverging orientations and their boundaries develop into new high angle boundaries [11-12]. Some grain orientations are particularly unstable during deformation, such that different parts of the grain rotate towards very different end orientations, and this can rapidly form highly misorientated boundaries.

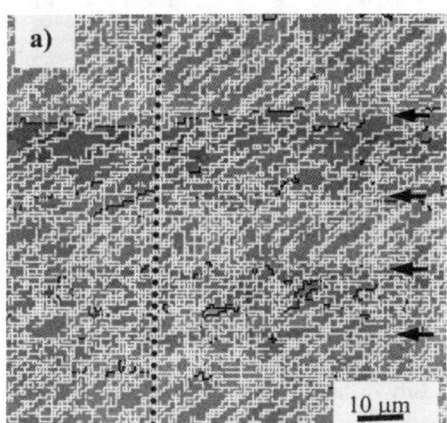

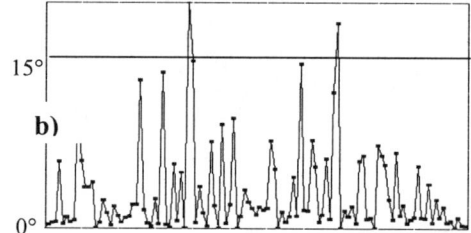

Figure 6 (a) An EBSD map of an Al-0.3%Mg alloy deformed by ECAE to an effective strain of 1.3 showing the early stages of grain subdivision. In (b) the boundary misorientations measured along the line indicated in (a) are shown.

In figure 6 an example is shown of the substructure within the centre of a grain obtained from an EBSD map of the Al-0.3%Mg alloy, deformed by ECAE to an effective strain of 1.4. From figure 6 it can be seen that the grain has started to subdivided on several scales. There are several coarse bands lying horizontal in the page (indicated by arrows in figure 6) and within some of these bands the low angle boundaries are all aligned at ~ 45° to the primary bands. The boundaries between these primary bands are already approaching 15° misorientation, however, the boundaries within the layers are typically of < 6° misorientations (see figure 6b).

As the strain increases, the high angle boundaries rotate towards the shear plane, to form a lamella structure. The change in the constraint on the grain will at the same time encourage some additional subdivision. In figure 7 the average subgrain size and high angle boundary spacing transverse to the shear plane are plotted as a function of strain. From figure 7 it can be seen that the subgrain size remains relatively constant, but the high angle boundary spacing reduces with strain and merges with the subgrain size at effective strains of ~ 10. The spacing of the lamella boundaries therefore reduces with strain until they become the order of the subgrain size [9]. It is less clear how the ribbon grains formed by the lamellae boundaries break up into submicron grain fragments. This process would be encouraged by heterogeneities in plastic flow caused by large (>1μm) second phase particles, or shear banding [27]. From figure 8 it can be seen that the fine grains develop very inhomogeneously, originating in bands within the deformed state. This behaviour is probably related to the strong relationship between a crystal's stability during deformation and its crystallographic orientation. As the strain increases these bands are pushed closer together and more high angle boundaries are formed, increasing the volume fraction of submicron grains. The submicron grains are therefore fist produced at relatively moderate strains (~ 3-5) but it is not until very high strains are reached (> 10) that the entire microstructure has been refined and a uniform submicron grain structure can be achieved (e.g. figure 2b).

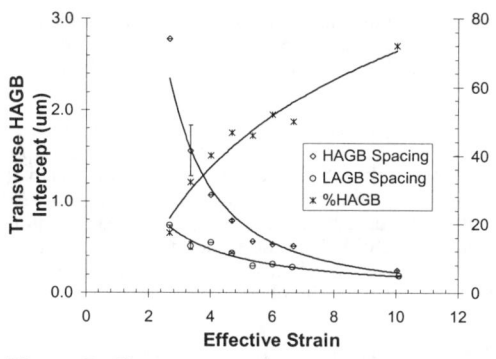

Figure 7 *The average subgrain size and high angle boundary separation transverse to the shear plane, as a function of total effective strain.*

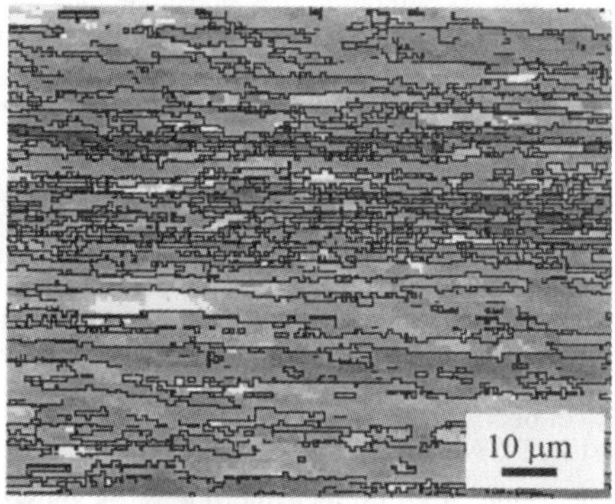

Figure 8 *An EBSD map of an Al-0.3%Mg alloy deformed by ECAE to an effective strain of 3.4 showing bands of submicron grains and lamella grain boundaries (only the HAGBs are shown).*

Effect of Strain Path

The evolution of the deformed state, when the strain path is changed in between each extrusion cycle, is currently not as well understood. With a cyclic strain path the strain per cycle is a more important factor than if the strain path is kept constant. The deformation behaviour will also be sensitive to the material's composition and the processing temperature. Unfortunately a systematic set of data does not yet exist for the effects of strain path, strain per-cycle, and material composition. However, the results obtained so far do illustrate some interesting trends that are discussed below.

With ECAE both the 180° rotation and +90° rotation routes lead to an overall redundant strain every second and fourth extrusion cycle respectively. Research carried out using torsion on an AA3003 alloy at 300°C has found that, on reversing the shear strain, not only were the original grain shapes reconstructed, but the initial texture was also recreated [9]. On deforming an Al-3%Mg-0.2Fe-0.2Zr alloys at 200°C by ECAE in a 120° die similar results have been observed, with respect to the grain shapes being reassembled after a redundant strain [22]. With redundant deformation, on reversal of the strain, it might be expected that if the deformation pattern and slip systems operating within a grain remain the same the deformation bands would tend to rotate back towards their starting orientations, and the misorientations across the cell block boundaries will reduce towards zero. With 180° rotation, which reverses the shear strain every second cycle, this would explain the observed very low rate of formation of new high angle boundaries within the grains (fig. 3b). With +90° rotation, despite the overall strain being redundant, a far higher level of grain refinement is generally found. In this case it can be argued [22] that it is more difficult for the deformation bands to rotate back towards their original orientations because shear occurs on two orthogonal planes and each shear is reversed out of sequence. In this case in the first extrusion cycle a deformation pattern will develop within each grain. On rotating the shear plane by 90°, in the next extrusion cycle, new deformation bands will become established, each deforming individually on a different set of slip systems to those used in the previous orientation. In the third extrusion cycle, on rotating through a further 90°, the first shear will now be reversed, but this will change the stress state acting on the grain compared to the previous cycle and it will now not be able to deform back with the same deformation pattern established during the immediately prior shear. The non-sequential reversal of the shears would therefore be expected to lead to a more complex deformation behaviour, involving intersecting deformation bands originating form the two shear planes. Under these conditions local boundary misorientations can still be built up in the deformation structure, despite the redundant strain, resulting in the formation of bands of submicron grains.

A further consideration with cyclic strain paths is that the shape of an element does not progressively change with strain. Relatively stable deformation patterns can therefore develop and it is more difficult to refine coarse deformation bands formed early in the process than with a constant strain path.

Recently, it has been suggested by Rauch [28] that changes in the strain path may promote shear localisation during ECAE processing. In two stage straining tests it has been shown that pre-straining a sheet and then rotating the tensile axis through 90° can lead to a reduction in tensile elongation in some materials, relative to that found in uninterrupted monotonic straining [29]. This effect is caused by a transient reduction in work hardening rate which can lead to strain localisation, and is thought to arise from instability in the dislocation structures, formed during the pre-strain, when the strain path is changed [29,30]. The significance of this behaviour has not yet been fully investigated in the context of the far higher strain levels per-cycle used in processes like ECAE.

In figure 9 an optical micrograph is shown of a Al-3%Mg alloy deformed at room temperature after the second extrusion, cycle through a 90° die, using the +90° rotation route. Intense shear bands can be seen within the grain leading to very high local shear strains. This behaviour first became obvious after the second extrusion cycle when the strain path was altered by 90°. At higher strains trans-granular macroscopic shear bands developed leading to the premature fracture of the billet after seven extrusion cycles. With a constant strain path such a dramatic shear localisation behaviour was not observed and the billet could be deformed to greater strains without fracture. Due to the strong interactions between dislocations and Mg, high Mg level alloys are particularly susceptible to this behaviour. However, in the lower solute alloys (e.g. the 0.3%Mn alloy), or high Mg alloys deformed at elevated temperatures (i.e. the 200°C used above), intense shear localisation was not seen.

Figure 9 An anodised optical micrograph of an Al-3%Mg alloy deformed twice at room temperature by ECAE through a 90° die using the +90° rotation route (B).

The intense shear localisation observed in the high solute 3%Mg alloys deformed at room temperature after a change in strain path leads to the majority of the strain being concentrated in narrow bands and leaves relatively undeformed regions between the bands. Although there appears little doubt that the intense strain gradients that occur across the shear bands will readily lead to the formation of new high angle boundaries, this will not result in a homogeneous submicron grain structure unless the entire volume of the material has been sheared by an intense deformation bands. This seems unlikely, given that the nature of the process is that the shear is localised, and in figure 9 the shear bands have a spacing of around 5 - 10 µm. In figure 5b after seven extrusion cycles (a total effective strain of 8, the maximum that could be obtained without fracture) the submicron grains within the Al-3%Mg sample were still inhomogeneously distributed, and it can be inferred that the bands of fine grains in figure 5b correspond to regions subjected to a higher localised shear strain. It thus appears that shear localisation may be an important factor in the grain refinement process in alloys deformed with cyclic strain paths, particularly with high solute levels at low temperatures, but that this does not generally result in uniform submicron grain structures.

CONCLUSIONS

The analysis of severely deformed alloys by high resolution EBSD has allowed the density and distribution of high angle boundaries within the deformed state to be quantified, as a function of the strain and processing conditions. This has revealed that even though very fine grain structures can be produced at high strains the deformed state is often inhomogeneous and contains a significant proportion of low angle boundaries. The highest fraction of HAGBs that has been achieved was ~ 0.8 at an effective strain of 16. The average grain size decreases with increasing solute levels and lower deformation temperatures. In room temperature deformation materials with a sub-micron average grain size can only be produced in Al-alloys containing a significant level of solute.

The grain structure produced is very sensitive to the strain path used. With a constant strain path the submicron grains develop inhomogeneously from strains of 3-5, leading to a banded structure, consisting of layers of submicron grains and larger fibrous grains. As the strain level increases, the volume fraction of submicron grains rises and the fibrous grains are progressively refined until eventually a relatively uniform ultra-fine grain structure can be produced at plastic strains greater than ten. The resultant materials consist of grains aligned in the shear direction with an aspect ratio of ~ 2 - 3.

In ECAE the microstructures produced by cyclic strain paths are substantially different to those developed with a constant strain path. The 180° rotation route, (route C) which gives rise to a cyclic redundant strain in only one shear plane, appears to be a very inefficient way of refining the grain size. Materials processed in this way were found to contain a comparatively low density of submicron grains. Alloys deformed by +90° rotation in ECAE (route B), where redundant shears take place non-sequentially on two orthogonal planes, have a far greater density of high angle boundaries and form a bimodal grain size distribution, of larger 4-6 µm grains surrounded by bands of submicron grains. However, they still tend to have a lower fraction of high angle grain boundaries than alloys deformed with a constant strain path. Shear localisation appears to be promoted by changes in strain path, particularly in high Mg content alloys deformed at room temperature, and is an important aspect of the development of their deformation structures.

ACKNOWLEDGEMENTS

The authors would like to acknowledge the support of the EPSRC (grants GR/L06997 and GR/L96779) Alcan International, British Steel, the DERA and GKN Westland Helicopters for finical support. We would also like to thank Prof. F.J. Humphreys and Dr. E.F. Rauch for helpful discussions and I. Brough for assistance with the EBSD measurements.

REFERENCES

1. M.V. Markushev, C.C. Bampton, M.Y. Murashkin, and D.A. Hardwick, Mat. Sci. Eng., **A234**, 927 (1997).
2. M. Kawazoe, T. Shibata, T. Mukai, and K. Higashi, Scripta Mater., **36**, 699 (1997).
3. S-Y. Yuan and J-W. Yeh, *in THERMEC'97*, **1**, edited by T. Chandre and T. Sakai, (TMS, Warendale, PA, 1997), pp. 1143-1149.
4. Y. Saito, N. Tsui, H. Utsunomiya, T. Saki, R-G. Hong, in 6th Int. Conf. on Aluminium Alloys, ICCA-6, **3**, edited by T. Sato, S. Kumai, T. Kobayashi, Y. Murakami, (JILM, Tokyo, Japan, 1998), pp. 1967-1973.
5. V. M. Segal, V. I. Reznikov, A. E. Drobyshevskiy, V. I. Kopylov, Russ. Metall. **1**, 99 (1981).
6. S. Ferrassee, V.M. Segal, K.T. Hartwig, R.E. Goforth, J. Mater. Res, **12**, 1253 (1997).
7. Y. Iwahashi, Z. Horita, M. Nemto, T.G. Langdon, Acta Mater, **46**, 3317 (1998).
8. A. Gholinia, J.R. Bowen, P.B. Prangnell, F.J. Humphreys, in 6th Int. Conf. on Aluminium Alloys, ICCA-6, **3**, edited by T. Sato, S. Kumai, T. Kobayashi, Y. Murakami, (JILM, Tokyo, Japan, 1998), pp. 577-583.
9. C. Harris, S.M. Roberts, P.B. Prangnell, F.J. Humphreys, ed. T. R. McNelley, in ReX'96, 3rd Int. Conf. on Recrystallization and Related Phenomena, Monteray, (MIAS, Montery, CA, 1997) p. 587-593.
10. J.G. Sevillano, P. van Houtte, E. Aernoudt, Prog. Mater. Sci., **25**, 69 (1980).
11. D.A. Hughes and N. Hansen, Acta Mater, **45**, 3886 (1997).
12. D.A. Hughes and A. Godfrey, ed. T.R. Bieler, L.A. Lali and S.R. MacEwen, in Hot Deformation of Al-Alloys II (TMS, Warendale PA, 1998) pp. 23-36.
13. K. Oh-Ishi, Z. Horita, M. Nemoto, M. Furukawa, and T.G. Langdon, Metall. and Materials Trans. **29A**, 2245 (1998).
14. H.E. Vatne, M.G. Mousavi, S. Benum, B. Ronning and E. Nes, Mat. Sci. Forum, **217-222**, 553 (1996).
15. A. Korznikov, O. Dimitrov, G. Korznikova, Annalas de Chime-Sci des Mat. **21**, 443 (1996)
16. L. Dupuy, E.F. Rauch and J.J. Blandin, NATO Workshop on Severe Deformation, (Moscow 1999) in press.
17. J. Wang, M. Furukawa, Z. Horita, M. Nemoto, R.Z. Valiev, T.G. Langdon, Mat. Sci. Eng., **A216**, 41 (1996).
18. J.R. Bowen, A. Gholinia, S.M. Roberts, P.B. Prangnell, Mat. Sci. Eng. in press.
19. D.J. Dingley and V. Randle, J. Mater. Sci, **27**, 4545, (1992).
20. F.J. Humphreys, I. Brough, J. Microscopy, **195**, 6, (1999).
21. O. Mishin (unpublished results).
22. A. Gholinia, P.B. Prangnell, and M.V. Markushev, Acta Mater. (1999) in press.
23. F.J. Humphreys and M. Hatherly, *Recrystallisation and Related Annealing Phenomena*, Pergamon Publishers, Oxford (1995).
24. J.R. Bowen, P.B. Prangnell, and F.J. Humphreys, in 20th Risø Symp, on Materials Science, editors J.B. Bilde-Størensen et al. (Risø National Labs, Roskilde, Denmark, 1999) pp. 269-276.
25. V.M. Segal, Mat. Sci. Eng., **A197**, 157 (1995).
26. K. Chihab, Y. Estrin, L.P Kubin, J. Vergnol, Scripta Metall. **21**, 203 (1987)
27. F.J. Humphreys, P.B. Prangnell, J.R. Bowen, A. Gholinia, and C. Harris, Phil. Trans. R. Soc. Lond. A **357**, 1663 (1999).
28. E.F. Rauch (private communication).
29. D.J. Lloyd and H. Sang, Metall. Trans., **10A**, 1767 (1979)
30. D.V. Wilson, M. Zanrahimi and W.T. Roberts, Acta metall. Mater, **38**, 215 (1990).

SPD PROCESSING AND SUPERPLASTICITY IN ULTRAFINE-GRAINED ALLOYS

R.Z. VALIEV*, R.K.ISLAMGALIEV**
*Institute of Physics of Advanced Materials, Ufa State Aviation Technical University,
12, K. Marks st., Ufa, 450000, Russia, RZValiev@mail.rb.ru
**Institute of Mechanics, Russian Academy of Science,
12, K. Marks st., Ufa, 450000, Russia, ris@mail.rb.ru

ABSTRACT

Severe plastic deformation (SPD), for example by intense plastic straining under high pressure, is an innovative technique for producing ultrafine-grained (UFG) metals and alloys. The SPD fabricated UFG structures can lead to enhanced superplasticity, which, however, depends strongly on processing parameters. The present paper focuses on examples of attaining enhanced superplasticity in several alloys, subjected to SPD and considers the relationship between processing – UFG structures - superplastic properties in SPD produced materials.

INTRODUCTION

Presently it is established that processing of ultrafine-grained (nano- and submicrocrystalline) materials by SPD methods, e.g. equal channel angular (ECA) pressing or severe plastic torsion straining (SPTS) under high pressure can provide a sharp increase in superplastic properties and obtaining of superplasticity at relatively low temperatures and/or high strain rates [1-6]. At the same time, attaining enhanced superplastic properties is a complex problem, which depends on different processing and microstructural parameters. It has been shown that for SPD materials it is typical a presence of not only very small grain size, but also specific defect structure and often a change of phase composition [7]. Moreover, SPD in multi-phase alloys leads to highly metastable states due to formation of supersaturated solid solutions and disordering of intermetallic phases. From the other hand, the obtained microstructural parameters are associated with the SPD processing conditions (e.g. processing routes, temperature and strain rates). For example, it was shown recently [8] that using various routes of ECA pressing it is possible to produce microstructures which differ very much in non-homogeneity and number of high angle grain boundaries. Therefore, it is important to analyze the relationship: processing – UFG structures - superplastic properties in SPD materials.

The present paper is devoted to studies of affects of regimes and SPD processing routes on superplastic behavior in several alloys and it focuses on requirements to microstructure in order to attain enhanced superplasticity in these alloys.

MATERIALS AND EXPERIMENTAL PROCEDURES

The light weight aluminum alloy 1420 (Al-5.5%Mg-2.2Li-0.12%Zr) [9], a boron doped Ni_3Al intermetallic (Ni-3.5%Al-7.8%Cr-0.6%Zr-0.02%B) [4] and a commercial Zn-22%Al binary alloy [2], where enhanced superplasticity was already demonstrated after the SPD processing, have been used for these investigations.
Two techniques of SPD have been applied for processing of UFG structures and their principles are illustrated in Fig. 1. During ECA pressing (Fig. 1a) the sample was pressed under a load P

through two channels of equal cross-section intersecting at the angle $\Psi=90°$ and the pressing was repeated to attain the required level of strain. This technique was also recently used to study the microstructure and superplasticity in Al 1420 processed by ECA pressing at temperatures decreasing from 400°C to 200°C [3,9]. However, this temperature interval coincided with the temperature of the DSC exotherms during heating of a quenched Al 1420 alloy with a rate of 10 K/min [10]. Therefore we used two other regimes of ECA pressing for processing the Al 1420 alloy: first – at 370°C for 10 passes, second – at 420°C for 10 passes. In these experiments the samples were prepared in the form of cylinders with diameters of 20 mm and length of about 80 mm.

During SPTS (Fig. 1b) disk samples were subjected to high plastic deformation at room temperature by torsion under imposed pressure of about 5 GPa. Typical samples processed by this technique were in a form of disks with a diameter of 12 mm and thickness of 0.3 mm.

Transmission electron microscopy (TEM) studies of the foils were conducted using a JEM–100B and a JEOL ARM 1000 (NCEM, Berckeley) electron microscopes. Microhardness was measured using a Vickers diamond pyramidal indenter under a load of 200 g. Tensile ECAP samples were machined parallel to the longitudinal axes of the as-pressed cylinders with gauge length of 5 mm and gauge cross-sections of 2.5×1 mm^2. These samples were pulled to failure in air at selected elevated temperatures using an Instron testing machine operating at a constant strain rate of cross-head displacement. The strain rate sensitivity value m was measured as a function of both strain and strain rate by jumping the strain rate to its higher value. Further details on processing and tests are also given in our previous works [2,7,9].

RESULTS

Al 1420 alloy

After quenching and SPTS (5 turns, P = 6 GPa) at room temperature the typical microstructure in Al 1420 alloy had the average grain size measured from the dark field image less than 70 nm [7, 10]. After heating up 300°C there was some grain growth, but the mean grain size was still less than 300 nm (Fig.3).

During the DSC study of the SPTS Al 1420 alloy two runs were made on the same specimen (Fig. 2). The specimen was first heated at a rate of 10K/min to 500°C, allowed to cool down to room temperature and then heated again. It can be noted that the specimen showed several peaks in the first run and these peaks are not prominent in the second run. There are four distinct peaks; A – endothermic at 70-120°C, B – exothermic at 140-170°C, C – exothermic at 180-250°C, and a broad D endothermic at 300-480°C. Following [10], the origin of the endothermic peak A is likely to be associated with the dissolution of metastable δ' – Al$_3$Li phase. On the other hand the appearance of two exothermic peaks B and C can be associated with the precipitation of T-Al$_2$LiMg phase and recovery of lattice defects (vacancies and dislocations). The broad D peak appears to be made of three individual peaks. The origin of D peak is likely to be a result of precipitate dissolution in the matrix.

The specimens after testing at various temperatures in the interval of 250-350°C show neck free elongation that is characteristic of superplastic flow. The alloy demonstrates superplasticity at very high strain rates, up to 5×10^{-1}s^{-1} at rather low temperature of 300°C (Fig.4). The maximum elongation of 900% is observed at a strain rate of 1×10^{-2} s^{-1} with a flow stress of 40 MPa. Elongation decreases slightly with increasing strain rate although it remains quite high and

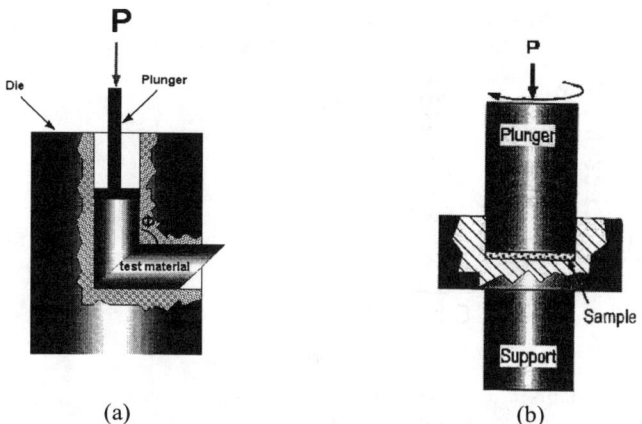

(a)   (b)

Fig. 1: Principles of SPD methods: (a) equal channel angular pressing, (b) severe plastic torsion straining.

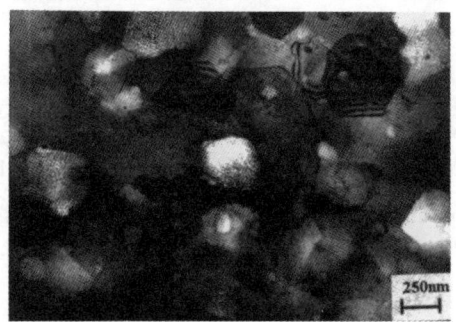

Fig.2. Microstructure of the Al 1420 alloy after SPTS and annealing at 300°C for 5 min

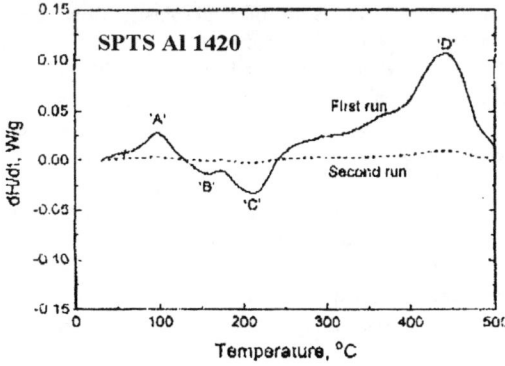

Fig.3. DSC curves of the SPTS Al 1420 alloy.

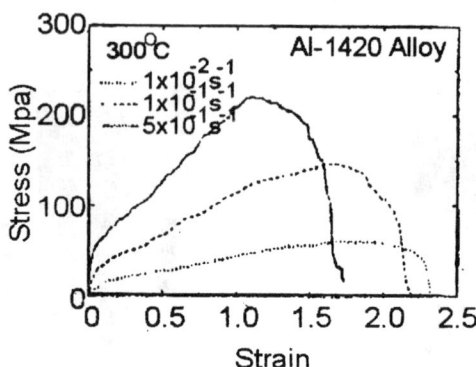

Fig.4.The variation true stress with strain as a function of strain rate for
the SPTS Al 1420 alloy.

significant strain hardening take place. Increasing in testing temperature up to 350°C at a strain
rate of 1×10^{-1} s^{-1} leads to a decrease in the flow stress to 16 MPa and the sample demonstrates
high elongation of 890%.

After ECA pressing at temperatures 370°C and 420°C the microstructure was characterized by a
mean grain size of about 1 μm and 3 μm, respectively. Moreover inspection of the
microstructures (Fig.5) reveals an essential difference in a phase composition, because the
specimens pressed at 370°C (Fig.5a) contain particles of the Ti-Al$_2$LiMg phase, 0.3-0.4 μm in
size, though there was no evidence of the presence of this phase in the specimens pressed at
420°C (Fig.5b). It should be noted that the same particles were also observed in the SPTS
samples (Fig.2) which demonstrated HSRS. These particles have a high volume fraction of
about 10-20%.

The samples after ECA pressing using the two different procedures described above were
pulled to failure at different testing temperatures. Some results of mechanical tests of the
specimens pressed by ECA pressing at 370°C are documented in Fig.6. It is apparent that the
samples exhibit the fundamental characteristics of superplastic flow: uniform deformation
within the gauge length and an absence of visible localized necking. The optimum of
superplasticity at 400°C is found at high strain rate of 1.2×10^{-2} s^{-1} (Fig.7) which is two orders of
magnitude higher that in the specimen with a grain size of about 5 μm [11]. Moreover the
elongation to failure of 1500% recorded in this material (Fig. 6) is highest elongation ever
reported for this commercial Al 1420 alloy.

Under these conditions the flow stress is not high and it reaches 20 MPa. During increasing the
strain rate up to $1{,}2*10^{-1}$ s^{-1} elongation to failure also exceeds the value of 1000%, being
1200%, and even at the highest strain rate 1.2 s^{-1} it amounts 340%. The view of this deformed
sample is typical for superplastic behavior (Fig.6). The flow stress in this case increases
noticeably, however, even at the highest strain rate 1.2s^{-1} it does not exceed 90 MPa (Fig.7).

The strain rate sensitivity of the flow stress, which was measured by jumping the strain rate, are
quite interesting (Fig.8). It is seen that the strain rate sensitivity in the ECA-pressed alloy at
400°C is not high and at the optimal strain rate reaches 0.40 - 0.45 depending on the strain

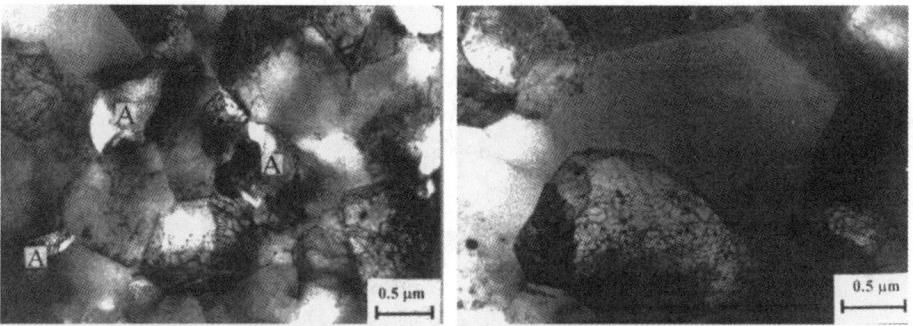

Fig.5. Microstructure of the Al 1420 alloy after ECA pressing at (a) 370°C and (b) 420°C. The particles (grains) of T-Al$_2$LiMg phase are marked by symbol A.

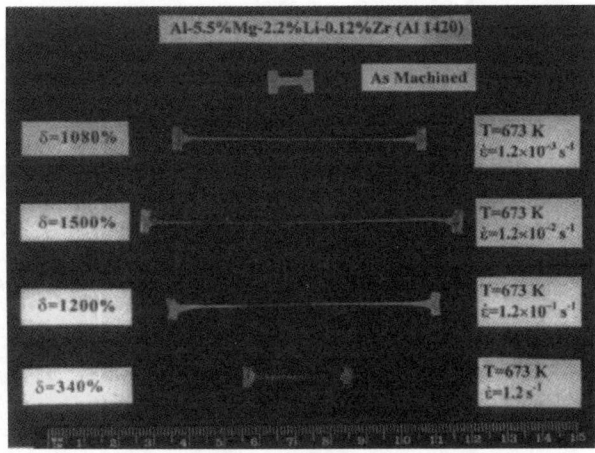

Fig.6. Appearance of specimens processed by ECA pressing at 370°C and pulled to failure.

of the sample. In other cases of strain rates the value of the strain rate sensitivity is less and does not exceed 0.25 whereas elongation to failure are high. Thus, the obtained values of the strain rate sensitivity in UFG 1420 alloy produced by SPD during high strain rate superplasticity are considerably lower compared to conventional superplasticity in the given alloy with a grain size of 5 μm, where the strain rate sensitivity was equal 0.5 for superplastic flow at 450°C and strain rate of 4×10^{-4} s^{-1} .

In the alloy subjected to ECA pressing high strain rate superplasticity was also observed at 350°C where elongation to failure at a strain rate of $1,2 \times 10^{-1}$ s^{-1} was 920%. In contrast to these results, when this alloy was subjected to ECA pressing at 420% it did not demonstrate enhanced superplasticity neither did it when processed at 350°C or 400°C. For example, its elongation was less 200% with a strain rate of $1,2 \times 10^{-2}$ s^{-1} and temperature 400°C.

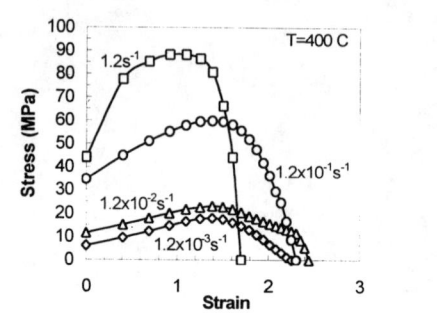

Fig.7. The variation in true stress with strain as a function of strain rate for the ECA-pressed at 370°C Al 1420 alloy.

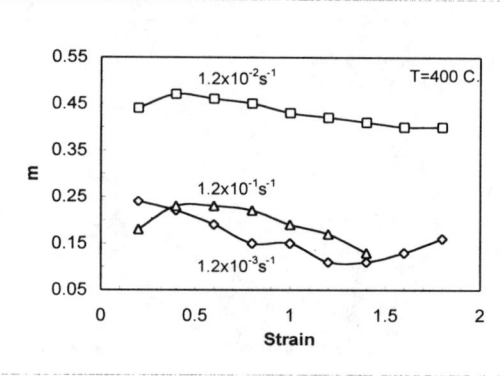

Fig.8. The dependence of the strain rate sensitivity of flow stress in the ECA-pressed (370°C) Al 1420 alloy on strain.

Zn-22%Al alloy

Another investigated material is the Zn-22%Al eutectoid alloy. It is a classic two-phase superplastic alloy capable of exhibiting tensile elongation exceeding 2000% under optimum conditions: temperature of 250°C and strain rate of $10^{-3} - 10^{-2}$ s^{-1} [12]. Superplasticity here is achieved in specimens with a grain size ranging 2-6 μm. In the present work, the samples from Zn-22%Al were quenched and subjected to SPTS. This processing led to formation of two-phase nanoduplex structure with a mean grain size of ~80 nm (Fig.10a) [13]. Superplasticity in

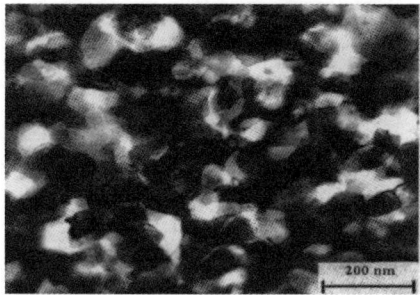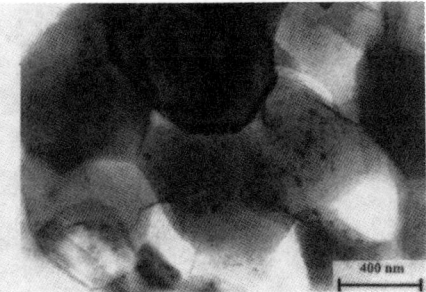

Fig. 9. TEM micrographs of the Zn-22%Al alloy, processed by (a) severe plastic torsion straining and (b) equal channel angular pressing.

this Zn-Al alloy subjected to SPTS was observed at 120°C and a strain rate of $1 \times 10^{-4} s^{-1}$, thus demonstrating superplasticity at a relatively low temperature. However the elongation to failure was equal to 280% only.

The Zn-22%Al alloy in the initial as-rolled state had a grain size of about 1-2 μm. ECA pressing at room temperature reduced the grain size in this two phase alloy to 0.5 μm by subjecting the samples to severe deformation (Fig.9 b). The ECAP processed Zn-22%Al alloy demonstrates very high tensile ductility at high strain rates of the order of 10^{-1} s^{-1} and temperature of 200°C. For example, elongations to failure were 1900% and 1540% at strain rates of $3.3 \times 10^{-2} s^{-1}$ and $3.3 \times 10^{-1} s^{-1}$, respectively [2,14]. This alloy demonstrated superplasticity at 120°C as well. Its elongation to failure was about of 600% at the strain rate of $1 \times 10^{-4} s^{-1}$, which is higher than for the SPTS specimen.

Ni₃Al intermetallic

After SPTS at room temperature a strong refinement of structure with a mean grain size of about 50 nm was found in a Ni_3Al intermetallic material. This value remained almost unchanged after annealing at T=650°C, and even after heating to 750°C the grain size did not exceed 100 nm [4,15]. TEM/HREM observations indicated that the grains are almost free from lattice dislocations (Fig. 10a), but the grain boundaries still retained their non-equilibrium character because they are not atomically flat and usually wavy and contained many defects, such as facets, steps and grain boundary dislocations (Fig.10 b). These are similar HREM observations for other metals after SPTS [16].

The nanostructured Ni_3Al revealed superplastic behavior at temperatures as low as 650°C. It is apparent from Fig.11 that the specimens exhibit very high tensile ductility (several hundred percent elongation to failure) without visible macroscopic necking. However, such low temperature superplastic behavior has several unusual features. In particular, the stress-strain curves for Ni_3Al at a strain rate of 10^{-3} s^{-1} demonstrated an extensive region of strain hardening at both temperatures 650°C and 725°C with the peak flow stress ranging from 0.9-1.5 GPa [4]. Special TEM/HREM investigations were performed on the thin foil prepared from a gauge section of Ni_3Al sample strained about 300% (see Fig.11). Although there is some grain growth during superplastic deformation, the grain size remained less than 100 nm. Grains were not

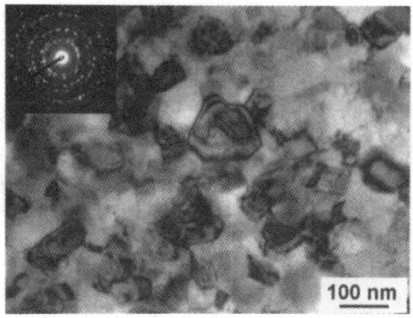 

Fig.10. Microstructure of Ni$_3$Al intermetallic after severe plastic torsion straining and annealing at 650 °C: (a) TEM micrograph; (b) HREM micrograph of typical grain boundary

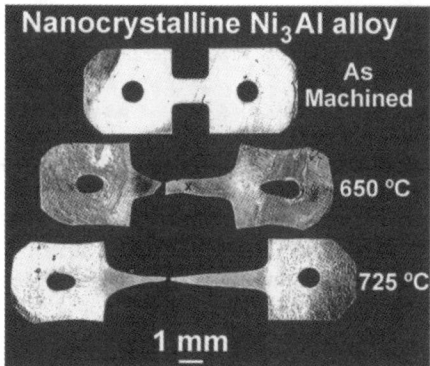

Fig. 11 (a) Appearance of nanostructured Ni$_3$Al samples prior and after tension at temperature 650°C, 1×10^{-3}, elongation 390% (a cross indicates the place where the foil was cut out for HREM/TEM) and at temperature 725°C 1×10^{-3}, elongation 560%.

Fig. 12. HREM micrograph of a typical grain boundary in Ni$_3$Al from a gage section of the specimen demonstrated superplastic elongation.

elongated and we could not find any evidence of large dislocation activity inside the grains. From the HREM observations it appears that grain boundaries still remain their narrow width and they are still atomically not flat (Fig.12). There were also some evidences of grain boundary dislocations.

DISCUSSION

It is known that all the three investigated alloys demonstrate superplastic behavior when they have a grain size of several microns [12,13,17]. Further decrease of the grain size to the submicron or nanocrystalline range in the alloys by SPD enhances their superplastic properties resulting in a decrease in temperature and/or increase of strain rate of superplastic deformation. A first manifestation of enhanced superplastic properties in SPD materials was demonstrated in the Al-based alloy from the Al-Cu-Zr-system [18]. Up to present time the effect has been revealed in a number of alloys and intermetallics [1-6,9,19 and others].

The results of the given work indicate that enhancement of superplasticity and attaining high strain rate superplasticity in the Al 1420 alloy can be provided both by ECA pressing and SPTS at room temperature. SPTS leads to formation of the very fine grain size of about 70 nm. However due to the grain growth during heating the samples produced by both techniques have similar microstructure and, therefore, it is clear why their properties at 350°C were very similar.

It is very important that the observed effect of high strain rate superplasticity in this case is conditioned by both the fine grain size (about 1 µm) and the presence of particles of the T-phase. As it is shown above, the dissolution of these particles during ECA pressing at 420°C results in sharp decrease in superplastic properties. The positive influence of a two-phase micro-duplex structure on superplasticity was shown in earlier studies with a number of alloys [11,12]. This effect can be atributed to both an influence of the second phase on microstructure stability during deformation and an enhancement of grain boundary sliding along interfacial boundaries. The data in the present work demonstrate that the submicron duplex structure contributes considerably to attaining high strain rate superplasticity in the Al1420 alloy.

Recent investigations showed that a very important requirement for producing high strain rate superplasticity is not only the strong refinement of microstructure by SPD techniques, but the formation of ultrafine-grained structure with high angle grain boundaries [3]. The latter can be achieved by increase of number of passes during ECA pressing. According to the data obtained it may be assumed that another requirement for enhancement of superplasticity is the optimization of the alloy phase composition, and this can be attained by optimal temperature and strain rate conditions in SPD processing.

Discussing the requirements of microstructure, let us consider in more detail the role of the grain refinement. The results of the present paper as well as previous works [2,3,9] indicate that grain refinement till a submicron range definitely contribute to enhancement of superplastic properties and attaining low temperature and/or high strain rate superplasticity. However, a further decrease in a grain size to the nanometer range does not necessarily lead to an additional enhancement of superplasticity. Moreover, as it is seen from the results of the Zn-22%Al alloy, the elongation to failure, in the case of the grain size of 80 nm, is even less than in the alloy with the grain size of 0.5 µm. Obviously, there exist a critical grain size within the range of 100-300 nm, and below this range any further grain refinement does not lead to additional enhancement of superplastic properties. The reason for the possible difficulty of superplasticity in nanocrystalline materials was recently discussed in [20], where it was supposed that this behavior is related to the difficulty in generation and motion of dislocations inside the very fine grains. The data on microstructure of superplastically deformed Ni_3Al presented above also confirm this assumption. The decrease of the strain rate sensitivity in UFG alloys till the values of 0.3 - 0.4, and even less, as compared to the values 0.5 - 0.6, which are typical for conventional superplasticity, indicate a possible change in deformation processes in UFG materials. In addition, it is important to note that in SPD alloys one can observe a visible strain hardening. Taking into account the mechanical approach by Hart [21] it may be assumed that enhanced ductility and good stability of superplastic flow under conditions of low temperature and/or high strain rate superplasticity in UFG alloys result from both the strain rate sensitivity of the flow stresses and strain hardening. This points to the importance of further investigations of the phenomenology and nature of the unusual superplasticity in SPD-processed materials.

SUMMARY

The investigation of the relationship between SPD processing, UFG structure, and enhanced superplasticity associated with low temperature and/or high strain rates was performed with the alloys Al 1420 and Zn-22%Al alloys as well as a Ni_3Al intermetallic. It was shown that the enhanced superplasticity in the Al 1420 alloy has important microstructural requirements, including both an UFG grain size and formation of the optimal phase composition. The greatest superplastic properties can be observed in UFG duplex-type structures, which can be attained due to optimal choice of temperature and strain rate of SPD processing. At the same time there is a "critical" grain size of 100-300 nm, below which the enhancement of superplastic properties becomes slower because of a possible change of the deformation mechanism.

ACKNOWLEDGEMENTS

This research on superplasticity in ultrafine-grained materials was supported in part by the NSF of the USA under grant NSF-CMS-96-34179 (P.I.: A.K. Mukherjee), in part the US ARO under grant 68171-98-M-5642 (in collaboration with T.G. Langdon), the INTAS grant N 97-1243, and Russian Foundation for Basic Research.

REFERENCES

1. R.Z. Valiev, Mater. Sci. Forum, **243-245**, 207 (1997).
2. R.Z. Valiev, R.K. Islamgaliev, Superplasticity and Superplastic Forming, edited by A.K. Ghosh and T.R. Bieler (The Minerals, Metals & Materials Society, 1998) pp.117-126.
3. P.B. Berbon, M. Furukawa, Z. Horita, M. Nemoto, N.K. Tsenev, R.Z. Valiev and T.G. Langdon, Phil. Mag. Let., **78**, 313 (1998).
4. R.S. Mishra, R.Z. Valiev, F.X. McFadden and A.K. Mukherjee, Mat. Sci. Eng. **A252**, 174 (1998).
5. M. Mabuchi and K.Higashi, Acta mater., **47**, 2047 (1999).
6. C.A.Salishchev, R.M.Galeev, S.P.Malisheva and O.R.Valiakhmetov, Mater. Sci. Forum, **243-245**, 585 (1997),
7. R.Z. Valiev, I.V. Alexandrov, and R.K Islamgaliev, NATO ASI, Eds G.M.Chow and N.I.Noskova. (KluwerPubl, 1998) pp.121–142.
8. Y. Iwahashi, Z. Horita, M. Nemoto and T.G. Langdon, Acta Mater., **46**, 3317 (1998).
9. R.Z. Valiev, D.A. Salimonenko, N.K. Tsenev, P.B. Berbon and T.G. Langdon, Scripta

Mater., **37**, 724 (1997).

10. R.S. Mishra, R.Z. Valiev, S.X. McFadden, R.K. Islamgaliev and A.K. Mukherjee, Phil.Mag., to be published.

11. O.A. Kaibyshev, *Superplasticity in Metals, Alloys and Intermetallics* (Dresden, Verlag, 1992), p.280.

12. T.G. Nieh, J. Wadsworth and O.D. Sherby, *Superplasticity in Metals and Ceramics* (Cambridge, Univer. Press, 1997), p.290.

13. R.S. Mishra, R.Z. Valiev and A.K. Mukherjee, Nanostruct. Mater., **2**, 430 (1997).

14. M. Furukawa, Z. Horita, M. Nemoto, R. Valiev and T.G. Langdon, J. Mater. Res. **11**, 2128 (1996).

15. R.Z. Valiev and R.K. Islamgaliev, Mater. Sci. Forum, **304-306**, 39 (1999).

16. Z. Horita, D.J. Smith, M. Nemoto, R.Z. Valiev and T.G. Langdon, J. Mater. Res. **13**, 446 (1998).

17. R.Z. Valiev, R.M. Gayanov, H.S. Yang and A.K. Mukherjee, Scripta Met. Mater. **25**, 1945 (1991).

18. R.Z. Valiev, O.A. Kaibyshev, R.I. Kuznetsov, R.Sh. Musalimov, N.K. Tsenev, DAN SSSR **301**, 4, 864 (1988).

19. .S.X. McFadden, R.S. Mishra, R.Z. Valiev, A.P. Zhilyaev and A.K. Mukherjee, Nature, **398**, 684 (1999).

20. R. S. Mishra, A. K. Mukherjee, Superplasticity and Superplastic Forming, edited by A.K. Ghosh and T.R. Bieler (The Minerals, Metals & Materials Society, 1998) pp.109-116.

21. J.Pilling and N.Ridley, Superplasticity in Crystalline Solids (Manchester, Inst. of Metals, 1992), p.214.

AN EXAMINATION OF THE DEFORMATION PROCESS
IN EQUAL-CHANNEL ANGULAR PRESSING

PATRICK B. BERBON*, MINORU FURUKAWA**, ZENJI HORITA***,
MINORU NEMOTO***, TERENCE G. LANGDON****
*Structural Metals Division, Rockwell Science Center, 1049 Camino Dos Rios,
Thousand Oaks, CA 91360, U.S.A. pbberbon@rsc.rockwell.com
**Department of Technology, Fukuoka University of Education, Munakata,
Fukuoka 811-4192, Japan
***Department of Materials Science and Engineering, Faculty of Engineering,
Kyushu University, Fukuoka 811-8581, Japan
****Departments of Materials Science and Mechanical Engineering,
University of Southern California, Los Angeles, CA 90089-1453, U.S.A.

ABSTRACT

In the past decade, there has been an important emphasis in materials science on the production and use of ultrafine-grained materials. These materials offer wide-ranging advantages such as an improvement in strength at low temperatures and enhanced superplastic behavior at high temperatures. New techniques to process these materials have been developed. Good properties have been achieved by powder metallurgy and by techniques of bulk processing. Equal-channel angular (ECA) pressing is a bulk-processing technique which has led to some remarkable achievements in the production of sub-microcrystalline materials with excellent superplastic properties at low temperatures and high strain rates. When considering the currently available physical and mechanical properties of these materials, it appears that some aspects cannot be explained solely by the very small grain size. In fact, the specific mode of deformation occurring in ECA pressing appears to have an influence on the final properties. This paper reviews the details of this metal working technique and then provides an explanation for the observed microstructures, their thermal stability and their remarkable superplastic properties.

INTRODUCTION

The equal-channel angular (ECA) pressing technique has been used successfully to produce sub-microcrystalline (SMC) materials and microstructures exhibiting superplastic deformation [1-3]. In this procedure, a billet of material is extruded from one channel into another channel of equal cross section lying at an angle to it of, typically, 90° [4]. The main stress component in the processed material is simple shear and the bulk of the deformation takes place at the intersection of the channels. Superposed on this shear stress is a high isostatic stress level. The deformation under this high compressive pressure allows the introduction of a very large amount of strain without fracture. In the currently investigated materials, namely aluminum and aluminum alloys, the deformation strain occurs through the creation of dislocations and their rearrangement into cell and grain boundaries (GBs). The processed materials have grain sizes in the range of 0.1 to 1 μm. When compared with other material in this same SMC range, there appear to be some unusual physical and mechanical properties in the ECA processed materials [5-7]. This paper examines the deformation in ECA pressing and proposes a possible explanation for the results observed on aluminum-based alloys.

347

THE ECA PRESSING TECHNIQUE

There are many parameters available when processing a material via ECA pressing and the most important are discussed in this section together with their influence on the deformation. The angle between the channels (angle Φ) is correlated to the amount of strain introduced in the material in one deformation pass and to the rate at which the deformation is imposed on the billet. The radius at the intersection of the channels (angle Ψ) is incorporated into the amount of strain introduced. The number of passes N through the die is the key parameter in determining the total amount of true strain applied to the billet. Calculations from geometric principles have shown the influence of N, Φ, and Ψ on the true strain ε_N [8]. For typical values of $\Phi = 90°$ and $\Psi = 20°$, the true strain introduced after N passes is $\varepsilon_N \approx N$. The effect of true strain on the material microstructure has been characterized [9,10] and will be discussed later. The pressing speed controls the rate at which the deformation occurs and it is important in determining which accommodation process is active after dislocations are introduced in large numbers in the initially large grains. It also influences the heat dissipation which may use up a large portion of the plastic work. The adiabatic heating is enhanced at higher pressing speeds and the time for heat loss to the die and environment is consequently shortened. This additive combination has an important influence on the resulting microstructures. The most important parameter in ECA pressing is the temperature of pressing which controls the evolution of the dislocation networks through the occurrence of dynamic recovery or dynamic recrystallization. The pressing temperature therefore affects the resulting microstructure. Another important parameter is the processing route followed. There are four different possible routes dependent upon whether the sample is rotated between consecutive pressings [11]. Results have shown that the route influences the final microstructure and it is important in the production of superplastic alloys [12]. The friction between the tool walls and the billet should be minimized in order to successfully press strong alloys. Annealing the billet between passes introduces static recovery or static recrystallization which may influence the evolution of the dislocation structures introduced during pressing.

The ECA processing method was described earlier [13]. The technique possesses numerous advantages over other procedures used to produce SMC microstructures:

The possibility of introducing very large strains: the geometrical dimensions of the billet are maintained in the procedure and therefore the same billet can be processed multiple times in order to reach very high strain levels. This is advantageous by comparison with standard rolling and extrusion processes.

Homogeneous deformation: a state of simple shear is created during deformation [4], as confirmed in microstructural investigations and through finite-element analysis [14].

No porosity or impurities: the deformation of the bulk sample occurs exclusively in the solid state. This constitutes a critical gain over powder metallurgy (P/M) techniques and rapid solidification.

Possibility of processing large billets: the billet size is only limited by the ECA die and the capacity of the press. It should be noted that even when using small dies (for example, a billet of 5 mm in diameter and 75 mm in length), the samples are large compared to P/M or torsion straining techniques.

Low tensile stresses: during deformation, the superimposed compressive isostatic stress keeps the tensile stresses very low, thereby reducing the structural damage to the billet.

There have been indications in the literature that some physical properties, such as the Young's modulus, the Debye temperature and the GB diffusion coefficient, may be changed by severe plastic deformation [5]. These results are controversial but intuitively it seems possible that the diffusion coefficient may be enhanced in an extremely strained structure because of differences in the atomic arrangement at the interfaces. Similarly, the dislocations created during the ECA pressing are very heterogeneous with a small number of clearly favored Burgers vectors. It is possible that long-range internal stresses remain after dislocation rearrangement into lower energy structures and therefore the materials may not follow the expectations for low energy dislocation structures (LEDS) [15].

EXPERIMENTAL DETAILS

Room temperature processing was performed on billets of pure aluminum [13] and Al-5% Zn [9]. Microstructural observations and mechanical testing were performed to assess the effect of processing parameters on the material strength and microstructure. These simple materials were convenient choices in order to follow the mechanism of deformation during ECA deformation in aluminum alloys.

On the first passage through the ECA die, many dislocations are introduced into the materials and the microstructure consists essentially of bands of subgrains. However, these subgrains evolve with increasing strain into grains separated by high angle grain boundaries. After a total of four passes, an ultrafine grain size develops in pure Al as shown in Fig. 1 for two different pressing speeds: these observations were recorded on the x plane perpendicular to the longitudinal axis of the sample and the selected area electron diffraction (SAED) patterns were recorded at selected points using a diameter of 12.3 μm [13]. The photomicrographs and the SAED patterns in Fig. 1 show that the measured grain size is essentially identical for both pressing speeds and equal to ~1.2 μm and, in addition, the grains are separated by boundaries having high angles of misorientation. Saturation essentially occurs at this amount of straining for the pure Al samples and there is very little additional evolution of the microstructure at higher strains. Information was given earlier on the evolution of the microstructure with the number of passes for Al-5% Zn and an example is shown in Fig. 2 after pressing for a total of 8 passes [9]. In this material, there was again an evolution of the microstructure with increasing numbers of passes, leading to an ultrafine-grained structure with high angle grain boundaries.

MODEL OF ECA DEFORMATION

To summarize the observations obtained in these investigations, sub-grains form in the first pass through the die and subsequently, with the advent of additional straining, these subgrains evolve into grains with high angle boundaries. The grain size is determined essentially at the first passage through the die and only the misorientation angles of the boundaries vary at the higher strains. A similar conclusion follows also from the detailed results on ECA pressing by Ferrasse et al. [16].

This type of deformation is therefore very much different from the classical observations arising from investigations dealing with LEDS theory or with models for the rolling of f.c.c. metals [15,17]. The difference in behavior arises because of specific and unique properties which are associated with the process of ECA pressing. These properties are examined in the following paragraphs.

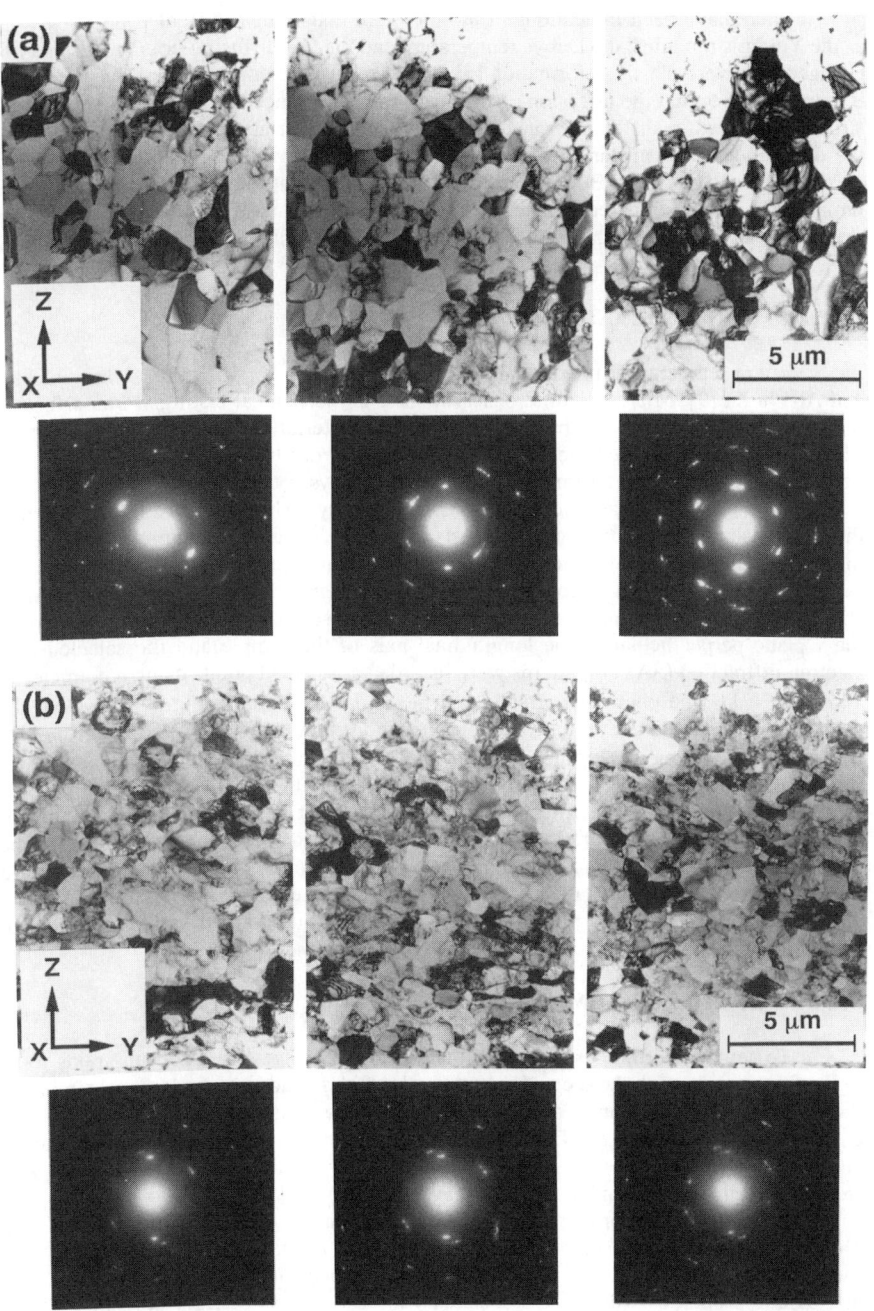

Fig. 1 Microstructures in pure Al after 4 passes at speeds of (a) 8.5×10^{-3} and (b) 7.6 mm s^{-1}.

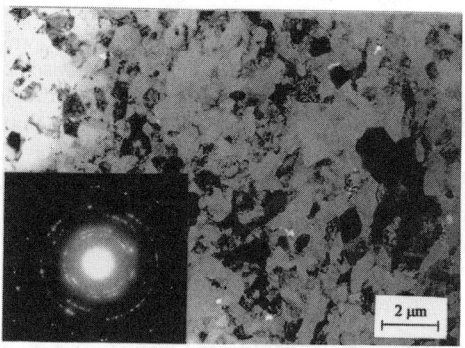

Fig. 2 Microstructure in Al-5% Zn after 8 passes.

<u>Large strain applied with no recovery time.</u> The most important feature of ECA pressing is
the rate at which the deformation is applied. For each pass, using a 90° die, a true strain of
~1 is introduced. The introduction of such a high strain is possible in this technique only
because the material is deformed under a high isostatic pressure: indeed, this would be
impossible using most classical metal working techniques. Comparing with rolling, it has
been shown that 4 minutes is insufficient for a complete reorganization of the dislocations
introduced by a strain of 0.1 in an Al-Mg alloy [17]. In ECA pressing, a strain ten times
larger is applied with no recovery time between the individual 1.0 true strain increments. It
may be anticipated that a state of equilibrium is never reached during the ECA pressing.
<u>Sub-grain formation completed in one pass.</u> The observations show that all of the steps
involved in the formation of cells and sub-grains are concentrated within the first pass.
This is consistent with the transition from cell boundaries to sub-grains which occurs at a
strain of ~ 0.1-0.22 in high purity Al [18] and at ~ 0.8 in low purity Al [19].
<u>GB pinning mechanism.</u> Again by analogy with rolling, it is known that Al-Mg precipitates
hold the cell boundaries in place retarding their rearrangement [17]. In pure Al and Al-5%
Zn in ECA pressing, the sub-GBs are also essentially pinned but there are no precipitates
in the microstructure. Therefore, another mechanism must be evoked.

A simple model may be proposed to account for these observations. During the
first ECA pass, shearing is concentrated in a short time and confined to the shear plane.
Only a limited number of slip systems is favorably oriented for slip in each grain and this
concentrates the deformation. This creates an unbalanced dislocation distribution with a
small number of Burgers vectors strongly favored. Recombination of these identical
dislocations leads not to the usual LEDS structure but rather to a situation where many
extrinsic dislocations remain in addition to the intrinsic dislocations necessary to form the
GB. A non-homogeneous structure therefore results. Dislocations in the sub-GBs will not
screen the long-range stresses well and the total dislocation density remains high. These
high stresses are probably responsible for stabilizing the boundaries and therefore they
essentially play the same role as precipitates in the Al-Mg alloys [12]. When the GBs are
pinned, the absorption of new lattice dislocations increases the mis-orientation angles.

SUMMARY AND CONCLUSIONS

This paper considers the deformation mode occurring in ECA pressing with reference to results obtained on pure Al and Al-5% Zn. A model is proposed containing attributes of the LEDS theory except that the long-range stresses due to the dislocations in the GBs are not well screened and there are long-range internal stresses in the material.

ACKNOWLEDGMENTS

This work was supported in part by the Light Metals Educational Foundation of Japan and in part by the National Science Foundation of the United States under Grants No. DMR-9625969 and INT-9602919.

REFERENCES

1. P.B. Berbon, M. Furukawa, Z. Horita, M. Nemoto, N.K. Tsenev, R.Z. Valiev and T.G. Langdon, Phil. Mag. Lett. **78**, 313 (1998).
2. P.B. Berbon, N.K. Tsenev, R.Z. Valiev, M. Furukawa, Z. Horita, M. Nemoto and T.G. Langdon, Metall. Mater. Trans. **29A**, 2237 (1998).
3. P.B. Berbon, S. Komura, A. Utsunomiya, Z. Horita, M. Furukawa, M. Nemoto and T.G. Langdon, Mater. Trans. JIM **40**, 772 (1999).
4. V.M. Segal, Mater. Sci. Eng. **A197**, 157 (1995).
5. R.Z. Valiev, N.A. Krasilnikov and N.K. Tsenev, Mater. Sci. Eng. **A137**, 35 (1991).
6. A.A. Nazarov, A.E. Romanov and R.Z. Valiev, Acta Metall. Mater. **41**, 1033 (1993).
7. A.A. Nazarov, A.E. Romanov and R.Z. Valiev, Nanostruct. Mater. **4**, 93 (1994).
8. Y. Iwahashi, J. Wang, Z. Horita, M. Nemoto and T.G. Langdon, Scripta Mater. **35**, 43 (1996).
9. P.B. Berbon, M. Furukawa, Z. Horita, M. Nemoto, N.K. Tsenev, R.Z. Valiev and T.G. Langdon in *Modeling the Mechanical Response of Structural Materials*, edited by E.M. Taleff and R.K. Mahidhara (The Minerals, Metals and Materials Society, Warrendale, PA, 1997), p.173.
10. Y. Iwahashi, Z. Horita, M. Nemoto and T.G. Langdon, Acta Mater. **46**, 3317 (1998).
11. M. Furukawa, Y. Iwahashi, Z. Horita, M. Nemoto and T.G. Langdon, Mater. Sci. Eng. **A257**, 328 (1998).
12. S. Lee, P.B. Berbon, M. Furukawa, Z. Horita, M. Nemoto, N.K. Tsenev, R.Z. Valiev and T.G. Langdon, Mater. Sci. Eng. **A272**, 63 (1999).
13. P.B. Berbon, M. Furukawa, Z. Horita, M. Nemoto and T.G. Langdon, Metall. Mater. Trans. **30A**, 1989 (1999).
14. C. Harris, S.M. Roberts, P.B. Prangnell and F.J. Humphreys in *The Third International Conference on Recrystallization and Related Phenomena*, edited by T.R. McNelley (Monterey Institute of Advanced Studies, Monterey, CA, 1997), p. 587.
15. N. Hansen and D.A. Hughes, Physica Status Solidi **A149**, 155 (1995).
16. S. Ferrasse, V.M. Segal, K.T. Hartwig and R.E. Goforth, J. Mater. Res. **12**, 1253 (1997).
17. S.J. Hales and T.R. McNelley in *Superplasticity in Aerospace*, edited by H.C. Heikkenen and T.R. McNelley (TMS, Warrendale, PA, 1988), p. 61.
18. N. Hansen, Trans. AIME **245**, 1305 (1969).
19. F. Schuh and M. von Heimendahl, Z. Metallk. **65**, 346 (1974).

SUPERPLASTIC PROPERTIES OF AN ALUMINUM-BASED ALLOY
AFTER EQUAL-CHANNEL ANGULAR PRESSING

JINGTAO WANG*, PATRICK B. BERBON**, YUZHONG XU*, LIZHONG WANG*, TERENCE G. LANGDON***
*Department of Metallurgy, Xi'an University of Architecture and Technology, Xi'an 710055, P.R. China
**Structural Metals Division, Rockwell Science Center, 1049 Camino Dos Rios, Thousand Oaks, CA 91360, U.S.A.
***Departments of Materials Science and Mechanical Engineering, University of Southern California, Los Angeles, CA 90089-1453, U.S.A. langdon@usc.edu

ABSTRACT

It is now recognized that superplasticity requires a very small grain size, typically <10 μm. A further reduction in grain size, to the submicrometer or nanometer level, offers the potential for attaining superplasticity at both faster strain rates and lower temperatures. This paper reports an investigation of the microstructure and tensile behavior of an Al-3% Mg-0.5% Zr alloy after processing by equal-channel angular pressing to an equivalent true strain of ~8.

INTRODUCTION

There are two fundamental requirements for achieving high tensile elongations and superplasticity [1]. First, the grain size must be very small and stable, typically less than ~10 μm. Second, the temperature must be reasonably high and typically above ~$0.5T_m$, where T_m is the absolute melting temperature of the material. In principle, small grain sizes may be obtained through appropriate thermo-mechanical processing but this leads to grain sizes which are generally in the range of ~1 – 10 μm. An alternative procedure for attaining an ultrafine grain size, often within the submicrometer or even the nanometer range, is to use either the processing procedure known as Equal-Channel Angular Pressing (ECAP) [2] or torsion straining under a high pressure [3]. In practice, ECAP is an especially attractive process because it has the potential for producing substantial grain refinement in large bulk samples whereas torsion straining is restricted to small disks with diameters of the order of ~1 cm.

This paper describes a series of experiments in which ECAP was applied to an Al-3% Mg-0.5% Zr alloy. This alloy was selected because earlier work established that it is possible to reduce the grain size of an Al-3% Mg solid solution alloy to ~0.2 – 0.3 μm through the use of ECAP [4-6] but these ultrafine grains are not stable at temperatures above ~500 K so that this alloy is not suitable for achieving superplastic ductilities. By contrast, there is a possibility of introducing additional thermal stability of the ultrafine grains through the addition of Zr to aluminum [7]. An example of this effect is provided by experiments on an Al-5.5% Mg-2.2% Li-0.12% Zr alloy where the ultrafine grain size introduced by ECAP was stable up to ~700 K [8] and the alloy exhibited superplastic characteristics [9-11].

EXPERIMENTAL MATERIAL AND PROCEDURES

The experiments were conducted using an Al-3% Mg alloy containing an addition of 0.5% Zr. The alloy was prepared by melting and casting into a water-cooled copper mold at a

Mat. Res. Soc. Symp. Proc. Vol. 601 © 2000 Materials Research Society

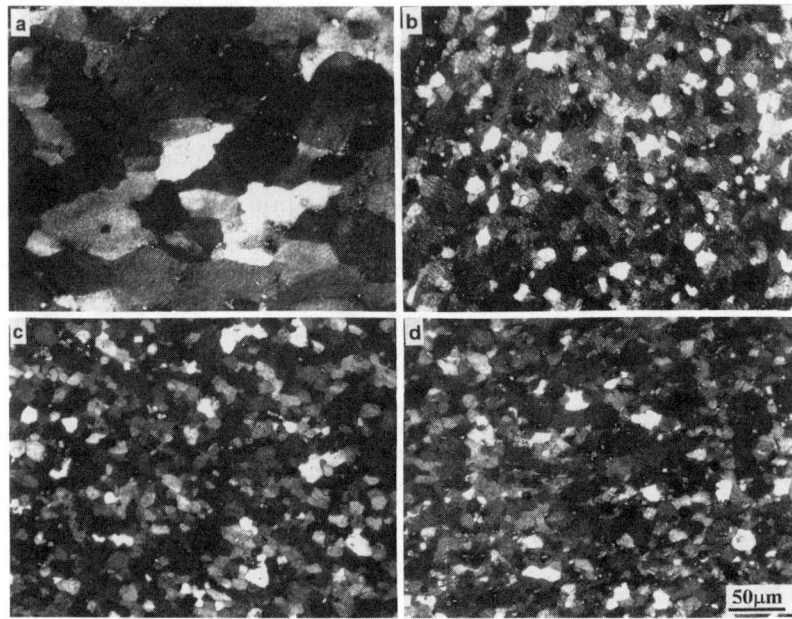

Fig. 1 Microstructures (a) before ECAP, and after annealing at 603 K for 60 minutes
following ECAP of (b) 2 passes, (c) 4 passes and (d) 8 passes.

temperature of 1083 K. The cast ingot had dimensions of $300 \times 200 \times 20$ mm and it was
homogenized for 24 hours at 723 K and then hot-rolled to a thickness of 15 mm. The ECAP was
conducted using bars, having dimensions of $15 \times 15 \times 60$ mm, which were cut from the rolled
plate in the transverse direction and annealed for 4 hours at 723 K to give a homogeneous grain
size of ~ 50 μm: an example of the microstructure prior to ECAP is shown in Fig. 1(a). The
ECAP was conducted at room temperature using processing route A in which the samples are not
rotated between consecutive passes through the die [12]. The die in these experiments had an
internal angle of 90° and the equivalent true strain introduced on each passage through the die is
~1 [13]. Samples were pressed up to a maximum of 8 passes, equivalent to a strain of ~8.

Following ECAP, tensile specimens were machined from the as-pressed bars with a
gauge length of 4 mm and a gauge section of 2×3 mm. These samples were pulled to failure in
an Instron testing machine operating at a constant rate of cross-head displacement and with
initial strain rates from 10^{-3} to 10^{-1} s^{-1} and testing temperatures from 723 to 773 K. The thermal
stability of the microstructure was investigated by annealing small samples for 1 hour at
temperatures up to 623 K and then quenching into water. The microstructures were examined
using both optical microscopy with polarized light and transmission electron microscopy (TEM).

EXPERIMENTAL RESULTS AND DISCUSSION

The microstructures of samples were examined after pressing through various numbers of
passes and then annealing at 603 K for 1 hour: Fig. 1 shows the resultant microstructures after

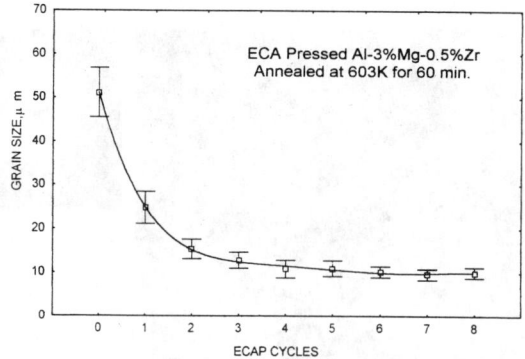

Fig. 2   Variation of grain size with number of passes after annealing at 603 K for 1 hour.

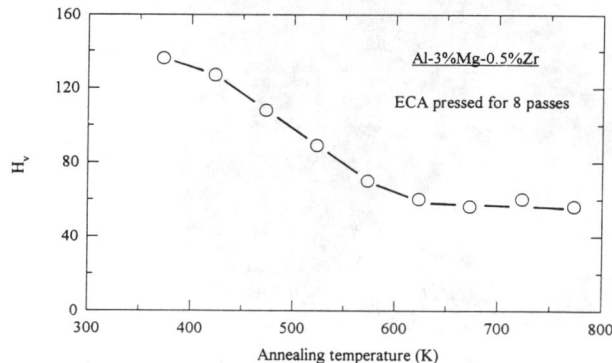

Fig. 3   Variation of hardness with annealing temperature after ECAP for 8 passes.

pressing through (b) 2 passes, (c) 4 passes and (d) 8 passes, respectively. Inspection shows the average grain sizes of these samples are in each case close to ~10 μm. More detailed information is given in Fig. 2 where the average grain size is plotted against the number of passes through the die for samples annealed after pressing for 1 hour at 603 K. Thus, the grain size is reduced after 1 pass to ~25 μm and to ~15 μm after 2 passes but the grain size is close to ~10 μm after 4 passes and it remains at this value for larger numbers of passes.

Figure 3 illustrates the influence of the annealing temperature on the Vickers microhardness, Hv, for samples subjected to ECAP for a total of 8 passes: each sample was annealed for 1 hour and then quenched in water prior to taking the hardness meassurements. These results show that the value of Hv decreases monotonically with increasing temperature up to ~600 K but thereafter the hardness remains reasonably constant.

When samples are subjected to ECAP, it is usual to define three orthogonal planes of sectioning for subsequent inspection: these are the x plane perpendicular to the pressing direction and the y and z planes parallel to the side and top faces of the sample at the point of exit from the die, respectively. Figure 4 illustrates the microstructure in the z plane in a sample subjected to a

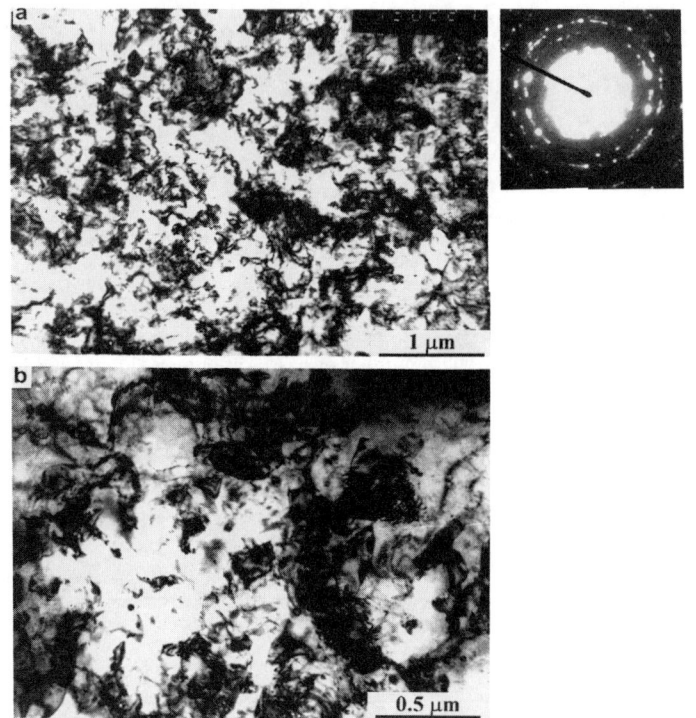

Fig. 4 Microstructure and associated SAED pattern after ECAP through 8 passes:
(a) evidence for an equiaxed microstructure and (b) details at a higher magnification.
The SAED pattern was taken from an area with a diameter of 2.5 μm.

total of 8 passes, where (a) shows evidence for an essentially equiaxed microstructure and (b) gives details at a higher magnification: the selected area electron diffraction (SAED) pattern was taken from an area having a diameter of 2.5 μm. It is apparent that a total of 8 passes through the die leads to a homogeneous array of grains and dislocation tangles, with an effective grain size under these conditions of the order of ~0.3 μm. Also, it is apparent that the SAED pattern exhibits rings, thereby indicating the presence of grains separated by grain boundaries having high angles of misorientation.

The mechanical properties were tested for samples subjected to ECAP through a total of 8 passes and therefore having similar microstructures to those shown in Fig. 4.

Typical plots of stress, σ, against strain, ε, are shown in Fig. 5 for samples tested at an absolute temperature, T, of 773 K and at three different initial strain rates. These curves reveal only limited strain hardening and with an increase in the ultimate tensile stress with increasing strain rate. Figure 6 shows the variation of the strain to failure, ε_f, with the testing temperature for samples pulled at the same initial strain rate of 1×10^{-3} s^{-1}. From this plot, there appears to be an optimum testing temperature in the vicinity of 770 K. The maximum tensile strain observed in these experiments was ~370%. Finally, Fig. 7 reveals that the elongations are dependent upon

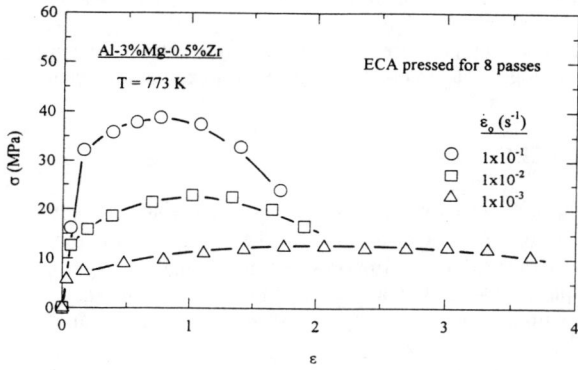

Fig. 5 Stress versus strain for tests conducted at 773 K after ECAP for 8 passes.

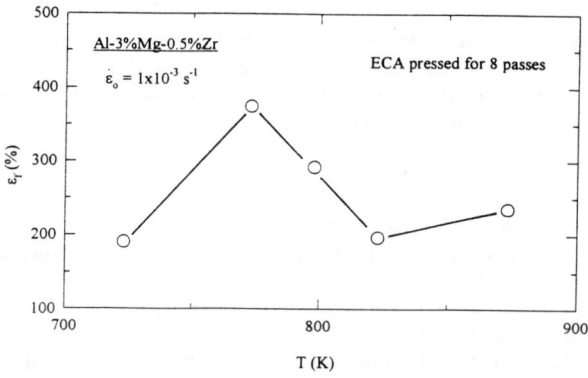

Fig. 6 Elongation to failure versus testing temperature for samples tested at an initial strain rate of 1×10^{-3} s^{-1}.

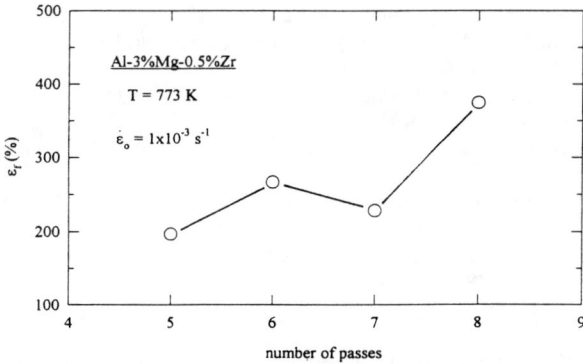

Fig. 7 Variation of elongation to failure with number of passes through the die.

the number of passes through the die, such that there are lower elongations to failure when the total number of passes is less than 8. This dependence on the number of passes through the die, and therefore upon strain, is a direct consequence of the higher strains which are needed in order to achieve reasonably homogeneous microstructures of equiaxed grains when Mg is added to an aluminum matrix [6].

SUMMARY AND CONCLUSIONS

1. An Al-3% Mg-0.5% Zr alloy was subjected to equal-channel angular pressing (ECAP) to a total of 8 passes, equivalent to a strain of ~8. The grain size was reduced to ~0.3 μm by ECAP but the grain size increased to ~10 μm on subsequent annealing for 1 hour at 603 K.

2. Some superplastic-like flow was recorded in this material when testing in tension at elevated temperatures, with a maximum elongation of ~370% achieved at 773 K with an initial strain rate of 1×10^{-3} s^{-1}.

ACKNOWLEDGMENTS

This work was supported in part by the National Science Foundation of China under Grant No. 59974018, in part by the Foundation of the National Key Laboratory of Metals Strength at Xi'an Jiaotong University and in part by the National Science Foundation of the United States under Grant No. DMR-9625969.

REFERENCES

1. T.G. Langdon, Metall. Trans. **13A**, 689 (1982).
2. V.M. Segal, V.I. Reznikov, A.E. Drobyshevskiy and V.I. Kopylov, Russian Metall. **1**, 99 (1981).
3. N.A. Smirnova, V.I. Levit, V.I. Pilyugin, R.I. Kuznetsov, L.S. Davydova and V.A. Sazonova, Fiz. Metall. Metalloved. **61**, 1170 (1986).
4. J. Wang, Y. Iwahashi, Z. Horita, M. Furukawa, M. Nemoto, R.Z. Valiev and T.G. Langdon, Acta Mater. **44**, 2973 (1996).
5. M. Furukawa, Z. Horita, M. Nemoto, R.Z. Valiev and T.G. Langdon, Acta Mater. **44**, 4619 (1996).
6. Y. Iwahashi, Z. Horita, M. Nemoto and T.G. Langdon, Metall. Trans. **29A**, 2503 (1998).
7. H. Hasegawa, S. Komura, A. Utsunomiya, Z. Horita, M. Furukawa, M. Nemoto and T.G. Langdon, Mater. Sci. Eng. **A265**, 188 (1999).
8. M. Furukawa, Y. Iwahashi, Z. Horita, M. Nemoto, N.K. Tsenev, R.Z. Valiev and T.G. Langdon, Acta Mater. **45**, 4751 (1997).
9. R.Z. Valiev, D.A. Salimonenko, N.K. Tsenev, P.B. Berbon and T.G. Langdon, Scripta Mater. **37**, 1945 (1997).
10. M. Furukawa, P.B. Berbon, Z. Horita, M. Nemoto, N.K. Tsenev, R.Z. Valiev and T.G. Langdon, Metall. Mater. Trans. **29A**, 169 (1998).
11. P.B. Berbon, M. Furukawa, Z. Horita, M. Nemoto, N.K. Tsenev, R.Z. Valiev and T.G. Langdon, Phil. Mag. Lett. **78**, 313 (1998).
12. M. Furukawa, Y. Iwahashi, Z. Horita, M. Nemoto and T.G. Langdon, Mater. Sci. Eng. **A257**, 328 (1998).
13. Y. Iwahashi, J. Wang, Z. Horita, M. Nemoto and T.G. Langdon, Scripta Mater. **36**, 143 (1996).

INFLUENCE OF EQUAL-CHANNEL ANGULAR PRESSING
ON THE SUPERPLASTIC PROPERTIES
OF COMMERCIAL ALUMINUM ALLOYS

SUNGWON LEE, TERENCE G. LANGDON
Departments of Materials Science and Mechanical Engineering
University of Southern California, Los Angeles, CA 90089-1453, U.S.A. langdon@usc.edu

ABSTRACT

Equal-channel angular (ECA) pressing was used to refine the microstructure in two commercial aluminum alloys, Al-2024 and the Supral-100 Al-2004 alloy. The ECA pressing was conducted at room temperature and at elevated temperatures for both alloys using several different processing routes. Tensile testing was carried out at elevated temperatures on both pressed and unpressed samples of each alloy in order to evaluate the effect of the pressing. This paper describes the influence of the ECA pressing on the subsequent mechanical properties of these two alloys. For both alloys, it is shown that the optimum superplastic conditions are influenced by the ECA pressing, and in practice there tends to be a decrease in the optimum temperature for superplasticity and a corresponding increase in the optimum strain rate. In addition, there was evidence for high strain rate superplasticity (HSR SP) in both alloys after the ECA pressing procedure.

INTRODUCTION

Much attention has focused recently on procedures which may be used to increase the strain rate and decrease the temperature for superplasticity in commercial aluminum-based alloys. This interest arises because of the relatively slow forming rates and high forming temperatures which are currently in use in conventional superplastic forming processes. Since superplasticity is dependent upon the grain size of the material, it may be possible to achieve these objectives by making a substantial reduction in the grain size [1]. At the present time, grain refinement is generally achieved through thermo-mechanical processing (TMP) but this has the limitation that different processing procedures must be developed separately for each alloy and, in addition, it is usually not possible to reduce the grain size below ~2 - 5 μm. As a result of these limitations, attention has been directed towards the possibility of using procedures in which materials are subjected to severe plastic deformation through either torsion straining [2] or equal-channel angular (ECA) pressing [3]. Both of these techniques are capable of producing materials with grain sizes at the submicrometer level [4,5] but in practice torsion straining is restricted to relatively small disks with diameters of ~1 cm whereas ECA pressing has a potential for use with large bulk samples.

In this investigation, attention was restricted exclusively to the effect of ECA pressing on the subsequent mechanical properties of two commercial aluminum-based alloys. Earlier reports demonstrated the potential for using severe plastic deformation to reduce the temperature for superplasticity [6,7] and also to achieve superplasticity at very rapid strain rates [8-10]. However, these earlier tests were conducted over only a limited range of strain rates and temperatures and with little or not attention given to the influence of different processing routes. The present investigation was therefore initiated in order to provide a more detailed appraisal of the factors influencing the mechanical characteristics of two representative aluminum-based alloys subjected to ECA pressing

Mat. Res. Soc. Symp. Proc. Vol. 601 © 2000 Materials Research Society

EXPERIMENTAL MATERIALS AND PROCEDURES

The experiments were conducted using two different commercial aluminum alloys: (1) an Al-2024 alloy with a chemical composition, in wt. %, of Al-3.8~4.9% Cu-1.2~1.8% Mg and (2) an Al-2004 Supral-100 Al alloy with a chemical composition of Al-6% Cu-0.4% Zr. The former alloy was annealed at 688 K for 3 hours to give plate-like grains with average dimensions of ~500 × 300 × 10 μm³ and the latter alloy was cast, homogenized for 5 hours at 648 K and hot rolled to give an initial grain size of ~100 μm..

Both alloys were subjected to ECA pressing using a die containing two channels, equal in cross-section, which intersected at an angle of 90° at the center of the die. Each sample was pressed several times through the die either at room temperature or at selected elevated temperatures. For these conditions, the strain introduced on each pass through the die is ~1 [11] and the samples were pressed up to a maximum of 12 passes corresponding to a total equivalent strain of ~12.

Several different processing routes have been defined for ECA pressing when samples are pressed repetitively through the die [12]. Specifically, route A denotes the situation where the specimen is pressed for more than one pass without any rotation, route B denotes the situation where the sample is rotated by 90° between each pass and route C denotes a rotation of 180° between the passes, where the two routes of B_A and B_C denote a rotation by 90° in the same direction between each pass and by 90° in alternate directions between each pass, respectively. It is also possible to combine different processing routes: for example, combinations of B_A-A and B_C-A were considered earlier [12]. In practice, the use of different processing routes has a significant influence on the microstructural development because different shearing patterns are introduced when the samples are rotated. It has been shown, for example, that route B_C is the optimum processing route for pure Al in order to most rapidly achieve a homogeneous microstructure of equiaxed grains separated by boundaries having high angles of misorientation [13]. In the present investigation, the Al-2024 alloy was pressed up to 8 passes at room temperature using either route B_C or new combinations of routes designated C-B_C and B_C-C, where these processing combinations are defined in Table I. These combinations were selected because route B_C was identified as the optimum for the pressing of pure Al [13] whereas route C restores a cubic element every second pass but without any deformation in one of the three orthogonal planes within the deformed sample [12].

Tensile specimens, with gauge lengths of 4 mm and cross-sections of 2 × 3 mm², were machined from the pressed samples parallel to the longitudinal axes after pressing and these samples were pulled to failure in air over a range of temperatures from 573 to 773 K. Similar tests were also conducted using specimens cut from the unpressed material. All of these tests were performed on an Instron testing machine operating at a constant rate of cross-head displacement.

Table I. Rotation angles and directions for two combination routes

Route	Number of pressings							
	1	2	3	4	5	6	7	8
C-B_C	N/R	180°↑	90°↓	180°↑	90°↓	180°↑	90°↓	180°↑
B_C-C	N/R	90°↓	180°↑	90°↓	180°↑	90°↓	180°↑	90°↓

* N/R corresponds to 'no rotation'

EXPERIMENTAL RESULTS AND DISCUSSION

Extensive ECA pressing was conducted on samples of the Al-2024 alloy using the three different processing routes and either room temperature or 373 K. Most of the subsequent tensile tests were performed with an initial strain rate of 1×10^{-3} s^{-1} but additional tests were conducted at selected temperatures where the ductilities were reasonably high.

Table II summarizes the more important results obtained in a detailed compilation of experimental data, including the testing temperature, the initial strain rate, the ultimate tensile strength (UTS) and the measured elongation to failure: the upper results show the data for unpressed samples and the remaining results cover samples subjected to a total of 8 passes under different ECA pressing conditions. The optimum superplastic condition may be defined as the combination of testing temperature and imposed strain rate which gives the maximum elongation to failure. Thus, inspection of Table II shows that the optimum condition in the unpressed Al-2024 alloy is at a temperature of 723 K with strain rates in the vicinity of $\sim 10^{-4}$ to 10^{-3} s^{-1}. After ECA pressing, however, it is evident that the optimum superplastic conditions tend to change to lower temperatures and possibly also to faster strain rates. For example, after ECA pressing through 8 passes at room temperature using route B$_C$, the optimum strain rate is again $\sim 10^{-3}$ s^{-1} but the optimum temperature is reduced to ~ 673 K. Furthermore, the elongations are higher after ECA pressing with a maximum recorded elongation of close to $\sim 500\%$ after pressing for 8 passes at 373 K using route B$_C$. It is also apparent from Table II that routes C-B$_C$ and B$_C$-C give similar elongations to failure but the ductilities are lower than with route B$_C$.

Table II. The superplastic properties of the unpressed and ECA pressed 2024 Al alloy

Processing condition	Test temp. (K)	strain rate (s^{-1})	UTS (MPa)	Elongation (%)
	673	1×10^{-3}	43	189
	723	1×10^{-2}	44	221
Unpressed	723	1×10^{-3}	25	280
	723	1×10^{-4}	13	267
	773	1×10^{-3}	12	189
	623	1×10^{-3}	40	344
	673	1	107	165
8 passes	673	1×10^{-1}	77	185
at room temp.	673	1×10^{-2}	52	304
via route B$_C$	673	3×10^{-3}	36	318
	673	1×10^{-3}	23	456
	673	1×10^{-4}	13	277
	723	1×10^{-3}	15	284
8 passes	573	1×10^{-3}	70	222
at room temp.	623	1×10^{-2}	66	233
via route C-B$_C$	623	1×10^{-3}	42	332
	673	1×10^{-3}	36	131
8 passes	573	1×10^{-3}	72	229
at room temp.	623	1×10^{-2}	73	225
via route B$_C$-C	623	1×10^{-3}	44	347
	673	1×10^{-3}	30	223
8 passes	623	1×10^{-3}	23	380
at 373 K	673	1×10^{-1}	58	288
via route B$_C$	673	1×10^{-2}	29	497
	673	1×10^{-3}	15	409

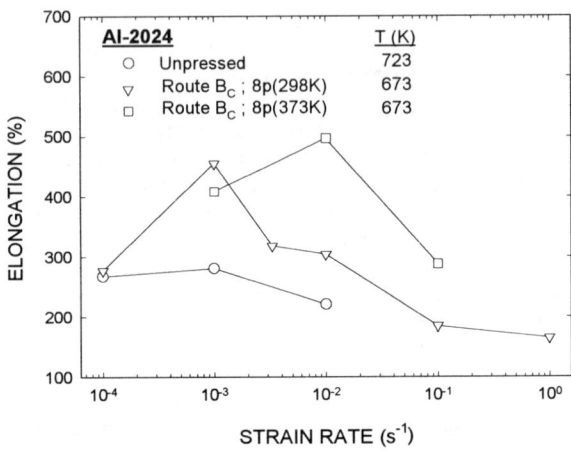

Fig. 1 Elongation versus strain rate under optimum superplastic conditions for Al-2024 alloy.

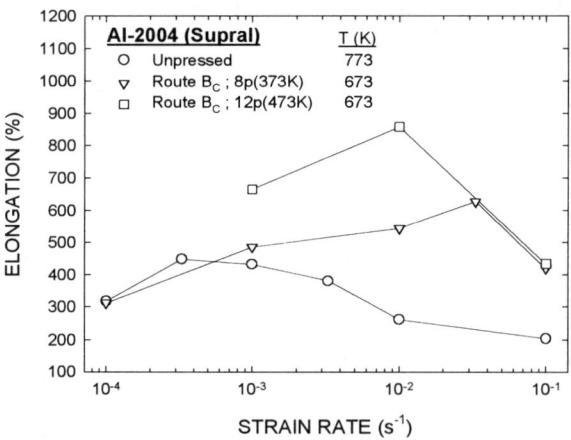

Fig. 2 Elongation versus strain rate under optimum superplastic conditions for Al-2004 alloy.

Figure 1 shows the optimum superplastic conditions for the Al-2024 alloy in the unpressed condition and after ECA pressing using route B_C at either room temperature or 373 K. This plot demonstrates the higher ductilities achieved in this alloy following the ECA pressing. It is also apparent that this alloy is essentially on the threshold of exhibiting high strain rate superplasticity (HSR SP) where high ductilities are achieved at strain rates above 10^{-2} s^{-1}.

Figure 2 shows a similar plot of optimum superplastic conditions for the Supral-100 Al-2004 alloy and a detailed set of experimental results is given in Table III. This material is more ductile than the Al-2024 alloy and the unpressed specimens exhibit elongations up to ~450% at a temperature of 773 K and strain rates from ~10^{-4} to 10^{-3} s^{-1}. It should be noted that this alloy is a standard superplastic material [14-17] widely used for commercial forming operations [18,19].

Table III. The superplastic properties of the unpressed and ECA pressed SUPRAL-100 Al alloy

Processing condition	Test temp. (K)	strain rate (s^{-1})	UTS (MPa)	Elongation (%)
	723	1×10^{-3}	21	244
	773	1×10^{-1}	43	204
	773	1×10^{-2}	25	261
Unpressed	773	3×10^{-3}	19	380
	773	1×10^{-3}	12	432
	773	3×10^{-4}	8	448
	773	1×10^{-4}	5	317
	623	1×10^{-1}	76	222
8 passes	673	1×10^{-1}	57	364
at room temp.	673	1×10^{-2}	30	567
via route B_C	693	1×10^{-1}	49	321
	723	1×10^{-2}	24	524
	623	1×10^{-3}	26	380
	673	1×10^{-1}	65	419
8 passes	673	3×10^{-2}	44	627
at 373 K	673	1×10^{-2}	30	544
via route B_C	673	1×10^{-3}	22	486
	673	1×10^{-4}	9	311
	723	1×10^{-3}	13	340
	623	1×10^{-2}	52	344
12 passes	623	1×10^{-3}	26	438
at 373 K	673	1×10^{-2}	37	437
via route B_C	673	1×10^{-3}	17	427
	723	1×10^{-3}	15	301
	623	1×10^{-2}	50	516
12 passes	623	1×10^{-3}	24	453
at 473 K	673	1×10^{-1}	63	434
via route B_C	673	1×10^{-2}	26	858
	673	1×10^{-3}	14	665

After ECA pressing, the elongations to failure in the Al-2004 alloy were significantly higher than in the unpressed condition. A maximum elongation of ~860% was attained at a strain rate of 1×10^{-2} s^{-1} with a testing temperature of 673 K after ECA pressing for 12 passes at 473 K using route B_C. It is very clear from Fig. 2 that the optimum conditions in this alloy are shifted to both a lower testing temperature and a higher imposed strain rate. Close inspection of Table III reveals, for example, that the optimum conditions after ECA pressing are consistently achieved at a temperature of 673 K which is 100° lower than in the unpressed condition. Furthermore, all samples after ECA pressing now exhibit HSR SP with good tensile ductilities at a strain rate of 10^{-2} s^{-1}. The elongations are also high at 673 K with a strain rate of 10^{-1} s^{-1}, ranging from ~360% after pressing at room temperature via route B_C to ~420% after pressing using the same route at 373 K. When this alloy is thermo-mechanically processed, the maximum elongation has been reported as ~1000% at 753 K with a strain rate of 3×10^{-3} s^{-1} [15]. Although the elongations attained in the present investigation are <1000% after ECA pressing, nevertheless it is instructive to note that ECA pressing is effective both in decreasing the optimum temperature and in increasing the optimum strain rate.

Finally, it is possible to conclude from the data in Table III that an increase in the temperature for ECA pressing gives the potential for a higher maximum elongation to failure when the total strain imposed by the ECA pressing remains the same. The same conclusion follows also from the data for the Al-2024 alloy shown in Table II.

SUMMARY AND CONCLUSIONS

Equal-channel angular (ECA) pressing was conducted on two different commercial alloys, Al-2024 and Supral-100 Al-2004. Following ECA pressing, the optimum superplastic conditions tended to occur at lower temperatures and faster strain rates.

ACKNOWLEDGMENTS

This work was supported by the National Science Foundation under Grant No. DMR-9625969.

REFERENCES

1. Y. Ma, M. Furukawa, Z. Horita, M. Nemoto, R.Z. Valiev and T.G. Langdon, Mater. Trans. JIM **37**, 336 (1996).
2. N.A. Smirnova, V.I. Levit, V.I. Pilyugin, R.I. Kuznetsov, L.S. Davydova and V.A. Sazonova, Fiz. Metall. Metalloved. **61**, 1170 (1986).
3. V.M. Segal, V.I. Reznikov, A.E. Drobyshevskiy and V.I. Kopylov, Russian Metall. (Metally) **1**, 99 (1981).
4. R.Z. Valiev, N.A. Krasilnikov and N.K. Tsenev, Mater. Sci. Eng. **A137**, 35 (1991).
5. R.Z. Valiev and N.K. Tsenev in *Hot Deformation of Aluminum Alloys*, edited by T.G. Langdon, H.D. Merchant, J.G. Morris and M.A. Zaidi (The Minerals, Metals and Materials Society, Warrendale, PA, 1991), p. 319.
6. R.Z. Valiev, O.A. Kaibyshev, R.I. Kuznetsov, R.Sh. Musalimov and N.K. Tsenev, Soviet Phys. Dokl. **33**, 626 (1988).
7. R.Z. Valiev, A.V. Korznikov and R.R. Mulyukov, Mater. Sci. Eng. **A168**, 141 (1993).
8. R.Z. Valiev, D.A. Salimonenko, N.K. Tsenev, P.B. Berbon and T.G. Langdon, Scripta Mater. **37**, 1945 (1997).
9. P.B. Berbon, M. Furukawa, Z. Horita, M. Nemoto, N.K. Tsenev, R.Z. Valiev and T.G. Langdon, Phil. Mag. Lett. **78**, 313 (1998).
10. S. Lee, P.B. Berbon, M. Furukawa, Z. Horita, M. Nemoto, N.K. Tsenev, R.Z. Valiev and T.G. Langdon, Mater. Sci. Eng. **A272**, 63 (1999).
11. Y. Iwahashi, J. Wang, Z. Horita, M. Nemoto and T.G. Langdon, Scripta Mater. **35**, 143 (1996).
12. M. Furukawa, Y. Iwahashi, Z. Horita, M. Nemoto and T.G. Langdon, Mater. Sci. Eng. **A257**, 328 (1998).
13. K. Oh-ishi, Z. Horita, M. Furukawa, M. Nemoto and T.G. Langdon, Metall. Mater. Trans. **29A**, 2011 (1998).
14. R. Grimes, C. Baker, M.J. Stowell and B.M. Watts, Aluminium **51**, 720 (1975)
15. R. Grimes, M.J. Stowell and B.M. Watts, Metals Tech. **3**, 154 (1976).
16. B.M. Watts, M.J. Stowell, B.L. Baikie and D.G.E. Owen, Metal Sci. **10**, 189 (1976).
17. B.M. Watts, M.J. Stowell, B.L. Baikie and D.G.E. Owen, Metal Sci. **10**, 198 (1976).
18. A.J. Barnes, Mater. Sci. Forum **170-172**, 701 (1994).
19. A.J. Barnes, Mater. Sci. Forum **304-306**, 785 (1999).

PROCESSING BY EQUAL-CHANNEL ANGULAR PRESSING: POTENTIAL FOR ACHIEVING SUPERPLASTICITY

MINORU FURUKAWA*, MINORU NEMOTO**†, ZENJI HORITA**,
TERENCE G. LANGDON***
*Department of Technology, Fukuoka University of Education, Munakata, Fukuoka 811-4192,
Japan
**Department of Materials Science and Engineering, Faculty of Engineering, Kyushu University,
Fukuoka 812-8581, Japan
***Departments of Materials Science and Mechanical Engineering, University of Southern
California, Los Angeles, CA 90089-1453, U.S.A. langdon@usc.edu
†Now at Sasebo National College of Technology, 1-1 Okishin-cho, Sasebo 857-1193, Japan

ABSTRACT

Equal-channel angular (ECA) pressing is a processing procedure whereby a very severe plastic
strain is imposed on a sample without any change in the cross-sectional dimensions of the material.
This processing method leads to a substantial grain refinement, producing grains which are within
the submicrometer or even the nanometer scale. This paper discusses the potential for using this
method to prepare materials for superplasticity. The results demonstrate that it is possible to achieve
superplastic deformation in selected materials subjected to ECA pressing and, in addition, there is
the possibility of extending the superplastic region so that it occurs at very rapid strain rates.

INTRODUCTION

Superplasticity is achieved in metals when samples with very small grain sizes, typically <10
μm, are tested in tension at temperatures at and above approximately one-half of the absolute
melting temperature. An important characteristic of superplasticity is that it occurs over a relatively
restricted range of strain rates. Thus, a plot of the elongation to failure against the imposed strain
rate invariably shows a peak at intermediate strain rates and reductions in the tensile elongations at
both slower and faster rates. This peak in the elongation curve occurs typically at strain rates in the
vicinity of ~10^{-3} - 10^{-2} s^{-1} [1].
    The high tensile ductilities exhibited by superplastic alloys make these materials very
attractive for use in industrial forming operations. Superplastic forming is now established as a
viable processing tool for the fabrication of a wide range of complex parts which are used in
applications associated with, for example, the aerospace, transportation, construction and sports
industries [2,3]. Despite this success, the slow strain rates associated with the occurrence of
optimum superplasticity has restricted the utilization of superplastic forming to limited production
runs in high-value markets. Efforts are therefore needed to extend the viability of superplastic
forming to a wider range of applications including, for example, in the automotive industry.
    It may be possible to achieve the goal of faster superplastic forming strain rates by making
a significant reduction in the grain size of the material since there is experimental evidence that the
range of optimum superplasticity is displaced to faster strain rates when the grain size is decreased
[1]. However, it is generally difficult to reduce the grain size of metallic alloys below ~1 μm
through standard thermo-mechanical processing. This difficulty led to the suggestion that it may
be feasible to attain high strain rate superplasticity by taking advantage of the process of equal-
channel angular (ECA) pressing in which a material is subjected to a very intense plastic strain [4].
This paper describes this process and the results obtained using two representative alloys.

Mat. Res. Soc. Symp. Proc. Vol. 601 © 2000 Materials Research Society

EXPERIMENTAL MATERIALS AND PROCEDURES

Equal-channel angular (ECA) pressing is a processing procedure developed by Segal and co-workers [5] in the Soviet Union in 1981. The principle of this technique is to subject a sample to simple shear within a confining channel so that a very severe plastic strain is introduced without any change in the cross-sectional dimensions of the sample. The lack of any changes in the cross-sectional dimensions is the important characteristic distinguishing this process from conventional industrial processes such as extrusion and rolling. The straining is achieved by making use of a die containing two channels, equal in cross-section, which intersect near the center of the die. The test sample is then machined so that it fits within the channel and it is pressed through the die using a plunger. Further details on this processing procedure are given elsewhere [6,7]. It has been shown that the use of ECA pressing provides an opportunity for producing materials with extremely fine grain sizes [8,9], generally within the submicrometer range. Therefore, this is a very useful procedure for producing grain sizes which are substantially smaller than those readily obtained through standard procedures.

The ECA pressing was conducted in the present investigation using a solid die containing two channels intersecting at an angle of 90°: this configuration was selected because it was shown in earlier work that an angle close to 90° is needed in order to attain a homogeneous microstructure of essentially equiaxed grains separated by boundaries having high angles of misorientation [10]. For this value of the angle between the channels, it can be shown that a single passage of the sample through the die leads to a strain of ~1 [11].

Two different materials will be used to illustrate the potential for achieving high strain rate superplasticity after ECA pressing. Initially, experiments were conducted using a light-weight high-strength aluminum alloy fabricated by casting and containing, in weight %, 5.5% Mg, 2.2% Li and 0.12% Zr. This alloy contains a fine dispersion of δ'-Al_3Li and β'-Al_3Zr precipitates and it was supplied in a hot-rolled non-superplastic condition with an initial grain size of ~400 μm. Subsequently, experiments were conducted using an Al-3% Mg-0.2% Sc alloy which was prepared by melting Al and Sc to form an Al-3% Sc alloy and then remelting with additional Al and Mg to form the required material. This alloy was prepared by casting, it was homogenized at 753 K for one day and then it was solution treated at 883 K for 1 hour to give a material with an initial grain size of ~200 μm.

When samples are subjected to ECA pressing, very high strains may be introduced by subjecting the same sample to a series of consecutive passages through the die. In practice, however, different shearing systems are introduced if the sample is rotated between each consecutive pass [12]. It has been shown that the optimum pressing condition occurs when a sample is rotated by 90° between each pass in the procedure known as route B_C [12] since this procedure leads to an equiaxed array of grains in the minimum number of passes through the die [13,14]. In the present work, the Al-Mg-Sc alloy was processed using route B_C but the Al-Mg-Li-Zr alloy was processed using route A where there is no rotation of the sample between consecutive pressings.

The Al-Mg-Li-Zr alloy was subjected to ECA pressing under two different conditions. In route 1, each sample was pressed for four passes to a strain of ~4 at a temperature of 673 K; in route 2, samples were pressed for eight passes at 673 K and then a further four passes at 473 K to give a total strain of ~12, where the lower temperature was selected for the final passes in order to minimize any grain growth. For the Al-Mg-Sc alloy, all samples were pressed for eight passes at room temperature giving a total strain of ~8. Further information is given elsewhere on the characteristics of the Al-Mg-Li-Zr alloy [15-18] and the Al-Mg-Sc alloy [19] after ECA pressing. Following the ECA pressing, tensile specimens were machined parallel to the longitudinal axes of the as-pressed with gauge lengths of either 4 mm(Al-Mg-Li-Zr) or 5 mm (Al-Mg-Sc) and these samples were pulled to failure in air at selected elevated temperatures.

366

EXPERIMENTAL RESULTS AND DISCUSSION

Figure 1 shows typical microstructures obtained after ECA pressing of (a) the Al-Mg-Li-Zr alloy processed by route 2 and (b) the Al-Mg-Sc alloy, together with selected area electron diffraction (SAED) patterns taken from regions having diameters of 12.3 μm. In both alloys, inspection shows there has been a very substantial refinement in the grain size, the microstructures are now reasonably homogeneous with equiaxed grains, and the grain sizes are ~1.2 μm for the Al-Mg-Li-Zr alloy and ~0.2 μm for the Al-Mg-Sc alloy, respectively. In addition, the SAED patterns demonstrate that the grains are separated by grain boundaries which have high angles of misorientation.

Despite this very substantial reduction in the grain size, these materials may not be suitable for attaining superplasticity because the grains may grow at the elevated temperatures which are needed to achieve superplastic flow. For example, earlier results showed it is possible to attain a grain size of ~0.2 μm through the ECA pressing of an Al-3% Mg solid solution alloy but these

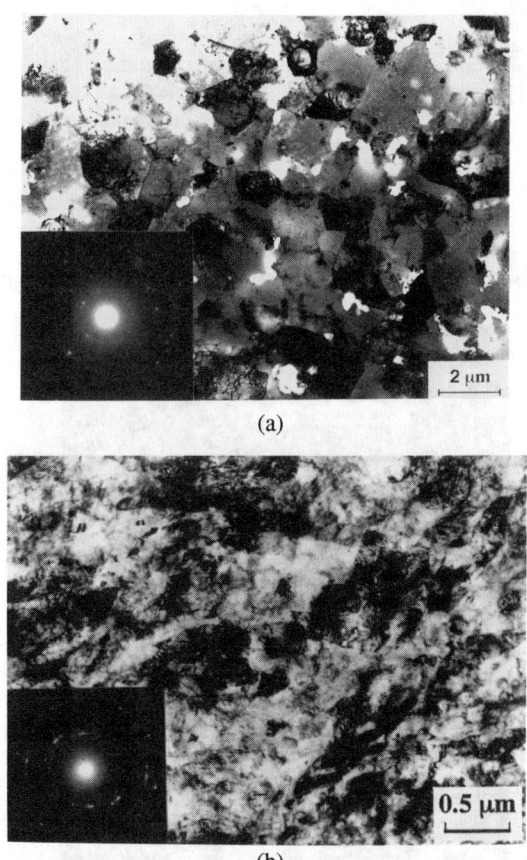

(a)

(b)

Fig. 1 Microstructures after ECA pressing of (a) the Al-Mg-Li-Zr alloy through route 2 and (b) the Al-Mg-Sc alloy, together with the SAED patterns.

grains grow rapidly when the material is heated to temperatures of ~500 K and higher [20,21]. To check on the thermal stability of these two alloys, samples were annealed for 1 hour at selected temperatures and the grain sizes were then measured. The result is shown in Fig. 2 which includes the earlier data for the Al-3% Mg alloy. It is apparent that the presence of precipitates in the present alloys gives a potential for retaining an ultrafine grain size in the materials up to temperatures of ~700 K and this suggests the possibility of achieving superplastic ductilities at temperatures in the range of ~500 - 700 K. An experimental example of these superplastic ductilities is shown in Fig. 3 for the Al-Mg-Li-Zr alloy processed via route 2.

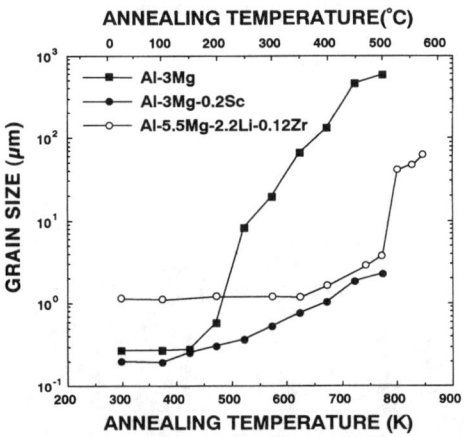

Fig. 2 Grain size versus annealing temperature for three different alloys after ECA pressing.

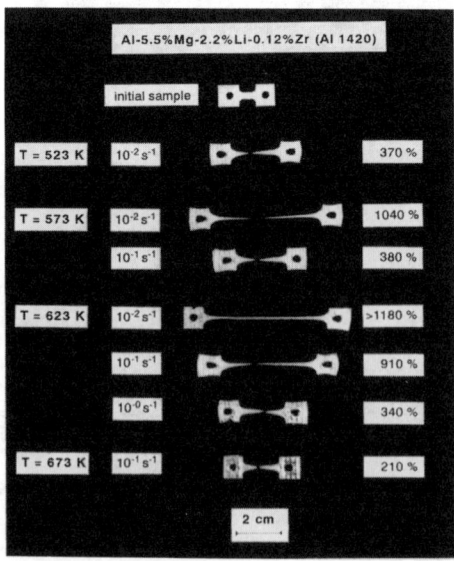

Fig. 3 Examples of superplastic ductilities in the Al-Mg-Li-Zr alloy after ECA pressing.

Figure 4 gives examples of the superplastic ductilities which may be attained through the use of ECA pressing where (a) is for the Al-Mg-Li-Zr alloy using both routes 1 and 2 and (b) is for the Al-Mg-Sc alloy: for both sets of data, additional results are also given for samples of the same or a similar alloy tested after appropriate thermo-mechanical treatments [22,23]. Inspection of Fig. 4(a) shows that ECA pressing through route 1 gives some improvement in the elongations to failure compared with standard processing where the grain size was ~7 μm [22] but this condition was not optimum because the as-pressed microstructure was heterogeneous with only ~60-70% of the grain boundaries having high angles of misorientation. By contrast, excellent superplastic ductilities are achieved in this alloy at high strain rates after processing through route 2 and even higher ductilities are recorded after ECA pressing of the Al-Mg-Sc alloy as shown in Fig. 4(b).

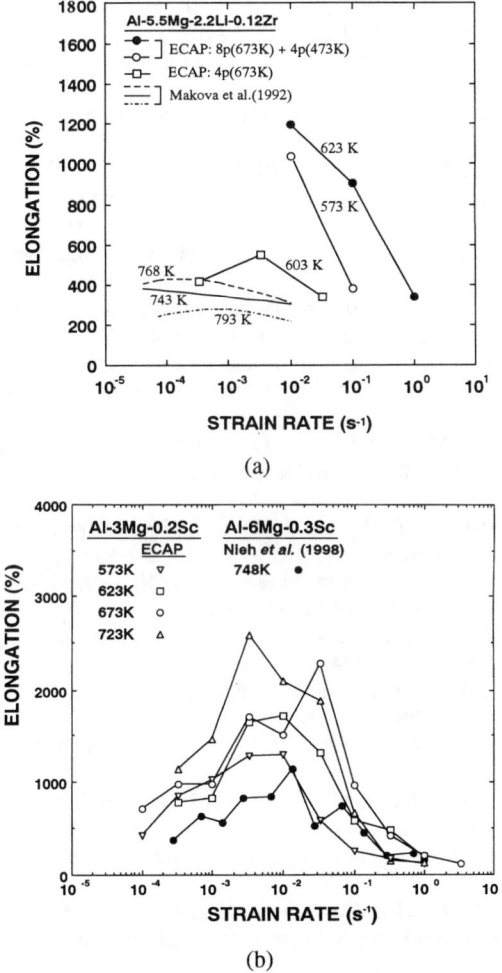

(a)

(b)

Fig. 4 Elongation to failure versus initial strain rate for (a) the Al-Mg-Li-Zr alloy and (b) the Al-Mg-Sc alloy: additional results after thermo-mechanical processing are also included [22,23].

ACKNOWLEDGMENTS

This work was supported in part by the Light Metals Educational Foundation of Japan, in part by a Grant-in-Aid for Scientific Research from the Ministry of Education, Science, Sports and Culture of Japan, in part by the Japan Society for the Promotion of Science and in part by the National Science Foundation of the United States under Grants No. DMR-9625969 and INT-9602919.

REFERENCES

1. F.A. Mohamed, M.M.I. Ahmed and T.G. Langdon, Metall. Trans. **8A**, 933 (1977).
2. A.J. Barnes, Mater. Sci. Forum **304-306**, 785 (1999).
3. D.G. Sanders, Mater. Sci. Forum **304-306**, 805 (1999).
4. Y. Ma, M. Furukawa, Z. Horita, M. Nemoto, R.Z. Valiev and T.G. Langdon, Mater. Trans. JIM **37**, 336 (1996).
5. V.M. Segal, V.I. Reznikov, A.E. Drobyshevskiy and V.I. Kopylov, Russian Metallurgy (Metally) **1**, 99 (1981).
6. V.M. Segal, Mater. Sci. Eng. **A197**, 157 (1995).
7. M. Nemoto, Z. Horita, M. Furukawa and T.G. Langdon, Metals and Materials **4**, 1181 (1998).
8. R.Z. Valiev and N.K. Tsenev in *Hot Deformation of Aluminum Alloys*, edited by T.G. Langdon, H.D. Merchant, J.G. Morris and M.A. Zaidi (The Minerals, Metals and Materials Society, Warrendale, PA, 1991), p. 319.
9. R.Z. Valiev, N.A. Krasilnikov and N.K. Tsenev, Mater. Sci. Eng. **A137**, 35 (1991).
10. K. Nakashima, Z. Horita, M. Nemoto and T.G. Langdon, Acta Mater. **46**, 1589 (1998).
11. Y. Iwahashi, J. Wang, Z. Horita, M. Nemoto and T.G. Langdon, Scripta Mater. **35**, 143 (1996).
12. M. Furukawa, Y. Iwahashi, Z. Horita, M. Nemoto and T.G. Langdon, Mater. Sci. Eng.. **A257**, 328 (1998).
13. Y. Iwahashi, Z. Horita, M. Nemoto and T.G. Langdon, Acta Mater. **46**, 3317 (1998).
14. K. Oh-ishi, Z. Horita, M. Furukawa, M. Nemoto and T.G. Langdon, Metall. Mater. Trans. **29A**, 2011 (1998).
15. M. Furukawa, Y. Iwahashi, Z. Horita, M. Nemoto, N.K. Tsenev, R.Z. Valiev and T.G. Langdon, Acta Mater. **45**, 4751 (1997).
16. M. Furukawa, P.B. Berbon, Z. Horita, M. Nemoto, N.K. Tsenev, R.Z. Valiev and T.G. Langdon, Metall. Mater. Trans. **29A**, 169 (1998).
17. P.B. Berbon, N.K. Tsenev, R.Z. Valiev, M. Furukawa, Z. Horita, M. Nemoto and T.G. Langdon, Metall. Mater. Trans. **29A**, 2237 (1998).
18. P.B. Berbon, M. Furukawa, Z. Horita, M. Nemoto, N.K. Tsenev, R.Z. Valiev and T.G. Langdon, Phil. Mag. Lett. **78**, 313 (1998).
19. S. Komura, P.B. Berbon, M. Furukawa, Z. Horita, M. Nemoto and T.G. Langdon, Scripta Mater. **38**, 1851 (1998).
20. J. Wang, Y. Iwahashi, Z. Horita, M. Furukawa, M. Nemoto, R.Z. Valiev and T.G. Langdon, Acta Mater. **44**, 2973 (1996).
21. M. Furukawa, Z. Horita, M. Nemoto, R.Z. Valiev and T.G. Langdon, Acta Mater. **44**, 4619 (1996).
22. O.B. Makova, V.K. Portnoy, L.I. Novikov and L.B. Khokhlatova in *Aluminum-Lithium*, edited by M. Peters and P.-J. Winkler (DGM Informationgesellschaft, Oberursel, Germany, 1992), vol. 2, p. 1133.
23. T.G. Nieh, L.M. Hsiung, J. Wadsworth and R. Kaibyshev, Acta Mater. **46**, 2789 (1998).

AUTHOR INDEX

SUBJECT INDEX

transmission electron microscopy, 25, 81, 169, 311

ultrafine-grained metals and alloys, 153, 283, 323, 335

viscosity, 141

weight reduction, 253

yield phenomena, 141

zirconia, 93, 99, 105
Zn-Al alloy, 37, 335